Application and Implementation of Finite Element Methods

COMPUTATIONAL MATHEMATICS AND APPLICATIONS

Series Editor
J. R. WHITEMAN

Institute of Computational Mathematics, Brunel University, England.

E. HINTON and D. R. J. OWEN: Finite Element Programming

M. A. JASWON and G. T. SYMM: Integral Equation Methods in Potential Theory and Elastostatics

J. R. CASH: Stable Recursions: with applications to the numerical solution of stiff systems

H. ENGELS: Numerical Quadrature and Cubature

L. M. DELVES AND T. L. FREEMAN: Analysis of Global Expansion Methods: Weakly Asymptotically Diagonal Systems

J. E. AKIN: Application and Implementation of Finite Element Methods

Application and Implementation of Finite Element Methods

J. E. AKIN

Department of Mechanical Engineering and Materials Science,
Rice University,
Houston, Texas, USA

ACADEMIC PRESS, INC.

(Harcourt Brace Jovanovich, Publishers)

London Orlando San Diego New York
Toronto Montreal Sydney Tokyo

ACADEMIC PRESS INC. (LONDON) LTD
24–28 Oval Road
London NW1

US edition published by
ACADEMIC PRESS, INC.
Orlando, Florida 32887

Reprinted with corrections 1984
First paperback edition published 1984

British Library Cataloguing in Publication Data

Akin, J. E.
Application and implementation of finite element methods.
1. Finite element method—Data processing
I. Title
515.3'53 TA347.F5

ISBN 0-12-047650-9
ISBN 0-12-047652-5 (Pbk)
LCCCN 81-69597

PRINTED IN THE UNITED STATES OF AMERICA

85 86 87 88 9 8 7 6 5 4 3 2

Preface

The finite element method has now become a well-established branch of computational mathematics and the theoretical foundations have been presented in several texts. However, the actual application of finite element procedures requires extensive programming effort. The present text has been developed to illustrate typical computational alogrithms and their applications. Whilst the theoretical discussions have been limited to the minimum required to introduce the topic, most of the computational procedures are discussed in detail. Many universities follow an introductory finite element course, with a course on the related computational procedures. This text should be well suited for such a course.

Numerous programs are presented and discussed. A particular controlling program and data structure have been included for completeness so that specific applications can be examined in detail. Emphasis has been placed on the use of isoparametric elements and numerical integration. The discussion of the programs for isoparametric elements, Chapter 5, and their example applications, Chapter 11, should clarify this important topic. Since practical problems often involve a large amount of data the subject of mesh generation is also examined. A very limited discussion of time integration procedures has been included. Also included is an Appendix which describes some subroutines of secondary interest, which are mentioned in Chapters 2 and 6, together with various sample applications and the input formats for MODEL.

Figures referred to in each Section are included at the end of the Section, followed by the Tables.

The text reflects the many studies and conversations on finite elements in which I was able to participate during a period in which I was on leave from the University of Tennessee. These studies were conducted at the University of Texas at Austin, Brunel University, and the California Institute of Technology. The support of a UK SRC Senior Visiting Fellow Grant and a US NSF Professional Development Grant is gratefully acknowledged. I would also like to acknowledge the support and encouragement of J. T. Oden, J. R. Whiteman, T. J. R. Hughes, E. B. Becker, and W. C. T. Stoddart.

Copies of the MODEL software are available either from the author, or from the Institute of Computational Mathematics at Brunel University.

University of Tennessee J. E. Akin
January 1982

To Pryntha, Jeff, and Chris

Contents

Preface v
Program Notation xi

1 Finite Element Concepts
1.1 Introduction 1
1.2 Foundation of finite element procedures 1
1.3 General finite element analysis procedure 4
1.4 Analytic example 13
1.5 Exercises 18

2 Control and Input Phase
2.1 Introduction 19
2.2 Control of major segments 22
2.3 Data input 32
2.4 Exercises 35

3 Pre-element Calculations
3.1 Introduction 37
3.2 Property retrieval 41
3.3 Effects of skyline storage 46
3.4 Exercises 49

4 Calculation of Element Matrices
4.1 Introduction 51
4.2 Square and column matrix considerations 51
4.3 Auxiliary calculations 53
4.4 Condensation of element's internal degrees of freedom . 59
4.5 Economy considerations in the generation of element matrices 62

5 Isoparametric Elements
5.1 Introduction 65
5.2 Fundamental theoretical concepts 65
5.3 Programming isoparametric elements 74

5.4 Simplex elements, a special case 77
5.5 Isoparametric contours 84
5.6 Exercises 90

6 Element Integration and Interpolation
6.1 Introduction 91
6.2 Exact integrals for triangular and quadrilateral geometries 91
6.3 Gaussian quadratures 94
6.4 Numerical integration over triangles 101
6.5 Minimal, optimal, reduced and selected integration . . 105
6.6 Typical interpolation functions 107
6.7 Interpolation enhancement for C^0 transition elements . 115
6.8 Special elements 122
6.9 Exercises 127

7 Assembly of Element Equations into System Equations
7.1 Introduction 129
7.2 Assembly programs 130
7.3 Example 137
7.4 Symbolic element assembly for quadratic forms . . 144
7.5 Frontal assembly and solution procedures . . . 148
7.6 Exercises 152

8 Application of Nodal Parameter Boundary Constraints
8.1 Introduction 153
8.2 Matrix manipulations 154
8.3 Constraints applied at the element level 160
8.4 Penalty modifications for nodal constraints . . . 160
8.5 Exercises 164

9 Solution and Result Output
9.1 Economical solution techniques for the system equations . 165
9.2 Output of results 175
9.3 Post-solution calculations within the elements . . . 180
9.4 Exercises 181

10 One-dimensional Applications
10.1 Introduction 183
10.2 Conductive and convective heat transfer 183
10.3 Plane truss structures 193
10.4 Slider bearing lubrication 205
10.5 Ordinary differential equations 212
10.6 Plane frame structures 225

11 Two-dimensional Applications

11.1 Introduction . . . 231
11.2 Plane stress analysis . . . 232
11.3 Heat conduction . . . 245
11.4 Viscous flow in straight ducts . . . 255
11.5 Potential flow . . . 259
11.6 Electromagnetic waveguides . . . 276
11.7 Axisymmetric plasma equilibria . . . 280
11.8 Exercises . . . 284

12 Three-dimensional Applications

12.1 Introduction . . . 285
12.2 Heat conduction . . . 285

13 Automatic Mesh Generation

13.1 Introduction . . . 295
13.2 Mapping functions . . . 299
13.3 Higher order elements . . . 302

14 Initial Value Problems

14.1 Introduction . . . 317
14.2 Parabolic equations . . . 318
14.3 Hyperbolic equations . . . 331
14.4 Exercises . . . 344

References and Bibliography . . . 345

Appendix—A Summary of Input Formats and Supporting Programs . . . 351

Subject and Author Index . . . 367

Subroutine Index . . . 371

Program Notation

AD = VECTOR CONTAINING FLOATING POINT VARIABLES
AJ = JACOBIAN MATRIX
AJINV = INVERSE JACOBIAN MATRIX

C = ELEMENT COLUMN MATRIX
CB = BOUNDARY SEGMENT COLUMN MATRIX
CC = COLUMN MATRIX OF SYSTEM EQUATIONS
CEQ = CONSTRAINT EQS COEFFS ARRAY
COORD = SPATIAL COORDINATES OF A SELECTED SET OF NODES
CP = PENALTY CONSTRAINT COLUMN MATRIX
CUTOFF IS SPECIFIED NUMBER FOR CUTTING OFF ITERATIONS

D = NODAL PARAMETERS ASSOCIATED WITH A GIVEN ELEMENT
DD = SYSTEM LIST OF NODAL PARAMETERS
DDOLD = SYSTEM LIST OF NODAL DOF FROM LAST ITERATION
DELTA = LOCAL DERIVATIVES OF INTERPOLATION FUNCTIONS H

ELPROP = ELEMENT ARRAY OF FLOATING POINT PROPERTIES

FLTEL = SYSTEM STORAGE OF FLOATING PT ELEMENT PROP
FLTMIS = SYSTEM STORAGE OF FLOATING PT MISC. PROP
FLTNP = SYSTEM STORAGE OF FLOATING PT NODAL PROP
FLUX = SPATIAL COMPONENTS OF SPECIFIED BOUNDARY FLUX

GLOBAL = GLOBAL DERIV.S OF INTERPOLATION FUNCTIONS H

H = INTERPOLATION FUNCTIONS FOR AN ELEMENT

ID = VECTOR CONTAINING FIXED POINT ARRAYS
IBC = NODAL POINT BOUNDARY RESTRAINT INDICATOR ARRAY
INDEX = SYSTEM DEGREE OF FREEDOM NUMBERS ARRAY
IF INRHS.NE.0 INITIAL VALUES OF CC ARE INPUT
IF IPTEST.GT.0 SOME PROPERTIES ARE DEFINED
IP1 = NUMBER OF ROWS IN ARRAY FLTNP
IP2 = NUMBER OF ROWS IN ARRAY NPFIX
IP3 = NUMBER OF ROWS IN ARRAY FLTEL
IP4 = NUMBER OF ROWS IN ARRAY LPFIX
IP5 = NUMBER OF ROWS IN ARRAY PRTLPT
ISAY = NO. OF USER REMARKS TO BE I/O

JP1 = NUMBER OF COLUMNS IN ARRAYS NPFIX AND NPROP
JP2 = NUMBER OF COLUMNS IN ARRAYS LPFIX AND LPROP
JP3 = NUMBER OF COLUMNS IN ARRAY MISFIX

KFIXED = ALLOCATED SIZE OF ARRAY ID
KFLOAT = ALLOCATED SIZE OF ARRAY AD
KODES = LIST OF DOF RESTRAINT INDICATORS AT A NODE
KP1 = NUMBER OF COL IN ARRAYS FLTNP & PRTLPT & PTPROP
KP2 = NUMBER OF COLUMNS IN ARRAYS FLTEL AND ELPROP
KP3 = NUMBER OF COLUMNS IN ARRAY FLTMIS
K1-K5 = NO. OF COLUMNS OF FLOATING PT CONSTRAINT DATA

LBN = NUMBER OF NODES ON AN ELEMENT BOUNDARY SEGMENT
IF LEMWRT = 0 LIST NODAL PARAMETERS BY ELEMENTS
IF LHOMO=1 ELEMENT PROPERTIES ARE HOMOGENEOUS
LNODE = THE N ELEMENT INCIDENCES OF THE ELEMENT
LPFIX = SYSTEM STORAGE ARRAY FOR FIXED PT ELEMENT PROP
LPROP = ARRAY OF FIXED POINT ELEMENT PROPERTIES
IF LPTEST.GT.0 ELEMENT PROPERTIES ARE DEFINED

```
M = NO. OF SYSTEM NODES
MAXBAN = MAX. HALF BANDWIDTH OF SYSTEM EQUATIONS
IF MAXTIM.GT.0 CALCULATE CPU TIMES OF MAJOR SEGMENTS
MAXACT = NO ACTIVE CONSTRAINT TYPES (<=MAXTYP)
MAXTYP = MAX NODAL CONSTRAINT TYPE (=3 NOW)
MISCFL = NO. MISC. FLOATING POINT SYSTEM PROPERTIES
MISCFX = NO. MISC. FIXED  POINT SYSTEM PROPERTIES
MISFIX = SYSTEM ARRAY OF MISC. FIXED POINT PROPERTIES
MTOTAL = REQUIRED SIZE OF ARRAY AD
M1 TO MNEXT = POINTERS FOR FLOATING POINT ARRAYS

N = NUMBER OF NODES PER ELEMENT
NCURVE = NO. CONTOUR CURVES CALCULATED PER PARAMETER
NDFREE = TOTAL NUMBER OF SYSTEM DEGREES OF FREEDOM
NDXC = CONSTRAINT EQS DOF NUMBERS ARRAY
NE = NUMBER OF ELEMENTS IN SYSTEM
NELFRE = NUMBER OF DEGREES OF FREEDOM PER ELEMENT
NG = NUMBER OF NODAL PARAMETERS (DOF) PER NODE
IF NHOMO=1 NODAL SYSTEM PROPERTIES ARE HOMOGENEOUS
NITER = NO. OF ITERATIONS TO BE RUN
NLPFIX = NO.    FIXED  POINT ELEMENT PROPERTIES
NLPFLO = NO. FLOATING POINT ELEMENT PROPERTIES
NNPFIX = NO. FIXED POINT NODAL PROPERTIES
NNPFLO = NO. FLOATING POINT NODAL PROPERTIES
NOCOEF = NO COEFF IN SYSTEM SQ MATRIX
NODES = ELEMENT INCIDENCES OF ALL ELEMENTS
NOTHER = TOTAL NO. OF BOUNDARY RESTRAINTS .GT. TYPE1
NPROP = NODAL ARRAY OF FIXED POINT PROPERTIES
IF NPTWRT = 0 LIST NODAL PARAMETERS BY NODES
NRANGE = ARRAY CONTAINING NODE NO.S OF EXTREME VALUES
NREQ = NO. OF CONSTRAINT EQS. OF EACH TYPE
NRES = NO. OF CONSTRAINT FLAGS OF EACH TYPE
NSEG = NO OF ELEM BOUNDARY SEGMENTS WITH GIVEN FLUX
NSPACE = DIMENSION OF SPACE
NTAPE1 = UNIT FOR POST SOLUTION MATRICES STORAGE
NTAPE2,3,4 = OPTIONAL UNITS FOR USER (USED WHEN > 0)
NTOTAL = REQUIRED SIZE OF ARRAY ID
IF NULCOL.NE.0  ELEMENT COLUMN MATRIX IS ALWAYS ZERO
NUMCE = NUMBER OF CONSTRAINT EQS
N1 TO NNEXT = POINTERS FOR FIXED POINT ARRAYS

PRTLPT = FLOATING PT PROP ARRAY OF ELEMENT'S NODES
PTPROP = NODAL ARRAY OF FLOATING PT PROPERTIES

RANGE: 1-MAXIMUM VALUE, 2-MINIMUM VALUE OF DOF

S = ELEMENT SQUARE MATRIX
SB = BOUNDARY SEGMENT SQUARE MATRIX
SS = 'SQUARE' MATRIX OF SYSTEM EQUATIONS

TIME = ARRAY STORING CPU TIMES FOR VARIOUS SEGMENTS
TITLE = PROBLEM TITLE

X = SPATIAL COORDINATES OF ALL NODES IN THE SYSTEM
XPT = SPATIAL COORDINATES OF A CONTOUR POINT

.................................................
```

1

Finite element concepts

1.1 Introduction

The finite element method has become an important and practical numerical analysis tool. It has found application in almost all areas of engineering and applied mathematics. The literature on finite element methods is extensive and rapidly increasing. Extensive bibliographies are available [59, 76] but even these are incomplete and are rapidly becoming outdated. Numerous texts are available which present the theory of various finite element procedures. Most of these relegate programming considerations to a secondary, or lower, level. One exception is the text by Hinton and Owen [42]. The present work takes a similar position in that it aims to provide a complete overview of typical programming considerations while covering only the minimum theoretical aspects. Of course, the theory behind the illustrated implementation procedures and selected applications is discussed.

This chapter begins the introduction of various *building block* programs for typical use in finite element analysis. These modular programs may be utilized in numerous fields of study. Specific examples of the application of the finite element method will be covered in later chapters.

1.2 Foundation of finite element procedures

From the mathematical point of view the finite element method is based on integral formulations. By way of comparison the older finite difference methods are usually based on differential formulations. Finite element models of various problems have been formulated from simple physical intuition and from mathematical principles. Historically, the use of physical intuition led to several early practical models. However, today there is increased emphasis on the now well established mathematical foundations of the procedure [10, 28, 61].

The mathematical rigor of the finite element method was lacking at first, but it is now a very active area of research. Modern finite element integral formulations are obtained by two different procedures: variational formulations and weighted residual formulations. The following sections will briefly review the common procedures for establishing finite element models. It is indeed fortunate that all of these techniques use the same *bookkeeping* operations to generate the final assembly of algebraic equations that must be solved for the unknown nodal parameters.

The earliest mathematical formulations for finite element models were based on variational techniques. Variational techniques still are very important in developing elements and in solving practical problems. This is especially true in the areas of structural mechanics and stress analysis. Modern analysis in these areas has come to rely on finite element techniques almost exclusively. Variational models usually involve finding the nodal parameters that yield a stationary (maximum or minimum) value of a specific integral relation known as a functional. In most cases it is possible to assign a physical meaning to the integral being extremized. For example, in solid mechanics the integral may represent potential energy, whereas in a fluid mechanics problem it may correspond to the rate of entropy production. Many physical problems have variational formulations that result in quadratic forms. These in turn yield algebraic equations for the system which are symmetric and positive definite. Another important practical advantage of variational formulations is that they often have error bound theorems associated with them. Numerous examples of variational formulations for finite element models can be found by examining the many texts available on the theory of variational calculus. Several applications of this type will be illustrated in the later chapters.

It is well known that the solution that yields a stationary value of the integral functional and satisfies the boundary conditions is equivalent to the solution of an associated differential equation, known as the Euler equation. If the functional is known, then it is relatively easy to find the corresponding Euler equation. Most engineering and physical problems are initially defined in terms of a differential equation. The finite element method requires an integral formulation. Thus, one must search for the functional whose Euler equation corresponds to the given differential equation (and boundary conditions). Unfortunately, this is generally a difficult, or impossible task. Therefore, there is increasing emphasis on the various weighted residual techniques that can generate an integral formulation directly from the original differential equations. Both the differential equation and integral form are defined in physical coordinates, say (x, y, z).

As a simple one-dimensional example of an integral statement, consider the functional

$$I = \tfrac{1}{2}\int_0^L [K(\mathrm{d}t/\mathrm{d}x)^2 + Ht^2]\,\mathrm{d}x,$$

which will be considered in later applications. Minimizing this functional is equivalent to satisfying the differential equation

$$K\mathrm{d}^2t/\mathrm{d}x^2 - Ht = 0.$$

In addition the functional satisfies the natural conditions of $\mathrm{d}t/\mathrm{d}x = 0$ at an end where the essential boundary condition, $t = t_0$, is not applied.

The generation of finite element models by the utilization of weighted residual techniques is a relatively recent development. However, these methods are increasingly important in the solution of differential equations and other non-structural applications. The weighted residual method starts with the governing differential equation

$$L(\phi) = Q$$

and avoids the often tedious search for a mathematically equivalent variational statement. Generally one assumes an approximate solution, say ϕ^*, and substitutes this solution into the differential equation. Since the assumption is approximate, this operation defines a residual error term in the differential equation

$$L(\phi^*) - Q = R.$$

Although one cannot force the residual term to vanish, it is possible to force a weighted integral, over the solution domain, of the residual to vanish. That is, the integral over the solution domain, Ω, of the product of the residual term and some weighting function is set equal to zero, so that

$$I = \int_\Omega RW\mathrm{d}\Omega = 0$$

Substituting interpolation functions for the approximate solution, ϕ^*, and the weighting function, W, results in a set of algebraic equations that can be solved for the unknown coefficients in the approximate solution. The choice of weighting function defines the type of weighted residual technique being utilized. To obtain the Galerkin criterion one selects

$$W = \phi^*,$$

while for a least squares criterion

$$W = \partial R/\partial \phi^*$$

gives the desired result. Similarly, selecting the Dirac delta function gives a point collocation procedure; i.e.

$$W = \delta.$$

Obviously, other choices of W are available and lead to alternate weighted residual procedures such as the subdomain procedure. The first two procedures seem to be most popular for finite element methods. Use of integration by parts with the Galerkin procedure usually reduces the continuity requirements of the approximating functions. If a variational procedure exists, the Galerkin criterion will lead to the same algebraic approximation. Thus it often offers optimal error estimates for the finite element solution.

For both variational and weighted residual formulations the following restrictions are now generally accepted as means for establishing convergence of the finite element model as the mesh refinement increases [87]:

1. (A necessary criterion) The element interpolation functions must be capable of modelling any constant values of the dependent variable or its derivatives, to the order present in the defining integral statement, in the limit as the element size decreases.
2. (A sufficient criterion) The element shape functions should be chosen so that at element interfaces the dependent variable and its derivatives, of one order less than those occurring in the defining integral statement, are continuous.

1.3 General finite element analysis procedure

1.3.1 Introduction

In the finite element method, the boundary and interior of the continuum (or more generally the solution domain) is subdivided (see Fig. 1.1—illustrations are arranged together at the end of the section in which they are mentioned) by imaginary lines (or surfaces) into a finite number of discrete sized subregions or *finite elements.* A discrete number of *nodal points* are established with the imaginary mesh that divides the region. These nodal points can lie anywhere along, or inside, the subdividing mesh lines, but they are usually located at intersecting mesh lines (or surfaces). Usually, the elements have straight boundaries and thus some geometric approximations will be introduced in the geometric idealization if the actual region of interest has curvilinear boundaries.

The nodal points are assigned identifying integer numbers (*node numbers*) beginning with unity and ranging to some maximum value, say M. Similarly, each element is assigned an identifying integer number. These *element numbers* also begin with unity and extend to a maximum value, say NE. As will be discussed later, the assignment of the nodal numbers and element numbers can have a significant effect on the solution time and storage requirements. The analyst assigns a number of (generalized) *degrees of freedom,* (dof), say NG, to each and every node. These are the (unknown) nodal parameters that have been chosen by the analyst to govern the formulation of the problem of interest. Common *nodal parameters* are pressure, velocity components, displacement components, displacement gradients, etc. The nodal parameters do not have to have a physical meaning, although they usually do. It will be assumed herein that each node in the system has the same number (NG) of nodal parameters. This is the usual case, but it is not necessary. A typical node, Fig. 1.1, will usually be associated with more than one element. The domains of influence of a typical node and typical element are also shown in Fig. 1.1. A typical element will have a number, say N, of nodal points associated with it located on or within its boundaries. It is assumed herein that every element has the same number (N) of nodes per element. This is the usual situation, but again it is not necessary in general.

This idealization procedure defines the total number of degrees of freedom associated with a typical node and a typical element. Obviously, the number of degrees of freedom in the system, say NDFREE, is the product of the number of nodes and the number of parameters per node, i.e. NDFREE = M* NG. Similarly, the number of degrees of freedom per element, say NELFRE, is defined by NELFRE = N* NG.

Recall that the total number of degrees of freedom of the system corresponds to the total number of nodal parameters. In general the system degree of freedom number, say NDF, associated with parameter number J at system node number I is defined (by induction) as:

$$\mathrm{NDF} = \mathrm{NG}^*(\mathrm{I} - 1) + \mathrm{J}, \tag{1.1}$$

where $1 \leqslant \mathrm{I} \leqslant \mathrm{M}$ and $1 \leqslant \mathrm{J} \leqslant \mathrm{NG}$ so that $1 \leqslant \mathrm{NDF} \leqslant \mathrm{NDFREE}$. This elementary equation forms the basis of the program "bookkeeping" method and thus is very important and should be clearly understood. Equation (1.1) is illustrated for a series of one-dimensional elements in Fig. 1.2, where a system with four line elements, five nodes and six degrees of freedom (dof) per node is illustrated (i.e. M = 5, NE = 4, NG = 6, N = 2). There are a total of thirty dof in the system. We wish to determine the dof number of the fifth parameter (J = 5) at system node number four (I = 4). Equation (1.1) shows that the required result is NDF = 6(4 − 1) + 5 = 23 for the system dof number. For element three this corresponds to local dof number 11 while for element four it is local dof number 5. Therefore we note that contributions to system equation number 23 comes from parts of two different element equation sets.

In addition to the above constants it is necessary to define the dimension of the space, say NSPACE, that is associated with the problem. As will be pointed out as the discussion proceeds, these quantities can be used to calculate the size of the storage requirements for the matrices to be generated in the analysis. The actual programs that read the problem data will be discussed in a later section.

Data must be supplied to define the spatial coordinates of each nodal point. This array of data, say X, will have the dimensions of M*NSPACE. It is common to associate an integer with each nodal point. The purpose of the code is to indicate which, if any, of the nodal parameters at the node have boundary constraints specified. This vector of data, say IBC, contains M integer coefficients. To accomplish this nodal boundary condition coding process recall that there are NG parameters per node. Thus, one can define an integer code, IBC, (right justified) to consist of NG digits. Let the ith digit be a single digit indicator corresponding to the ith parameter at that node. If the indicator equals j where $0 \leq j \leq 9$ then this is defined to mean that the ith parameter has a boundary constraint of type j. If the single digit indicator is zero, this means that there is no boundary constraint on that parameter. As will be discussed later, the present program allow several common types of nodal parameter boundary constraints. Figure 1.3 illustrates a set of boundary condition codes for a typical set of nodes with six parameters per node (NG = 6). This code is also considered in the example in Section 7.3 and all the applications.

An important concept is that of *element connectivity*, ie the list of global node numbers that are attached to an element. The element connectivity data defines the topology of the mesh. Thus for each element it is necessary to input (in some consistent order) the N node numbers that are associated with that particular element. This array of data, say NODES, has the dimensions of NE*N. The list of node numbers connected to a particular element is usually referred to as the *element incident list* for that element. The identification of these data is important to the use of Eqn. (1.1).

1.3.2 Approximation of element behaviour and equations

It is assumed that the variable(s) of interest, and perhaps its derivatives, can be uniquely specified throughout the solution domain by the nodal parameters associated with the nodal points of the system. These nodal parameters will be the unknown parameters of the problem. It is assumed that the parameters at a particular node influence only the values of the quantity of interest within the elements that are connected to that particular node. Next, an interpolation function is assumed for the purpose of relating the quantity of interest within the element in terms of the values of the nodal parameters at the nodes that are connected to that particular element. Figure 1.4 illustrates a common interpolation associated with that particular element. Figure 1.4 illustrates a common interpolation function and its constituent parts defined in terms of the nodal coordinates (x_i, y_i), the element area A^e, the spatial location (x, y) and the nodal parameters T_i, T_j and T_k.

After the element behaviour has been assumed, the remaining fundamental problem is to establish the element matrices $\mathbf{S}^e$ and $\mathbf{C}^e$. Generally, they involve substituting the interpolation functions into the governing integral form. Historically these matrices have been called the *element stiffness matrix* and *load vector,* respectively. Although these matrices can sometimes be developed from physical intuition, they are usually formulated by the minimization of a functional or by the method of weighted residuals. These procedures are described in several texts, and will be illustrated in detail in later chapters.

Almost all element matrix definitions involve some type of defining properties, or coefficients. A few finite element problems require the definition of properties at the nodal points. For example, in a stress analysis one may wish to define variable thickness elements by specifying the thickness of the material at each node point. The finite element method is very well suited to the solution of non-homogeneous problems; therefore, most finite element programs also require the analyst to assign certain properties to each element. It is usually desirable to have any data that are common to every element (or every node) stored as miscellaneous system data. The analyst must decide which data are required and how best to input and recover them.

1.3.3 Assembly and solution of equations

Once the element equations have been established the contribution of each element is added to form the *system equations.* The programming details of the assembly procedure will be discussed in Section 7.2. The system equations resulting from a finite element analysis will usually be of the form

$$\mathbf{S}\,\mathbf{D} = \mathbf{C}, \tag{1.2}$$

where the square matrix **S** is NDFREE*NDFREE in size and the vectors **D** and **C** contain NDFREE coefficients each. The vector **D** will contain the unknown nodal parameters and the matrices **S** and **C** are obtained by assembling (as described later) the element matrices, $\mathbf{S}^e$ and $\mathbf{C}^e$, respectively. Matrices $\mathbf{S}^e$ and $\mathbf{C}^e$ are NELFRE*NELFRE and NELFRE*1 in size. In the majority of problems $\mathbf{S}^e$, and thus **S**, will be symmetric. Also, the system, **S**, is usually banded about the diagonal. It will be assumed herein that the system equations are banded and symmetric. Thus, the half-bandwidth, including the diagonal, is an important quantity to be considered in any finite element analysis. From consideration of the technique used to approximate the element behaviour, it is known that the half-bandwidth, say IBW, of the system equations due to a typical element "e" is defined by

$$\mathrm{IBW} = \mathrm{NG}^* (\mathrm{NDIFF} + 1), \tag{1.3}$$

where NDIFF is the absolute value of the maximum difference in node numbers of the nodes connected to the element. Equation (1.3) will be derived later. The maximum half-bandwidth, say MAXBAN, of the system is the largest value of IBW that exists in the system. That is,

$$\mathrm{MAXBAN} = \mathrm{IBW}_{\mathrm{maximum}}. \tag{1.4}$$

The quantity MAXBAN is one of the important quantities which govern the storage requirements and solution time of the system equations. Thus, although the assignment of node numbers is arbitrary, the analyst, in practice, should try to minimize the maximum difference in node numbers (and the bandwidth) associated with a typical element. The assembly process is illustrated in Fig. 1.5 for a four element mesh consisting of three-node triangles with one parameter per node. The top of the figure shows an assembly of the system **S** and **C** matrices that is coded to denote the sources of the contributing terms but not their values. A hatched area in these indicates a term that was added in from an element that has the hash code. For example, the load vector term C(6) is seen to be the sum of contributions from elements 2 and 3 which are hatched with horizontal (−) and oblique (/) lines, respectively. By way of comparison the term C(1) has only a contribution from element 2.

In closing, it should be noted that several efficient finite element codes do not utilize a banded matrix solution technique. Instead, they may employ a frontal solution or a sparse matrix solution. These important topics will be covered in some detail.

After the system equations have been assembled, it is necessary to apply the boundary constraints before solving for the unknown nodal parameters. The most common types of nodal parameter boundary constraints are

(1) defining explicit values of the unknowns on the boundary, and
(2) defining constraint equations that are linear combinations of the nodal quantities.

Methods for accomplishing both of these conditions will be presented in Section 8.2. Another type of boundary condition is the type that involves a flux or traction on the boundary of one or more elements. These element boundary constraints contribute additional terms to the element square and/or column matrices for the element(s) on which the constraints were placed. Thus, although these (Neumann type) conditions do enter into the system equations, their presence may not be obvious at the system level. Flux boundary conditions are considered in Sections 4.2. and 11.2. There are other applications where the analyst may need to specify and input the initial terms in the system equations column vector. For example, in structural mechanics, any applied concentrated nodal forces would be read directly into the system column matrix. Any additional loading terms calculated within the program would then be added to the initial set of specified loads.

After all the above conditions are satisfied, the system equations are solved by means of procedures which account for the sparse, symmetric nature of the problem. This can greatly reduce (by 90% or more) the number of calculations which would normally be required to solve the equations. The three most common methods for solving the system equations are by (1) Gauss Elimination, (2) Choleski Factorization, (3) Gauss–Choleski Factorization.

The relative merits of all three methods will be discussed in Section 9.1. The present program utilizes an *in core* solution method. However, the programming concepts of out of core techniques, that must be used for very large systems of equations, are also discussed in the literature. The comparison of full and compact band storage modes for **S** are shown in Fig. 1.6. where D's, C's, R's, and dots denote non-zero entries. Note that the rows, R, remain as rows but that columns, C, do not. The last dot in each row locates the *bandwidth* of that row. The largest bandwidth in the system defines the size of the array that will store the half-bandwidth of **S**.

After the Eqns. (1.2) have been solved for the unknown nodal parameters, it is usually necessary to output the parameters, **D**. In some cases the problem would be considered completed at this point, but in others it is necessary to use the calculated values of the nodal parameters to calculate other quantities of interest. For example, in stress analysis one uses the calculated nodal displacements to solve for the stresses and strains. Of course, these secondary quantities must also be output in some manner. Techniques of this type are usually called *post processing*.

1.3.4 Other considerations

The above discussion has been in terms of problems which lead to linear algebraic equations. Other computational problems, which can arise in a typical finite element analysis, include eigenvalue problems, time integration algorithms, and nonlinear problems. Entire books have been written on each of these subjects, but only limited discussion will be included herein.

One related task that also often arises with the finite element technique is the problem of data generation. One disadvantage of finite element procedures is that they can require large amounts of data. For regions with complex geometries, the requirements of closely approximating the boundaries together with the desirability of a fine mesh make it desirable to use a large number of nodal points and elements. The specification of the locations of such a large number of nodal points and elements is a time consuming job in which there is a high probability of human error. Data preparation and evaluation can require as much as thirty to forty percent of the total cost and time involved in the solution of large practical problems. To minimize the data preparations time and the probability of error several schemes for the automatic generation of much of the required data have been developed. However, no optimum technique has yet been found. Chapter 13 presents one useful technique.

Fig. 1.1 Influence domains

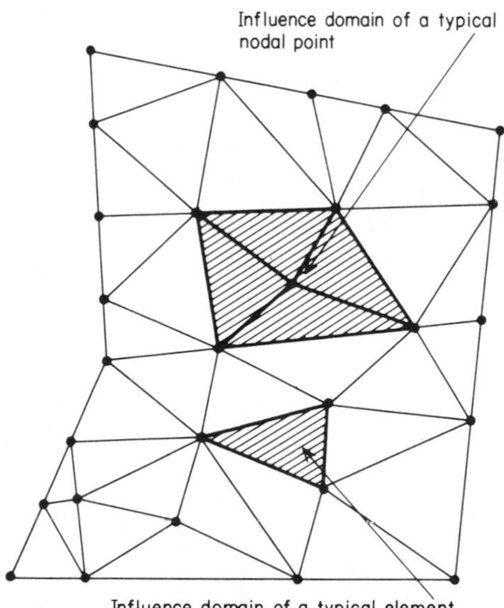

Fig. 1.2 Calculation of degree of freedom numbers

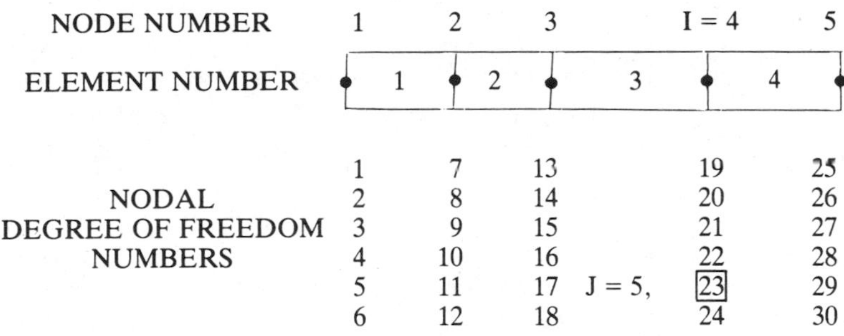

Fig. 1.3 Interpretation of nodal boundary code

Node	IBC	Boundary code for nodal parameter number, J J = 1	2	3	4	5	
1	111111	1	1	1	1	1	1
2	0	0	0	0	0	0	0
3	220000	2	2	0	0	0	0
4	100	0	0	0	1	0	0
5	11000	0	1	1	0	0	0

Fig. 1.4 Common interpolation components

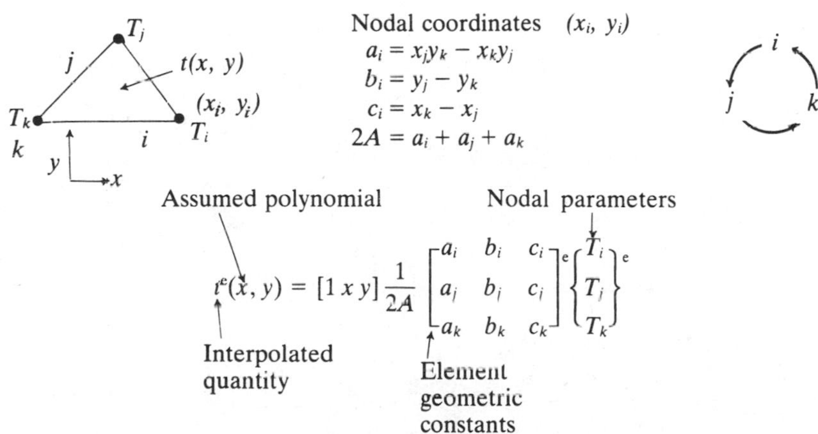

or

$$t^e(x, y) = H_i^e(x, y)\, T_i + H_j^e(x, y)\, T_j + H_k^e(x, y)\, T_k$$

$$t^e(x, y) = \boldsymbol{H}^e(x, y)\, \boldsymbol{T}^e$$

Fig. 1.5 Assembled contributions, **S D** = **C**; **(a)** System matrices, and **(b)** contributing elements

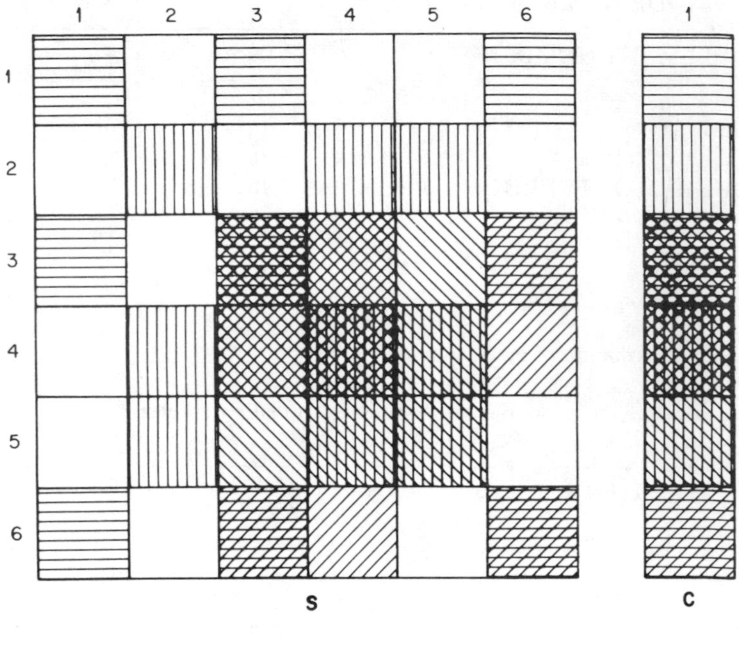

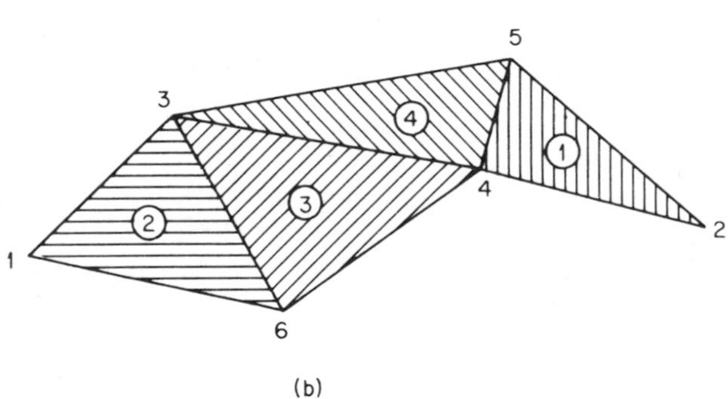

Fig. 1.6 Storage of system matrix, **X**

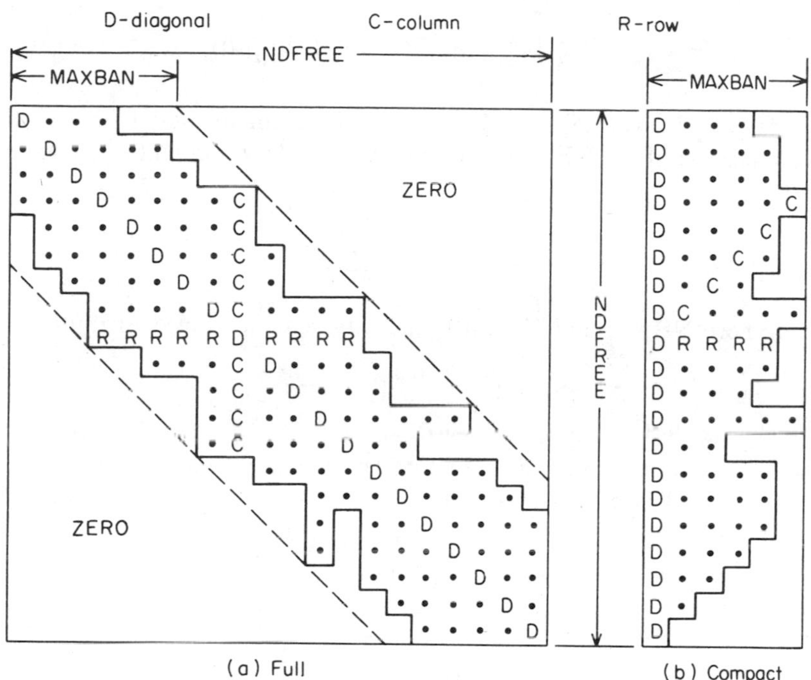

1.4 Analytic example

In order to review the previous introductory concepts consider the following model problem which will serve as an analytic example. The differential equation of interest is

$$\frac{d^2t}{dx^2} + Q = 0 \tag{1.5}$$

on the closed domain, $x \in [0, L]$, and is subjected to the boundary conditions

$$t(L) = t_0 \quad \text{and} \quad \frac{dt}{dx}(0) = q. \tag{1.6}$$

The corresponding governing integral statement to be used for the finite element model is obtained from a *variational statement*. In this case the variational formulation states that the function, t, which satisfies the essential condition, $t(L) = t_0$, and minimizes

$$I = \frac{1}{2}\int_0^L [(\mathrm{d}t/\mathrm{d}x)^2 - 2tQ]\,\mathrm{d}x + qt(0), \tag{1.7}$$

also satisfies Eqns (1.5) and (1.6). In a finite element model I is assumed to be the sum of the NE element and NB boundary segment contributions so that

$$I = \sum_{e=1}^{\mathrm{NE}} I^e + \sum_{b=1}^{\mathrm{NB}} I^b, \tag{1.8}$$

where here NB = 1 and $I^b = qt(0)$ and a typical element contribution is

$$I^e = \frac{1}{2}\int_{l^e} [(\mathrm{d}t^e/\mathrm{d}x)^2 - 2Q^e t^e]\,\mathrm{d}x,$$

where l^e is the length of the element. To evaluate such a typical element contribution it is necessary to introduce a set of interpolation functions, **H**, such that

$$t^e(x) = \mathbf{H}^e(x)\,\mathbf{T}^e = \mathbf{T}^{eT}\,\mathbf{H}^{eT}$$

and

$$\frac{\mathrm{d}t^e}{\mathrm{d}x} = \frac{\mathrm{d}\mathbf{H}^e}{\mathrm{d}x}\mathbf{T}^e = \mathbf{T}^{eT}\frac{\mathrm{d}\mathbf{H}^{eT}}{\mathrm{d}x}, \tag{1.9}$$

where $\mathbf{T}^e$ denotes the nodal values of t for element e. Thus a typical element contribution is

$$I^e = \frac{1}{2}\mathbf{T}^{eT}\,\mathbf{S}^e\,\mathbf{T}^e - \mathbf{T}^{eT}\,\mathbf{C}^e, \tag{1.10}$$

where

$$\mathbf{S}^e \equiv \int_{l^e} \frac{\mathrm{d}\mathbf{H}^{eT}}{\mathrm{d}x}\frac{\mathrm{d}\mathbf{H}^e}{\mathrm{d}x}\,\mathrm{d}x \tag{1.11}$$

and

$$\mathbf{C}^e \equiv \int_{l^e} Q^e\,\mathbf{H}^{eT}\,\mathrm{d}x.$$

If the domain, $[0, L]$, contains M nodes such that $t(0) = T_1$ and $t(L) = T_M$, then the non-essential (natural) boundary term is

$$I^b \equiv \mathbf{T}^{bT}\,\mathbf{C}^b = T_1 q.$$

Clearly, both the element degrees of freedom, $\mathbf{T}^e$, and the boundary degrees of freedom, $\mathbf{T}^b$, are subsets of the total vector of unknown parameters, $\mathbf{T}$. That is, $\mathbf{T}^e \subseteq \mathbf{T}$ and $\mathbf{T}^b \subset \mathbf{T}$. Of course, the $\mathbf{T}^b$ are usually a subset of the $\mathbf{T}^e$ (i.e., $\mathbf{T}^b \subset \mathbf{T}^e$ and in higher dimensional problems $\mathbf{H}^b \subset \mathbf{H}^e$). The main point here is that $I = I(\mathbf{T})$ and that fact must be considered in the summation in Eqn. (1.8) and in the minimization. The consideration of the subset relations is merely a bookkeeping problem. As will be shown later this allows Eqn. (1.8) to be written as

$$I = \frac{1}{2}\mathbf{T}^T \mathbf{S}\,\mathbf{T} - \mathbf{T}^T \mathbf{C},$$

where

$$\mathbf{S} = \sum_{e=1}^{NE} \mathbf{B}^{eT} \mathbf{S}^e \mathbf{B}^e,$$

$$\mathbf{C} = \sum_{e=1}^{NE} \mathbf{B}^{eT} \mathbf{C}^e + \mathbf{B}^{bT} \mathbf{C}^b$$

and where $\mathbf{B}$ denotes a set of symbolic bookkeeping operations. The combination of the summations and bookkeeping is commonly referred to as the assembly process. It is often simply (incorrectly) written as a summation. No uniformly accepted symbol has been developed to represent these operations.

It is easily shown that minimizing $I = I(\mathbf{T})$ leads to

$$\partial I/\partial \mathbf{T} = \mathbf{0} = \mathbf{S}\,\mathbf{T} - \mathbf{C},$$

as the governing algebraic equations to be solved for the unknown nodal parameters, $\mathbf{T}$. To be specific, consider a linear interpolation element with $N = 2$ nodes per element. If the element length is $l^e = x_2 - x_1$ then the element interpolation is

$$\mathbf{H}^e(x) = [(x_2 - x)/l^e \quad (x - x_1)/l^e],$$

so

$$\frac{d\mathbf{H}^e}{dx} = [-1/l^e \qquad 1/l^e]. \tag{1.12}$$

Therefore the element square matrix is simply

$$\mathbf{S}^e = \frac{1}{l^e}\begin{bmatrix} 1 & -1 \\ -1 & 1 \end{bmatrix}, \tag{1.13}$$

while the element column matrix is

$$\mathbf{C}^e = \int_{l^e} Q^e \begin{Bmatrix} (x_2 - x) \\ (x - x_1) \end{Bmatrix} dx.$$

Assuming that $Q = Q_0$, a constant, this simplifies to

$$\mathbf{C}^e = \frac{Q_0 l^e}{2}\begin{Bmatrix}1\\1\end{Bmatrix}. \tag{1.14}$$

The exact solution of the original problem for $Q = Q_0$ is

$$t(x) = t_0 + q(x - L) + Q_0(L^2 - x^2)/2. \tag{1.15}$$

Since for $Q_0 \neq 0$ the exact value is quadratic and the selected element is linear, our finite element model can give only an approximate solution. However, for the homogeneous problem $Q_0 = 0$, the model can (and does) give an exact solution. To compare a finite element solution with the exact one, select a two element model. Let the elements be of equal length, $l^e = L/2$. Then the element matrices are the same for both elements.

The assembly process yields, $\mathbf{S}\,\mathbf{T} = \mathbf{C}$, as

$$\frac{2}{L}\begin{bmatrix}1 & -1 & 0\\ -1 & (1+1) & -1\\ 0 & -1 & 1\end{bmatrix}\begin{Bmatrix}T_1\\T_2\\T_3\end{Bmatrix} = \frac{Q_0 L}{4}\begin{Bmatrix}1\\(1+1)\\1\end{Bmatrix} - \begin{Bmatrix}q\\0\\0\end{Bmatrix}.$$

However, these equations do not yet satisfy the essential boundary condition of $t(L) = T_3 = t_0$. That is, $\mathbf{S}$ is singular. After applying this condition the reduced equations are

$$\frac{2}{L}\begin{bmatrix}1 & -1\\ -1 & 2\end{bmatrix}\begin{Bmatrix}T_1\\T_2\end{Bmatrix} = \frac{Q_2 L}{4}\begin{Bmatrix}1\\2\end{Bmatrix} - \begin{Bmatrix}q\\0\end{Bmatrix} + \frac{2t_0}{L}\begin{Bmatrix}0\\1\end{Bmatrix},$$

or $\mathbf{S}_r\,\mathbf{T}_r = \mathbf{C}_r$. Solving for $\mathbf{T}_r = \mathbf{S}_r^{-1}\,\mathbf{C}_r$ where

$$\mathbf{S}_r^{-1} = \frac{L}{2}\begin{bmatrix}2 & 1\\ 1 & 1\end{bmatrix}$$

yields

$$\mathbf{T}_r = \begin{Bmatrix}T_1\\T_2\end{Bmatrix} = \frac{Q_0 L^2}{8}\begin{Bmatrix}4\\3\end{Bmatrix} - \frac{qL}{2}\begin{Bmatrix}2\\1\end{Bmatrix} + t_0\begin{Bmatrix}1\\1\end{Bmatrix}. \tag{1.16}$$

These are the exact nodal values as can be verified by evaluating Eqn. (1.15) at $x = 0$ and $x\ L/2$, respectively. Thus our finite element solution is giving an *interpolate* solution. That is, it interpolates the solution exactly at the node points and is approximate at all other points. For the homogeneous problem, $Q_0 = 0$, the finite element solution is exact at all points. These properties are common to other finite element problems. The exact and finite element solutions are illustrated in Fig. 1.7. Note that the derivatives are also exact at least at one point in each element. The optimal derivative sampling points will be considered in a later section. For this differential operator it can be shown that the centre point gives a derivative estimate accurate to $0(l^{e2})$ while all other points are only $0(l^{e})$ accurate. For $Q = Q_0$ the centre point derivatives are exact in the example.

Fig. 1.7 A two element solution

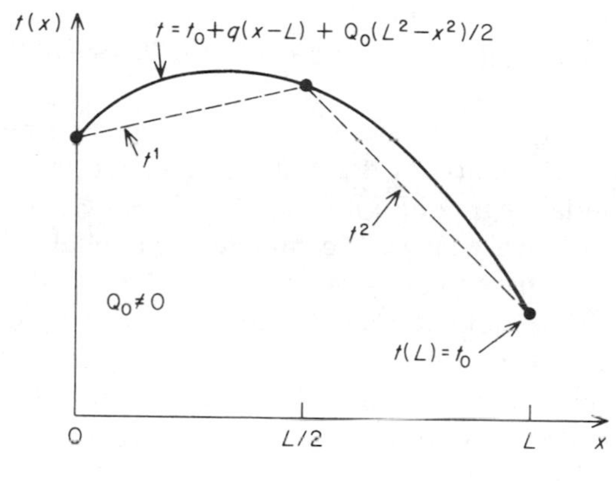

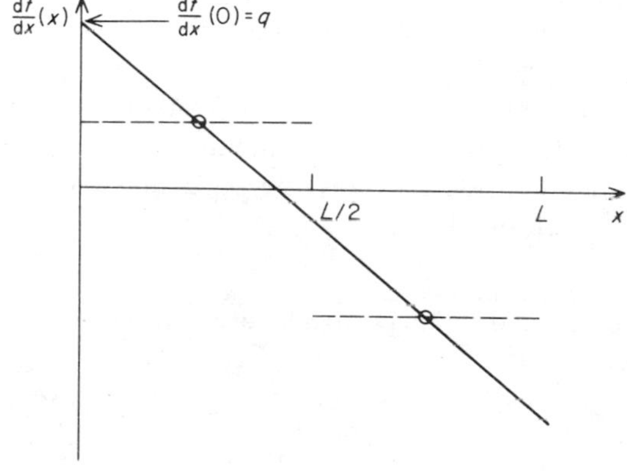

1.5 Exercises

1. Derive Eqn. (1.1) by induction.
2. An alternate way of assigning degree of freedom numbers is to number the ith parameters at all nodes before numbering the $(i + 1)$th parameters. Derive by induction the resulting equation that would be analogous to Eqn. (1.1). Does Eqn. (1.3) remain the same?
3. Repeat the graphical assembly in Fig. 1.5 after adding mid-side nodes to the elements.
4. Redraw Fig. 1.2 using NG = 3. Calculate the degree of freedom number at I = 4, J = 2.
5. Determine the maximum and minimum element degree (i.e. number of connecting elements) of the nodes in Fig. 1.1. Also determine the maximum nodal degree of a node.
6. Construct the Pascal triangle of complete polynomial terms corresponding to the three triangular elements in Fig. 1.8.
7. List likely nodal parameters and required properties data for a problem involving: (a) linear stress analysis, (b) heat conduction, (c) potential flow, (d) fluid dynamics, and (e) magnetic fields.
8. Employ a local coordinate system, $0 \leqslant r \leqslant 1$, so that Eqn. (1.12) is replaced by $H_1(r) = (1 - r)$ and $H_2(r) = r$, where $x(r) = H_1 x_1 + H_2 x_2$. Evaluate Eqn. (1.11) using this system.
9. Evaluate the element matrices $\mathbf{S}^e$ and $\mathbf{C}^e$ when Q varies linearly over the element.
10. Suggest how to evaluate numerically the convergence rate of the centre point element derivatives as the element sizes decrease uniformly.
11. Replot Fig. 1.7 considering only the particular part of the exact and finite element solutions.

Fig. 1.8 First three members of the triangular element family

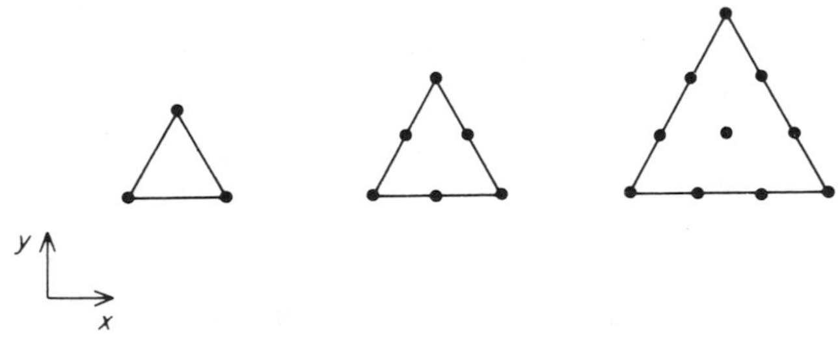

2

Control and input phase

2.1 Introduction

The practical application of the finite element method requires the use of a computer and a sizeable amount of programming effort. If a potential user of the finite element method were forced to program the entire code from scratch, he would be faced with a long, difficult, and often impractical task. It is therefore essential that the programming effort and knowledge of previous workers be utilized as much as possible. A number of large finite element computer codes have been developed. Several of the early codes were developed at a great cost in both money and labour, yet they were highly specialized, machine dependent, and difficult to modify; thus they rapidly became obsolete.

The most desirable types of general purpose finite element codes are those that are designed for comprehension, modification, and updating by the user. It is also desirable for the user to be familiar with the internal functions of the computer code if full advantage is to be taken of its basic potential and if inappropriate applications are to be avoided. These desirable objectives are most easily met if the program consists of several *building blocks* or *modules*. Generally, any specific function performed with some degree of repetition (or having a particular educational value) is separated into a distinct subroutine (or module) with a conscious effort on the part of the original programmer to make obvious the subroutine's function. By applying this concept one can establish a program *library* that contains many such building blocks. Then, for a given application, it is possible for the user to assemble the appropriate subroutines from the library plus any that he has written to meet his special needs. Of course, such a system of finite element codes generally requires a simple main program whose function is to call the various required subroutines in the appropriate order.

This modular approach to programming a finite element code has a number of important advantages. From the educational point of view, it is useful in that the student of the finite element method has only to master one elementary concept at a time. Therefore, it is easier to master the general programming concepts associated with the finite element method. The modular approach also makes it easier to apply the finite element technique to various fields of analysis such as heat transfer, fluid flow, and solution of differential equations. It also reduces to a minimum the amount of computer codes that the user must write when undertaking an analysis in a new field of application. Of still greater importance, however, is the fact that the modularity of such a code allows someone other than the original author to keep a program from becoming obsolete by incorporating new types of elements, solution techniques, input/output routines and other facilities as they become available. The various building block programs to be presented herein were developed by the author with these objectives in mind. These programs have been kept as short and independent as is practical in order to make their application as general as possible. The author has also attempted to reduce the solution time and storage requirements.

A typical finite element program will require the storage of a large number of matrices. These matrices include the spatial coordinates of the nodes, the element incidences, the element matrices, the system equations, etc. In general, the size of a typical matrix can be defined by the control parameters (number of nodes, number of elements, etc.) and/or the input data (element nodal incidences, etc.). This fact allows the programmer to utilize efficiently the available in-core storage. Unfortunately, most programs do not take advantage of this option. Instead they specify, once and for all, the maximum size of each array. That is, the program sets limits on the maximum number of nodal points, elements, boundary conditions, etc. Usually a typical problem will at most reach one or two of these limits so that the sizes of the other matrices are larger than required and storage is wasted. Of course, the size of the limits can be changed, but this usually involves changing several dimension statements in the program and its various subroutines. A procedure is described which allows for the semi-dynamic allocation of storage for these matrices. The first step is to dimension a single vector of a size large enough to contain all the matrices needed in an average problem. Next, the size of each matrix is calculated, for the problem of interest, by using the control parameters and the input data. Then, as illustrated in Fig. 2.1, a set of *pointers* is defined to locate the position, in the storage vector, of the first coefficient of each matrix of interest. By using this approach, the user has only one limit about which to worry. That limiting number is the dimension of the single storage vector and that limit in turn is usually governed by the size of the computer system to which the user has access. Only one dimension card has to be changed to modify the in-core storage capacity of the present building block program. The disadvantage is that one must do the extra "bookkeeping" necessary to define the pointers to locate the matrices.

In some cases, an additional storage saving is possible through overwriting. Referring to Fig. 2.1 for three representative matrices, **A** (3×3), **B** (1×5), **C** (2×1), assume that the matrix **B** is generated, stored and completely utilized before matrix **C** is generated and stored. Thus, to save storage space one could define the pointer for matrix **C** as I3 = I2. This would result in the coefficients of matrix **C**, and the following matrices, being stored in the now unneeded space that matrix **B** occupied.

It is possible to write a FORTRAN code so as to obtain true dynamic storage allocation abilities, see [17]. However, these extra details will not be considered here.

Fig. 2.1 An efficient matrix storage technique

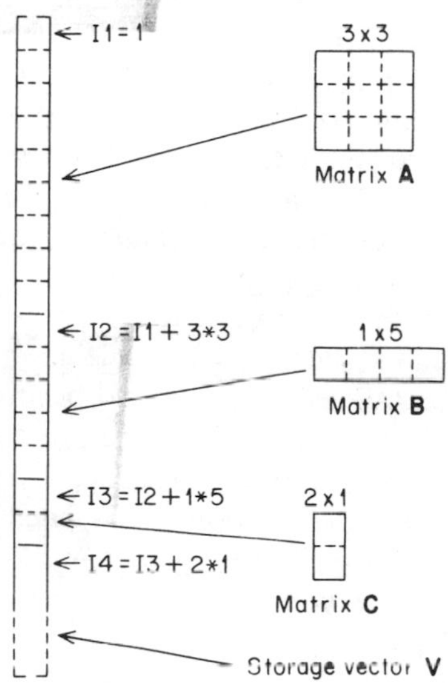

2.2 Control of major segments

The controlling main program and the data input routines are very important to the user. They significantly affect the ease of program use as well as its range of analysis capabilities. However, they vary greatly from one program to the next. For example, large commercial packages such as PAFEC [39], tend to have special control macros, free-format input, and extensive graphics to simplify the ease of use. Other small systems may use special macros to simplify the order of data input. An example of this type of code is the one presented by Taylor [69] in the appendix of the text by Zienkiewicz. Other programs often use the more classical type of fixed format data input. That is the procedure utilized here in the MODEL program.

The major point of the discussions of the routines in MODEL is to illustrate the typical calculations that go on within a finite element analysis system. The details of interfacing with the user are given lesser importance here. However, when considering a large commercial code a user should probably reverse the order of the above priorities.

Having made these choices for MODEL the details of how the controlling program and input routines work is probably not important to the reader at this time. However, for the sake of completeness the input formats and the subroutines employed for these phases are given in Appendix A.

Figure 2.2 presents a flow chart that illustrates the major steps that fall under the control of the main program, MODEL. It is important that one understand the relationships between these operations. The first half of the steps, down to point 1, deal with initializing the system, determining the input options to be utilized, reading and printing the input data, and assigning storage locations that will hold the information to be generated in later phases. For future reference, Fig. 2.3 lists the names of the major subroutines that control the corresponding segments listed in Fig. 2.2.

The steps from points 1 to 2, in Fig. 2.2, deal with the generation of element matrices, setting up data to be used for later secondary calculations, assemblying the system matrices, and applying the essential boundary conditions before solving the equations. Between points 2 and 3 the system equations are solved (factorized) to yield the unknown nodal parameters. If additional calculations are desired then the required auxilary data are recovered from storage and the post-solution calculations are executed.

A problem with a material or geometric nonlinearity would have an additional loop such as the one shown between points 1 and 3 in Fig. 2.2. For example, if a temperature calculation involves temperature dependent thermal conductivities then it is necessary to estimate initially the conductivities in each element. After the nodal temperatures have been computed a more accurate value of the conductivity is obtained and compared with the previous estimate. If all element conductivities are reasonably close to the estimate then the calculated temperatures are acceptable, otherwise a new estimate must be made and the solution repeated.

A linear time dependent problem would involve a loop through points 2 and 3 at each time step. Other combinations are possible such as a nonlinear dynamic analysis that requires material property iterations at the end of each time step. Thus the main controlling program can have complicated logic if it is designed to be very general.

The major subroutines ASYMBL, APLYBC, FACTOR, SOLVE, and POST that are listed in Fig. 2.3 will be considered in more detail in a later chapter. Their names and comparison of Figs. 2.2 and 2.3 should suggest their major functions. The assembly segment control program, ASYMBL, uses several of the subroutines to be discussed in Chapters 3 and 4. It will be discussed in detail in Chapter 7. However, Fig. 2.4 is included here to outline its major operations and to identify the relative functional positions of some of the routines to be presented in the next few chapters. Chapter 4 will discuss typical procedures utilized in the generation of element matrices. Additional specific details are given later beginning in Chapter 10. Figure 2.5 illustrates the flow of operations in the element generation control segment, GENELM, and identifies some of the subroutines it utilizes.

Since the generation of the popular isoparametric elements requires special programming considerations they are given detailed attention in Chapter 5. The major steps that occur in a program, say ISOPAR, that controls the generation of isoparametric elements is shown in Fig. 2.6. It also identifies typical routines to be presented later. From Fig. 2.6 it is noted that isoparametric element generation usually requires the knowledge of element interpolation, or shape, equations and numerical integration. The latter two topics are included in Chapter 6.

Following the presentation of the assembly procedures in Chapter 7 the applications of the algorithms for applying constraints to the nodal dof is discussed in Chapter 8. A flow chart for the operations in a controlling routine, say APLYBC in Fig. 2.3, is given in Fig. 2.7. As discussed later, the type of constraint is defined by the number of dof occurring in the linear constraint equation given in the input data.

Chapter 9 will give specific details of the equation solution process and the output of the calculated nodal dof, that is the operations of FACTOR, SOLVE, WRTPT, and WRTELM in Fig. 2.3. It also gives the typical procedures for executing any post-solution calculations defined at the element level and stored at assembly time. These are listed in Fig. 2.8.

Very detailed examples of common applications of finite element analysis methods are given in Chapters 10, 11, and 12. These include the element matrices, typical post-solution quantities of practical importance, and specific input and output data. To aid in generating larger and more practical meshes, Chapter 13 is included to present an algorithm for automatic mesh generation. The last two chapters are included to demonstrate the implementation aspects of selected time dependent finite element solutions. However, since there are at least thirty time integration algorithms the ones presented may not be typical of those procedures employed in the more advanced applications.

As the reader proceeds through the following chapters it will probably be useful to refer back to flow charts given previously. If the reader is interested in searching in the text for specific programming details, then the Index of Subroutines along with these flow charts should allow the text to also serve as a quick reference book.

Fig. 2.2 A typical analysis sequence

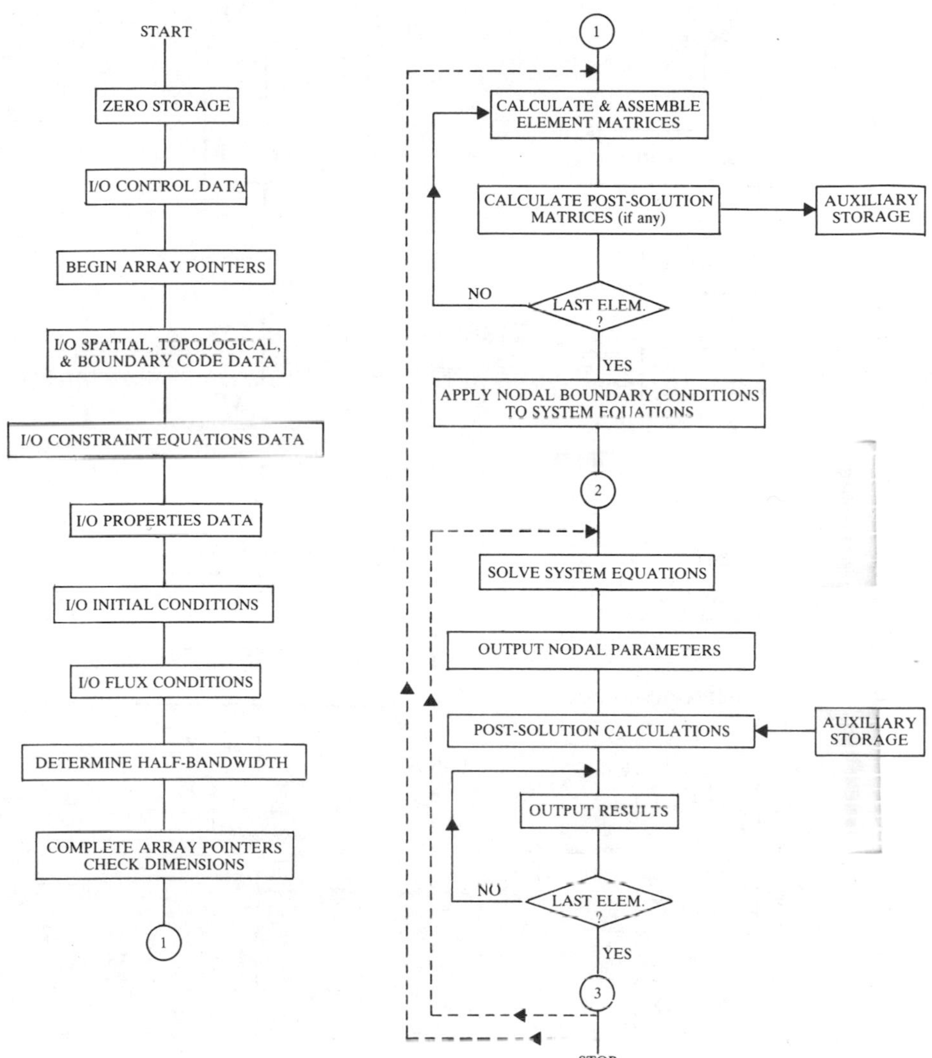

Fig. 2.3 Subroutines related to the segments in Fig. 2.2

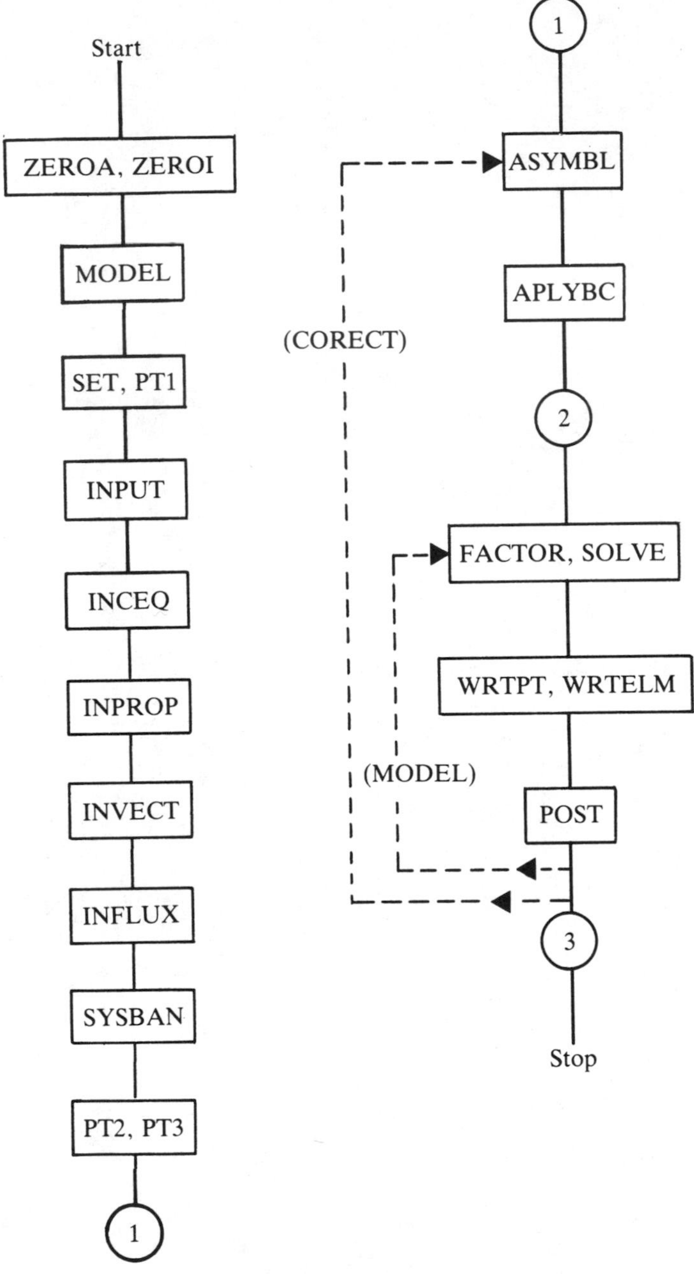

Fig. 2.4 A typical assembly flow chart

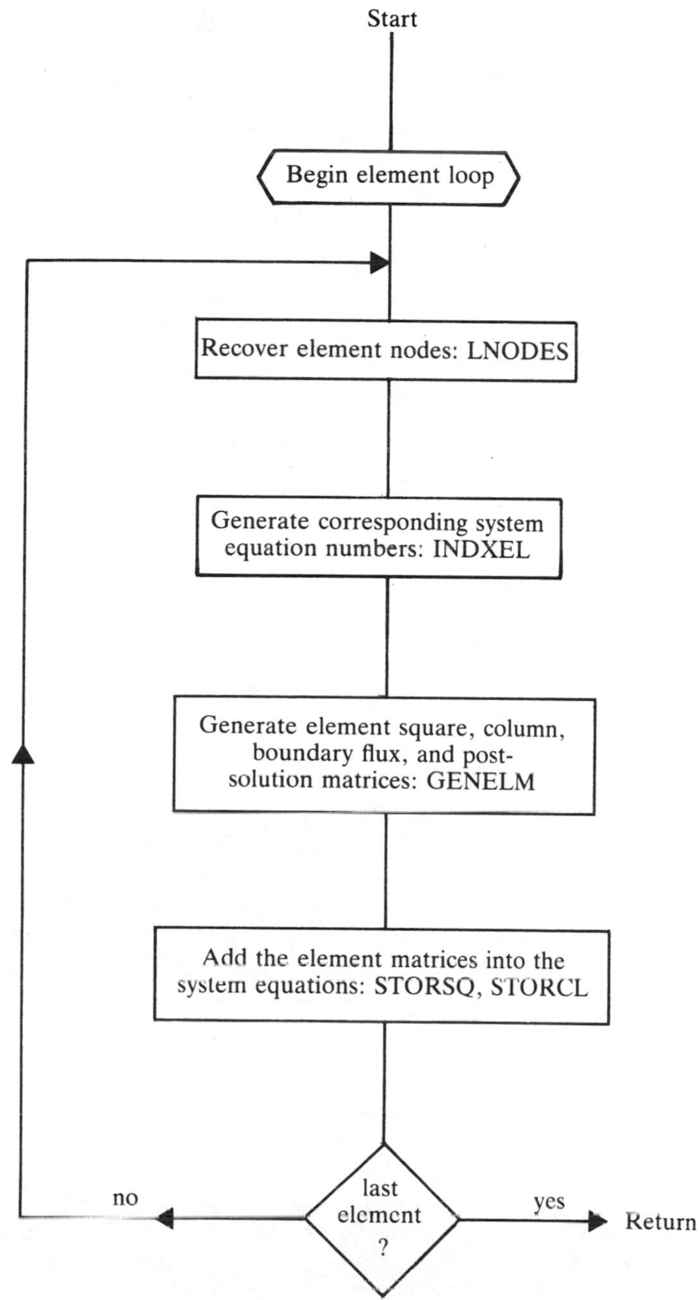

Fig. 2.5 A typical element generation procedure

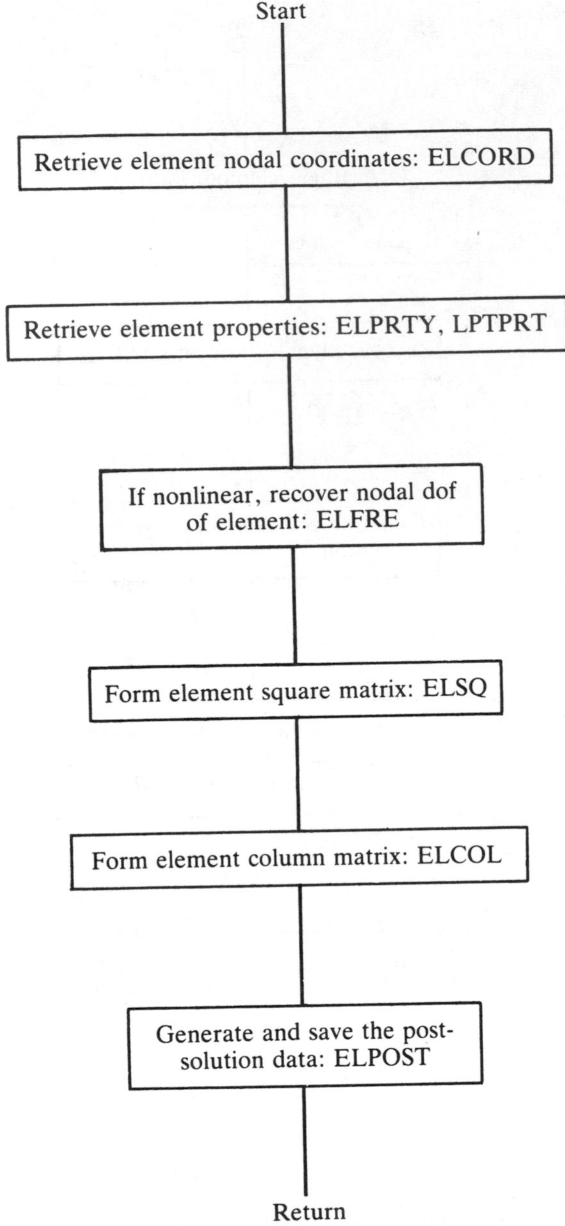

Fig. 2.6 Steps in generating matrices for isoporametric elements

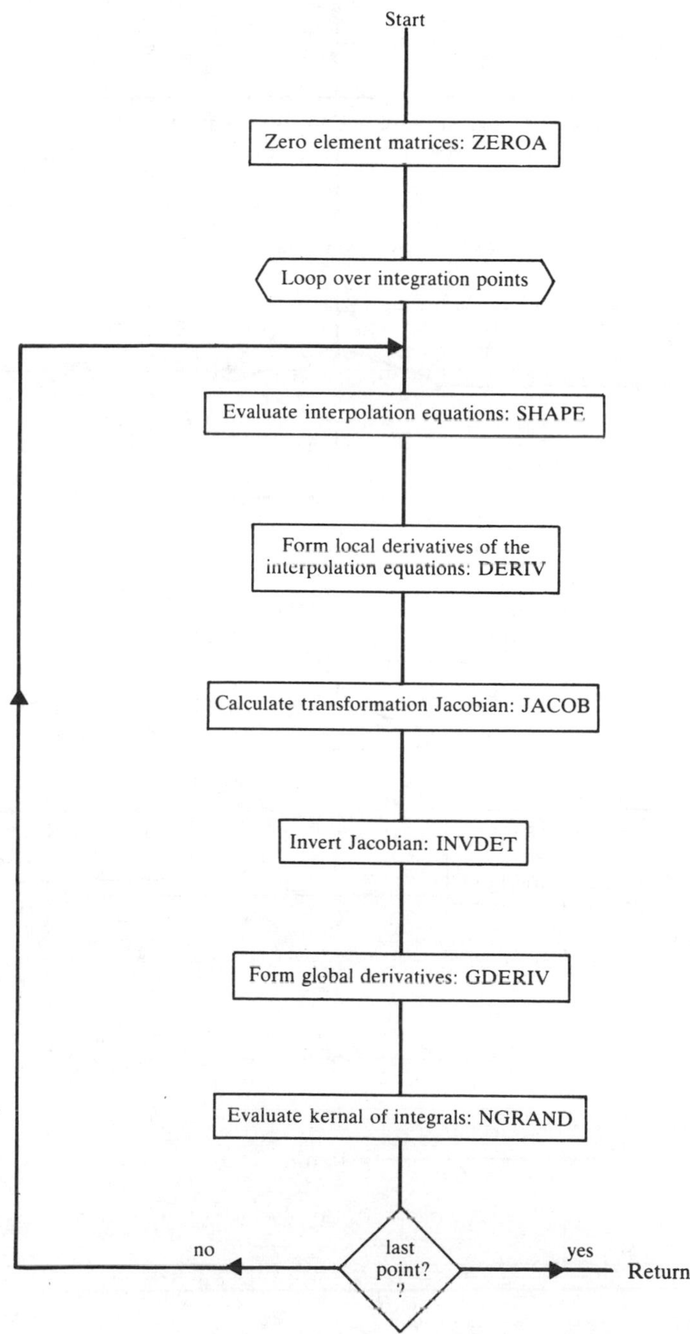

Fig. 2.7 Applying nodal constraints to the system equations

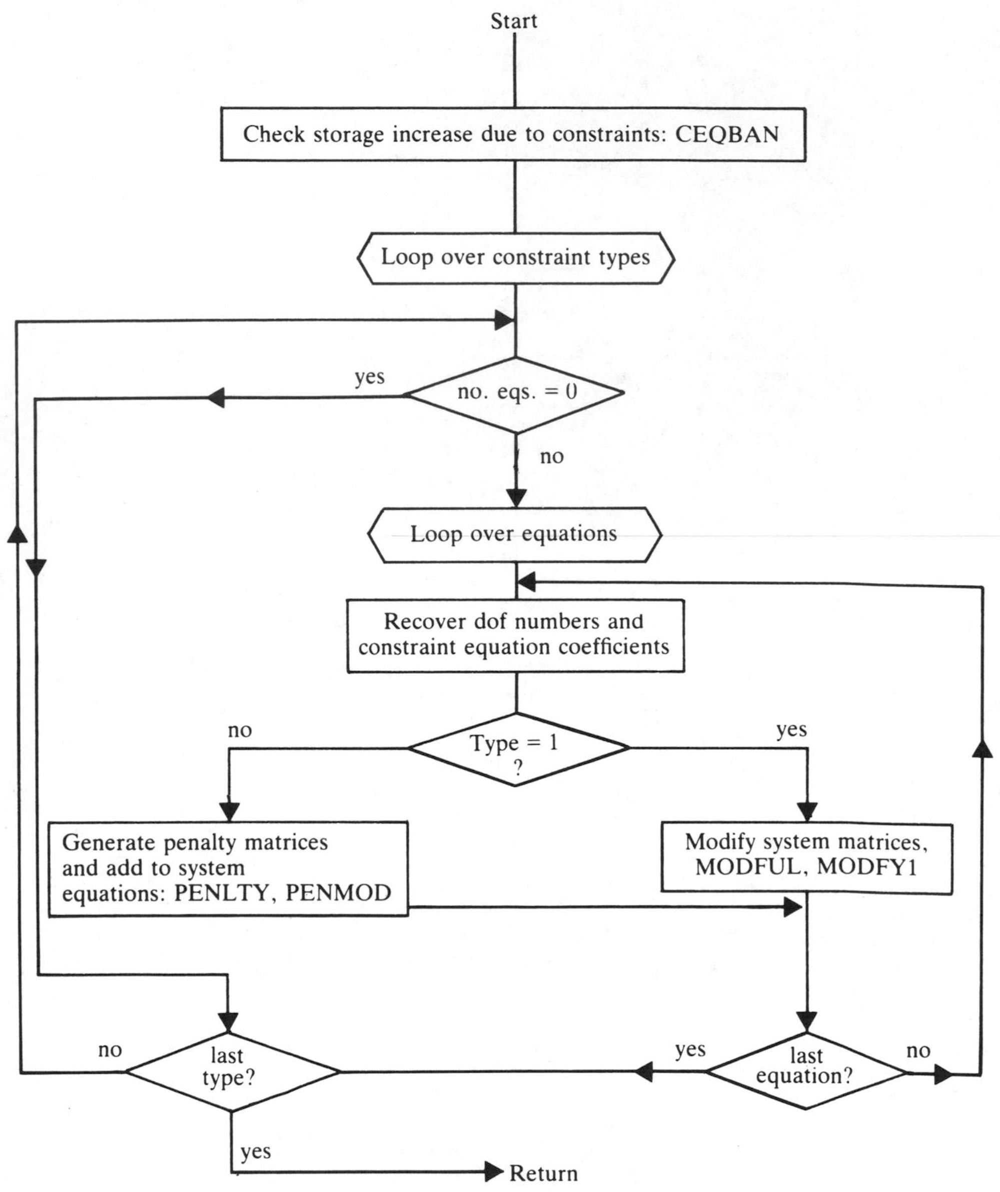

Fig. 2.8 Evaluating element's post-solution variables

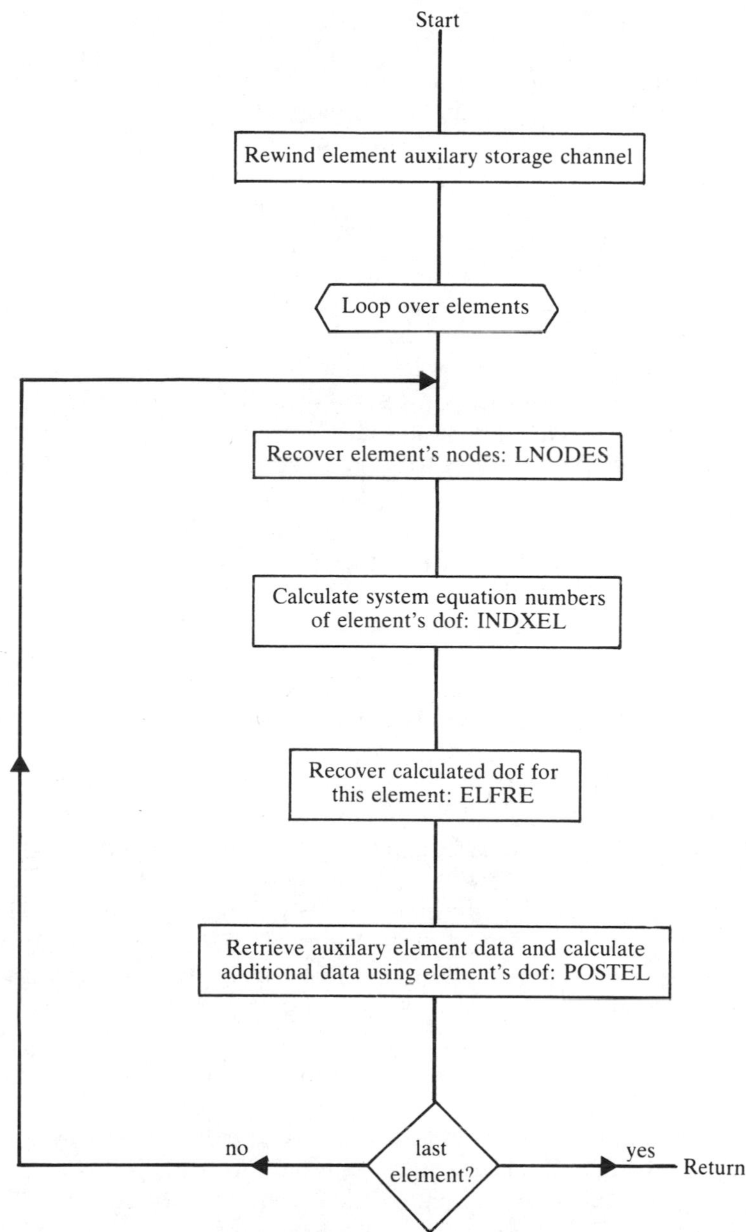

2.3 Data input

The basic data dealing with the nodal points and elements are read by subroutine INPUT. This routine reads the node number, boundary code indicator and the spatial coordinates for every node in the system. Next it reads the element number and the element incidences (node numbers of all nodal points associated with the element) for every element in the system. These data are also printed for the purpose of checking the input data. Given the number of nodes, number of elements, the number of nodes per element, number of parameters per node, and the dimensions of the space, this subroutine sets aside storage for the spatial coordinates, element incidences, and the nodal point boundary code indicator. This routine and its input instructions are shown in the Appendix.

The nodal point boundary code indicator serves the important task of indicating the number and type of nodal parameter boundary conditions occurring in the system. Several pieces of data are included in this code. Subroutine PTCODE extracts these data. This program takes the integer boundary code indicator (consisting of NG digits, right justified) and extracts NG single digit integer codes. These single digit codes range from zero to nine and they identify the type of constraint equation (defined by the analyst) that the parameter must satisfy. A zero code implies no constraint. This allows the analyst to write programs to handle up to nine types of boundary constraint conditions. The present set of building block programs use only the single digit values of 0, 1, 2, and 3.

After subroutine PTCODE has extracted the NG single digit integer codes associated with the NG parameters at a node, these codes must be scanned to count the number of different types of nodal parameter boundary conditions encountered in the system. This is done by subroutine CCOUNT which utilizes the nodal point indicators, KODES, to count the number of non-zero constraints of each type that occur in the system. Subroutine CCOUNT calls subroutine PTCODE, and both are given in the Appendix. For the example shown in Fig. 1.3, the constraint indicator counts are nineteen zeros, nine ones and two twos. As discussed in Section 1.3, these data imply nine Type 1 nodal parameter constraints and one Type 2 constraint.

Once the number and type of nodal parameter boundary conditions have been determined, it is necessary to supply the constraint equation data. These nodal parameter constraint data are read by subroutine INCEQ (see Appendix). The nodal parameter constraint equations will be discussed in detail in Section 8.1. At this point it is only necessary to note that the present program allows the following three types of linear equations:

$$\begin{aligned} &\text{TYPE 1} \quad D_j = A \\ &\text{TYPE 2} \quad AD_j + BD_k = C \qquad (2.1) \\ &\text{TYPE 3} \quad AD_j + BD_k + CD_l = E \end{aligned}$$

where A, B, C, and E are constants and j, k, and l denote system degree of freedom numbers associated with the nodal parameters (D). The program can easily be expanded to allow for nine types of linear constraint equations. Regardless of which type of constraint equation one has, it is still necessary for the analyst to input the constant(s) and identify the degree of freedom number(s). The degree of freedom numbers are identified indirectly by specifying the node where the parameter is located and its local parameter number. The coefficients in the equation are specified directly. Subroutine INCEQ reads two sets of data for each constraint equation of each type. These data include identification (node number, parameter number) of the constrained degree of freedom numbers and the corresponding coefficients of the constraint equation. The input formats for this routine are shown at the beginning of the Appendix.

As outlined in the above section, the present program specifies three allowable types of linear constraint equations. It can easily be expanded to allow for nine (or more) types of linear constraint equations. For any given type of linear constraint equation it is necessary to define an integer array, NDX, and a floating point array, CEQ. Both of these arrays will have the same number of rows. The number of rows is of course equal to the maximum number of degrees of freedom in the constraints. Each row of the integer array contains the system degree of freedom numbers of the nodal parameters occurring in a particular constraint equation. The floating point array contains the corresponding set of coefficients occurring in the constraint equation. The number of columns equals the total number of nodal constraint equations.

The degree of freedom numbers will be calculated from the input data by using Eqn. (1.1) and the input coefficients will be modified so that the coefficient associated with the first parameter in the equation is defined to be unity. For example, a TYPE 2 constraint equation $AD_j + BD_k = C$ becomes $D_j + aD_k = b$, where $a = B/A$ and $b = C/A$. Thus, it will be necessary to store only two integers (j and k) and two floating point numbers (a and b) for each equation of this type. In general, the number of rows in these two arrays equals the constraint type number. The use of these data will be shown in Section 8.1.

Subroutine INPROP (Appendix) inputs the various properties necessary to define the problem. If there are nodal point properties, then the node number and corresponding properties are read for each nodal point.

Similarly, if element properties are to be input, the element number and corresponding properties are read for each element. Finally, any miscellaneous system properties are also read. The user is allowed the option of specifying the nodal point and element properties to be homogeneous. This greatly reduces the required amount of data in several problems. The specific input instructions are presented in Appendix A. Later applications will illustrate the use of these properties.

In some applications, the initial forcing vector (system column matrix) terms will be known to be non-zero. In such cases, the non-zero initial contributions must be input by the analyst. Subroutine INVECT reads these contributions and then prints the entire initial load vector. The subroutine is also shown in the Appendix, as are the corresponding input instructions. In subroutine INVECT each specified coefficient has an identifying parameter number and system node number associated with it. It is necessary to convert the latter two quantities to the corresponding system degree of freedom number. This calculation is defined by Eqn. (1.1), which is included in subroutine DEGPAR for possible uses elsewhere.

There are numerous applications with boundary conditions which do not directly involve the nodal parameters. Usually these conditions involve a flux across, or traction on, a specified segment of an element boundary. For example, in a heat transfer problem one may specify the components of the heat flux across a portion of the boundary, while in stress analysis one may have an applied pressure acting on a portion of the structure's surface. Generally, these types of boundary conditions lead to contributions to the system matrices. The calculations of these contributions are closely related to the calculation of the element's square matrix and column matrix. The latter two matrices are discussed in Sections 4.2 and 5.3. The addition of the flux contributions to the system matrices is closely related to the procedures to be given in Section 7.2.

Subroutine INFLUX (see Appendix together with its input formats) reads the flux type boundary condition data for each boundary segment. The actual calculation of the flux contributions is carried out by a problem dependent subroutine called BFLUX which is discussed later. If the specified segment has LBN nodes associated with it and NG parameters per node, then subroutine BFLUX would utilize the NG components of FLUX to calculate the contributions of the flux to the NFLUX = LBN * NG system degrees of freedom on that segment. Since the actual calculations in BFLUX would vary from one application to another the analyst could choose to interpret the specified flux components to be normal and tangential to the segment instead of being components in each of the spatial directions.

2.4 Exercises

1. Several programs test for gaps in the input of sequential data. Such gaps are filled by generating supplemental data by linear interpolation between the bounding data sets. Write a supplementing input subroutine that will:
   (a) assign missing coordinates by linear interpolation and set the boundary code equal to the last value before the gap; and
   (b) assign missing element incidences lists by incrementing consistently from the two lists bounding the gap.
2. Structural analysis can require solutions for several forcing vectors (system column matrices). Modify INVECT to account for NRHS such vectors. That is, expand **CC** to a rectangular array. What corresponding changes do you think would occur in the control variables and storage pointers?
3. With quadratic (or cubic) elements it is desirable for omitted mid-side (or third point) coordinates to lie on a straight line between the boundary corner points. Write a routine to calculate these coordinates. Would changes in INPUT be required?
4. The properties' definitions do not specifically include a definition of a material number and a number of properties per material. Later applications will show how this can be done. Write the changes necessary to directly include the input of such properties.
5. It is common for Type 1 nodal constraints to assign zero nodal values. Write a version of INCEQ that will assign a default value of zero to any omitted type 1 constraint. If necessary, include IBC or other quantities in the argument list.
6. A number of codes allow only Type 1 nodal boundary codes. Thus for parameter I at a node KODE(I) is zero or unity. Only dof with a zero code need to have an equation number. Write a routine to generate an array, say IDEQ(NG, M), that contains the equation number of parameter I of node J for all $I \leq NG$ and $J \leq M$. The equation number IDEQ(I, J) is zero if the parameter is specified with a Type 1 constraint.
7. Use the IDEQ array to generate an element location array, say LM(NG, N, NE), where LM(I, J, K) contains the equation number of dof I of local node J on element K.

3

Pre-element calculations

3.1 Introduction

After the problem control parameters and properties have been input, some useful preliminary calculations can be executed. One important step is the calculation of the half-bandwidth size of the system equations. This information is necessary to calculate the amount of storage to be required by the system equations. It is used to test the feasibility of proceeding with the calculations. This type of calculation requires the identification of the node numbers (element incidences) connected to a typical element. The latter operation is performed by subroutine LNODES. This subroutine extracts the N element incidences of a specified element from the element incidences list for the entire system. As shown in Fig. 3.1, the N element incidences are stored in a column matrix, called LNODE.

This elementary operation is also very important to the bookkeeping involved in various parts of the program. The operation is illustrated in Fig. 3.2. Once the element incidences of the specified element have been extracted (by subroutine LNODES) from the input data, they are scanned to determine the maximum difference in these node numbers. This difference in node numbers is then substituted into Eqn. (1.2) to determine the half-bandwidth associated with that particular element, without regard to the effect of constraints. These elementary operations are executed by subroutine ELBAND, which is shown in Fig. 3.3, and they are based on Fig. 3.4.

The simple but important task of determining the value of the maximum half-bandwidth of the system equations is performed by subroutine SYSBAN, Fig. 3.5. The half-bandwidth of each element is checked and the largest value found in the system is retained. The system equations usually occupy a very large percentage of the total storage. Knowing the system equation's half-bandwidth, determined above, and the total number of degrees of freedom in the system, the storage requirements for these banded, symmetric equations are easily determined. It is possible for nodal parameter constraint equations to increase the half-bandwidth of the system equations. This depends on how the constraints are implemented.

There are certain basic operations that must be carried out before one actually generates the element matrices for the particular problem under study. For simplicity it will be assumed that the problem under study is linear and involves a quadratic functional. One must always use the spatial coordinates of the nodes associated with the element. Recall that the spatial coordinates of all the nodal points and the element incidences of all elements are available as input data. Thus to define an array containing the spatial coordinates of the nodes associated with an element one simply extracts its element incidences with subroutine LNODES and then extracts the spatial coordinates of each node in the list of incidences. The present program for this purpose, ELCORD, is shown in Fig. 3.6. A similar program, PTCORD, for extracting the spatial coordinates of a single node is shown in Fig. 3.7.

Fig. 3.1 Subroutine LNODES

```
      SUBROUTINE   LNODES (LID,NE,N,NODES,LNODE)
C     * * * * * * * * * * * * * * * * * * * * * * * * *
C            EXTRACT NODES ASSOCIATED WITH ELEMENT LID
C     * * * * * * * * * * * * * * * * * * * * * * * * *
      DIMENSION NODES(NE,N), LNODE(N)
C     NE = NUMBER OF ELEMENTS IN SYSTEM
C     N = NUMBER OF NODES PER ELEMENT
C     LID = ELEMENT NUMBER
C     NODES = NODAL INCIDENCES OF ALL ELEMENTS
C     LNODE = THE N NODAL INCIDENCES OF THE ELEMENT
      DO 10  I = 1,N
   10 LNODE(I) = NODES(LID,I)
      RETURN
      END
```

Fig. 3.2 Typical mesh data

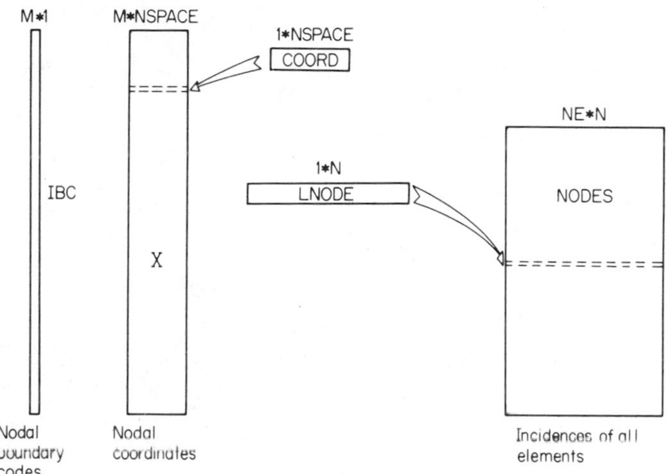

Fig. 3.3 Subroutine ELBAND

```
      SUBROUTINE  ELBAND (N,NG,IBW,LNODE)
C     * * * * * * * * * * * * * * * * * * * * * * * * *
C     ELEMENT BANDWIDTH CALCULATION
C     * * * * * * * * * * * * * * * * * * * * * * * * *
      DIMENSION  LNODE(N)
C     LNODE = ELEMENT INCIDENCES
C     N = NUMBER OF NODES PER ELEMENT
C     NG = NUMBER OF PARAMETERS PER NODE
C     IBW = HALF BANDWIDTH INCLUDING THE DIAGONAL
C        OF A SYMMETRICAL SYSTEM
      IBW = 1
      NLESS = N - 1
      DO 20 I = 1,NLESS
      II = I + 1
      LNI = LNODE(I)
      IF ( LNI.LT.1 )  GO TO 20
      DO 10  J = II,N
      NEW = NG*(IABS(LNODE(J)-LNI) + 1)
      IF ( NEW.GT.IBW ) IBW = NEW
   10 CONTINUE
   20 CONTINUE
      RETURN
      END
```

Fig. 3.4 Element bandwidth and column sizes

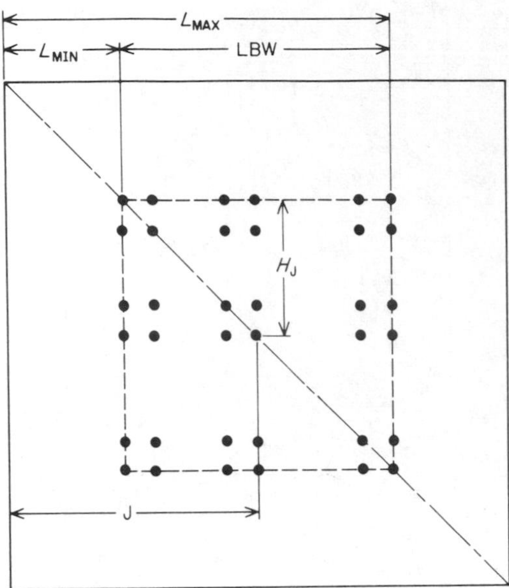

Fig. 3.5 Subroutine SYSBAN

```
      SUBROUTINE SYSBAN (NE,N,NG,IBW,NODES,LNODE,LMAX)
C     * * * * * * * * * * * * * * * * * * * * * * * * * * *
C     DETERMINE UPPER HALF BAND WIDTH OF SYSTEM
C     * * * * * * * * * * * * * * * * * * * * * * * * * * *
      DIMENSION  NODES(NE,N),  LNODE(N)
C     NE = NUMBER OF ELEMENTS IN SYSTEM
C     N = NUMBER OF NODES PER ELEMENT
C     NG = NUMBER OF PARAMETERS (DOF) PER ELEMENT
C     IBW = MAXIMUM HALF BANDWIDTH = LBW MAX
C     NODES = NODAL INCIDENCES OF ALL ELEMENTS
C     LNODE = ELEMENT INCIDENCES LIST
C     LBW = ELEMENT HALF BANDWIDTH
C     LMAX = LAST ELEMENT CAUSING LBW
      LMAX = 1
      IBW = 1
      DO 20  I=1,NE
      CALL  LNODES (I,NE,N,NODES,LNODE)
      CALL  ELBAND (N,NG,LBW,LNODE)
      IF ( LBW-IBW )  20,20,10
   10 IBW = LBW
      LMAX = I
   20 CONTINUE
      RETURN
      END
```

Fig. 3.6 Subroutine ELCORD

```
      SUBROUTINE ELCORD (M,N,NSPACE,X,COORD,LNODE)
C     * * * * * * * * * * * * * * * * * * * * * * * * *
C     DETERMINE COORDINATES OF NODES ON ELEMENT
C     * * * * * * * * * * * * * * * * * * * * * * * * *
CDP   IMPLICIT REAL*8(A-H,O-Z)
      DIMENSION X(M,NSPACE), COORD(N,NSPACE), LNODE(N)
C     M = NUMBER OF NODES IN SYSTEM
C     NSPACE = DIMENSION OF SPACE
C     N = NUMBER OF NODES PER ELEMENT
C     X = COORDINATES OF SYSTEM NODES
C     COORD = COORDINATES OF ELEMENT NODES
C     LNODE = N ELEMENT INCIDENCES OF ELEMENT
      DO 20  K = 1,NSPACE
      DO 10  I = 1,N
      IF ( LNODE(I).LT.1 )  GO TO 10
      COORD(I,K) = X(LNODE(I),K)
 10   CONTINUE
 20   CONTINUE
      RETURN
      END
```

Fig. 3.7 Subroutine PTCORD

```
      SUBROUTINE  PTCORD (IPT,M,NSPACE,X,COORD)
C     * * * * * * * * * * * * * * * * * * * * * * * * *
C     EXTRACT COORDINATES OF POINT NUMBER IPT
C     * * * * * * * * * * * * * * * * * * * * * * * * *
CDP   IMPLICIT REAL*8(A-H,O-Z)
      DIMENSION  X(M,NSPACE), COORD(1,NSPACE)
C     X = SPATIAL COORDINATES OF ALL NODES IN SYSTEM
C     COORD = SPATIAL COORDINATES OF THE NODE
C     M = TOTAL NUMBER OF NODES IN SYSTEM
C     N = NUMBER OF NODES PER ELEMENT
C     NSPACE = DIMENSION OF THE SPACE
      DO 10  J = 1,NSPACE
 10   COORD(1,J) = X(IPT,J)
      RETURN
      END
```

3.2 Property retrieval

Recall that the input data include the element properties for each element in the system. Thus it is a simple matter to scan these data and extract the properties for any particular element. This operation is executed by subroutine ELPRTY. After the element properties have been extracted they can be passed to the subroutine(s) that calculate the element matrices. The above discussion assumed that the element properties do not vary spatially over the element. If this is not the case then it would be necessary to introduce an alternative procedure that could define a spatial variation of element properties. This could be accomplished by inputting the properties at the nodal points and using the element interpolation functions to also define the spatial variations of the properties.

Subroutine LPTPRT can be used to establish an array containing the nodal point properties of all nodes connected to a particular element. The routine uses the element incidence list to guide the extraction of the proper data from the input nodal properties array. The properties data associated with the element incidences are stored in array PRTLPT and are passed to the element matrices routines. Subroutines ELPRTY and LPTPRT are shown in Fig. 3.8 and Fig. 3.9, respectively. The ability of the finite element methods to treat many different types of properties is one of their major practical advantages. Examples of the types of properties (or differential equation coefficients) that one commonly encounters are the thermal conductivity in a heat conduction analysis and the nodal point temperatures in a thermal stress analysis. Several specific examples of typical property data will be given in the example applications beginning in Chapter 10.

Consider a typical nodal point in the system and recall that there are NG parameters associated with each node. Thus at a typical node there will be NG local degree of freedom numbers ($1 \leq J \leq NG$) and a corresponding set of system degree of freedom numbers. If I denotes the system node number of the point, then the NG corresponding system degree of freedom numbers are calculated by utilizing Eqn. (1.1). These elementary calculations are carried out by subroutine INDXPT. The program assigns NG storage locations for the vector, say INDEX, containing the system degree of freedom numbers associated with the specified nodal point—see Table 3.1 and Fig. 3.10 for the related computations.

Recall from the previous section that in order to calculate the system degree of freedom numbers at a point it is necessary to know the system node number of the point. Similarly, it is necessary to know the system node numbers (element incidences) associated with a particular element if one wishes to calculate the system degree of freedom numbers associated with the element. By utilizing the list of N nodal points (element incidences) associated with a specific element, it is possible to again apply Eqn. (1.1) and calculate the system degree of freedom numbers that correspond to the N*NG parameters assigned to the specific element. These calculations, illustrated in Table 3.2, are executed by subroutine INDXEL as shown in Fig. 3.11. Storage locations are established for the N element incidences (extracted by subroutine LNODES) and the N*NG system degree of freedom numbers, in vector INDEX, associated with the element.

If the properties depend on the nodal parameters, as they often do in an iterative solution, then one may need to recover the previous nodal parameters associated with an element. Subroutine ELFRE, Fig. 3.12, uses the above element INDEX vector to extract those data. We will see typical uses of the above pre-element calculations in later sections.

Fig. 3.8 Subroutine ELPRTY

```
      SUBROUTINE   ELPRTY (LID,LHOMO,IP3,IP4,JP2,KP2,
     1                     LPFIX,FLTEL,LPROP,ELPRCP)
C     * * * * * * * * * * * * * * * * * * * * * * * * * *
C     EXTRACT PROPERTIES OF A ELEMENT,LID, FROM TCTAL
C        PROPERTIES ARRAYS
C     * * * * * * * * * * * * * * * * * * * * * * * * * *
CDP   IMPLICIT REAL*8(A-H,O-Z)
      DIMENSICN   FLTEL(IP3,KP2), ELPROP(KP2),
     1            LPFIX(IP4,JP2), LPROP(JP2)
C     LPFIX = SYSTEM ARRAY OF FIXED PT ELEM PROPERTIES
C     LPROP = ELEM FIXED PT PROPERTIES ARRAY
C     FLTEL = SYS ARRAY OF FLOATING PT NCDAL PROP
C     ELPRCP = ELEM FLOATING PT PROPERTIES ARRAY
C     IF LHOMO=1 ELEMENT PROPERTIES ARE HOMOGENEOUS
      I = LID
      IF ( LHOMO.EQ.1 )  I = 1
C     FLOATING POINT PROPERTIES
      DO 10  J = 1,KP2
   10 ELPROP(J) = FLTEL(I,J)
C     FIXED POINT PROPERTIES
      DO 20  J = 1,JP2
   20 LPRCP(J) = LPFIX(I,J)
      RETURN
      END
```

Fig. 3.9 Subroutine LPTPRT

```
      SUBROUTINE   LPTPRT (N,IP1,IP5,KP1,FLTNP,PRTLPT,
     1                     LNODE,NHOMO)
C     * * * * * * * * * * * * * * * * * * * * * * * * * *
C     EXTRACT FLOATING POINT PROPERTIES AT NCDAL PCINTS
C                       OF AN ELEMENT
C     * * * * * * * * * * * * * * * * * * * * * * * * * *
CDP   IMPLICIT REAL*8 (A-H,O-Z)
      DIMENSION  FLTNP(IP1,KP1), PRTLPT(IP5,KP1), LNODE(N)
C     FLTNP = FLOATING POINT PROP. ARRAY OF SYSTEM NODES
C     PRTLPT= FLOATING POINT PROP. ARRAY OF ELEMENT NCDES
C     LNODE = ELEMENT INCIDENCES ARRAY CF THE ELEMENT
C     N = NUMBER OF NCDES PER ELEMENT
C     IF NHOMO=1 NCDAL PROPERTIES ARE HOMOGENEOUS
      DO 20  I = 1,N
      IROW = LNODE(I)
      IF ( NHOMO.EQ.1 )   IROW = 1
      DO 10  J = 1,KP1
   10 PRTLPT(I,J)  = FLTNP(IROW,J)
   20 CCNTINUE
   30 CCNTINUE
      RETURN
      END
```

Fig. 3.10 Subroutine INDXPT

```
      SUBROUTINE   INDXPT (IPT,NG,INDEX)
C     * * * * * * * * * * * * * * * * * * * * * * * * * *
C     DETERMINE DEGREES OF FREEDOM NUMBERS AT NOCF
C     * * * * * * * * * * * * * * * * * * * * * * * * * *
      DIMENSION  INDEX(NG)
C     IPT = SYSTEM NODE NUMBER
C     NG = NUMBER OF PARAMETERS (DOF) PER NCDE
C     INDEX = SYSTEM DOF NOS OF NODAL DOF
      NGIM1 = NG*(IPT-1)
      DO 10  J = 1,NG
C     INDEX(J) = NG*(IPT-1) + J
   10 INDEX(J) = NGIM1 + J
      RETURN
      END
```

Fig. 3.11 Subroutine INDXEL

```
      SUBROUTINE INDXEL (N,LEMFRE,NG,LNODE,INDEX)
C     * * * * * * * * * * * * * * * * * * * * * * * * * * * *
C     DETERMINE DEGREES OF FREEDOM NUMBERS OF ELEMENT
C     * * * * * * * * * * * * * * * * * * * * * * * * * * * *
      DIMENSION INDEX(LEMFRE), LNODE(N)
C     N = NUMBER OF NODES PER ELEMENT
C     NG = NUMBER OF PARAMETERS (DOF) PER NODE
C     LEMFRE = N*NG = NUMBER OF DOF PER ELEMENT
C     LNODE = NODAL INCIDENCES OF THE ELEMENT
C     INDEX = SYSTEM DOF NOS OF ELEMENT DOF
      DO 20  K = 1,N
      IDOF = -NG
      IF ( LNODE(K).GT.0 ) IDOF = IDOF + NG*LNODE(K)
      NGKM1 = NG*(K-1)
      DO 10  IG = 1,NG
      IELM = NGKM1 + IG
C     INDEX(NG*(K-1)+IG) = NG*(LNODE(K)-1)+IG
   10 INDEX(IELM) = IDOF + IG
   20 CONTINUE
      RETURN
      END
```

Fig. 3.12 Subroutine ELFRE

```
      SUBROUTINE  ELFRE (NDFREE,NELFRE,D,DD,INDEX)
C     * * * * * * * * * * * * * * * * * * * * * * * * * * * *
C     EXTRACT ELEMENT DEGREES OF FREEDOM FROM SYSTEM DOF
C     * * * * * * * * * * * * * * * * * * * * * * * * * * * *
CDP   IMPLICIT REAL*8(A-H,O-Z)
      DIMENSION  D(NELFRE), DD(NDFREE), INDEX(NELFRE)
C     D = NODAL PARAMETERS ASSOCIATED
C     DD = SYSTEM ARRAY OF NODAL PARAMETERS
C     INDEX = ARRAY OF SYSTEM DEGREE OF FREEDOM NUMBERS
C     NELFRE = NUMBER OF DEGREES OF FREEDOM PER ELEMENT
C     NDFREE = TOTAL NUMBER OF SYSTEM DEGREES OF FREEDOM
      DO 10  I = 1,NELFRE
      D(I) = 0.D0
      IF ( INDEX(I).GT.0 ) D(I) = DD(INDEX(I))
   10 CONTINUE
      RETURN
      END
```

Table 3.1

Degree of freedom numbers at node I_s	
Local	System*
1	INDEX(1)
2	INDEX(2)
.	.
.	.
.	.
J	INDEX(J)
.	.
.	.
.	.
NG	INDEX(NG)

INDEX(J) = NG(I_s − 1) + J

Table 3.2

Local node I_L	Parameter number J	System node $I_s = \mathrm{LNODE}(I_L)$	Element degree of freedom numbers: Local $NG*(I_L-1)+J$	Element degree of freedom numbers: System $NG*(I_s-1)+J$
1	1	LNODE(1)	1	NG*(LNODE(1) − 1) + 1
1	2	LNODE(1)	2	
⋮	⋮	⋮		
1	NG	LNODE(1)	⋮	⋮
2	1	LNODE(2)		
⋮	⋮	⋮		
K	IG	LNODE(K)	NG*(K − 1) + IG	NG*(LNODE(K) − 1) + IG
⋮	⋮	⋮		
N	1	LNODE(N)	⋮	⋮
⋮	⋮	⋮		
N	NG	LNODE(N)	N*NG	NG*(LNODE(N) − 1) + NG

3.3 Effects of skyline storage

Up to this point we have been assuming that a banded storage mode will be used. While this is the case in most of the discussions, here it should be noted that other important storage modes can also be treated easily. For example, the *skyline* or profile storage is suggested by Taylor [69] and Bathe and Wilson [14]. To use these procedures we must know the maximum height (out to the last non-zero term) of each column of the assembled system equations.

One easy way to find these data is to find the contribution of each element to the column heights and retain the maximum encountered for each column in the system square matrix. Thus we can have routines similar to ELBAND and SYSBAN. From Fig. 3.4 we note that the typical column height, H_J, of an element is $H_J = J - L_{min} + 1$. These simple calculations are executed by subroutine ELHIGH and the system maximums are tabulated by routine SKYHI which loops over all of the elements and calls ELHIGH.

Other information of importance to skyline solution procedures is the location of the diagonal elements and the total number of coefficients to be stored (which is always less than or equal to the band storage). If we decide to store the columns from the top down, then these calculations are simple, as shown in subroutine SKYDIA. Note that the location of the last diagonal coefficient also corresponds to the total number of coefficients to be stored. The last three programs are shown in Figs. 3.13 to 3.15. An example of typical equation heights due to various element contributions is shown in Fig. 3.16. Also shown is the assembled form of the system equations for the mesh in the figure. Figure 3.17 shows a typical system square matrix. It indicates which columns would be stored, and the location of the diagonal coefficients.

The above discussion has shown how to reduce storage requirements and problem costs by using efficient storage techniques. More importantly we have a better understanding of how to relate the element and system equation numbers, and how to recover basic input data needed for calculating the element matrices. In the next chapter we will outline additional important concepts associated with the generation of various element matrices. Very specific and extensive details of these concepts will be presented in the chapters on typical applications.

Fig. 3.13 Subroutine ELHIGH

```
      SUBROUTINE  ELHIGH (NELFRE,INDEX,LHIGH)
C     * * * * * * * * * * * * * * * * * * * * * * * * * * * * *
C      FIND SYSTEM COLUMN HEIGHTS OF AN ELEMENT
C     * * * * * * * * * * * * * * * * * * * * * * * * * * * * *
      DIMENSION  INDEX(NELFRE), LHIGH(NELFRE)
C     NELFRE = NO OF DEGREES OF FREEDOM OF ELEMENT
C     INDEX= SYSTEM DOF NOS OF ELEMENT PARAMETERS
C     LHIGH(I) = COLUMN HEIGHT FOR EQUATION INDEX(I)
      MIN = INDEX(1)
C     FIND MINIMUM INDEX
      DO 10  I = 1,NELFRE
      LHIGH(I) = 0
      NDX = INDEX(I)
      IF ( NDX.LT.1 )  GO TO 10
      IF ( NDX.LT.MIN )  MIN = NDX
   10 CONTINUE
C     CONVERT TO COLUMN HEIGHTS
      MIN = MIN - 1
      DO 20  I = 1,NELFRE
      NDX = INDEX(I)
      IF ( NDX.LT.1 )  GO TO 20
      LHIGH(I) = NDX - MIN
   20 CONTINUE
      RETURN
      END
```

Fig. 3.14 Subroutine SKYHI

```
      SUBROUTINE  SKYHI (NDFREE,NE,N,NG,NELFRE,NODES,
     1                   LNODE,INDEX,LHIGH,IDOFHI)
C     * * * * * * * * * * * * * * * * * * * * * * * * * * * * * *
C     FIND COLUMN HEIGHTS OF SYSTEM EQUATIONS
C     SYMMETRIC SKYLINE STORAGE MODE
C     * * * * * * * * * * * * * * * * * * * * * * * * * * * * * *
      DIMENSION  NODES(NE,N), LNODE(N), INDEX(NELFRE),
     1           LHIGH(NELFRE), IDOFHI(NDFREE)
C     NDFREE = TOTAL NO OF SYSTEM DOF
C     NELFRE = NO OF ELEMENT PARAMETERS (DOF)
C     NE = NUMBER OF ELEMENTS
C     N = NUMBER OF NODES PER ELEMENT
C     NG = NO OF PARAMETERS PER NODE
C     NODES = NODAL INCIDENCES OF ALL ELEMENTS
C     LNODE = ELEMENT NODAL INCIDENCES
C     INDEX(I) = SYS DOF NO OF ELEMENT DOF I
C     IDOFHI(I) = COL HEIGHT OF SYS DOF I
C
C     ZERO HEIGHTS
      CALL  ZEROI (NDFREE,IDOFHI)
C     LOOP OVER ELEMENTS
      DO 20  IE = 1,NE
C     EXTRACT NODES, FIND DOF NOS
      CALL LNODES (IE,NE,NODES,LNODE)
      CALL INDXEL (N,NELFRE,NG,LNODE,INDEX)
C     FIND ELEMENT COLUMN HEIGHTS
      CALL ELHIGH (NELFRE,INDEX,LHIGH)
C     COMPARE WITH CURRENT MAXIMUMS
      DO 10  J = 1,NELFRE
      NDX = INDEX(J)
      IF ( NDX.LT.1 )  GO TO 10
      IF ( IDOFHI(NDX).LT.LHIGH(J)) IDOFHI(NDX)=LHIGH(J)
   10 CONTINUE
   20 CONTINUE
      RETURN
      END
```

Fig. 3.15 Subroutine SKYDIA

```
      SUBROUTINE  SKYDIA (NDFREE,IDOFHI,IDIAG)
C     * * * * * * * * * * * * * * * * * * * * * * * * *
C     USE COLUMN HEIGHTS TO FIND DIAGONAL COEFFICIENTS
C     FOR SYMMETRIC SKYLINE STORAGE MODE
C     * * * * * * * * * * * * * * * * * * * * * * * * *
C     ASSUMING SYMMETRIC COLS STORED FROM TOP DOWN
      DIMENSION  IDOFHI(NDFREE), IDIAG(NDFREE)
C     NDFREE = TOTAL NO OF SYSTEM EQUATIONS
C     IDOFHI(I) = COL HEIGHT OF EQ I, WITH DIAG
C     IDIAG(I) = LOCATION OF DIAG OF I-TH EQ
C     NO COEFF IN UPPER TRIANGLE = IDIAG(NDFREE)
      IPOINT = 0
      DO 10  I = 1,NDFREE
      IPOINT = IPOINT + IDOFHI(I)
   10 IDIAG(I) = IPOINT
      RETURN
      END
```

Fig. 3.16 Sample element and system column heights

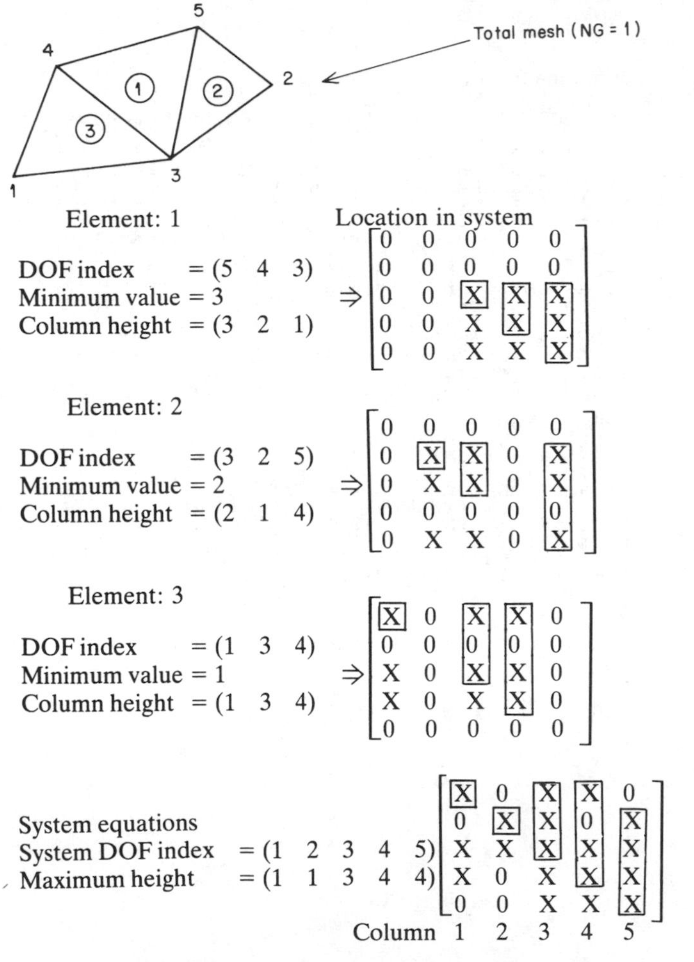

Fig. 3.17 A typical skyline storage of the system square matrix

$$
S = \begin{bmatrix}
S_{11} & S_{12} & 0 & S_{14} & 0 & 0 & 0 & 0 \\
 & S_{22} & S_{23} & 0 & 0 & 0 & 0 & 0 \\
 & & S_{33} & 0 & 0 & S_{36} & 0 & 0 \\
 & & & S_{44} & S_{45} & S_{46} & 0 & 0 \\
 & & & & S_{55} & S_{56} & 0 & S_{58} \\
 & & & & & S_{66} & S_{67} & S_{68} \\
 & & & & & & S_{77} & 0 \\
\text{sym.} & & & & & & & S_{88}
\end{bmatrix}
$$

(a) Actual system square matrix

$$
S \Leftrightarrow \begin{bmatrix}
1 & 2 & - & 6 & - & - & - & - \\
 & 3 & 4 & 7 & - & - & - & - \\
 & & 5 & 8 & - & 12 & - & - \\
 & & & 9 & 10 & 13 & - & - \\
 & & & & 11 & 14 & - & 18 \\
 & & & & & 15 & 16 & 19 \\
 & & & & & & 17 & 20 \\
 & & & & & & & 21
\end{bmatrix}
\qquad
\begin{matrix} 1 \\ 3 \\ 5 \\ 9 \\ 11 \\ 15 \\ 17 \\ 21 \end{matrix}
$$

(b) Corresponding vector locations (c) Diagonal location

3.4 Exercises

1. Refer to Fig. 3.18. Determine the bandwidth if the elements are: (a) four node quadrilaterals; and (b) two node lines.
2. Renumber the nodes in Fig. 3.18 by numbering across line 1–18 first and then proceeding to the right. How much is the bandwidth reduced? Determine a node numbering that gives a smaller bandwidth. What is the element bandwidth in each of the three cases for the quadrilateral in the bottom left corner?
3. Calculate the element column heights in the bottom left quadrilateral in Fig. 3.18. What are the heights if there are two degrees of freedom per node?
4. Consider Fig. 3.19. If the elements are two node line elements sketch the skyline of the assembled equations; what is the system bandwidth? Repeat these operations with the node numbers alternating from side to side (that is, 6 becomes 3, etc.).
5. Utilize the arrays, IDEQ and LM, developed in exercises 6 and 7 of Section 2.4 to develop alternative versions of the programs INDXPT and INDXEL.

Fig. 3.18 Problem mesh 1

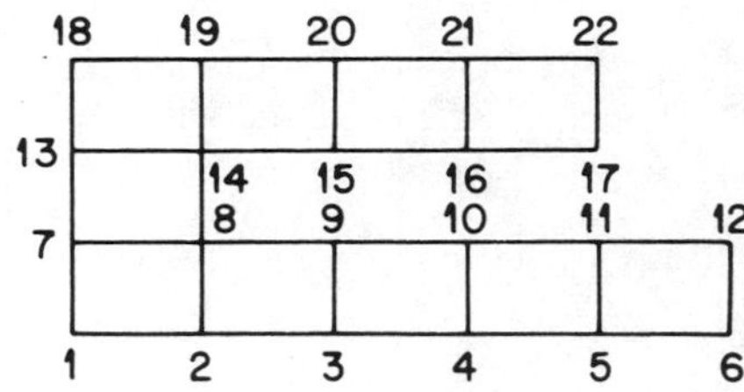

Fig. 3.19 Problem mesh 2

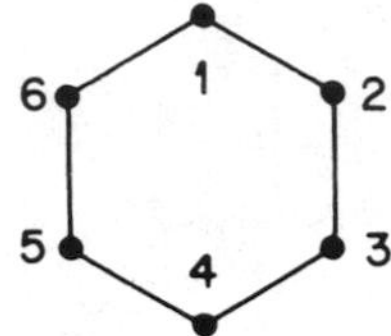

4

Calculation of element matrices

4.1 Introduction

In this chapter typical computations related to problem dependent element calculations will be outlined. These correspond to the boxes after point 1 in the flowchart of Fig. 2.2. Much more detail will be given when example applications are considered later. The major subjects of interest here are typical data requirements for the element square matrix, element column matrix, auxilary element calculations and boundary flux contributions. The absolute minimum problem dependent calculation involves the generation of an element square matrix. Thus our discussions begin with this.

4.2 Square and column matrix considerations

A typical definition of an element square matrix $\boldsymbol{S}$, frequently called the element stiffness matrix, is often of the form

$$\boldsymbol{S}^e = \int_{V^e} \left(P_x \frac{\partial \boldsymbol{H}^T}{\partial x} \frac{\partial \boldsymbol{H}}{\partial x} + P_y \frac{\partial \boldsymbol{H}^T}{\partial y} \frac{\partial \boldsymbol{H}}{\partial y} + P_z \frac{\partial \boldsymbol{H}^T}{\partial z} \frac{\partial \boldsymbol{H}}{\partial z} \right) \mathrm{d}v.$$

This can be considered as a three-dimensional extension of Eqn. (1.11). The integral is over the volume of the element, thus the necessity of defining the spatial orientations of the element; i.e., the nodal coordinates. The integrand involves typical element properties (P_x, P_y, P_z) and the element interpolation functions, $\boldsymbol{H}$, and/or their spatial derivatives, $\partial \boldsymbol{H}/\partial x$, as shown in Section 1.2. The evaluation of such an integral can be difficult or impossible in closed form with the result that numerical integration is often used. Numerical integration techniques will be reviewed in Chapter 6. In the present program the element square matrix is calculated in subroutine ELSQ, Fig. 4.4 and the element column matrix is calculated in ELCOL, Fig. 4.5 (see Section 4.3). These subroutines are two of the few programs that change from one typical problem to another. Several specific examples are given in Chapters 10, 11, and 12.

The following parameters are included in the argument lists to the routines ELSQ and ELCOL which are used to calculate the problem dependent element square and column matrices:

N	= number of nodes per element
NELFRE	= number of element degrees of freedom
NSPACE	= dimension of the space
COORD(N,NSPACE)	= spatial coordinates of the element's nodes
D(NELFRE)	= nodal parameters associated with the element from the previous iteration (if any)
ELPROP(KP2)	= floating point properties of the element (if any)
LPROP(JP2)	= fixed point properties of the element (if any)
PRTLPT(IP5,KP1)	= floating point properties (if any) associated with the nodes of the element
FLTMIS(KP3)	= miscellaneous floating point properties common to the entire system (if any)
S(NELFRE,NELFRE)	= element square matrix (output)
C(NELFRE)	= element column matrix (output)

In addition, the property arrays' dummy dimension parameters IP5, JP2, KP1, KP2, KP3 are also included in the argument list. If any array is not needed, then the corresponding dummy dimensions are set equal to one (in the main program) so that only one storage location is wasted. The array $\boldsymbol{D}$ is usually utilized only in problems requiring iterations. Other problem dependent data are often included in additional dimension or common statements within the subroutines ELSQ and ELCOL. In the present building block programs COMMON statements are utilized only in the problem dependent routines ELSQ, ELCOL, ELPOST, and POSTEL, see Section 4.3. All of these routines also have in their argument list four auxiliary storage unit numbers, NTAPE1, NTAPE2, NTAPE3, and NTAPE4. The latter three units are available for use by the analyst if they are assigned positive unit numbers. The value of NTAPE1 (when > 0) represents the auxiliary unit on which data generated by ELPOST are to be stored.

A typical column matrix, $\boldsymbol{C}$, frequently called the load or force vector, is often defined by an integral of the form

$$C^e = \int_{V^e} g(x, y, z)\, \boldsymbol{H}^T\, dv,$$

where g is some known quantity (e.g. rate of heat generation) that varies over the element. In the present program the calculation of the element column matrix is carried out in subroutine ELCOL. Several specific examples will be presented in Chapters 11 and 12. In some cases the programming will be most efficient if both the element square and column matrices are generated in a single subroutine, say ELSQCL. This is particularly true if the integrals are evaluated numerically.

4.3 Auxiliary calculations

There are two other problem dependent subroutines that are closely related to ELSQ. They are ELPOST, Fig. 4.6, and POSTEL, Fig. 4.7, and they are both related to any post-solution calculations to be performed once the nodal parameters have been calculated. Subroutines ELPOST and POSTEL, respectively, generate and process element data that are required in any post-solution calculations. For example, if the nodal parameters represent temperatures, these routines could perform the calculations for the temperatures and temperature gradients at specific points within the elements once the temperatures (nodal parameters) have been calculated. Subroutine ELPOST usually shares selected element data with subroutine ELSQ by means of a problem dependent COMMON statement. It utilizes these data and/or other information to relate "secondary" quantities of interest to the, as yet unknown, nodal parameters. Subroutine ELPOST contains the input parameter NTAPE1 in its argument list. If NTAPE1 is greater than zero, it represents the external unit number on which any element data associated with the post-solution calculations are to be stored. Of course, the parameter NTAPE1 must be greater than zero or this subroutine is not called.

Subroutine POSTEL is called (if NTAPE1 > 0) after the nodal parameters have been calculated. This program reads, from NTAPE1, the data that were generated by ELPOST. Its argument list contains the following parameters:

NTAPE1 = the external unit containing data generated by ELPOST
NELFRE = number of degrees of freedom per element
D(NELFRE) = array of nodal parameters associated with element IE
IE = element number
NTAPE2,3,4 = additional scratch units

It combines the data stored by ELPOST and the calculated nodal parameters, $\boldsymbol{D}$, to determine secondary quantities of interest within element IE.

The element square and column matrices must be generated for each and every element in the system. These calculations can represent a large percentage of the total computing time and so these calculations should be programmed as efficiently as possible. Some economy considerations associated with generating the element matrices will be presented in Section 4.5.

In an integral formulation involving a Neumann-type boundary condition the boundary flux contributions usually define a column matrix, say $\boldsymbol{C}^b$, with NFLUX coefficients and a corresponding square matrix. These matrices are often defined by integrals of the form

$$\boldsymbol{C}^b = \int_\Gamma \boldsymbol{H}_b^T \boldsymbol{f} \, \mathrm{d}s,$$

$$\boldsymbol{S}^b = \int_\Gamma \mathrm{H}_b^T \mathrm{H}_b P \, \mathrm{d}s,$$

where $\boldsymbol{f}$ denotes the specified flux components, $P = P(\boldsymbol{f})$ is a coefficient defined on the boundary, and $\boldsymbol{H}_b$ represents the element interpolation functions along that boundary segment. If the flux components and the parameter of interest both vary linearly along the element boundary segment, one obtains (as shown later) a boundary column matrix contribution such as the one shown in Fig. 4.1, where l_{ij} is the line segment and there is a linear variation in the flux (f_x, f_y).

Similar contributions for two flux components on a quadratic segment are shown in Fig. 4.2. The calculation of boundary segment matrices for typical segments, such as Fig. 4.3, are carried out by the problem dependent subroutine BFLUX.

Typical examples of the problem dependent arguments in ELSQ, ELCOL, ELPOST, POSTEL, and BFLUX are given in Figs. 4.4 to 4.8. The specific applications will not be considered until later, beginning with Chapter 10. Future references to these routines will include specific problem dependent calculations.

Most of the previous operations that are required to generate the element problem dependent calculations can be combined into a single routine. Figure 4.9 shows subroutine GENELM which is typical of an element generation control routine. This was discussed in Chapter 2. Some programs also store the element matrices on auxiliary storage for later use.

Fig. 4.1 Typical boundary column matrix **C** for a segment, l_{ij}, with linear variation of the flux (f_x, f_y) and parameter of interest

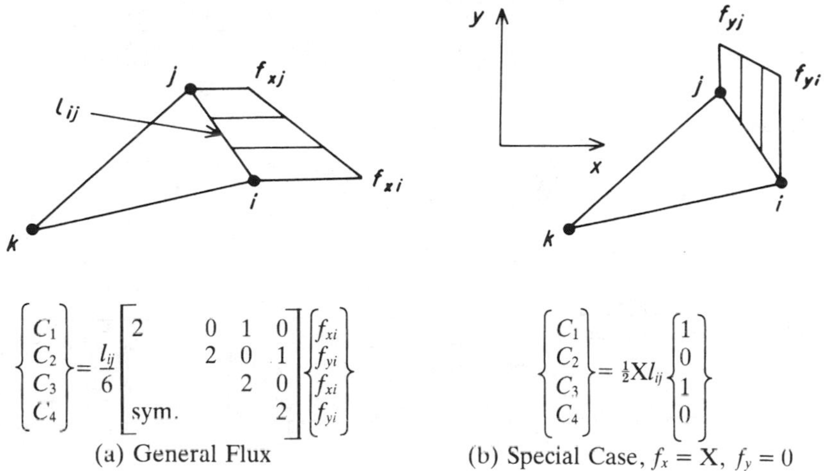

Fig. 4.2 Typical boundary column matrix for a quadratic variation

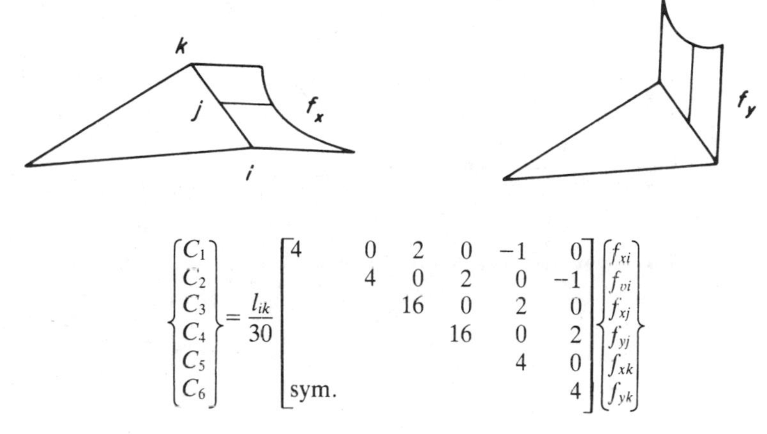

k
j
i
f_x
f_y
$$\begin{Bmatrix} C_1 \\ C_2 \\ C_3 \\ C_4 \\ C_5 \\ C_6 \end{Bmatrix} = \frac{l_{ik}}{30} \begin{bmatrix} 4 & 0 & 2 & 0 & -1 & 0 \\ & 4 & 0 & 2 & 0 & -1 \\ & & 16 & 0 & 2 & 0 \\ & & & 16 & 0 & 2 \\ & & & & 4 & 0 \\ \text{sym.} & & & & & 4 \end{bmatrix} \begin{Bmatrix} f_{xi} \\ f_{vi} \\ f_{xj} \\ f_{yj} \\ f_{xk} \\ f_{yk} \end{Bmatrix}$$

Fig. 4.3 Typical boundary segments

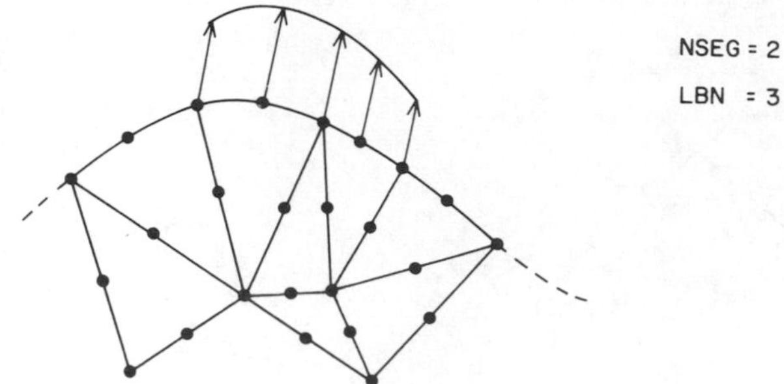

Fig. 4.4 Subroutine ELSQ

```
      SUBROUTINE  ELSQ  (N,NELFRE,NSPACE,IP5,JP2,KP1,KP2,
     1                KP3,COORD,D,ELPROP,LPROP,PRTLPT,FLTMIS,
     2                S,NTAPE1,NTAPE2,NTAPE3,NTAPE4)
C     * * * * * * * * * * * * * * * * * * * * * * * * *
C                 GENERATE ELEMENT SQUARE MATRIX
C     * * * * * * * * * * * * * * * * * * * * * * * * *
CDP   IMPLICIT REAL*8(A-H,O-Z)
      DIMENSION COORD(N,NSPACE), D(NELFRE), ELPROP(KP2),
     1          LPROP(JP2), S(NELFRE,NELFRE),
     2          PRTLPT(IP5,KP1), FLTMIS(KP3)
C     N = NUMBER OF NODES PER ELEMENT
C     ELFRE = NUMBER OF DEGREES OF FREEDOM PER ELEMENT
C     SPACE = DIMENSION OF SPACE
C     P5 = NUMBER OF ROWS IN ARRAY PRTLPT
C     P2 = NUMBER OF COLUMNS IN ARRAYS LPFIX ANC LPROP
C     P1 = NUMBER OF COLUMNS IN FLTNP & PRTLPT & PTPROP
C     P2 = NUMBER OF COLUMNS IN ARRAYS FLTEL AND ELPRCP
C     KP3 = NUMBER OF COLUMNS IN ARRAY FLTMIS
C     COORD = SPATIAL COORDINATES OF ELEMENT'S NCDES
C     D = NODAL PARAMETERS ASSOCIATED WITH AN ELEMENT
C     ELPROP = ELEMENT ARRAY OF FLOATING PT PROPERTIES
C     LPROP = ARRAY OF FIXED POINT ELEMENT PROPERTIES
C     PRTLPT = FLOATING POINT PROP FOR ELEMENT'S NODES
C     FLTMIS = SYSTEM STORAGE OF FLOATING PT MISC PROP
C     S = ELEMENT SQUARE MATRIX
C     NTAPE1 = UNIT FOR POST SOLUTION MATRICES STCRAGE
C     NTAPE2,3,4 = OPTIONAL UNITS FOR USER (USED WHEN > 0)
C     ..........................................................
C     *** ELSQ PROBLEM DEPENDENT STATEMENTS FOLLCW ***
C     ..........................................................
      RETURN
      END
```

Fig. 4.5 Subroutine ELCOL

```
      SUBRCUTINE  ELCOL (N,NELFRE,NSPACE,IP5,JP2,KP1,KP2,
     1            KP3,COORD,D,ELPROP,LPRCP,PRTLPT,FLTMIS,
     2            C,NTAPE1,NTAPE2,NTAPE3,NTAPE4)
C     * * * * * * * * * * * * * * * * * * * * * * * * * *
C               GENERATE ELEMENT COLUMN MATRIX
C     * * * * * * * * * * * * * * * * * * * * * * * * * *
CDP   IMPLICIT REAL*8(A-H,O-Z)
      DIMENSION COORD(N,NSPACE), D(NELFRE), ELPRCP(KP2),
     1          LPROP(JP2), C(NELFRE), PRTLPT(IP5,KP1),
     2          FLTMIS(KP3)
C     N = NUMBER CF NODES PER ELEMENT
C     NELFRE = NUMBER OF DEGREES OF FREEDOM PER ELEMENT
C     NSPACE = DIMENSION OF SPACE
C     IP5 = NUMBER OF ROWS IN ARRAY PRTLPT
C     JP2 = NUMBER OF COLUMNS IN ARRAYS LPFIX ANC LPROP
C     KP1 = NUMBER OF COLUMNS IN  FLTNP & PRTLPT & PTPROP
C     KP2 = NUMBER OF COLUMNS IN ARRAYS FLTEL AND ELPROP
C     KP3 = NUMBER OF COLUMNS IN ARRAY FLTMIS
C     COORD = SPATIAL COORDINATES OF ELEMENT'S NCDES
C     D = NODAL PARAMETERS ASSOCIATED WITH A ELEMENT
C     ELPROP = ELEMENT ARRAY OF FLOATING PCINT PROPERTIES
C     LPRCP = ARRAY OF FIXED POINT ELEMENT PROPERTIES
C     PRTLPT = FLOATING POINT PROP OF ELEMENT'S NODES
C     FLTMIS = SYSTEM STORAGE OF FLOATING PCINT MISC. PROP
C     C - ELEMENT COLUMN MATRIX
C     NTAPE1 = UNIT FOR POST SOLUTION MATRICES STCRAGE
C     NTAPE2,3,4 = OPTIONAL UNITS FOR USER  (USED WHEN > 0)
C     ...........................................................
C     *** ELCOL PROBLEM DEPENDENT STATEMENTS FCLLOW ***
C     ...........................................................
      RETURN
      END
```

Fig. 4.6 Subroutine ELPOST

```
      SUBROUTINE ELPOST (NTAPE1,NTAPE2,NTAPE3,NTAPE4)
C     * * * * * * * * * * * * * * * * * * * * * * * * * *
C          GENERATE DATA FOR POST SOLUTICN CALCULATIONS
C     * * * * * * * * * * * * * * * * * * * * * * * * * *
CDP   IMPLICIT REAL*8(A-H,O-Z)
C     NTAPE1 = UNIT FOR POST SOLUTION MATRICES STORAGE
C     NTAPE2,3,4 = OPTIONAL UNITS FOR USER (WHEN > 0)
C     ...........................................................
C     *** ELPOST PROBLEM DEPENDENT STATEMENTS FCLLCW ***
C     ...........................................................
      RETURN
      END
```

Fig. 4.7 Subroutine POSTEL

```
      SUBROUTINE POSTEL (NTAPE1,NELFRE,D,IE,NTAPE2,
     1                   NTAPE3,NTAPE4,IT,NITER,NE,M)
C     * * * * * * * * * * * * * * * * * * * * * * * * * *
C          ELEMENT LEVEL POST SOLUTION CALCULATICNS
C     * * * * * * * * * * * * * * * * * * * * * * * * * *
C     NTAPE1 = UNIT FOR POST SOLUTION MATRICES STCRAGE
C     NELFRE = NUMBER OF DEGREES OF FREEDOM PER ELEMENT
C     D = NODAL PARAMETERS ASSOCIATED WITH THE ELEMENT
C     IE = ELEMENT NUMBER
C     NTAPE2,3,4 = OPTIONAL UNITS FOR USER (USED WHEN > 0)
C     IT = CURRENT ITERATION NUMBER
C     NITER = MAXIMUM NUMBER OF ITERATICNS
C     NE = TOTAL NUMBER OF ELEMENTS
C     M = TOTAL NUMBER OF NODES
CDP   IMPLICIT REAL*8 (A-H,O-Z)
      DIMENSION  D(NELFRE)
C     ...........................................................
C     *** POSTEL PROBLEM DEPENDENT STATEMENIS FOLLCW ***
C     ...........................................................
      RETURN
      END
```

Fig. 4.8 Subroutine BFLUX

```
      SUBROUTINE  BFLUX (FLUX,COORD,LBN,NSPACE,NFLUX,
     1                   NG,C,S,IOPT)
C     * * * * * * * * * * * * * * * * * * * * * * * * * * *
C         PROBLEM DEPENDENT BOUNDARY FLUX CONTRIBUTIONS
C     * * * * * * * * * * * * * * * * * * * * * * * * * * *
CDP   IMPLICIT REAL*8 (A-H,O-Z)
      DIMENSION COORD(LBN,NSPACE), FLUX(LBN,NG), C(NFLUX),
     1          S(NFLUX,NFLUX)
C     FLUX = SPECIFIED BOUNDARY FLUX COMPONENTS
C     COORD = SPATIAL COORDINATES OF SEGMENT NODES
C     LBN = NO. OF NODES ON AN ELEMENT BOUNDARY SEGMENT
C     NSPACE = DIMENSION OF SOLUTION SPACE
C     NFLUX = LBN*NG = MAXIMUM NUMBER OF FLUX CONTRIBUTIONS
C     C = BOUNDARY FLUX COLUMN MATRIX CONTRIBUTIONS
C     S = BOUNDARY FLUX SQUARE MATRIX
C     NG = NUMBER OF PARAMETERS PER NODE POINT
C     IOPT = PROBLEM MATRIX REQUIREMENTS
C          = 1, CALCULATE C ONLY
C          = 2, CALCULATE S ONLY
C          = 3, CALCULATE BOTH C AND S
C     ..................................................
C       ** BFLUX PROBLEM DEPENDENT STATEMENTS FOLLOW **
C     ..................................................
      IOPT = 0
      RETURN
      END
```

Fig. 4.9 Subroutine GENELM

```
      SUBROUTINE  GENELM (IE,M,N,NSPACE,NELFRE,
     1            NDFREE,NITER,LPTEST,LHOMO,NNPFLO,IP1,
     2            IP3,IP4,IP5,JP2,KP1,KP2,NTAPE1,
     3            LNODE,INDEX,LPFIX,LPROP,X,FLTNP,FLTEL,
     4            FLTMIS,ELPROP,COORD,S,C,D,DDOLD,NHOMO,
     5            PRTLPT,KP3,NULCOL,NTAPE2,NTAPE3,NTAPE4)
C     * * * * * * * * * * * * * * * * * * * * * * * * * *
C      GENERATE ELEMENT MATRICES AND POST SOLUTION DATA
C     * * * * * * * * * * * * * * * * * * * * * * * * * *
CDP   IMPLICIT REAL*8 (A-H,O-Z)
      DIMENSION X(M,NSPACE), FLTNP(IP1,KP1), FLTEL(IP3,KP2),
     1          FLTMIS(KP3), ELPROP(KP2), COORD(N,NSPACE),
     2          S(NELFRE,NELFRE), C(NELFRE), DDOLD(NDFREE),
     3          D(NELFRE), PRTLPT(IP5,KP1), LNODE(N),
     4          INDEX(NELFRE), LPFIX(IP4,JP2), LPROP(JP2)
C     IE = ELEMENT NO, M = NO OF SYSTEM NODES
C     N = NO OF NODES PER ELEM, NSPACE = DIM OF SPACE
C     NELFRE = NUMBER OF DEGREES OF FREEDOM PER ELEMENT
C     NDFREE = TOTAL NUMBER OF SYSTEM DEGREES OF FREEDOM
C     NITER = NO. OF ITERATIONS TO BE RUN (USUALLY 1)
C     IF LPTEST.GT.0  ELEMENT PROPERTIES HAVE BEEN DEFINED
C     IF LHOMO=1 ELEMENT PROPERTIES ARE HOMOGENEOUS
C     NNPFLO = NO.  FLOATING POINT NODAL PROPERTIES
C     IP,JP,KP = PROPERTIES ARRAYS DIMENSION DATA
C     IF NHOMO=1 NODAL   PROPERTIES ARE HOMOGENEOUS
C     NTAPE1 = UNIT FOR POST SOLUTION MATRICES STORAGE
C     NTAPE2,3,4 = OPTIONAL UNITS FOR USER (USED WHEN > 0)
C     LNODE = THE N ELEMENT INCIDENCES OF THE ELEMENT
C     INDEX = SYSTEM DOF NUMBERS ASSOCIATED WITH ELEMENT
C     LPFIX = SYSTEM ARRAY OF FIXED PT ELEM PROP
C     LPROP = ELEM ARRAY OF FIXED PT ELEM PROP
C     X = COORDINATES OF SYSTEM NODES
C     FLTNP, FLTEL, FLTMIS = FLOATING POINT PROPS OF SYSTEM
C     COORD = COORDINATES OF ELEMENT NODES
C     D = ARRAY OF ELEMENT DEGREES OF FREEDOM
C     DDOLD = SYSTEM NODAL PARAMETERS FROM LAST ITERATION
C     PRTLPT= FLOATING POINT PROP. ARRAY OF ELEMENT NODES
C     IF NULCOL.NE.0  ELEMENT COLUMN MATRIX IS ALWAYS ZERO
C
C-->  EXTRACT NODAL COORDINATES
      CALL  ELCORD (M,N,NSPACE,X,COORD,LNODE)
C     EXTRACT NODAL POINT PROPERTIES (IF ANY)
      IF ( NNPFLO.GT.0 )  CALL  LPTPRT (N,IP1,IP5,KP1,FLTNP,
     1                          PRTLPT,LNODE,NHOMO)
```

Fig. 4.9—continued

```
C       EXTRACT NODAL PARAMETERS FROM LAST ITERATICN (IF ANY)
        IF ( NITER.GT.1 )  CALL  ELFRE (NDFREE,NELFRE,D,DDOLD,
     1                                  INDEX)
C-->    EXTRACT ELEMENT MATERIAL PROPERTIES (IF ANY)
        IF ( LPTEST.GT.0 )  CALL ELPRTY (IE,LHOMO,IP3,IP4,JP2,
     1                          KP2,LPFIX,FLTEL,LPROP,ELPROP)
C-->    GENERATE ELEMENT SQUARE AND COLUMN MATRICES
        CALL  ELSQ (N,NELFRE,NSPACE,IP5,JP2,KP1,KP2,KP3,COORD,
     1             D,ELPROP,LPROP,PRTLPT,FLTMIS,S,NTAPE1,
     2             NTAPE2,NTAPE3,NTAPE4)
        IF ( NULCOL.EQ.0 )  CALL ELCOL (N,NELFRE,NSPACE,IP5,
     1                      JP2,KP1,KP2,KP3,COORD,D,ELPROP,
     2                      LPROP,PRTLPT,FLTMIS,C,NTAPE1,
     3                      NTAPE2,NTAPE3,NTAPE4)
C-->    STORE DATA FOR POST SOLUTION CALCULATICNS (IF ANY)
        IF ( NTAPE1.GT.0 )  CALL  ELPOST (NTAPE1,NTAPE2,
     1                                   NTAPE3,NTAPE4)
C       SYS PROP UPDATE COULD BE DONE HERE
        RETURN
        END
```

4.4 Condensation of element's internal degrees of freedom

As one becomes more experienced with the finite element method one finds it desirable at times to consider elements with internal degrees of freedom. Here the word internal is used to mean that the quantity of interest is not shared with any other element (or master element). Figure 4.10 illustrates the concept of distinguishing between internal and shared degrees of freedom. Internal degrees of freedom can be introduced to improve the accuracy of the interpolating functions and/or as an artifice to reduce the amount of input data. The most common use is illustrated in part (d) of Fig. 4.10. This shows a *master* quadrilateral element with four nodes which is input by the user. In the subroutines that define the element matrices for the quadrilateral the following procedure is utilized. First an internal node is defined to be at some convenient location such as the centroid of the quadrilateral. This internal node is used to define four triangular sub-elements. The element matrices of these triangular elements are generated and assembled together to represent the matrices of the quadrilateral element with five nodes. However, this combined element would have more degrees of freedom (5*NG) than the system definition indicates it should have (4*NG). Therefore, it is desirable to condense out the element's internal degrees of freedom. To do this it is necessary, in effect, to require the governing functional to also be minimized over the element. This allows one to write the element equations as:

$$\left[\begin{array}{c:c} S_{AA} & S_{AB} \\ \hdashline S_{BA} & S_{BB} \end{array}\right] \left\{\begin{array}{c} D_A \\ \hdashline D_B \end{array}\right\} = \left\{\begin{array}{c} C_A \\ \hdashline C_B \end{array}\right\}, \tag{4.1}$$

where the element equations have been partitioned to distinguish between shared degrees of freedom, $\mathbf{D}_A$, and internal degrees of freedom, $\mathbf{D}_B$. From these equations it is possible to relate $\mathbf{D}_B$ to $\mathbf{D}_A$; that is

$$\mathbf{D}_B = \mathbf{S}_{BB}^{-1}(\mathbf{C}_B - \mathbf{S}_{BA}\mathbf{D}_A). \tag{4.2}$$

Therefore, the element equations can be rewritten in condensed form as

$$\mathbf{S}_{AA}^{*}\mathbf{D}_A = \mathbf{C}_A^{*}, \tag{4.3}$$

where $\mathbf{S}^*$ and $\mathbf{C}^*$ are defined in terms of the original matrices $\mathbf{S}$ and $\mathbf{C}$. The use of internal degrees of freedom has been shown to improve accuracy and reduce the number of system equations that must be solved in many cases.

The operations, illustrated for a numerical example in Fig. 4.11, required to condense out the internal degrees of freedom are executed by subroutine CONDSE. Since matrix $\mathbf{S}_{BB}$ can be relatively large for some elements, it is desirable to avoid the actual inversion of this matrix. Therefore, the routine uses Gaussian elimination, which is computationally much more efficient. The elimination begins with the last degree of freedom to be condensed and works backwards towards the first one. Subroutine CONDSE is shown in Fig. 4.12.

Fig. 4.10 Typical internal 'nodes'

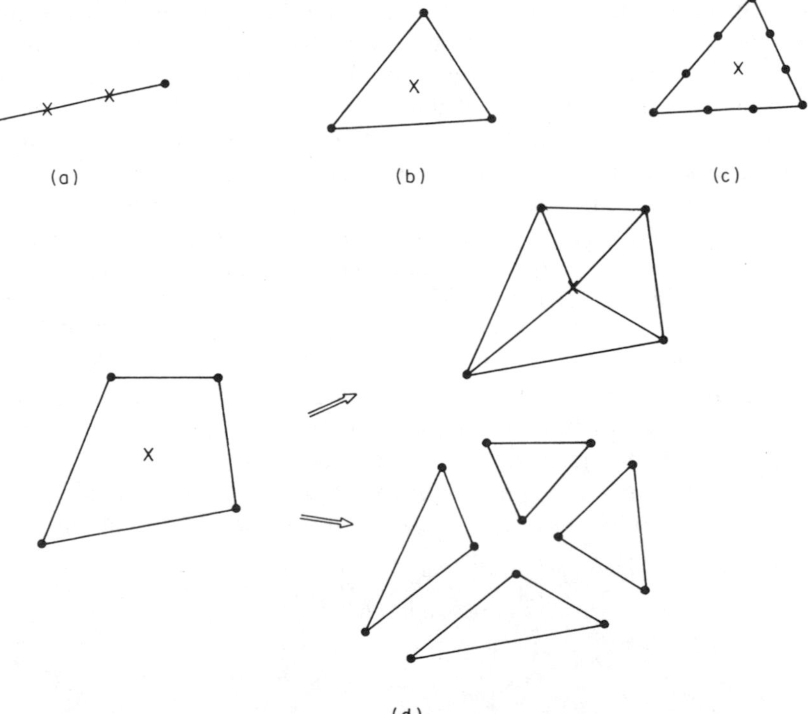

Fig. 4.11 Example condensation procedure

$$\begin{bmatrix} S_{AA} & S_{AB} \\ S_{BA} & S_{BB} \end{bmatrix} \begin{bmatrix} D_A \\ D_B \end{bmatrix} = \begin{bmatrix} C_A \\ C_B \end{bmatrix} \quad \begin{array}{l} B \rightarrow \text{internal dof} \\ A \rightarrow \text{shared dof} \end{array}$$

$$\left[\begin{array}{cccc:c} 8 & 1 & 0 & 1 & 2 \\ 1 & 8 & 1 & 0 & 2 \\ 0 & 1 & 8 & 1 & 2 \\ 1 & 0 & 1 & 8 & 2 \\ \hdashline 2 & 2 & 2 & 2 & 16 \end{array}\right] \left[\begin{array}{c} D_1 \\ D_2 \\ D_3 \\ D_4 \\ \hdashline D_5 \end{array}\right] = \left[\begin{array}{c} 24 \\ 37 \\ 40 \\ 46 \\ \hdashline 110 \end{array}\right]$$

(a) Original Element Equations

$$\mathbf{C}_B = \{110\}, \qquad \mathbf{S}_{BB} = [16], \qquad \mathbf{S}_{BB}^{-1} = [1/16]$$

$$\mathbf{S}_{AA}^{*} = \begin{bmatrix} 8 & 1 & 0 & 1 \\ 1 & 8 & 1 & 0 \\ 0 & 1 & 8 & 1 \\ 1 & 0 & 1 & 8 \end{bmatrix} - \frac{1}{4} \begin{bmatrix} 1 & 1 & 1 & 1 \\ 1 & 1 & 1 & 1 \\ 1 & 1 & 1 & 1 \\ 1 & 1 & 1 & 1 \end{bmatrix} = \begin{bmatrix} 7.75 & 0.75 & -0.25 & 0.75 \\ 0.75 & 7.75 & 0.75 & -0.25 \\ -0.25 & 0.75 & 7.75 & 0.75 \\ 0.75 & -0.25 & 0.75 & 7.75 \end{bmatrix}$$

$$\mathbf{C}_A^{*} = \begin{bmatrix} 24 \\ 37 \\ 40 \\ 46 \end{bmatrix} - 13.75 \begin{bmatrix} 1 \\ 1 \\ 1 \\ 1 \end{bmatrix} = \begin{bmatrix} 10.25 \\ 23.25 \\ 26.25 \\ 32.25 \end{bmatrix}$$

(b) Condensed Element Equation

Fig. 4.12 Subroutine CONDSE

```
      SUBROUTINE  CONDSE (NTOTAL,NELFRE,S,C)
C     * * * * * * * * * * * * * * * * * * * * * * * * * * *
C     CONDENSATION OF ELEMENT MATRICES TO REMOVE
C                 INTERNAL DEGREES OF FREEDOM
C     * * * * * * * * * * * * * * * * * * * * * * * * * * *
CDP   IMPLICIT REAL*8(A-H,O-Z)
      DIMENSION  S(NTOTAL,NTOTAL), C(NTOTAL)
C     INTERNAL DEGREES OF FREEDOM *MUST* COME LAST
C         |    SAA   SAB   | |  DA  |     |  CA  |
C         | ............   | | .... |  =  | .... |
C         |    SBA   SBB   | |  DB  |     |  CB  |
C     ENTER FULL ; RETURN CONDENSED  IN SAA AND CA
C     DIMENSION  SAA(NELFRE,NELFRE), CA(NELFRE)
C     SAA* = (SAA) - (SAB)*(SBB)I*(SAB)T
C     CA* = (CA) - (SAB)*(SBB)I*(CB)
C     NTOTAL = ORIG. NO. OF D.O.F. OF ELEMENT
C     NELFRE = FINAL NO. OF D.O.F. OF ELEMENT
C     S = SQUARE ELEMENT MATRIX, C = ELEMENT COLUMN MATRIX
      NELIM = NTOTAL - NELFRE
      ZERO = 0.D0
      DO 30  I = 1,NELIM
      J = NTOTAL - I
      K = J + 1
      SKK = S(K,K)
      CK = C(K)
      IF ( SKK.EQ.ZERO )  GO TO 30
      C(K) = CK/SKK
      DO 20  L = 1,J
      SLKSKK = S(L,K)/SKK
      S(L,K) = SLKSKK
      DO 10  M = L,J
      S(L,M) = S(L,M) - S(K,M)*SLKSKK
   10 S(M,L) = S(L,M)
      C(L) = C(L) - CK*SLKSKK
   20 CONTINUE
   30 CONTINUE
      RETURN
      END
```

4.5 Economy considerations in the generation of element matrices

The calculation of the element matrices can represent an expensive operation. This is especially true for the more advanced element types such as a sixty degree of freedom elastic solid element or a higher order shell element. In such cases the programmer may find it necessary to take special steps to obtain an economical solution. The present building block routines do not have these features but could be easily modified to implement the following procedures.

For simplicity, assume that the element properties are homogeneous throughout the system. In that case, the element matrices would depend on certain geometrical parameters such as its size, shape, and orientation relative to the global coordinate system. That is, if these parameters are the same for a group of elements then the element matrices of that group will be identical. Consider the mesh illustrated in Fig. 4.13(a), which contains twenty-four triangular elements. Note that elements 1 to 12 are geometrically identical as are the group of elements from 13 to 24. Thus it is only necessary to evaluate the element matrices for element number 1 and then to use these matrices for elements 2 to 12. Of course the matrices for element 13 would be evaluated and used for the remaining elements. This approach would only require the evaluation of two sets of element matrices instead of twenty-four. Such a reduction can represent a major economy if the elements have a relatively high number of degrees of freedom.

To take advantage of this concept, one must generally have a uniform grid spacing over most of the mesh area. Assuming this is the case, it is still necessary to identify clearly those elements which are the same. Some programs allow the user to input a special integer element code to indicate if an element is geometrically the same as the previous element. This is a simple and effective procedure but it places the full burden on the user. Another method is to have a special program segment in the routine for evaluating the element square matrix that checks the important geometrical parameters of the element. If any of these parameters differ from those of the previous element, then a new element matrix is calculated; otherwise, the subroutine simply returns with the matrix calculated for the previous element. The latter approach is relatively inexpensive since many, if not all, of the geometrical parameters are required for later use in calculating the element matrices. Also, elements of different geometries are often quickly identified. For example, if their areas differ, a new element matrix will be required. This approach to more economical matrix generation has an added advantage in that the average user does not have to know of its existence. Of course, the user should be requested to number (or input) geometrically similar elements in a consecutive manner.

Figure 4.13(b) illustrates another economy that some authors recommend, that is to use exact formulations of rectangular elements where possible and mix them with numerically integrated curved elements when it is necessary to match curved boundaries. The matrices for a single rectangular element can be used again and again when multiplied by a scaling factor. This approach is especially attractive when the problem is three-dimensional. The numerically integrated elements are shown hashed in Fig. 4.13.

Fig. 4.13 Element economy considerations

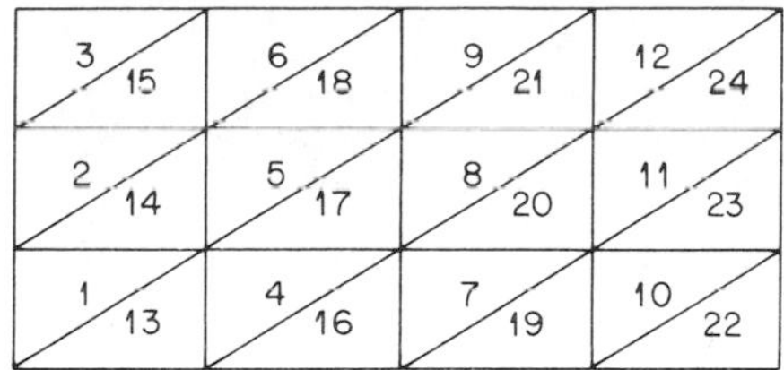

(a) Groups of repeated elements

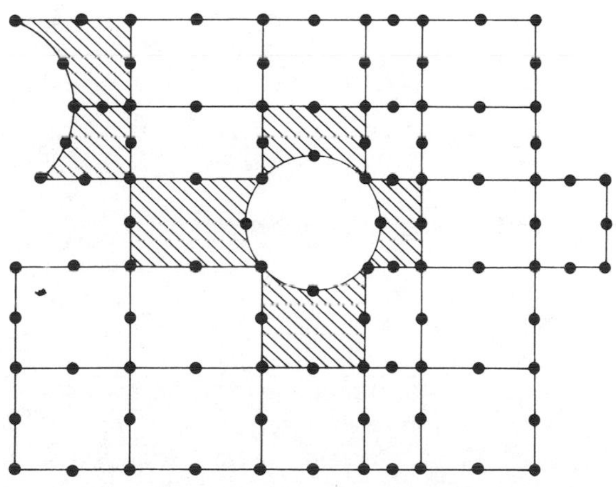

(b) Combining exact and numerical integration

5

Isoparametric elements

5.1 Introduction

In modern finite element methods the use of numerically integrated *isoparametric elements* is becoming increasingly common. In order to understand the programming concepts associated with isoparametric elements, it is necessary to understand some of their basic theoretical foundations. Since detailed treatments of isoparametric elements are lacking, or incomplete, in most theoretical texts, the basic foundations will be reviewed in the next section before their programming aspects are considered.

5.2 Fundamental theoretical concepts

Isoparametric elements utilize a local coordinate system to formulate the element matrices. The local coordinates, say r, s, and t, are usually dimensionless and range from 0 to 1, or from -1 to 1. The latter range is usually preferred since it is directly compatible with the definition of abscissa utilized in numerical integration by Gaussian quadratures. The elements are called isoparametric since the same (iso) local coordinate parametric equations (interpolation functions) used to define any quantity of interest within the elements are also utilized to define the global coordinates of any point within the element in terms of the global spatial coordinates of the nodal points. Let the global spatial coordinates again be denoted by x, y, and z. Let the number of nodes per element be N. For simplicity, consider a single scalar quantity of interest, say $V(r, s, t)$. The value of this variable at any local point (r, s, t) within the element is assumed to be defined by the values at the N nodal points of the element (V_i^e, $1 \leqslant i \leqslant$

N), and a set of interpolation functions ($H_i(r, s, t)$, $1 \leq i \leq$ N). That is,

$$V(r, s, t) = \sum_{i=1}^{N} H_i(r, s, t) V_i^e,$$

or symbolically, as illustrated in Fig. 1.4,

$$V(r, s, t) = \mathbf{H}\mathbf{V}^e, \tag{5.1}$$

where **H** is a row vector. Generalizing this concept, the global coordinates are defined as

$$x(r, s, t) = \mathbf{H}\mathbf{x}^e, y = \mathbf{H}\mathbf{y}^e, z = \mathbf{H}\mathbf{z}^e.$$

Programming considerations make it desirable to write the last three relations in a partitioned form

$$[x : y : z] = \mathbf{H}[\mathbf{x}^e : \mathbf{y}^e : \mathbf{z}^e], \tag{5.2}$$

where the last matrix simply contains the spatial coordinates (COORD in the building block code) of the N nodal points incident with element e. To illustrate a typical two-dimensional isoparametric element, consider a quadrilateral element with nodes at the four corners, as shown in Fig. 5.1. The global coordinates and local coordinates of a typical corner, i, are (x_i, y_i) and (r_i, s_i), respectively. The following local coordinate interpolation functions (shape functions) have been developed for this element:

$$H_i(r, s) \equiv \tfrac{1}{4}(1 + rr_i)(1 + ss_i), \qquad 1 \leq i \leq 4. \tag{5.3}$$

Recall that

$$V(r, s) = \mathbf{H}(r, s)\mathbf{V}^e$$

$$= [H_1\, H_2\, H_3 H_4] \begin{Bmatrix} V_1 \\ V_2 \\ V_3 \\ V_4 \end{Bmatrix}$$

and

$$x(r, s) = \mathbf{H}(r, s)\mathbf{x}^e,$$

$$y(r, s) = \mathbf{H}(r, s)\mathbf{y}^e.$$

Note that along an edge of the element ($r = \pm 1$, or $s = \pm 1$) these interpolation functions become linear and thus any of these three quantities can be uniquely defined by the two corresponding nodal values on that edge. If the adjacent element is of the same type (linear on the boundary), then these quantities will be continuous between elements since their values are uniquely defined by the shared nodal values on that edge. Since the variable of interest, V, varies linearly on the *edge* of the element, it is called the linear isoparametric quadrilateral although the interpolation functions are bilinear inside the element. The shape functions for this element are presented in subroutine SHP4Q and are given for future reference in Fig. 5.2. It has been shown that, in general, the isoparametric shape functions must satisfy the relations

$$\sum_{i=1}^{N} H_i(r, s) = 1$$

and (5.4)

$$H_i(r_j, s_j) = \begin{cases} 1 & \text{if} \quad i = j \\ 0 & \text{if} \quad i \neq j \end{cases}$$

at the node points. The above two relations can be utilized to check numerically subroutines such as SHP4Q. The latter condition should be checked at each node and the first relation should be checked at random points within the element. A typical routine, SCHECK, for checking shape function routines is shown in Fig. 5.3.

For future reference, note that if one can define the interpolation functions in terms of the local coordinates then one can also define their partial derivatives with respect to the local coordinate system. For example, the local derivatives of the shape functions of the above element are

$$\begin{aligned} \frac{\partial H_i(r, s)}{\partial r} &\equiv \tfrac{1}{4} r_i (1 + s s_i), \\ \frac{\partial H_i(r, s)}{\partial s} &\equiv \tfrac{1}{4} s_i (1 + r r_i). \end{aligned} \tag{5.5}$$

In the three dimensions, let the array containing the local derivatives of the interpolation functions be denoted by $\mathbf{\Delta}$, a 3 by N matrix, where

$$\mathbf{\Delta}(r, s, t) = \begin{bmatrix} \dfrac{\partial}{\partial r}\mathbf{H} \\ \cdots\cdots \\ \dfrac{\partial}{\partial s}\mathbf{H} \\ \cdots\cdots \\ \dfrac{\partial}{\partial t}\mathbf{H} \end{bmatrix}.$$

The expressions in Eqn. (5.5) have been incorporated into subroutine DER4Q, which is shown in Fig. 5.4. These partial derivatives are necessary to establish certain relationships between the two coordinate systems. The functions in subroutine DER4Q (or similar programs) can be numerically checked in several ways. They could be checked at random points within the element by comparing with finite difference approximations based on the use of subroutine SHP4Q. However, a simpler check is available. Taking the local derivatives of Eqn. (5.4), we obtain

$$\sum_{i=1}^{N} \frac{\partial H_i(r,s)}{\partial r} = 0, \qquad \sum_{i=1}^{N} \frac{\partial H_i(r,s)}{\partial s} = 0. \tag{5.6}$$

Checking these two equations at random points in the element would give a numerical check of the functions programmed in DER4Q. A checking routine, DCHECK, of the latter type is shown in Fig. 5.5. To have a valid set of shape functions, it is necessary that they satisfy the checks in SCHECK and DCHECK.

Although x, y, and z can be defined in an isoparametric element in terms of the local coordinates, r, s, and t, a unique inverse transformation is not needed. Thus one usually does not define r, s, and t in terms of x, y, and z. What one must have, however, are the relations between derivatives in the two coordinate systems. From calculus, it is known that the derivatives are related by the *Jacobian*. Recall that from the chain rule of calculus one can write, in general,

$$\frac{\partial}{\partial r} = \frac{\partial}{\partial x}\frac{\partial x}{\partial r} + \frac{\partial}{\partial y}\frac{\partial y}{\partial r} + \frac{\partial}{\partial z}\frac{\partial z}{\partial r}$$

with similar expressions for $\partial/\partial s$ and $\partial/\partial t$. In matrix form these become

$$\begin{Bmatrix} \dfrac{\partial}{\partial r} \\ \dfrac{\partial}{\partial s} \\ \dfrac{\partial}{\partial t} \end{Bmatrix} = \begin{bmatrix} \dfrac{\partial x}{\partial r} & \dfrac{\partial y}{\partial r} & \dfrac{\partial z}{\partial r} \\ \dfrac{\partial x}{\partial s} & \dfrac{\partial y}{\partial s} & \dfrac{\partial z}{\partial s} \\ \dfrac{\partial x}{\partial t} & \dfrac{\partial y}{\partial t} & \dfrac{\partial z}{\partial t} \end{bmatrix} \begin{Bmatrix} \dfrac{\partial}{\partial x} \\ \dfrac{\partial}{\partial y} \\ \dfrac{\partial}{\partial z} \end{Bmatrix}, \tag{5.7}$$

where the square matrix is called the Jacobian. Symbolically, one can write the derivatives of a quantity, such as $V(r, s, t)$, which for convenience is written as $V(x, y, z)$ in the global coordinate system, in the following manner

$$\boldsymbol{\partial}_{\mathrm{l}}\mathbf{V} = \mathbf{J}(r,s,t)\,\boldsymbol{\partial}_{\mathrm{g}}\mathbf{V},$$

where $\mathbf{J}$ is the Jacobian matrix and where the subscripts l and g have been introduced to denote local and global derivatives, respectively. Similarly, the inverse relation is

$$\boldsymbol{\partial}_g \mathbf{V} = \mathbf{J}^{-1} \boldsymbol{\partial}_l \mathbf{V}. \tag{5.8}$$

Thus, to evaluate global and local derivatives, one must be able to establish $\mathbf{J}$ and $\mathbf{J}^{-1}$. In practical application, these two quantities usually are evaluated numerically. Consider the first row of the Jacobian in Eqn. (5.7). Since x, y, and z can be defined, for isoparametric elements, by Eqn. (5.2) it is noted that

$$\left[\frac{\partial x}{\partial r} \vdots \frac{\partial y}{\partial r} \vdots \frac{\partial z}{\partial r}\right] = \left[\frac{\partial}{\partial r}\mathbf{H}\right] [\mathbf{x}^e \,\vdots\, \mathbf{y}^e \,\vdots\, \mathbf{z}^e].$$

Generalizing this observation yields

$$\begin{bmatrix} \frac{\partial x}{\partial r} & \frac{\partial y}{\partial r} & \frac{\partial z}{\partial r} \\ \frac{\partial x}{\partial s} & \frac{\partial y}{\partial s} & \frac{\partial z}{\partial s} \\ \frac{\partial x}{\partial t} & \frac{\partial y}{\partial t} & \frac{\partial z}{\partial t} \end{bmatrix}_{3\times 3} = \begin{bmatrix} \frac{\partial}{\partial r}\mathbf{H} \\ \cdots\cdots \\ \frac{\partial}{\partial s}\mathbf{H} \\ \cdots\cdots \\ \frac{\partial}{\partial t}\mathbf{H} \end{bmatrix}_{3\times N} [\mathbf{x}^e \,\vdots\, \mathbf{y}^e \,\vdots\, \mathbf{z}^e]_{N\times 3}, \tag{5.9}$$

or, in symbolic form,

$$\mathbf{J}(r, s, t) = \boldsymbol{\Delta}(r, s, t)\,[\mathbf{x}^e \,\vdots\, \mathbf{y}^e \,\vdots\, \mathbf{z}^e]. \tag{5.10}$$

This numerically defines the Jacobian matrix, $\mathbf{J}$, at a local point inside a typical element in terms of the spatial coordinates of the element's nodes, $[\mathbf{x}^e \,\vdots\, \mathbf{y}^e \,\vdots\, \mathbf{z}^e]$ which is referenced by the name **COORD** in the subroutines, and the local derivatives, $\boldsymbol{\Delta}$, of the interpolation functions, $\mathbf{H}$. Thus, at any point (r, s, t) of interest, such as a numerical integration point, it is possible to define the values of $\mathbf{J}$, $\mathbf{J}^{-1}$, and the determinant of the Jacobian, $|\mathbf{J}|$. Figure 5.6 illustrates a subroutine, JACOB, which evaluates numerically the Jacobian matrix at a point. A simple program to invert and find the determinant of a one-, two-, or three-dimensional matrix Jacobian is shown in Fig. 5.7.

The integral definitions of the element matrices usually involve the global derivatives of the quantity of interest. From Eqn. (5.1) it is seen that the local derivatives of V are related to the nodal parameters by

$$\begin{Bmatrix} \frac{\partial V}{\partial r} \\ \frac{\partial V}{\partial s} \\ \frac{\partial V}{\partial t} \end{Bmatrix} = \begin{bmatrix} \frac{\partial}{\partial r}\mathbf{H} \\ \cdots\cdots \\ \frac{\partial}{\partial s}\mathbf{H} \\ \cdots\cdots \\ \frac{\partial}{\partial t}\mathbf{H} \end{bmatrix} \mathbf{V}^e,$$

or, symbolically,

$$\boldsymbol{\partial}_1 V(r, s, t)_{3\times 1} = \boldsymbol{\Delta}(r, s, t)_{3\times N}\mathbf{V}^e_{N\times 1}. \tag{5.11}$$

To relate the global derivatives to the nodal parameters one substitutes the above expression into Eqn. (5.8) to obtain

$$\boldsymbol{\partial}_g V = \mathbf{J}^{-1}_{3\times 3}\boldsymbol{\Delta}_{3\times N}\mathbf{V}^e_{N\times 1},$$

or

$$\boldsymbol{\partial}_g V(r, s, t)_{3\times 1} = \mathbf{d}(r, s, t)_{3\times N}\mathbf{V}^e_{N\times 1}, \tag{5.12}$$

where

$$\mathbf{d}(r, s, t) \equiv \mathbf{J}(r, s, t)^{-1}\boldsymbol{\Delta}(r, s, t). \tag{5.13}$$

The matrix **d** is very important since it relates the global derivatives of the quantity of interest to the quantity's nodal values. For the sake of completeness, note that **d** can be partitioned as

$$\mathbf{d}(r, s, t) = \begin{bmatrix} d_x \\ \cdots \\ d_y \\ \cdots \\ d_z \end{bmatrix} = \begin{bmatrix} \dfrac{\partial}{\partial x} \mathbf{H} \\ \cdots\cdots \\ \dfrac{\partial}{\partial y} \mathbf{H} \\ \cdots\cdots \\ \dfrac{\partial}{\partial z} \mathbf{H} \end{bmatrix}, \tag{5.14}$$

so that each row represents a derivative of the interpolation functions with respect to a global coordinate direction. In practice the **d** matrix exists in numerical form at selected points. Clearly, at such a point one would evaluate the matrix $\boldsymbol{\Delta}$ (call DER4Q), calculate the Jacobian (call JACOB) and its inverse, and numerically evaluate the product in Eqn. (5.13). Equation (5.13) is evaluated by subroutine GDERIV, which is shown in Fig. 5.8.

Fig. 5.1 A four node isoparametric element

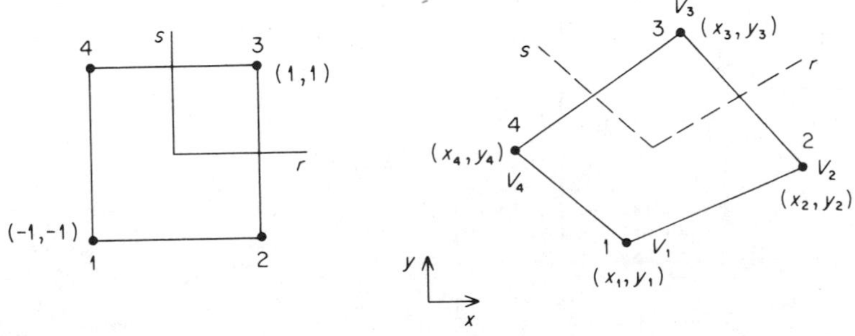

Fig. 5.2 Subroutine SHP4Q

```
      SUBROUTINE  SHP4Q  (R,S,H)
C     * * * * * * * * * * * * * * * * * * * * * * * * * *
C     SHAPE FUNCTIONS OF A 4 NODE ISOPARAMETRIC QUAD
C     * * * * * * * * * * * * * * * * * * * * * * * * * *
CDP   IMPLICIT REAL*8 (A-H,O-Z)
      DIMENSION H(4)
C     (R,S) IS A POINT IN THE LOCAL COORDS        4--3
C     H = LOCAL COORD INTERPOLATION FUNCTIONS     I  I
C     HERE H(I) = 0.25*(1+R*R(I))*(1+S*S(I))      I  I
C     R(I) = LOCAL R-COORDINATE OF NODE I         1--2
C     LOCAL COORDS, 1=(-1,-1)    3=(+1,+1)
      RP = 1. + R
      RM = 1. - R
      SP = 1. + S
      SM = 1. - S
      H(1) = 0.25*RM*SM
      H(2) = 0.25*RP*SM
      H(3) = 0.25*RP*SP
      H(4) = 0.25*RM*SP
      RETURN
      END
```

Fig. 5.3 Subroutine SCHECK

```
      SUBROUTINE  SCHECK (H,N)
C     * * * * * * * * * * * * * * * * * * * * * * * * * *
C     NUMERICAL CHECKING OF N SHAPE FUNCTIONS, H,T
C        LOCAL POINT IN AN ISOPARAMETRIC ELEMENT
C     * * * * * * * * * * * * * * * * * * * * * * * * * *
CDP   IMPLICIT REAL*8 (A-H,O-Z)
      DIMENSION  H(N)
C     H = LOCAL COORDINATE INTERPOLATION FUNCTIONS
C     N = NUMBER OF SHAPE FUNCTIONS
      ONE = 1.D0
      SUM = 0.D0
      DO 10  I = 1,N
   10 SUM = SUM + H(I)
      IF ( SUM.NE.ONE )  WRITE (6,5000)
 5000 FORMAT ( ' SUPPLIED SHAPE FUNCTIONS INCORRECT')
      RETURN
      END
```

Fig. 5.4 Subroutine DER4Q

```
      SUBROUTINE  DER4Q (R,S,DELTA)
C     * * * * * * * * * * * * * * * * * * * * * * * * * *
C     LOCAL DERIVATIVES OF THE SHAPE FUNCTIONS FOR AN
C     ISOPARAMETRIC QUADRILATERAL WITH FOUR NODES
C     * * * * * * * * * * * * * * * * * * * * * * * * * *
CDP   IMPLICIT REAL*8 (A-H,O-Z)
      DIMENSION  DELTA(2,4)
C     DELTA(1,I) = DH/DR , DELTA(2,I) = DH/DS
C     H = LOCAL COORDINATE INTERPOLATION FUNCTIONS
C     HERE D(H(I))/DR = 0.25*R(I)*(1+S*S(I)), ETC.
C     (R,S) IS A POINT IN THE LOCAL COORDINATES
C     R(I) = LOCAL R-COORDINATE OF NODE I
      RP = 1. + R
      RM = 1. - R
      SP = 1. + S
      SM = 1. - S
      DELTA(1,1) = -0.25*SM
      DELTA(1,2) =  0.25*SM
      DELTA(1,3) =  0.25*SP
      DELTA(1,4) = -0.25*SP
      DELTA(2,1) = -0.25*RM
      DELTA(2,2) = -0.25*RP
      DELTA(2,3) =  0.25*RP
      DELTA(2,4) =  0.25*RM
      RETURN
      END
```

Fig. 5.5 Subroutine DCHECK

```
      SUBROUTINE  DCHECK (DELTA,N,NSPACE)
C     * * * * * * * * * * * * * * * * * * * * * * * * * *
C      CHECKING OF THE LOCAL COORDINATE DERIVATIVES OF
C          THE N SHAPE FUNCTIONS AT A LOCAL POINT
C               FOR AN ISOPARAMETRIC ELEMENT
C     * * * * * * * * * * * * * * * * * * * * * * * * * *
CDP   IMPLICIT REAL*8 (A-H,O-Z)
      DIMENSION  DELTA(NSPACE,N)
C     DELTA = LOCAL DERIVATIVES OF SHAPE FUNCTIONS
C     N = NUMBER OF SHAPE FUNCTIONS
C     NSPACE = DIMENSION OF LOCAL SPACE
      ZERO = 0.D0
      DO 20  J = 1,NSPACE
      SUM = ZERO
      DO 10  I = 1,N
   10 SUM = SUM + DELTA(J,I)
      IF ( SUM.NE.ZERO )  WRITE (6,5000)
 5000 FORMAT (' SUPPLIED DERIVATIVES ARE INCORRECT')
   20 CONTINUE
      RETURN
      END
```

Fig. 5.6 Subroutine JACOB

```
      SUBROUTINE  JACOB (N,NSPACE,DELTA,COORD,AJ)
C     * * * * * * * * * * * * * * * * * * * * * * * * * *
C      CALCULATE THE JACOBIAN MATRIX AT A LOCAL POINT
C     * * * * * * * * * * * * * * * * * * * * * * * * * *
CDP   IMPLICIT REAL*8 (A-H,O-Z)
      DIMENSION  DELTA(NSPACE,N), COORD(N,NSPACE),
     1           AJ(NSPACE,NSPACE)
C     N = NUMBER OF NODES PER ELEMENT
C     NSPACE = DIMENSION OF SPACE
C     DELTA = LOCAL  DERIVATIVES OF N INTERPOLATION
C         FUNCTIONS. EVALUATED AT POINT OF INTEREST.
C     COORD = SPATIAL COORDINATES OF ELEMENT'S NODES
C     AJ = JACOBIAN MATRIX = DELTA*COORD
      DO 30  I = 1,NSPACE
      DO 20  J = 1,NSPACE
      SUM = 0.0
      DO 10  K = 1,N
      SUM = SUM + DELTA(I,K)*COORD(K,J)
   10 CONTINUE
      AJ(I,J) = SUM
   20 CONTINUE
   30 CONTINUE
      RETURN
      END
```

Fig. 5.7 **(a)** Subroutine INVDET

```
      SUBROUTINE  INVDET (AJ,AJINV,DET,NSPACE)
C     * * * * * * * * * * * * * * * * * * * * * * * * * *
C     FIND INVERSE AND DETERMINANT OF JACOBIAN
C     * * * * * * * * * * * * * * * * * * * * * * * * * *
CDP   IMPLICIT REAL*8 (A-H,O-Z)
      DIMENSION  AJ(NSPACE,NSPACE), AJINV(NSPACE,NSPACE)
C     NSPACE = NUMBER OF SPATIAL DIMENSIONS
C     AJ = JACOBIAN MATRIX AT A POINT
C     AJINV = INVERSE OF AJ
C     DET = DETERMINANT OF AJ
      GO TO (100,200,300,400), NSPACE
C-->  1-D
  100 DET = AJ(1,1)
      AJINV(1,1) = 1.D0/DET
      RETURN
C-->  2-D
  200 CALL  I2BY2 (AJ,AJINV,DET)
      RETURN
C-->  3-D
  300 CALL  I3BY3 (AJ,AJINV,DET)
      RETURN
C     UNSUPPORTED OPTION
  400 WRITE (6,1000)
 1000 FORMAT (' INVALID DATA SUBR INVDET')
      RETURN
      END
```

Fig. 5.7 **(b)** Subroutine I2BY2

```
      SUBROUTINE  I2BY2 (A,AINV,DET)
C     * * * * * * * * * * * * * * * * * * * * * * * * * *
C     CALCULATE THE DETERMINANT AND INVERSE OF A(2,2)
C     * * * * * * * * * * * * * * * * * * * * * * * * * *
CDP   IMPLICIT REAL*8 (A-H,O-Z)
      DIMENSION  A(2,2), AINV(2,2)
      DET = A(1,1)*A(2,2) - A(1,2)*A(2,1)
      AINV(1,1) = A(2,2)/DET
      AINV(1,2) = -A(1,2)/DET
      AINV(2,1) = -A(2,1)/DET
      AINV(2,2) = A(1,1)/DET
      RETURN
      END
```

Fig. 5.7 **(c)** Subroutine I3BY3

```
      SUBROUTINE  I3BY3 (A,AINV,DET)
C     * * * * * * * * * * * * * * * * * * * * * * * * * *
C     FIND INVERSE AND DETERMINANT OF MATRIX A(3,3)
C     * * * * * * * * * * * * * * * * * * * * * * * * * *
CDP   IMPLICIT REAL*8 (A-H,O-Z)
      DIMENSION A(3,3), AINV(3,3)
      AINV(1,1) = A(2,2)*A(3,3) - A(3,2)*A(2,3)
      AINV(2,1) = -A(2,1)*A(3,3) + A(3,1)*A(2,3)
      AINV(3,1) = A(2,1)*A(3,2) - A(3,1)*A(2,2)
      AINV(1,2) = -A(1,2)*A(3,3) + A(3,2)*A(1,3)
      AINV(2,2) = A(1,1)*A(3,3) - A(3,1)*A(1,3)
      AINV(3,2) = -A(1,1)*A(3,2) + A(3,1)*A(1,2)
      AINV(1,3) = A(1,2)*A(2,3) - A(2,2)*A(1,3)
      AINV(2,3) = -A(1,1)*A(2,3) + A(2,1)*A(1,3)
      AINV(3,3) = A(1,1)*A(2,2) - A(2,1)*A(1,2)
      DET = A(1,1)*AINV(1,1) + A(1,2)*AINV(2,1)
     1    + A(1,3)*AINV(3,1)
      DO 20  J = 1,3
      DO 10  I = 1,3
   10 AINV(I,J) = AINV(I,J)/DET
   20 CONTINUE
      RETURN
      END
```

Fig. 5.8 Subroutine GDERIV

```
      SUBROUTINE  GDERIV (NSPACE,N,AJINV,DELTA,GLOBAL)
C     * * * * * * * * * * * * * * * * * * * * * * * * * *
C     NSPACE GLOBAL DERIVATIVES OF N INTERPOLATION
C         FUNCTIONS AT A LOCAL POINT.
C     * * * * * * * * * * * * * * * * * * * * * * * * * *
CDP   IMPLICIT REAL*8 (A-H,O-Z)
      DIMENSION  AJINV(NSPACE,NSPACE), DELTA(NSPACE,N),
     1           GLOBAL(NSPACE,N)
C     NSPACE = DIMENSION OF SPACE
C     N = NUMBER OF NODES PER ELEMENT
C     AJINV = INVERSE JACOBIAN MATRIX AT LOCAL POINT
C     DELTA = LOCAL COORD DERIV OF N INTERPOLATION
C      FUNCTIONS. EVALUATED AT POINT OF INTEREST.
C     GLOBAL = GLOBAL DERIVATIVES MATRIX AT LOCAL POINT
C             GLOBAL = AJINV*DELTA
      DO 30  I = 1,NSPACE
      DO 20  J = 1,N
      SUM = 0.D0
      DO 10  K = 1,NSPACE
      SUM = SUM + AJINV(I,K)*DELTA(K,J)
 10   CONTINUE
      GLOBAL(I,J) = SUM
 20   CONTINUE
 30   CONTINUE
      RETURN
      END
```

Fig. 5.9 Subroutine CALPRT

```
      SUBROUTINE  CALPRT (N,NNPFLO,H,PRTLPT,VALUES)
C     * * * * * * * * * * * * * * * * * * * * * * * * * *
C     CALCULATE NNPFLO PROPERTIES AT A LOCAL PT USING
C     ELEMENT'S NODAL PROPERTIES, PRTLPT, AND THE N
C     INTERPOLATION FUNCTIONS, H, AT THE POINT
C     * * * * * * * * * * * * * * * * * * * * * * * * * *
CDP   IMPLICIT REAL*8 (A-H,O-Z)
      DIMENSION  H(N), PRTLPT(N,NNPFLO), VALUES(NNPFLO)
C                VALUES = H*PRTLPT
C     N = NUMBER OF NODES PER ELEMENT
C     NNPFLO = NO. FLOATING POINT NODAL PROPERTIES
C     H = INTERPOLATION FUNCTIONS FOR AN ELEMENT
C     PRTLPT = FLOATING PT PROPS OF ELEMENT'S NODES
C     VALUES = LOCAL VALUES OF PROPERTIES
      DO 20  I = 1,NNPFLO
      SUM = 0.D0
      DO 10  J = 1,N
      SUM = SUM + H(J)*PRTLPT(J,I)
   10 CONTINUE
      VALUES(I) = SUM
   20 CONTINUE
      RETURN
      END
```

5.3 Programming isoparametric elements

The element matrices, $\mathbf{S}^e$ or $\mathbf{C}^e$, are usually defined by integrals of the symbolic form

$$\mathbf{A}^e = \iiint_{\Omega^e} \mathbf{B}^e(x, y, z)\, dx\, dy\, dz,$$

where $\mathbf{B}^e$ is usually the sum of products of other matrices involving the element interpolation functions and problem properties. With the element formulated in terms of the local (r, s, t) coordinates, where $\mathbf{B}^e(x, y, z)$ is transformed into $\hat{\mathbf{B}}^e(r, s, t)$, this expression must be rewritten as

$$\mathbf{A}^e = \int_{-1}^{1}\int_{-1}^{1}\int_{-1}^{1} \hat{\mathbf{B}}^e(r, s, t)\,|J(r, s, t)|\,dr\, ds\, dt.$$

In practice, one would use numerical integration to obtain

$$[\mathbf{A}^e] = \sum_{i=1}^{\text{NIP}} W_i \hat{\mathbf{B}}^e(r_i, s_i, t_i)\,|J(r_i, s_i, t_i)|$$

where $\mathbf{B}^e$ and $|\mathrm{J}|$ are evaluated at each of the NIP integration points and where (r_i, s_i, t_i) and W_i denote the tabulated abscissae and weights, respectively. The general procedure for programming an isoparametric element is outlined below:

(A) Initial Programming Steps

1. Define a subroutine (such as SHP4Q) to evaluate the N interpolation functions $\mathbf{H}(r, s, t)$ for any given values of the local coordinates (r, s, t).

2. Define a subroutine (such as DER4Q) to evaluate the NSPACE local derivatives of the interpolation functions $\mathbf{\Delta}(r, s, t)$ for any given values of the local coordinates.
3. Tabulate or calculate the number of integration points, NIP, and the weight coefficients W_i, and local coordinate abscissae (r_i, s_i, t_i) of each integration point.

(B) Steps Preceding the Element Matrix Subroutines
1. Extract the global coordinates, $[\mathbf{x}^e : \mathbf{y}^e : \mathbf{z}^e]$, **COORD**, of the nodes of the element.
2. Extract any constant element properties (ELPROP and/or LPROP), and tabulate the nodal point values (LPTPRT) of any properties which vary with local coordinates.

(C) Steps Within the Element Matrix Routine
1. Establish storage for any problem dependent variables ($\mathbf{H}$, $\mathbf{\Delta}$, $\mathbf{d}$, $\mathbf{J}$, $\mathbf{J}^{-1}$, etc.).
2. Zero the element matrices.
3. Perform the numerical integration (see Part D).

(D) Steps for Each Integration Point
1. Extract, or calculate, the weight and abscissae of the integration point.
2. For the given abscissae (r_i, s_i, t_i) evaluate the interpolation functions, $[\mathbf{H}]$, (call SHP4Q) and the local derivatives of the interpolation functions, $[\mathbf{\Delta}]$, (call DER4Q).
3. Calculate the Jacobian matrix at the point (call JACOB):

$$\mathbf{J}(r_i, s_i, t_i) = \mathbf{\Delta}(r_i, s_i, t_i)\ \mathbf{COORD}.$$

4. Calculate the inverse, $\mathbf{J}(r_i, s_i, t_i)^{-1}$, and determinant $|\mathrm{J}|$ of the Jacobian matrix (e.g., call I2BY2).
5. Calculate the first order global derivatives, $\mathbf{d}$, of the interpolation functions (call GDERIV):

$$\mathbf{d} = \mathbf{J}^{-1}\mathbf{\Delta}.$$

6. If the problem properties vary with the local coordinates, use the nodal values of the properties and the interpolation functions to calculate the properties at the integration point (call CALPRT, Fig. 5.9).
7. Execute the matrix operations defining the matrix integrand. In general, this involves the sum of products of element properties, $\mathbf{H}$, and $\mathbf{d}$.
8. Multiply the resulting matrix by the weighting coefficient and the determinant of the Jacobian and add it to the previous contributions to the element matrix.
9. If any of the above data are to be utilized in post-solution calculations, write them on auxiliary storage.

A typical arrangement of the latter operations is illustrated in subroutine

ISOPAR which is shown in Fig. 5.10. Note that the global coordinates of the integration point are also determined since they are often used for later output of the derivatives. Of course, in an axisymmetric analysis the global radial coordinate is required in the integrand. For most practical problems the Jacobian inverse can be easily obtained with a routine such as INVDET in Fig. 5.7.

The interpolation function routine, SHAPE, and the corresponding local derivative routine, DERIV, will be considered later. Numerical integration is also reviewed in more detail in a later section. The control section of subroutine NGRAND is shown in Fig. 5.11. Its purpose is to evaluate the actual problem dependent integrands, such as (grad **H** . grad **H**) which occurs in a Poisson problem. As we will see later it is sometimes desirable to use different integration rules for different parts of the integrand (see Section 6.5). Thus two or more versions of NGRAND might be required in a typical application. Therefore, in calling ISOPAR the name NGRAND is treated as a dummy EXTERNAL variable so that the user passes in the name of the subroutine that is actually required for the problem. Specific examples will be given in Chapters 10 and 11.

Fig. 5.10 Subroutine ISOPAR

```
      SUBROUTINE  ISOPAR (N,NSPACE,NELFRE,NIP,SQ,COL,QPT,
     1                    QWT,H,DLH,DGH,COORD,XPT,AJ,
     2                    AJINV,NTAPE1,NGRAND)
C     * * * * * * * * * * * * * * * * * * * * * * * * * *
C     NUMERICAL INTEGRATION IN AN ISOPARAMETRIC ELEMENT
C     * * * * * * * * * * * * * * * * * * * * * * * * * *
CDP   IMPLICIT REAL*8 (A-H,O-Z)
      DIMENSION COL(NELFRE),SQ(NELFRE,NELFRE),QWT(NIP),
     1          QPT(NSPACE,NIP),H(N),DLH(NSPACE,N),
     2          DGH(NSPACE,N),COORD(N,NSPACE),XPT(NSPACE),
     3          AJ(NSPACE,NSPACE), AJINV(NSPACE,NSPACE)
C     N = NO NODES PER ELEMENT
C     NSPACE = NO OF SPATIAL DIMENSIONS
C     NELFRE = NO OF ELEMENT DEGREES OF FREEDOM
C     NIP = NUMBER OF INTEGRATION POINTS
C     QPT, QWT = QUADRATURE PT COORDS, WEIGHT
C     SQ,COL = PROB DEPENDENT SQ AND COL INTEGRANDS
C     H = ELEMENT INTERPOLATION FUNCTIONS
C     DLH, DGH = LOCAL, AND GLOBAL, DERIVATIVES OF H
C     COORD = GLOBAL COORD OF NODES OF ELEMENT
C     XPT = GLOBAL COORD OF QUADRATURE POINT
C     AJ,AJINV, DET = JACOBIAN, ITS INVERSE, DETERMINANT
C     NTAPE1 = STORAGE UNIT FOR POST SOLUTION DATA
C     NGRAND = 'EXTERNAL' PROB DEP INTEGRAND ROUTINE
C-->  ZERO INTEGRANDS
      CALL  ZEROA (NELFRE,COL)
      LSQ = NELFRE*NELFRE
      CALL  ZEROA (LSQ,SQ)
C-->  BEGIN INTEGRATION
      DO 100  IP = 1,NIP
C     EVALUATE INTERPOLATION FUNCTIONS
      CALL  SHAPE (QPT(1,IP),H,N,NSPACE)
C     FIND GLOBAL COORD, XPT = H*COORD
      CALL  MMULT (H,COORD,XPT,1,N,NSPACE)
C     FIND LOCAL DERIVATIVES
      CALL  DERIV (QPT(1,IP),DLH,N,NSPACE)
C     FIND JACOBIAN AT THE PT
      CALL  JACOB (N,NSPACE,DLH,COORD,AJ)
C     FORM INVERSE AND DETERMINANT OF JACOBIAN
      CALL  INVDET (AJ,AJINV,DET,NSPACE)
C     EVALUATE GLOBAL DERIVATIVES
      CALL  GDERIV (NSPACE,N,AJINV,DLH,DGH)
C     *** FORM PROBLEM DEPENDENT INTEGRANDS ***
      CALL  NGRAND (QWT(IP),DET,H,DGH,XPT,N,NSPACE,
     1              NELFRE,COL,SQ,NTAPE1)
  100 CONTINUE
      RETURN
      END
```

Fig. 5.11 Subroutine NGRAND

```
      SUBROUTINE  NGRAND (WT,DET,H,DGH,XPT,N,NSPACE,
     1                    NELFRE,COL,SQ,NTAPE1)
C     * * * * * * * * * * * * * * * * * * * * * * * * *
C     PROBLEM DEPENDENT INTEGRAND EVALUATION IN
C     AN ISOPARAMETRIC ELEMENT
C     * * * * * * * * * * * * * * * * * * * * * * * * *
CDP   IMPLICIT REAL*8 (A-H,O-Z)
      DIMENSION COL(NELFRE), SQ(NELFRE,NELFRE),
     1          H(N), DGH(NSPACE,N), XPT(NSPACE)
C     N = NUMBER OF NODES PER ELEMENT
C     NSPACE = NUMBER OF SPATIAL DIMENSIONS
C     NELFRE = NO OF ELEMENT DEGREES OF FREEDOM
C     H = ELEMENT INTERPOLATION FUNCTIONS
C     DGH = GLOBAL DERIVATIVES OF H
C     XPT = GLOBAL COORDS OF THE POINT
C     WT = QUADRATURE WEIGHT AT POINT
C     DET = JACOBIAN DETERMINATE AT POINT
C     COL = PROB DEP COLUMN MATRIX INTEGRAND
C     SQ = PROB DEP SQUARE MATRIX INTEGRANC
C     NTAPE1 = STORAGE UNIT FOR POST SOLUTION DATA
C     ..........................................................
C     *** NGRAND PROBLEM DEPENDENT STATEMENTS FOLLOW ***
C     ..........................................................
      RETURN
      END
```

5.4 Simplex elements, a special case

A complete discussion of isoparametric elements should also include the special case of *simplex* elements. These are the simplest of all finite elements and they have the following properties:

1. The number of nodes per element, N, is one greater than the dimension of the space, i.e. N = NSPACE + 1.
2. The element interpolation functions, H_i, are complete linear functions of the local coordinates.
3. The global interpolation functions are continuous over the entire region of the problem.

The first three members of this family of elements are the line element, triangular element, and tetrahedral element. The first two are frequently utilized in current finite element programs, but the latter has been found to be computationally inefficient and has generally been displaced by the hexahedral isoparametric element family.

As shown in Fig. 5.12, all these elements can be defined in terms of local non-dimensional coordinates, say (r, s, t), which range from zero to unity. Note that the triangle could be thought of as a face of the tetrahedron and the line element as an edge of the tetrahedron (or triangle). The introduction of the non-dimensional local coordinates (r, s, t) makes it possible to define element interpolation functions, $H_i(r, s, t)$, which are independent of the geometry of the element in the global system. For example, the two line element interpolation functions are

$$\begin{aligned} H_1(r) &= 1 - r \\ H_2(r) &= r \end{aligned} \tag{5.15}$$

and for the triangle one has

$$\begin{aligned} H_1(r,s) &= 1 - r - s \\ H_2(r,s) &= \quad\;\; r \\ H_3(r,s) &= \qquad\quad s. \end{aligned} \tag{5.16}$$

Note that line interpolation is a subset of that of the triangle, etc. The generalization of this subset concept for complete polynomials is illustrated in Fig. 5.13.

Recall that the definitions of element (or boundary segment) matrices involve integrals of the interpolation functions, and/or their derivatives. Thus typical coefficients in these matrices can be defined in terms of integrals of the local coordinates (*r, s, t*). The main importance of the simplex elements is the fact that the integrals, over the element, of their local coordinates can be expressed in closed form. This occurs because the Jacobian of the transformation from local to global coordinates is constant. For example, consider an integration over a triangular element. It is known that

$$I^e \equiv \int_{A^e} r^m s^n da = \frac{2A^e\Gamma(m+1)\Gamma(n+1)}{\Gamma(3+m+n)}, \tag{5.17}$$

where Γ denotes the Gamma function. Restricting consideration to positive integer values of the exponents, m and n, yields

$$I^e \equiv 2A^e \frac{m!n!}{(2+m+n)!} = A^e/K_{mn}, \tag{5.18}$$

where ! denotes the factorial and where K_{mn} is an integer constant given in Table 5.1 for common values of m and n. Similarly for the line elements

$$I^e \equiv \int_{L^e} r^m \, dl = L^e \frac{m!}{(1+m)!} = L^e/(1+m) \tag{5.19}$$

and for the tetrahedron

$$I^e \equiv \int_{V^e} r^m s^n t^p \, dv = 6V^e \frac{m!n!p!}{(3+m+n+p)!}. \tag{5.20}$$

Thus, one notes that common integrals of this type can be evaluated by simply multiplying the element characteristic (i.e. global length, area, or volume) by known constants which could be stored in a data statement. These exact integrals can also be utilized for higher order elements if their Jacobians remain constant. That is, if their nodes are in the same relative locations as on the parent elements (i.e. at mid-sides for quadratics, third points for cubic, etc.). These element families are shown in Fig. 5.13. Figure 5.14 summarizes the closed form of integrals for polynomials and Table 5.1 gives commonly used constants.

To illustrate the application of these equations in evaluating element matrices, we consider the following examples, where $\mathbf{H}(r, s, t) = [H_1\, H_2 \ldots H_N]$:

$$\mathbf{I}_1 \equiv \int_{L^e} \mathbf{H}^T \mathrm{d}l = \int_{L^e} \begin{Bmatrix} (1-r) \\ r \end{Bmatrix} \mathrm{d}l = \begin{Bmatrix} L^e - L^e/2 \\ L^e/2 \end{Bmatrix} = \frac{L^e}{2} \begin{Bmatrix} 1 \\ 1 \end{Bmatrix}, \tag{5.21}$$

$$\mathbf{I}_2 \equiv \int_{A^e} \mathbf{H}^T \,\mathrm{d}a = \int_{A^e} \begin{Bmatrix} (1-r-s) \\ r \\ s \end{Bmatrix} \mathrm{d}r = \begin{Bmatrix} A^e - A^e/3 - A^e/3 \\ A^e/3 \\ A^e/3 \end{Bmatrix} = \frac{A^e}{3} \begin{Bmatrix} 1 \\ 1 \\ 1 \end{Bmatrix}, \tag{5.22}$$

and

$$\mathbf{I}_3 \equiv \int_{L^e} \mathbf{H}^T\mathbf{H} \,\mathrm{d}l = \int_{L^e} \begin{bmatrix} (1-2r+r^2) & (r-r^2) \\ \text{sym.} & r \end{bmatrix} \mathrm{d}l$$

$$= \begin{bmatrix} (L^e - 2L^e/2 + L^e/3) & (L^e/2 - L^e/3) \\ \text{sym.} & L^e/3 \end{bmatrix} = \frac{L^e}{6} \begin{bmatrix} 2 & 1 \\ 1 & 2 \end{bmatrix}. \tag{5.23}$$

Many element matrix definitions involve global derivatives of the interpolation functions. Recall from Eqn. (5.14) that these can be represented as

$$\mathbf{d}(r, s, t) \equiv \begin{bmatrix} d_x \\ \cdots \\ d_y \\ \cdots \\ d_z \end{bmatrix} \equiv \begin{bmatrix} \partial \mathbf{H}/\partial x \\ \cdots\cdots \\ \partial \mathbf{H}/\partial y \\ \cdots\cdots \\ \partial \mathbf{H}/\partial z \end{bmatrix} \begin{bmatrix} \mathbf{H}_{,x} \\ \mathbf{H}_{,y} \\ \mathbf{H}_{,z} \end{bmatrix}$$

where, from Eqn. (5.13),

$$\mathbf{d} = \mathbf{J}^{-1}\mathbf{\Delta},$$

in which $\mathbf{J}$ is the Jacobian of the transformation and $\mathbf{\Delta}$ is the local derivative matrix given in Eqn. (5.10). For simplex elements all three of these matrices are constant. Thus they can be moved outside some integrals. For example, consider the integral of the x derivative of the interpolation function $\mathbf{H}$, i.e. of $\mathbf{H}_{,x}$,

$$\mathbf{I} = \int_{L^e} \mathbf{H}_{,x}^T \, dl = \mathbf{d}_x^T \int_{L^e} dl = \frac{1}{L^e} \begin{Bmatrix} -1 \\ 1 \end{Bmatrix} L^e = \begin{Bmatrix} -1 \\ 1 \end{Bmatrix}.$$

At this point it is interesting to compare the exact and numerical integration procedures. Consider a typical term such as

$$\mathbf{I} = \int_{L^e} \mathbf{H}^T q \, dl,$$

where q is a known variable scalar quantity. Assume that one defines the value of q within the element by interpolating to its known nodal values, i.e. $q(r) = \mathbf{H}(r)\mathbf{Q}^e$. Integrating this expression numerically for a constant Jacobian one obtains:

$$\mathbf{I} \approx |\mathrm{J}| \sum_{i=1}^{\mathrm{NP}} W_i \mathbf{H}(r_i)^T q(r_i), \tag{5.24}$$

where W_i denotes the tabulated weights, r_i denotes the tabulated integration point and where the scalar $q(r_i) = \mathbf{H}(r_i)\mathbf{Q}^e$. By way of comparison the exact form would become

$$\mathbf{I} = \int_{L^e} \mathbf{H}^T q \, dl = \left(\int_{L^e} \mathbf{H}^T \mathbf{H} \, dl \right) \mathbf{Q}^e = \frac{L^e}{6} \begin{bmatrix} 2 & 1 \\ 1 & 2 \end{bmatrix} \mathbf{Q}^e. \tag{5.25}$$

Of course, the expressions (5.24) and (5.25) are identically equal for a properly selected numerical integration order, but in this simple case the exact form is probably faster to compute. However, it is possible for additional terms to occur in the integrand. For example, assume that the integral contains an additional variable, v, which is also defined by its nodal values, $\mathbf{V}^e$. The numerical integration is easily extended to include any number of additional terms. For example,

$$\mathbf{I} = \int_{L^e} \mathbf{H}^T v q \, dl \approx |\mathrm{J}| \sum_{i=1}^{\mathrm{NP}} W_i \mathbf{H}_i^T v_i q_i,$$

where the scalar $v_i = \mathbf{H}_i \mathbf{V}^e$. The exact method is less easily extended. For example,

$$\mathbf{I} = \int_{L^e} \mathbf{H}^T v q \, dl = \int_{L^e} v \mathbf{H}^T q \, dl,$$

$$\mathbf{I} = \mathbf{V}^{eT} \left[\int_{L^e} \mathbf{H}^T \mathbf{H}^T \mathbf{H} \, dl \right] \mathbf{Q}^e,$$

where the quantity in the square brackets must be considered as a hyper-matrix. A hyper-matrix is a matrix whose coefficients are themselves matrices. This concept is not usually covered in most engineers' formal education. Also, as more terms appear in the integrand the resulting hyper-matrices become more and more complicated and difficult to evaluate. The resulting hyper-matrix multiplications rapidly become costly as the number of terms increases. Since the numerical integration does not have this added complexity, and also allows one to generate curved elements, it is recommended as the best procedure for the beginner. Of course, one should utilize the known closed form of the simple integrals if it is more efficient.

Before leaving the subject of simplex integrations one should give consideration to the common special case of axisymmetric geometries, with coordinates (ρ, z). Recall from calculus that the Theorem of Pappus relates a differential volume and surface area to a differential area and length in the (ρ, z) plane of symmetry, respectively. That is, $dv = 2\pi\rho\, dA$ and $dS = 2\pi\rho dl$, where ρ denotes the radial distance to the differential element. Thus, typical axisymmetric volume and surface integrals reduce to

$$\begin{aligned}
\mathbf{I}_V &= \int_V \mathrm{H}^T\, dv = 2\pi \int_{A^e} \mathbf{H}^T \rho\, da \\
&= 2\pi \left(\int_{A^e} \mathbf{H}^T\mathbf{H}\, da \right) \boldsymbol{\rho}^e \\
&= \frac{2\pi A^e}{12} \begin{bmatrix} 2 & 1 & 1 \\ 1 & 2 & 1 \\ 1 & 1 & 2 \end{bmatrix} \boldsymbol{\rho}^e
\end{aligned}$$

and

$$\begin{aligned}
\mathbf{I}_S &= \int_S \mathbf{H}^T\, dS = 2\pi \int_{L^e} \mathbf{H}^T \rho\, dl \\
&= 2\pi \left(\int_{L^e} \mathbf{H}^T\mathbf{H}\, dl \right) \boldsymbol{\rho}^e \\
&= \frac{2\pi L^e}{6} \begin{bmatrix} 2 & 1 \\ 1 & 2 \end{bmatrix} \boldsymbol{\rho}^e,
\end{aligned}$$

since $\rho = \mathbf{H}\boldsymbol{\rho}^e$. Many workers like to omit the 2π term and work on a per-unit-radian basis so that they can more easily do both two-dimensional and axisymmetric calculations with a single program.

Fig. 5.12 The simplex element family

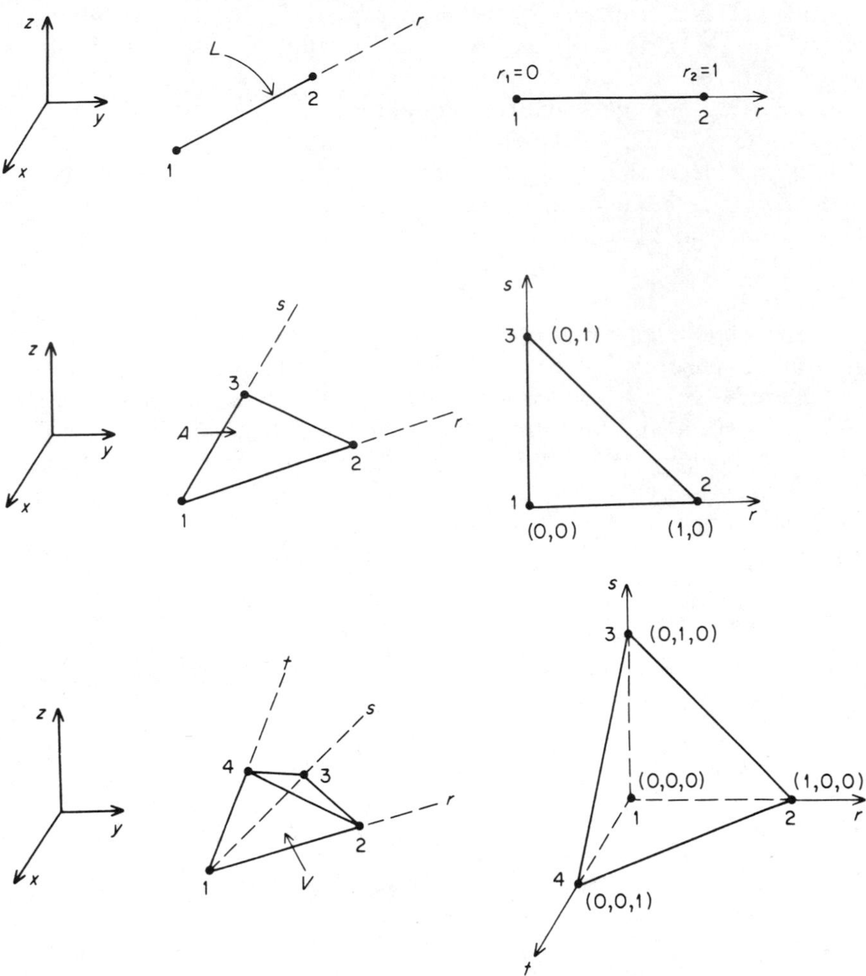

Fig. 5.13 Complete polynomial families

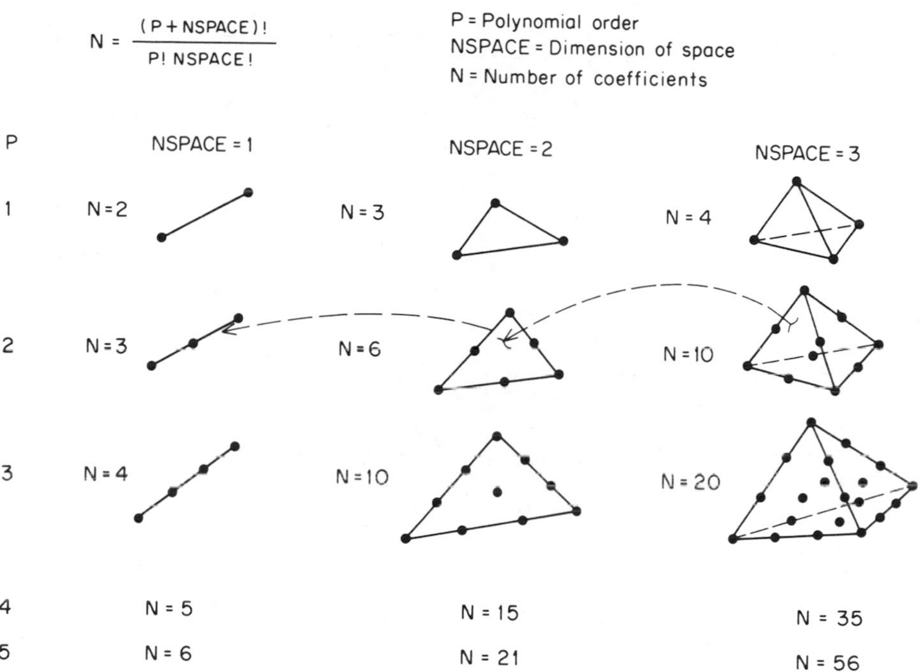

Fig. 5.14 Exact polynomial integrals for constant Jacobian elements in unit coordinates

NSPACE = 1 $\lvert J \rvert = L^e$	$\int_{L^e} r^m \, dl = \frac{L^e}{(1+m)}$
NSPACE = 2 $\lvert J \rvert = 2A^e$	$\int_{A^e} r^m s^n \, da = 2A^e \frac{m!n!}{(2+m+n)!}$
NSPACE = 3 $\lvert J \rvert = 6V^e$	$\int_{V^e} r^m s^n t^p \, dv = 6V^e \frac{m!n!p!}{(3+m+n+p)!}$

Table 5.1 Exact Integrals in Unit Triangle Coordinates

$A = \int dA, I = \int r^a s^b dA \equiv A/N$

a	*b*	N
0	0	1
1	0	3
0	1	3
2	0	6
1	1	12
0	2	6
3	0	10
2	1	30
1	2	30
0	3	10
4	0	15
3	1	60
2	2	90
1	3	60
0	4	15

5.5 Isoparametric contours

Contour plotting on two- or three-dimensional isoparametric surfaces is a problem facing many finite element code developers. Most procedures use some type of linear interpolation to approximate the contour locations. In general, the use of linear interpolation to locate contour lines on higher order isoparametric finite elements destroys much of the inter-element information contained within the results of an analysis. Without adequate means of visually showing the results of higher order finite element calculations, numerous developers of finite element codes have been reluctant to include these elements in their element libraries. Akin and Gray [6] have presented an interpolation procedure for contouring on any isoparametric finite element surface. This procedure is based upon the isoparametric interpolation function itself and as such does not destroy or invent any information that was not already contained within the results. The work presented here extends the procedure with regard to accuracy in following the contour lines.

Consider any isoparametric element formulated in local coordinates, r and s as shown in Fig. 5.15. At any local point the value of the quantity of interest is as in Section 5.2:

$$t(r, s) = \mathbf{H}(r, s)\mathbf{T}^e, \tag{5.26}$$

where $\mathbf{H}$ and $\mathbf{T}^e$ again denote the interpolation functions and the nodal values, respectively. If a contour of t can be defined in the local space, then it can easily be converted to the global space.

Assume that the contour is given by $t = K$ where K is the contour level, and also assume that the local coordinates (r_0, s_0) of one point on the contour are known. The equation of such a curve is given by $\mathrm{d}t = 0$. In local coordinates, this becomes

$$0 = \frac{\partial t}{\partial r}\,\mathrm{d}r + \frac{\partial t}{\partial s}\,\mathrm{d}s, \tag{5.27}$$

where $\mathrm{d}r$ and $\mathrm{d}s$ are components of a displacement $\mathrm{d}L$ which is tangential to the contour. Select a constant step size, $\mathrm{d}L$, to be used in tracing the contour path. As shown in Fig. 5.16, if the contour segment makes an angle of θ with the r axis, then $\mathrm{d}r = \mathrm{d}L \cos\theta$ and $\mathrm{d}s = \mathrm{d}L \sin\theta$. Clearly, the value of $\tan\theta$ can be obtained from Eqn. (5.27) using the values of $\partial t/\partial r$ and $\partial t/\partial s$, which are obtained easily for any isoparametric element by utilizing the local derivatives of Eqn. (5.26). That is,

$$\partial t/\partial r = \partial \mathbf{H}/\partial r T^e,$$

$$\partial t/\partial s = \partial \mathbf{H}/\partial s T^e,$$

or symbolically

$$\boldsymbol{\partial}_l t = \boldsymbol{\Delta} T^e, \tag{5.28}$$

where the subscript l denotes local derivatives. Thus, given a starting point on an element contour line, all one needs to trace the line through the element are the nodal values of the function, the local coordinate derivatives of the interpolation equations, and an assumed step increment $\mathrm{d}L$. Using this assumption, the formulae which predict the position of the contour line's next (or new) location with respect to the previous (or old) location are

$$g \equiv \left[\left(\frac{\partial t}{\partial r}\right)^2 + \left(\frac{\partial t}{\partial s}\right)^2\right]^{\frac{1}{2}},$$

$$r_{\text{new}} \equiv r_{\text{old}} + \left(\frac{\mathrm{d}L}{g}\right)\frac{\partial t}{\partial s}, \tag{5.29}$$

$$s_{\text{new}} \equiv s_{\text{old}} - \left(\frac{\mathrm{d}L}{g}\right)\frac{\partial t}{\partial r}.$$

Clearly the accuracy and cost of such a procedure depend on the size of the local increment, dL. This procedure will tend to deviate from the true contour path because of the numerical errors involved, that is, at the end of a single step the new contour value is not exactly equal to the one which was supposed to be followed. The gradient of the new contour will be used to calculate another position which is more inaccurate than the previous one, etc. In practice, however, this accumulation of error is not perceptible visually except when dealing with a closed contour. Because of the above inaccuracies, the contour will not close; instead it will spiral until an element boundary is reached, or the maximum number of allowable points upon a contour has been exceeded.

The above procedure can be modified to make the accumulated error acceptably small without excessive increase in computational cost by taking a simple correction step normal to the contour line to reduce the amount of deviation. In our terminology, the corrected local coordinates may be calculated from the equations

$$r_{cor} \equiv r_{new} - \left(\frac{1}{g}\right)^2 \frac{\partial t}{\partial r} [t(r_{new}, s_{new}) - K],$$

$$s_{cor} \equiv s_{new} - \left(\frac{1}{g}\right)^2 \frac{\partial t}{\partial s} [t(r_{new}, s_{new}) - K],$$

where K is the correct contour value. This simple correction produces contour lines which are in most cases more accurate than it is possible to record on a graphics display device.

To terminate the contour, one checks the coordinates at the end of each new segment to see if they remain within the element. For higher order elements it is likely that some contours will be closed loops, and in this case the test would fail to stop the calculations. Thus, it is necessary to have a maximum number of allowable segments on any one contour. The required value is $\text{MAX} \leq P/\mathrm{d}L$, where P is the element perimeter length.

Once a contour has been completely traced in the local space, all points on the contour are converted to the corresponding global coordinates and returned for plotting. The global values would generally be transformed again for isometric or perspective plotting.

The selection of the local coordinate starting point can involve two options: one for conveying quick and inexpensive engineering information, and another for detailed quality plots of specific contour values. In the first case, one could plot contours beginning at specific points instead of those which have specific values. For example, one could plot all contours beginning at the mid or third points of each edge of each element. This would give contour segments that are continuous across two elements and then terminate. This eliminates the need for expensive contour locating logic but is not desirable for final reports.

The application of the second and more common procedure is more expensive. One must test each element to determine if a specified contour value lies within the element. The local coordinates of a contour intersecting the edge of a linear, quadratic, or cubic element are easily found. However, a given curve will usually intersect more than one edge. Thus, to avoid plotting such curves twice, it is necessary to compare the calculated terminal point of one curve to see if it lies within a distance of dL of any other initial point on other edges of the element. For quadratic and cubic elements, it is necessary to test for closed loops not intersecting the edge. This is more difficult. One could check the value of t on lines of $r =$ const. and $s =$ const. that pass through the centroid of the element to determine an internal starting point in local coordinates.

Reference [6] shows a subroutine, LCONTC, for tracing the contour curve in a single element. It calls subroutine SHAPE to extract the interpolation functions $\mathbf{H}$, subroutine DERIV to extract the local derivatives, $\mathbf{\Delta}$ of $\mathbf{H}$, and a standard matrix multiplication subroutine, MMULT. Figure 5.17 shows typical results of the application of subroutine LCONTC to contour calculation in a quadratic triangle. For comparison, Fig. 5.18 shows a similar problem except that the corrector was not used. Notice that in Fig. 5.18 the interior contour line spirals, while in Fig. 5.17 the contour line closes upon itself, graphically demonstrating the difference between contour line calculation without and with a correction step normal to the contour line, respectively.

Figure 5.19 shows typical results obtained from a quadratic quadrilateral. This figure demonstrates the successful calculation of contours when the nodal contour function contains a *separatrix*. Two continuous curves are drawn to represent the separatrix. The first starts at the right-hand mid-point node and proceeds to the bottom mid-point node. The second starts at the left-hand mid-point node and proceeds to the top mid-point node.

Fig. 5.15 An isoparametric element formulated in local coordinates *r* and *s*

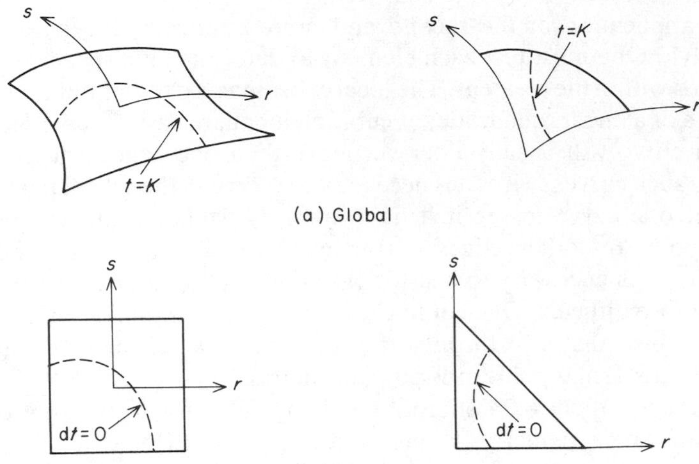

Fig. 5.16

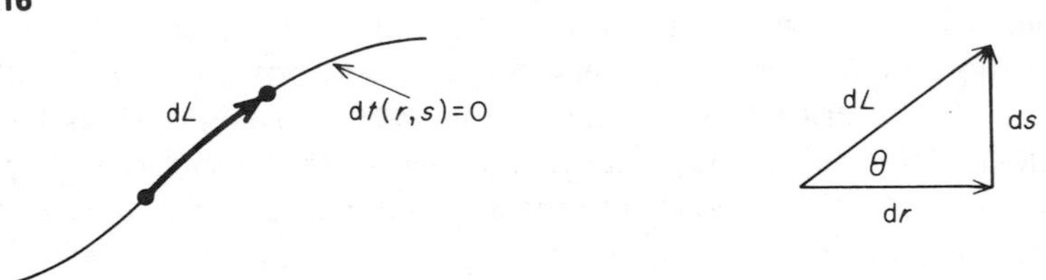

Fig. 5.17 Typical results of the application of subroutine LCONTC to contour calculation in a quadratic triangle

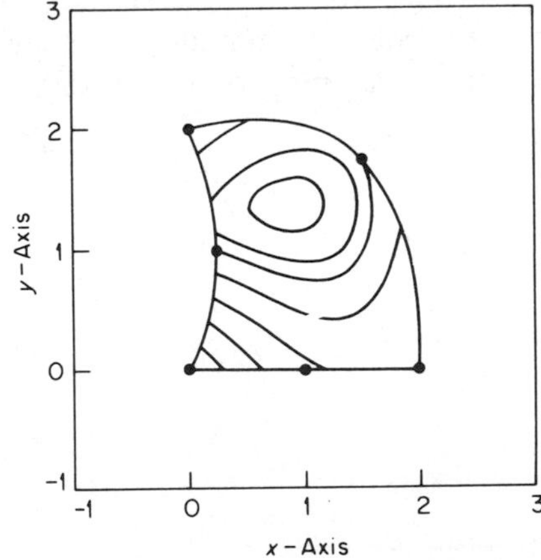

Fig. 5.18 As Fig. 17, but without use of corrector

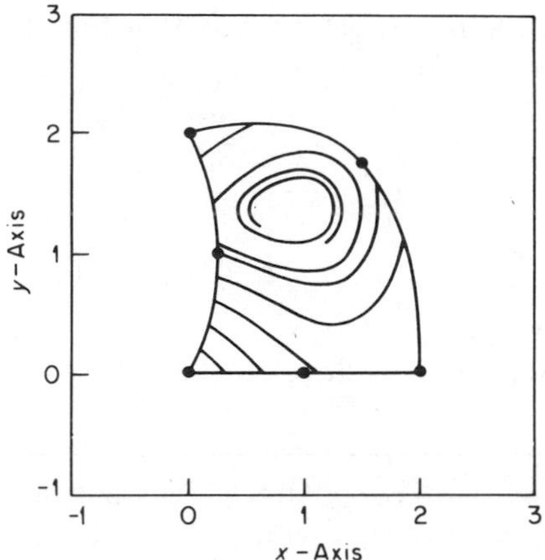

Fig. 5.19 Typical results obtained from a quadratic quadrilateral

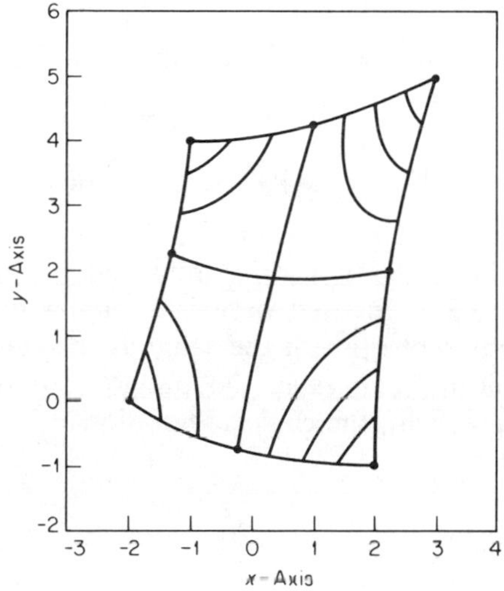

5.6 Exercises

1. For a two-dimensional problem verify that the inverse Jacobian can be written in closed form as

$$\mathbf{J}^{-1} = \frac{1}{|J|}\begin{bmatrix} \dfrac{\partial y}{\partial s} & -\dfrac{\partial y}{\partial r} \\ -\dfrac{\partial x}{\partial s} & \dfrac{\partial x}{\partial r} \end{bmatrix}$$

   where $x = x(r, s)$, and $y = y(r, s)$.
2. It is common (when sufficient storage is available) to evaluate the interpolation functions and their local derivatives only once at each integration point. The arrays for each point are stored for later use. That is, the array H becomes H(N,NQP) and the local derivatives, say DLH, are DLH(NSPACE,N,NQP) where NQP denotes the number of quadrature points. Write a routine to generate and store H and DLH in this form. Rewrite subroutine ISOPAR to use the arrays in this form.
3. Rewrite subroutine JACOB to allow for the case where the local space dimension, say LSPACE, is not the same as the global space dimension, NSPACE.
4. For a two-dimensional problem show that at any typical point

$$\mathrm{H}_{i,x} = [\mathrm{H}_{i,r}\mathrm{Y}_{,s} - \mathrm{H}_{i,s}\mathrm{Y}_{,r}]/|\mathrm{J}|$$

$$\mathrm{H}_{i,y} = [-\mathrm{H}_{i,r}\mathrm{X}_{,s} + \mathrm{H}_{i,s}\mathrm{X}_{,r}]/|\mathrm{J}|$$

   where $|\mathrm{J}|$ is the Jacobian determinant.
5. Show that the integral $\int \mathrm{H}_{i,x}\mathrm{H}_i dv$ is independent of $|\mathrm{J}|$.
6. The Jacobian definition used here is common to the finite element literature. Verify that it is not the same as the standard mathematical definition. Indicate changes in the previous routines that would be necessary to agree with the classical notation.

6

Element integration and interpolation

6.1 Introduction

Recall that the finite element analysis techniques are always based on an integral formulation. Therefore the accuracy of the integration of the element matrices is quite important. In many cases it is impossible or impractical to integrate the expression in closed form and numerical integration must therefore be utilized. If one is using sophisticated elements, it is almost always necessary to use numerical integration. Similarly, if the application is complicated, e.g. the solution of a non-linear ordinary differential equation, then even simple one-dimensional elements can require numerical integration. Many analysts have found that the use of numerical integration simplifies the programming of the element matrices. This results from the fact that lengthy algebraic expressions are avoided and thus the chance of algebraic and/or programming errors is reduced. There are many numerical integration methods available. Only those methods commonly used in finite element applications will be considered here.

6.2 Exact integrals for triangular and quadrilateral geometries

Most simple finite element matrices for two-dimensional problems are based on the use of linear triangular or quadrilateral elements. Since a quadrilateral can be divided into two or more triangles (see Fig. 4.8), only exact integrals over arbitrary triangles will be considered here. Integrals over triangular elements commonly involve integrands of the form

$$I = \iint_A x^m y^n \mathrm{d}x\mathrm{d}y, \tag{6.1}$$

where A is the area of a typical triangle shown in Fig. 6.1. When $0 \leqslant (m+n) \leqslant 2$, the above integral can easily be expressed in closed form in terms of the spatial coordinates of the three corner points. For a right-handed coordinate system, the corners must be numbered in counter-clockwise order. In this case, the above integral becomes

$$
\begin{aligned}
&\text{(a) } m=0, n=0, A=\iint \mathrm{d}A = \frac{1}{2}[x_1(y_2-y_3)+x_2(y_3-y_1)+x_3(y_1-y_2)],\\
&\text{(b) } m=0, n=1, I=\iint y\,\mathrm{d}A = A\bar{y},\\
&\text{(c) } m=1, n=0, I=\iint x\,\mathrm{d}A = A\bar{x},\\
&\text{(d) } m=0, n=2, I=\iint y^2\,\mathrm{d}A = \frac{1}{12}A(y_1^2+y_2^2+y_3^2+9\bar{y}^2),\\
&\text{(e) } m=1, n=1, I=\iint xy\,\mathrm{d}A = \frac{1}{12}A(x_1y_1+x_2y_2+x_3y_3+9\bar{x}\bar{y}),\\
&\text{(f) } m=2, n=0, I=\iint x^2\,\mathrm{d}A = \frac{1}{12}A(x_1^2+x_2^2+x_3^2+9\bar{x}^2),
\end{aligned}
\tag{6.2}
$$

where

$$
\bar{x} = \frac{1}{3}(x_1+x_2+x_3),
$$

$$
\bar{y} = \frac{1}{3}(y_1+y_2+y_3).
$$

The integrals (6.2) should be recognized as the area, and first and second moments of the area. If one had a volume of revolution that had a triangular cross-section in the (ρ, z)-plane, then one should recall that

$$
I = \iiint_V \rho f(\rho, z)\,\mathrm{d}\rho\,\mathrm{d}z\,\mathrm{d}\theta = 2\pi \iint_A \rho f(\rho, z)\,\mathrm{d}\rho\,\mathrm{d}z
$$

so that similar expressions could be used to evaluate the volume integrals. The above closed form integrals are included in subroutine TRGEOM, which is shown in Fig. 6.2. One enters this routine with the values of the integers M and N, the corner coordinates, and returns with the desired area (or volume) integral. Similar operations for quadrilaterals are performed by splitting the quadrilateral into two triangles and making two calls to TRGEOM to evaluate the integral.

Fig. 6.1 A typical triangular element

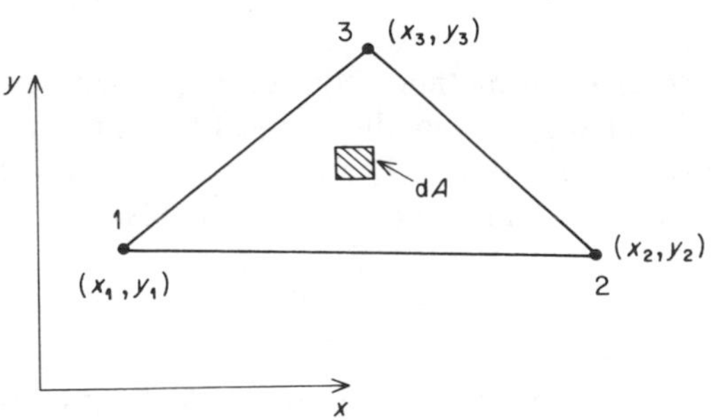

Fig. 6.2 Subroutine TRGEOM

```
      SUBROUTINE  TRGEOM (NCODE,COORD,VALUE)
C     * * * * * * * * * * * * * * * * * * * * * * * *
C     EXACT GEOMETRIC PROPERTIES OF AN ARBITRARY TRIANGLE
C     * * * * * * * * * * * * * * * * * * * * * * * *
CDP   IMPLICIT REAL*8(A-H,O-Z)
      DIMENSION  COORD(3,2)
CDP   ABS(Z) = DABS(Z)
C     VALUE = INTEGRAL OF (X**M)*(Y**N)*DA (0<=M,N<=2)
C     VALUE BY NCODE: 1-AREA, 2-VOL REV ABOUT X,
C     3-VOL REV ABOUT Y, 4-FIRST MOMENT ABOUT X,
C     5-FIRST MOMENT ABOUT Y, 6-SEC MOMENT ABOUT X,
C     7-SEC MOMENT WRT X-Y AXES, 8-SEC MOMENT ABOUT Y
      PI = 3.1415927D0
      XI = COORD(1,1)
      XJ = COORD(2,1)
      XK = COORD(3,1)
      YI = COORD(1,2)
      YJ = COORD(2,2)
      YK = COORD(3,2)
      AREA = 0.5*(XI*(YJ-YK)+XJ*(YK-YI)+XK*(YI-YJ))
      XB = (XI+XJ+XK)/3.D0
      YB = (YI+YJ+YK)/3.D0
      GO TO (10,20,30,40,50,60,70,80), NCODE
C     M=0, N=0,   AREA
   10 VALUE = AREA
      RETURN
C     M=0, N=0, VOLUME OF REVOLUTION ABOUT X-AXIS
   20 VALUE = 2.D0*PI*AREA*YB
      IF ((ABS(YI)+ABS(YJ)+ABS(YK)) .NE. ABS(YI+YJ+YK))
     1   WRITE (6,5000)
 5000 FORMAT ('0INVALID DATA FOR VOLUME OF REVOLUTION.')
      RETURN
C     M=0, N=0, VOLUME OF REVOLUTION ABOUT Y-AXIS
   30 VALUE = 2.D0*PI*AREA*XB
      IF ((ABS(XI)+ABS(XJ)+ABS(XK)) .NE. ABS(XI+XJ+XK))
     1   WRITE (6,5000)
      RETURN
C     M=0, N=1, FIRST MOMENT ABOUT X-AXIS
   40 VALUE = AREA*YB
      RETURN
C     M=1, N=0, FIRST MOMENT ABOUT Y-AXIS
   50 VALUE = AREA*XB
      RETURN
C     M=0, N=2, SECOND MOMENT ABOUT X-AXIS
   60 VALUE = AREA*(YI*YI+YJ*YJ+YK*YK+9.D0*YB*YB)/12.D0
      RETURN
C     M=1, N=1, SECOND MOMENT W.R.T. X-Y AXES
   70 VALUE = AREA*(XI*YI+XJ*YJ+XK*YK+9.D0*XB*YB)/12.D0
      RETURN
C     M=2, N=0, SECOND MOMENT ABOUT Y-AXIS
   80 VALUE = AREA*(XI*XI+XJ*XJ+XK*XK+9.D0*XB*XB)/12.D0
      RETURN
      END
```

6.3 Gaussian quadratures

Since the finite element method requires a large amount of integration, it is imperative that one obtain the greatest possible accuracy with the minimum cost (computer time). The most accurate numerical method in ordinary use is the *Gauss quadrature formula*. Consider the definite integral

$$I = \int_a^b f(x)\,dx,$$

which is to be computed numerically from a given number, n, of values of $f(x)$. Gauss considered the problem of determining which values of x should be chosen in order to get the greatest possible accuracy. In other words, how shall the interval (a, b) be subdivided so as to give the best possible results? Gauss found that the "n" points in the interval should not be equally spaced but should be symmetrically placed with respect to the mid-point of the interval. Some results of Gauss's work are outlined below. Let y denote $f(x)$ in the integral to be computed. Define a change of variable

$$x(r) = \tfrac{1}{2}(b - a)r + \tfrac{1}{2}(b + a) \tag{6.3}$$

so that the non-dimensionalized limits of integration of r become -1 and $+1$. The new value of $y(r)$ is

$$y = f(x) = f[\tfrac{1}{2}(b - a)r + \tfrac{1}{2}(b + a)] \equiv \phi(r). \tag{6.4}$$

Noting from Eqn. (6.3) that $dx = \tfrac{1}{2}(b - a)dr$, the original integral becomes

$$I = \tfrac{1}{2}(b - a)\int_{-1}^{1} \phi(r)\,dr. \tag{6.5}$$

Gauss showed that the integral in Eqn. (6.5) is given by

$$\int_{-1}^{1} \phi(r)\,dr = \sum_{i=1}^{n} W_i\phi(r_i), \tag{6.6}$$

where W_i and r_i represent tabulated values of the *weight functions* and *abscissae* associated with the n points in the non-dimensional interval $(-1, 1)$. Thus the final result is

$$I = \tfrac{1}{2}(b - a)\sum_{i=1}^{n} W_i\phi(r_i) = \sum_{i=1}^{n} f(x(r_i))W_i. \tag{6.7}$$

Gauss showed further that this equation will exactly integrate a polynomial of degree $(2n-1)$, e.g. $x^{(2n-1)}$. For a higher number of space dimensions (which range from -1 to $+1$) one obtains a multiple summation. For example, consider a typical integration in three dimensions

$$I = \int_{-1}^{1}\int_{-1}^{1}\int_{-1}^{1} \phi(r, s, t)\, \mathrm{d}r\mathrm{d}s\mathrm{d}t$$

or

$$I \approx \sum_{i=1}^{n}\sum_{j=1}^{n}\sum_{k=1}^{n} W_i W_j W_k \phi(r_i, s_j, t_k)$$

which could also be written as a single summation over n^3 points. That is,

$$I = \sum_{l=1}^{n^3} W_l \phi(r_l, s_l, t_l), \tag{6.8}$$

where

$$\begin{aligned} l &= i + (j-1)n + (k-1)n^2, \\ r_l &= \alpha_i, \\ s_l &= \alpha_j, \\ t_l &= \alpha_k, \\ W_l &= W_i W_j W_k, \end{aligned}$$

and where the W's and α's are the tabulated one-dimensional Gauss coefficients. Since the time required, and thus the cost, is a function of n^{NSPACE} it is desirable to keep n as small as possible. Of course, it is not always necessary to use the same number of points in each space dimension.

Most texts on numerical analysis contain tabulated data on the weights and abscissae required for integration by Gaussian quadratures, e.g. Ref. [14]. Table 6.1 contains a set of typical one-dimensional data. In a special purpose program, these data would probably be stored in a BLOCK DATA or COMMON statement. In the present building block system, these data are stored in a subroutine called GAUSCO. One enters this routine, which is shown in Fig. 6.3, with the order of the integration, n, and returns with the n values of the one-dimensional weights and abscissae. Data for higher order dimensional spaces are obtained from the above one-dimensional data by utilizing Eqn. (6.8). These quadrature data for two- and three-dimensional data are presented in Tables 6.2 and 6.3, respectively. These two-dimensional data were generated by subroutine GAUS2D as shown in Fig. 6.4.

As an example of Gaussian quadratures, consider the following one-dimensional integral:

$$\mathbf{I} = \int_1^2 \begin{bmatrix} 2 & 2x \\ 2x & (1 + 2x^2) \end{bmatrix} dx = \int_1^2 [F(x)]\, dx.$$

If two Gauss points are selected, then the tabulated values give $W_1 = W_2 = 1$; and $A_1 = 0.57735 = -A_2$. The change of variable gives $x(r) = (r + 3)/2$, so that $x(A_1) = 1.788675$ and $x(A_2) = 1.211325$. Therefore from Eqn. (6.7)

$$\mathbf{I} = \tfrac{1}{2}(2 - 1)\{W_1[F(x(A_1))] + W_2[F(x(A_2))]\}$$

$$= \tfrac{1}{2}(1)\left\{(1)\begin{bmatrix} 2 & 2(1.788675) \\ \text{sym.} & 1 + 2(1.788675)^2 \end{bmatrix} + (1)\begin{bmatrix} 2 & 2(1.211325) \\ \text{sym.} & 1 + 2(1.211325)^2 \end{bmatrix}\right\},$$

or

$$\mathbf{I} = \begin{bmatrix} 2.00000 & 3.00000 \\ 3.00000 & 5.666667 \end{bmatrix},$$

which is easily shown to be in good agreement with the exact solution. Similarly, evaluating the two-dimensional integral

$$I = \int_{-1}^{1} \int_{-1}^{1} (r^2 + s)\, dr\, ds$$

by using the data in Table 6.2 for two Gauss points in each dimension results in

$$I = W_1(r_1^2 + s_1) + W_2(r_2^2 + s_2) + W_3(r_3^2 + s_3) + W_4(r_4^2 + s_4)$$

or

$$I = 1[(+0.5773)^2 + (+0.5773)] + 1[(+0.5773)^2 + (-0.5773)]$$
$$+ 1[(-0.5773)^2 + (+0.5773)] + 1[(-0.5773)^2 + (-0.5773)]$$

so that $I = 1.33333$. This agrees well with the exact value of $I = 4/3$.

The above Gaussian data are well suited to line elements, quadrilateral elements and hexahedral elements. However, if numerical integration over a triangular or wedge-shaped element is required then other techniques are usually utilized. Some of these methods are presented in the next section.

Fig. 6.3 Subroutine GAUSCO

```
      SUBROUTINE  GAUSCO (N,A,W)
C     * * * * * * * * * * * * * * * * * * * * * * * * * *
C     GAUSSIAN QUADRATURE ABSCISSAE AND WEIGHT COEFFS
C     * * * * * * * * * * * * * * * * * * * * * * * * * *
CDP   IMPLICIT REAL*8(A-H,O-Z)
      DIMENSION  A(N), W(N)
      DATA  NMAX/4/
C     N = NO. OF GAUSS POINTS IN ONE DIMENSION
C     A = ABSCISSAE, W = WEIGHTS  OF GAUSS POINTS
C     NMAX = MAX. NO. OF POINTS TABULATED HEREIN
      NLES1 = N - 1
      GO TO (10,20,30,40), NLES1
C     N = 2
   10 A(1) = 0.577350269189626D0
      A(2) = -A(1)
      W(1) = 1.D0
      W(2) = 1.D0
      RETURN
C     N = 3
   20 A(1) = 0.774596669241483D0
      A(2) = 0.D0
      A(3) = -A(1)
      W(1) = 0.555555555555556D0
      W(2) = 0.888888888888889D0
      W(3) = W(1)
      RETURN
C     N = 4
   30 A(1) = 0.861136311594053D0
      A(2) = 0.339981043584856D0
      A(3) = -A(2)
      A(4) = -A(1)
      W(1) = 0.347854845137454D0
      W(2) = 0.652145154862546D0
      W(3) = W(2)
      W(4) = W(1)
      RETURN
   40 WRITE (6,5000) NMAX
 5000 FORMAT ('ON >',I3,' COEFFICIENTS NOT TABULATED')
      RETURN
      END
```

Fig. 6.4 Subroutine GAUS2D

```
      SUBROUTINE  GAUS2D (NQP,GPT,GWT,NIP,PT,WT)
C     * * * * * * * * * * * * * * * * * * * * * * * * * *
C     USE 1-D GAUSSIAN DATA TO GENERATE
C     QUADRATURE DATA FOR A QUADRILATERAL
C     * * * * * * * * * * * * * * * * * * * * * * * * * *
CDP   IMPLICIT REAL*8 (A-H,O-Z)
      DIMENSION GPT(NQP), GWT(NQP), PT(2,1), WT(1)
C     DIM PT(2,NIP), WT(NIP)
C     NQP = NO. TABULATED 1-D POINTS, NIP = NQP*NQP
C     GPT, GWT = TABULATED 1-D QUADRATURE RULES
C     PT, WT = CALCULATED COORDS AND WEIGHTS IN QUAD
      CALL  GAUSCO (NQP,GPT,GWT)
      NIP = NQP*NQP
      K = 0
      DO 20  I = 1,NQP
      DO 10  J = 1,NQP
      K = K + 1
      WT(K) = GWT(I)*GWT(J)
      PT(1,K) = GPT(J)
   10 PT(2,K) = GPT(I)
   20 CONTINUE
      RETURN
      END
```

Table 6.1 Abscissae and Weight Coefficients of the Gaussian Quadrature Formula

$$\int_{-1}^{1} f(x)\,dx = \sum_{j=1}^{n} W_j f(r_j)$$

$\pm r_i$		W_j
	$n = 2$	
0.57735 02691 89626		1.00000 00000 00000
	$n = 3$	
0.77459 66692 41483		0.55555 55555 55556
0.00000 00000 00000		0.88888 88888 88889
	$n = 4$	
0.86113 63115 94053		0.34785 48451 37454
0.33998 10435 84856		0.65214 51548 62546
	$n = 5$	
0.90617 98459 38664		0.23692 68850 56189
0.53846 93101 05683		0.47862 86704 99366
0.00000 00000 00000		0.56888 88888 88889
	$n = 6$	
0.93246 95142 03152		0.17132 44923 79170
0.66120 93864 66265		0.36076 15730 48139
0.23861 91860 83197		0.46791 39345 72691
	$n = 7$	
0.94910 79123 42759		0.12948 49661 68870
0.74153 11855 99394		0.27970 53914 89277
0.40584 51513 77397		0.38183 00505 05119
0.00000 00000 00000		0.41795 91836 73469
	$n = 8$	
0.96028 98564 97536		0.10122 85362 00376
0.79666 64774 13627		0.22238 10344 53374
0.52553 24099 16329		0.31370 66458 77887
0.18343 46424 95650		0.36268 37833 78362

Table 6.2 Weights and Abscissae for Integration Over a Quadrilateral by Gaussian Quadrature

Order	Point	r	s	Weight
2	1	$+A_1$	$+A_1$	W_1
	2	$+A_1$	$-A_1$	W_1
	3	$-A_1$	$+A_1$	W_1
	4	$-A_1$	$-A_1$	W_1
3	1	$+A_2$	$+A_2$	W_2
	2	$+A_2$	0	W_3
	3	$+A_2$	$-A_2$	W_2
	4	0	$+A_2$	W_3
	5	0	0	W_4
	6	0	$-A_2$	W_3
	7	$-A_2$	$+A_2$	W_2
	8	$-A_2$	0	W_3
	9	$-A_2$	$-A_2$	W_2
4	1	$+A_3$	$+A_3$	W_5
	2	$+A_3$	$+A_4$	W_6
	3	$+A_3$	$-A_4$	W_6
	4	$+A_3$	$-A_3$	W_5
	5	$+A_4$	$+A_3$	W_6
	6	$+A_4$	$+A_4$	W_7
	7	$+A_4$	$-A_4$	W_7
	8	$+A_4$	$-A_3$	W_6
	9	$-A_4$	$+A_3$	W_6
	10	$-A_4$	$+A_4$	W_7
	11	$-A_4$	$-A_4$	W_7
	12	$-A_4$	$-A_3$	W_6
	13	$-A_3$	$+A_3$	W_5
	14	$-A_3$	$+A_4$	W_6
	15	$-A_3$	$-A_4$	W_6
	16	$-A_3$	$-A_3$	W_5

$A_1 = 0.577350269189626$ $\quad W_1 = 1.0$
$A_2 = 0.774596669241483$ $\quad W_2 = 0.3086419753086425$
$A_3 = 0.861136311594053$ $\quad W_3 = 0.4938271604938276$
$A_4 = 0.339981043584856$ $\quad W_4 = 0.7901234567901236$
$W_5 = 0.1210029932856021$
$W_6 = 0.2268518518518519$
$W_7 = 0.4252933030106941$

Table 6.3 Gaussian Quadrature Weights and Abscissae for Integration Over a Cube

Order	Point	r	s	t	Weight
2	1	$+A_1$	$+A_1$	$+A_1$	W_1
	2	$+A_1$	$+A_1$	$-A_1$	W_1
	3	$+A_1$	$-A_1$	$+A_1$	W_1
	4	$+A_1$	$-A_1$	$-A_1$	W_1
	5	$-A_1$	$+A_1$	$+A_1$	W_1
	6	$-A_1$	$+A_1$	$-A_1$	W_1
	7	$-A_1$	$-A_1$	$+A_1$	W_1
	8	$-A_1$	$-A_1$	$-A_1$	W_1
3	1	$+A_2$	$+A_2$	$+A_2$	W_2
	2	$+A_2$	$+A_2$	0	W_3
	3	$+A_2$	$+A_2$	$-A_2$	W_2
	4	$+A_2$	0	$+A_2$	W_3
	5	$+A_2$	0	0	W_4
	6	$+A_2$	0	$-A_2$	W_3
	7	$+A_2$	$-A_2$	$+A_2$	W_2
	8	$+A_2$	$-A_2$	0	W_3
	9	$+A_2$	$-A_2$	$-A_2$	W_2
	10	0	$+A_2$	$+A_2$	W_3
	11	0	$+A_2$	0	W_4
	12	0	$+A_2$	$-A_2$	W_3
	13	0	0	$+A_2$	W_4
	14	0	0	0	W_5
	15	0	0	$-A_2$	W_4
	16	0	$-A_2$	$+A_2$	W_3
	17	0	$-A_2$	0	W_4
	18	0	$-A_2$	$-A_2$	W_3
	19	$-A_2$	$+A_2$	$+A_2$	W_2
	20	$-A_2$	$+A_2$	0	W_3
	21	$-A_2$	$+A_2$	$-A_2$	W_2
	22	$-A_2$	0	$+A_2$	W_3
	23	$-A_2$	0	0	W_4
	24	$-A_2$	0	$-A_2$	W_3
	25	$-A_2$	$-A_2$	$+A_2$	W_2
	26	$-A_2$	$-A_2$	0	W_3
	27	$-A_2$	$-A_2$	$-A_2$	W_2

$A_1 = 0.577350269189626$
$A_2 = 0.774596669241483$
$A_3 = 0.861136311594053$
$A_4 = 0.339981043584856$

$W_1 = 1.0$
$W_2 = 0.171467764060357$
$W_3 = 0.274348422496571$
$W_4 = 0.438957475994513$
$W_5 = 0.702331961591221$

$W_6 = 0.042091477490532$
$W_7 = 0.078911515795071$
$W_8 = 0.147940336056781$
$W_9 = 0.277352966953913$

6.4 Numerical integration over triangles

As noted earlier, the use of standard Gaussian quadrature data over triangular domains is uncommon. Integration over triangles usually requires a special coordinate system. Many finite element texts present triangular elements in terms of area or natural coordinates. However, the author feels it is simpler to introduce the local coordinate system for the unit triangle shown in Fig. 5.12. If the element has sides defined by the three corner points, then it is easily shown that

$$\begin{aligned} x(r, s) &= x_1 + (x_2 - x_1)r + (x_3 - x_1)s, \\ y(r, s) &= y_1 + (y_2 - y_1)r + (y_3 - y_1)s, \end{aligned} \tag{6.9}$$

where the x_i and y_i denote the global coordinates of the corners. The Jacobian (see Section 5.2) for the unit triangle transformation then becomes

$$\mathbf{J} = \begin{bmatrix} (x_2 - x_1) & (y_2 - y_1) \\ (x_3 - x_1) & (y_3 - y_1) \end{bmatrix} \tag{6.10}$$

which is constant, so that its determinant

$$|\mathrm{J}| = 2A = (x_2 - x_1)(y_3 - y_1) - (x_3 - x_1)(y_2 - y_1)$$

is also constant. As before, one approximates an integral of $\mathrm{f}(x, y) \equiv \mathrm{F}(r, s)$ over a triangle by

$$I = \int \mathrm{f}(x, y)\, \mathrm{d}x\mathrm{d}y \approx \sum_{i=1}^{n} W_i \mathrm{F}(r_i, s_i)\, |\mathrm{J}_i|.$$

For triangular regions the weights, W_i, and abscissae (r_i, s_i), are less well known so several different rules on the unit triangle are tabulated in Figs. 6.5 and 6.6. Figure 6.5 presents rules that yield points that are symmetric with respect to all corners of the triangle. Figure 6.6 presents rules that yield points that are symmetric with respect to the line $r = s$. They are known as Radau rules.

As a simple example of integration over a triangle, consider the integral

$$I = \iint y \mathrm{d}x\mathrm{d}y$$

over a triangle with its three vertices at (0, 0), (3, 0) and (0, 6), respectively, in (x, y) coordinates. Then $A = 9$ and $|\mathrm{J}| = 18$ and for a three point quadrature rule the integral is thus given by

$$I = \sum_{i=1}^{3} W_i y_i |\mathrm{J}_i|$$

where

$$y_i = y_1 + (y_2 - y_1)r_i + (y_3 - y_1)s_i = 6s_i.$$

Substituting from the three point rule in Fig. 6.5(b), and factoring out the constant $|\mathbf{J}|$,

$$I = 18[(1/6)(6)(0) + (1/6)(6)(1/2) + (1/6)(6)(1/2)] = 18,$$

which is the exact solution. The above data can be related to the so-called *area coordinates* (A_1, A_2, A_3), see [87], by noting that $r = A_2$, $s = A_3$, and $1 = A_1 + A_2 + A_3$. It is also useful to be aware of the numerical check that the sum of the weights must equal one half.

Fig. 6.5 Symmetric quadrature rules for the unit triangle

(a) $n = 1$,

i	r_i	s_i	W_i
1	1/3	1/3	1/2

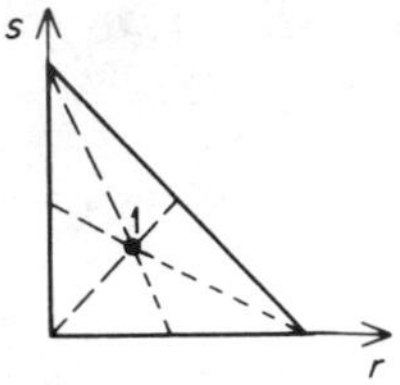

(b) $n = 3$,

i	r_i	s_i	W_i
1	1/2	0	1/6
2	1/2	1/2	1/6
3	0	1/2	1/6

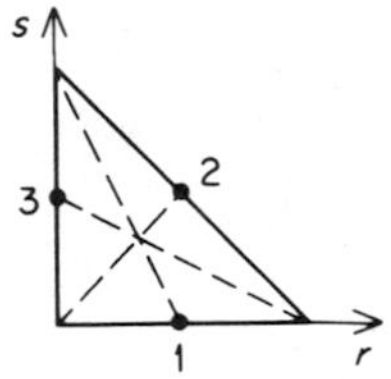

(c) $n = 7$,

i	r_i	s_i	W_i
1	0	0	1/40
2	1/2	0	1/15
3	1	0	1/40
4	1/2	1/2	1/15
5	0	1	1/40
6	0	1/2	1/15
7	1/3	1/3	9/40

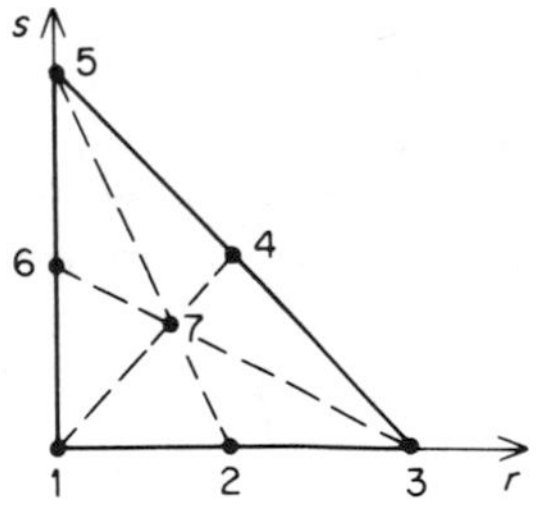

(d) $n = 7$,

i	r_i	s_i	W_i
1	A_2	A_2	W_1
2	A_3	A_1	W_2
3	A_4	A_2	W_1
4	A_3	A_3	W_2
5	A_2	A_4	W_1
6	A_1	A_3	W_2
7	1/3	1/3	W_3

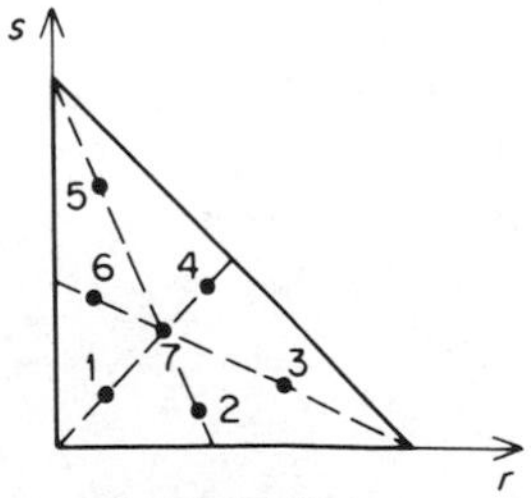

$A_1 = 0.05971587$ $\quad A_3 = 0.47014206$ $\quad W_1 = 0.06296959$
$A_2 = 0.10128651$ $\quad A_4 = 0.79742699$ $\quad W_2 = 0.06619708$
$W_3 = 0.1125$

Fig. 6.6 **(a)** Gauss–Radau quadrature rules for the unit triangle

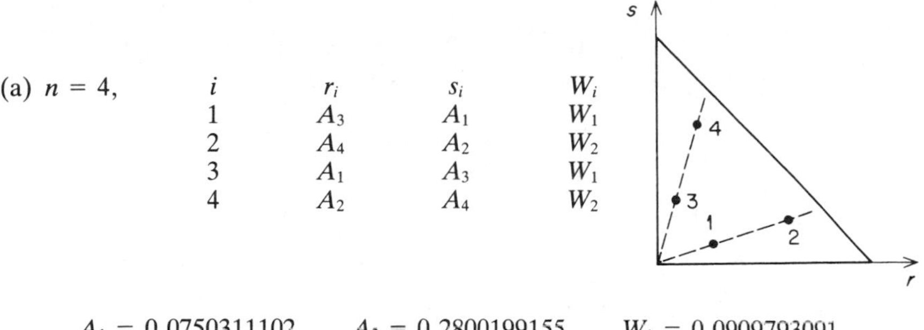

(a) $n = 4$,

i	r_i	s_i	W_i
1	A_3	A_1	W_1
2	A_4	A_2	W_2
3	A_1	A_3	W_1
4	A_2	A_4	W_2

$A_1 = 0.0750311102 \quad A_3 = 0.2800199155 \quad W_1 = 0.0909793091$

$A_2 = 0.1785587283 \quad A_4 = 0.6663902460 \quad W_2 = 0.1590206909$

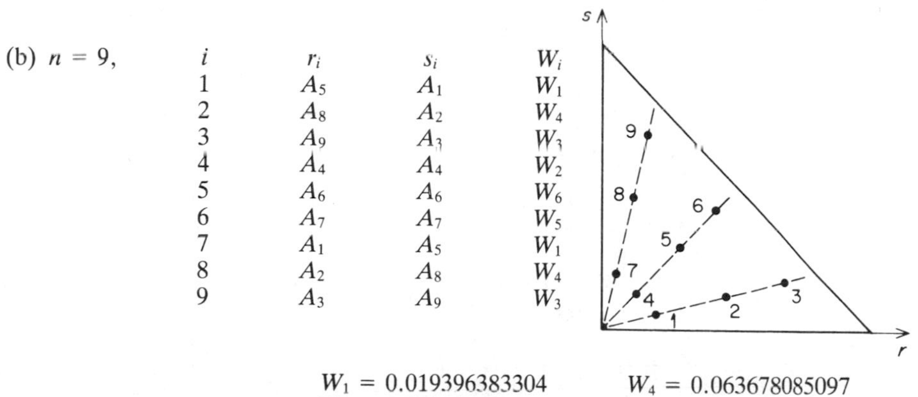

(b) $n = 9$,

i	r_i	s_i	W_i
1	A_5	A_1	W_1
2	A_8	A_2	W_4
3	A_9	A_3	W_3
4	A_4	A_4	W_2
5	A_6	A_6	W_6
6	A_7	A_7	W_5
7	A_1	A_5	W_1
8	A_2	A_8	W_4
9	A_3	A_9	W_3

$W_1 = 0.019396383304 \quad W_4 = 0.063678085097$

$W_2 = 0.031034213285 \quad W_5 = 0.089303072783$

$W_3 = 0.055814420490 \quad W_6 = 0.101884936154$

$A_1 = 0.02393113229 \quad A_4 = 0.10617026910 \quad A_7 = 0.45570602025$

$A_2 = 0.06655406786 \quad A_5 = 0.188409405913 \quad A_8 = 0.52397906774$

$A_3 = 0.10271765483 \quad A_6 = 0.29526656780 \quad A_9 = 0.80869438567$

Fig. 6.6 (b) Gauss–Radau quadrature rules

(c) $n = 16$,

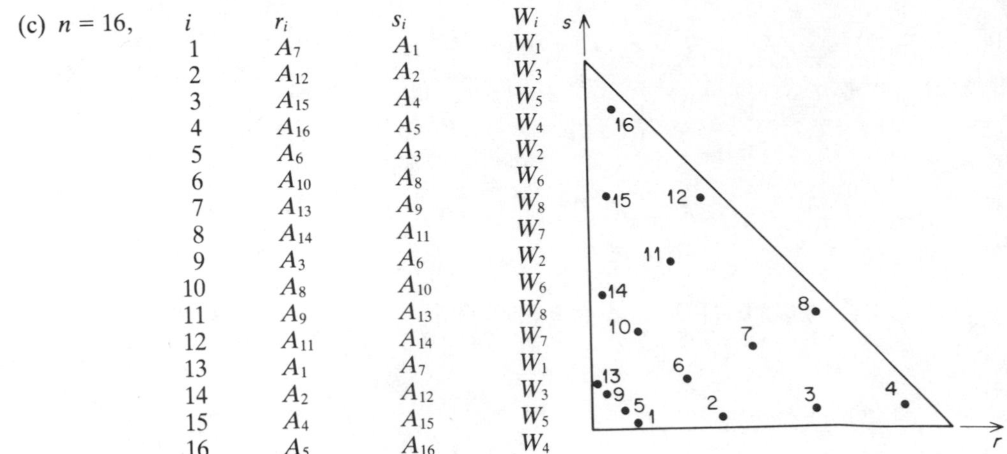

i	r_i	s_i	W_i
1	A_7	A_1	W_1
2	A_{12}	A_2	W_3
3	A_{15}	A_4	W_5
4	A_{16}	A_5	W_4
5	A_6	A_3	W_2
6	A_{10}	A_8	W_6
7	A_{13}	A_9	W_8
8	A_{14}	A_{11}	W_7
9	A_3	A_6	W_2
10	A_8	A_{10}	W_6
11	A_9	A_{13}	W_8
12	A_{11}	A_{14}	W_7
13	A_1	A_7	W_1
14	A_2	A_{12}	W_3
15	A_4	A_{15}	W_5
16	A_5	A_{16}	W_4

$A_1 = 0.00970378512$
$A_2 = 0.02891208422$
$A_3 = 0.04612207989$
$A_4 = 0.05021012321$
$A_5 = 0.06546699455$
$A_6 = 0.09363778441$
$A_7 = 0.13005607918$
$A_8 = 0.13741910412$
$W_1 = 0.005423225910$
$W_2 = 0.010167259561$
$W_3 = 0.022584049287$
$W_4 = 0.023568368199$

$A_0 = 0.23864865974$
$A_{10} = 0.27899046348$
$A_{11} = 0.31116455224$
$A_{12} = 0.38749748338$
$A_{13} = 0.48450832666$
$A_{14} = 0.63173125166$
$A_{15} = 0.67294686319$
$A_{16} = 0.87742880935$
$W_5 = 0.035388067900$
$W_6 = 0.042339724518$
$W_7 = 0.044185088522$
$W_8 = 0.066344216093$

6.5 Minimal, optimal, reduced and selected integration

Since the numerical integration of the element square matrix can represent a large part of the total cost it is desirable to use low order integration rules. Care must be taken when selecting the *minimal order* of integration. Usually the integrand will contain global derivatives so that in the limit, as the element size h approaches zero, the integrand can be assumed to be constant, and then only the integral $I = \int dv = \int |J| dr ds dt$ remains to be integrated exactly. Such a rule could be considered the minimal order. However, the order is often too low to be practical since it may lead to a rank deficient element (and system) square matrix, if the rule does not exactly integrate the equations. Typical integrands involve terms such as the strain energy density: $\mathbf{B}^T\mathbf{D}\mathbf{B}/2$ (see Chapter 10).

Let NQP denote the number of element integration points while NI represents the number of independent relations at each integration point; then the rank of the element is NQP*NI. Generally, NI corresponds to the number of rows in $\mathbf{B}$ in the usual symbolic integrand $\mathbf{B}^T\mathbf{D}\mathbf{B}$. For a typical element we want (NQP*NI-NC) ≥ NELFRE, where NC represents the number of element constraints, if any. For a non-singular system matrix a similar expression is NE*(NQP*NI-NC) ≥ NDFREE-NR, where NR denotes the number of nodal parameter restraints (NR ≥ 1). These relations can be used as guides in selecting a minimal value of NQP.

Consider a problem involving a governing integral statement with m^{th} order derivatives. If the interpolation (trial) functions are complete polynomials of order p then to maintain the theoretical convergence rate NQP should be selected [83] to give accuracy of order $0(h^{2(p-m)+1})$. That is, to integrate polynomial terms of order $2(p-m)$ exactly.

It has long been known that a finite element model gives a stiffness which is too high. Using reduced integration so as to underestimate the element stiffness has been accepted as one way to improve the results. These procedures have been investigated by several authors including Zienkiewicz [87], Zienkiewicz and Hinton [85], Hughes, Cohen and Haroun [44] and Malkus and Hughes [50]. Reduced integration has been especially useful in problems with constraints, such as material incompressibility.

A danger of low order integration rules is that *zero energy modes* may arise in an element. That is, $\mathbf{D}^{eT}\mathbf{S}^e\mathbf{D}^e = 0$ for $\mathbf{D}^e \neq \mathbf{0}$. Usually these zero energy modes, $\mathbf{D}^e$, are incompatible with the same modes in an adjacent element. Thus the assembly of elements may have no zero energy modes (except for the standard *rigid* modes). Cook [25] illustrates that an eigen-analysis of the element can be used as a check since zero eigenvalues correspond to zero energy modes.

The integrand usually involves derivatives of the function of interest. Many solutions require the post-solution calculation of these derivatives for auxilary calculations. Thus a related question is which points give the most accurate estimates for those derivatives. These points are often called *optimal points* or Barlow points. Their locations have been derived by Barlow [12] and Moan [55]. The optimal points usually are the common quadrature points. For low order elements the optimal points usually correspond to the minimal integration points. This is indeed fortunate.

As discussed in Chapter 1, it is possible in some cases to obtain exact derivative estimates from the optimal points. Barlow considered line elements, quadrilaterals and hexahedra while Moan considered the triangular elements. The points were found by assuming that the pth order polynomial solution, in a small element, is approximately equal to the $(p + 1)$ order exact polynomial solution. The derivatives of the two forms were equated and the coordinates of points where the identity is satisfied were determined. For triangles the optimal rules are the symmetric rules involving 1, 4, 7, and 13 points. For machines with small word lengths the 4 and 13 point rules may require higher precision due to the negative centroid weights.

Generally, all interior point quadrature rules can be used to give accurate derivative estimates. The derivatives of the interpolation functions are least accurate at the nodes. Hinton and Campbell [40] have presented relations for extrapolating the optimal values to the nodal points.

For element formulations involving element constraints, or penalties, it is now considered best to employ selective integration rules [44, 50]. For penalty formulations it is common to have equations of the form

$$(\mathbf{S}_1 + \alpha \mathbf{S}_2)\, \mathbf{D} = \mathbf{C}$$

where the constant $\alpha \to \infty$ in the case where the penalty constraint is exactly satisfied.

In the limit as $\alpha \to \infty$ the system degenerates to $\mathbf{S}_2\, \mathbf{D} = \mathbf{0}$, where the solution approaches the trivial result, $\mathbf{D} = \mathbf{0}$. To obtain a non-trivial solution in this limit it is necessary for $\mathbf{S}_2$ to be singular. Therefore, the two contributing element parts $\mathbf{S}_1^e$ and $\mathbf{S}_2^e$ are *selectively* integrated. That is, $\mathbf{S}_2^e$ is under integrated so as to be rank deficient (singular) while $\mathbf{S}_1^e$ is integrated with a rule which renders $\mathbf{S}_1$ non-singular. Typical applications of selective integration were cited above and include problems such as plate bending where the bending contributions are in $\mathbf{S}_1^e$ while the shear contributions are in $\mathbf{S}_2^e$.

6.6 Typical interpolation functions

The practical application of the finite element method depends on the use of various interpolation functions and their derivatives. Most of the interpolation functions for C^0 and C^1 continuity elements are now well known. The C^0 functions are continuous across an inter-element boundary while the C^1 functions also have their first derivatives continuous across the boundary. Some of these functions have already been introduced. For example, the bilinear isoparametric quadrilateral (Q4), its interpolation functions and their local derivatives were presented in Figs. 5.2 and 5.4, respectively.

The one-dimensional n^{th} order Lagrange interpolation polynomial is the ratio of two products. For an element with $(n+1)$ nodes, $r_i, i = 1, 2, \ldots, (n+1)$, the interpolation function is

$$L_k^n(r) = \prod_{\substack{i=1 \\ i \neq k}}^{n+1} (r - r_i) \Big/ \prod_{\substack{i=1 \\ i \neq k}}^{n+1} (r_k - r_i) \tag{6.11}$$

This is a complete n^{th} order polynomial in one dimension. It has the property that $L_k(r_i) = \delta_{ik}$.

For local coordinates given on the domain $[-1, 1]$ a typical term for equally spaced nodes is

$$L_2^3(r) = \frac{(r-(-1))(r-1)}{(0-(-1))(0-1)} = (1 - r^2)$$

Similarly $L_1^3(r) = r(r-1)/2$ and $L_3^3 = r(r+1)/2$ and $\sum_{k=1}^{3} L_k^3(r) = 1$. Now consider the two-dimensional extension. The first member of the quadrilateral family has already been introduced in Chapter 5. It is recognized as a product of the linear Lagrange interpolation polynomial in each direction. The four node quadrilateral is commonly called the bilinear quadrilateral. When using m points in each spatial dimension the standard Lagrangian quadrilateral will have m^2 nodes.

Another popular group of quadrilaterals are the *Serendipity elements*. The best known eight node, Q8, element has interpolation functions, for the element in Fig. 6.7,

$$H_i(r,s) = (1 + r_i r)(1 + s_i s)(r_i r + s_i s - 1)/4, 1 \leq i \leq 4, \tag{6.12}$$

at the corners, while at the mid-sides

$$H_i(r,s) = r_i^2(1 - s^2)(1 + r_i r)/2 + s_i^2(1 - r^2)(1 + s_i s)/2, 5 \leq i \leq 8. \tag{6.13}$$

The interpolation functions for the eight node Serendipity element are given in subroutine SHP8Q and the local derivatives are in DER8Q, see Fig. 6.8. By adding a central node this element can be extended to become a Lagrangian biquadratic element (Q9). It is also possible to add the same centre function to convert the Q4 element to a Q5 element. These internal nodes are usually omitted but they can become useful for problems involving element constraints such as an incompressibility condition.

The interpolation functions for the bilinear quadrilateral are easily generalized to a trilinear hexadral solid. In local coordinates (r, s, t) the interpolation functions for this eight node element, H8, are

$$H_i(r, s, t) = (1 + r_i r)(1 + s_i s)(1 + t_i t)/8 \tag{6.14}$$

where

i	r_i	s_i	t_i	i	r_i	s_i	t_i
1	1	1	1	5	−1	1	1
2	1	−1	1	6	−1	−1	1
3	1	−1	−1	7	−1	−1	−1
4	1	1	−1	8	−1	1	−1

are the local coordinates of the nodes. Figures 6.9 and 6.10 give the subroutines for evaluating these functions and their local derivatives, respectively.

The performance of an element depends on the polynomial terms included in the interpolation functions. Complete polynomials are the most desirable. However, they are exactly included only in the simplex elements and their generalizations. Figure 6.11 illustrates typical polynomial terms in two-dimensional elements. The Pascal triangle of complete polynomials is included for comparison. Note that the selected terms are symmetric relative to the centre of the Pascal triangle. The figure shows that the term x^2y^2 is included in the Q9 element but omitted from the Q8 given in Fig. 6.8.

The Serendipity elements have been defined on the intervals $[-1, 1]$ in r, s and, t. The use of unit coordinates $[0, 1]$ was presented in Section 5.4 for the simplex elements. The interpolation functions for some of the simplex elements were given there also. These included the two node line element (L2) and the common three node triangle (T3). The interpolation functions and local derivatives for the latter element are given in Fig. 6.12. The extension of the interpolation functions to the complete quadratic triangle, T6, are given in Fig. 6.13 along with the corresponding derivatives. Figure 6.14 shows a procedure that can be employed to derive the element interpolation functions for the six node triangle. It is not uncommon to define the family of complete triangular elements formulated in the so-called *natural* or *area* coordinates. However, those procedures are not used here.

We want to generate an interpolation function, H_i, that vanishes at the jth node when $i \neq j$. Such a function can be obtained by taking the products of the equations of selected curves through the nodes on the element. For example, let $H_1(r, s) = C_1\Gamma_1\Gamma_2$ where the Γ_j are shown in Fig. 6.14, and where C_1 is a constant chosen so that $H_1(r_1, s_1) = 1$. This yields $H_1 = (1 - 3r - 3s + 2r^2 + 4rs + 2s^2)$. Similarily, letting $H_4 = C_4\Gamma_1\Gamma_3$ gives $C_4 = 4$ and $H_4 = 4r(1 - r - s)$. This type of procedure is usually quite straightforward. However, there are times when there is not a unique choice of products, and then care must be employed to select the proper products. The results should always be checked against Eqn. (5.4).

It is common for finite element codes to have an element library that selects the desired element according to some set procedure. Such a subroutine, SHAPE, is shown in Fig. 6.15. It was referenced by ISOPAR in Section 5.3. This particular subroutine uses two integer keys to select the desired element. The first key is the dimension of the space, NSPACE, while the second is the number of nodes per element, N. A similar procedure is used in DERIV to establish the local derivatives of the interpolation functions.

It may be better to use a single element type number as a key for the element library. This makes it easier to select both C^0 and C^1 elements and avoids confusion when the choice is not unique when using two keys (as in selecting Q9 and T9 for two-dimensional problems). For simplicity, C^1 elements will be only briefly considered in Section 10.5.

Fig. 6.7 Quadratic quadrilateral element

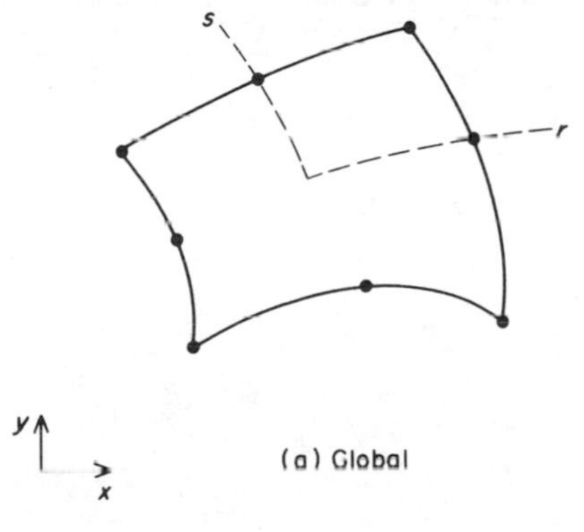

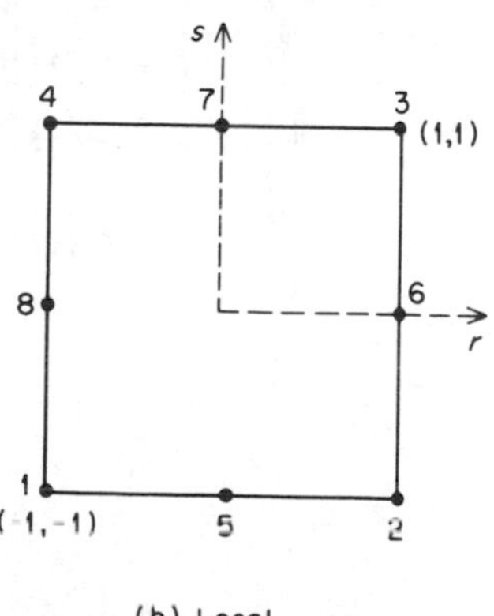

Fig. 6.8 **(a)** Subroutine SHP8Q, and **(b)** subroutine DER8Q

```
      SUBROUTINE  SHP8Q (S,T,H)
C     * * * * * * * * * * * * * * * * * * * * * * * * * *
C     SHAPE FUNCTIONS OF 8 NODE ISOPARAMETRIC QUADRILATERAL
C     * * * * * * * * * * * * * * * * * * * * * * * * * *
CDP   IMPLICIT REAL*8 (A-H,O-Z)
      DIMENSION H(8)
C     NODAL ORDER SHOWN TO RIGHT.                  4 - 7 - 3
C     S,T = LOCAL COORDINATES OF POINT             |   T   |
C     H = SHAPE FUNCTION ARRAY                     8   *S  6
C     NODE 1 AT (-1,-1)                            |       |
C     NODE 3 AT (1,1)                              1 - 5 - 2
      SP = 1. + S
      SM = 1. - S
      TP = 1. + T
      TM = 1. - T
      H(1) = 0.25*SM*TM*( SM + TM - 3. )
      H(2) = 0.25*SP*TM*( SP + TM - 3. )
      H(3) = 0.25*SP*TP*( SP + TP - 3. )
      H(4) = 0.25*SM*TP*( SM + TP - 3. )
      H(5) = 0.5*TM*( 1. - S*S )
      H(6) = 0.5*SP*( 1. - T*T )
      H(7) = 0.5*TP*( 1. - S*S )
      H(8) = 0.5*SM*( 1. - T*T )
      RETURN
      END
```

(a)

```
      SUBROUTINE  DER8Q (S,T,DH)
C     * * * * * * * * * * * * * * * * * * * * * * * * * *
C     FIND LOCAL DERIVATIVES OF SHAPE FUNCTIONS FOR AN
C     EIGHT NODE ISOPARAMETRIC QUADRILATERAL ELEMENT
C     * * * * * * * * * * * * * * * * * * * * * * * * * *
CDP   IMPLICIT REAL*8 (A-H,O-Z)
      DIMENSION DH(2,8)
C     S,T = LOCAL COORDINATES OF POINT
C     DH = LOCAL DERIVATIVES OF SHAPE FUNCTIONS, H
C     DH(1,J) = DH(J)/DS,   DH(2,J) = DH(J)/DT
C     H = SHAPE FUNCTION ARRAY
      SP = 1. + S
      SM = 1. - S
      TP = 1. + T
      TM = 1. - T
      DH(1,1) = -0.25*TM*( SM + SM + TM - 3. )
      DH(2,1) = -0.25*SM*( TM + SM + TM - 3. )
      DH(1,2) =  0.25*TM*( SP + SP + TM - 3. )
      DH(2,2) = -0.25*SP*( TM + SP + TM - 3. )
      DH(1,3) =  0.25*TP*( SP + SP + TP - 3. )
      DH(2,3) =  0.25*SP*( TP + SP + TP - 3. )
      DH(1,4) = -0.25*TP*( SM + SM + TP - 3. )
      DH(2,4) =  0.25*SM*( TP + SM + TP - 3. )
      DH(1,5) = -S*TM
      DH(2,5) = -0.5*( 1. - S*S )
      DH(1,6) =  0.5*( 1. - T*T )
      DH(2,6) = -T*SP
      DH(1,7) = -S*TP
      DH(2,7) =  0.5*( 1. - S*S )
      DH(1,8) = -0.5*( 1. - T*T )
      DH(2,8) = -T*SM
      RETURN
      END
```

(b)

Fig. 6.9 Subroutine SHP8H

```
      SUBROUTINE  SHP8H (R,S,T,H)
C     * * * * * * * * * * * * * * * * * * * * * * * * * *
C     SHAPE FUNCTIONS OF 8 NODE ISOPARAMETRIC HEXAHEDRON
C     * * * * * * * * * * * * * * * * * * * * * * * * * *
CDP   IMPLICIT REAL*8 (A-H,O-Z)
      DIMENSION  H(8)
C     R,S,T = LOCAL COORDS OF PT                 | T
C     H = ELEM SHAPE FUNCTIONS                 6/--/5
C     NODES ORDERED BY RHR                     /  / |
C     ABOUT THE R-AXIS                       2/--/1 /|--S
C     LOCAL COORD:1=(1,1,1)                   |  | /  8
C     4=(1,1,-1)  7=(-1,-1,-1)             R 3|--|/ 4
      RP = 1. + R
      RM = 1. - R
      SP = 1. + S
      SM = 1. - S
      TP = 1. + T
      TM = 1. - T
      H(1) = 0.125*RP*SP*TP
      H(2) = 0.125*RP*SM*TP
      H(3) = 0.125*RP*SM*TM
      H(4) = 0.125*RP*SP*TM
      H(5) = 0.125*RM*SP*TP
      H(6) = 0.125*RM*SM*TP
      H(7) = 0.125*RM*SM*TM
      H(8) = 0.125*RM*SP*TM
      RETURN
      END
```

Fig. 6.10 Subroutine DER8H

```
      SUBROUTINE  DER8H (R,S,T,DH)
C     * * * * * * * * * * * * * * * * * * * * * * * * * *
C     LOCAL DERIVATIVES FOR EIGHT NODE HEXAHEDRON
C     * * * * * * * * * * * * * * * * * * * * * * * * * *
CDP   IMPLICIT REAL*8 (A-H,O-Z)
      DIMENSION  DH(3,8)
C     R,S,T = LOCAL COORDINATES OF THE POINT
C     DH(1,K)=DH/DR, DH(2,K)=DH/DS, DH(3,K)=DH/DT
C     H = ELEMENT SHAPE FUNCTIONS
      RP = 1. + R
      RM = 1. - R
      SP = 1. + S
      SM = 1. - S
      TP = 1. + T
      TM = 1. - T
      DH(1,1) =  0.125*SP*TP
      DH(2,1) =  0.125*RP*TP
      DH(3,1) =  0.125*RP*SP
      DH(1,2) =  0.125*SM*TP
      DH(2,2) = -0.125*RP*TP
      DH(3,2) =  0.125*RP*SM
      DH(1,3) =  0.125*SM*TM
      DH(2,3) = -0.125*RP*TM
      DH(3,3) = -0.125*RP*SM
      DH(1,4) =  0.125*SP*TM
      DH(2,4) =  0.125*RP*TM
      DH(3,4) = -0.125*RP*SP
      DH(1,5) = -0.125*SP*TP
      DH(2,5) =  0.125*RM*TP
      DH(3,5) =  0.125*RM*SP
      DH(1,6) = -0.125*SM*TP
      DH(2,6) = -0.125*RM*TP
      DH(3,6) =  0.125*RM*SM
      DH(1,7) = -0.125*SM*TM
      DH(2,7) = -0.125*RM*TM
      DH(3,7) = -0.125*RM*SM
      DH(1,8) = -0.125*SP*TM
      DH(2,8) =  0.125*RM*TM
      DH(3,8) = -0.125*RM*SP
      RETURN
      END
```

Fig. 6.11 Typical two-dimensional polynominal terms

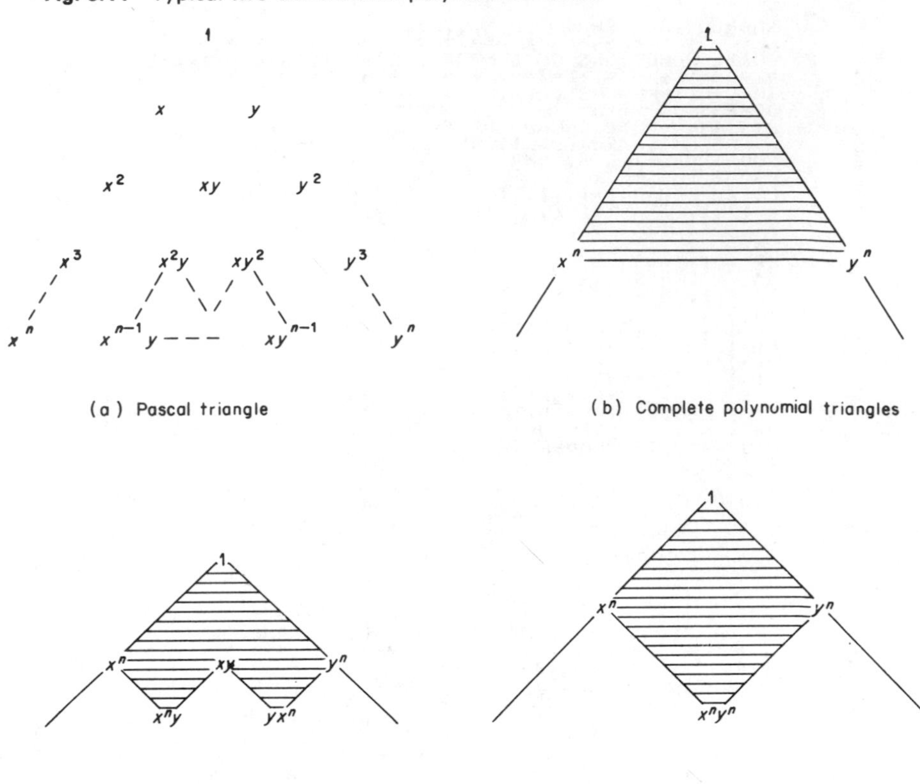

Fig. 6.12 **(a)** Subroutine SHP3T, and **(b)** subroutine DER3T

```
      SUBRCUTINE  SHP3T (S,T,H)
C     * * * * * * * * * * * * * * * * * * * * * * * * * *
C     SHAPE FUNCTICNS FOR A THREE NODE UNIT TRIANGLE
C     * * * * * * * * * * * * * * * * * * * * * * * * * *
CDP   IMPLICIT REAL*8 (A-H,O-Z)
      DIMENSION  H(3)
C     S,T = LOCAL COORDINATES OF THE POINT      3     T
C     H = SHAPE FUNCTIONS                       . .   .
C     NODAL COORDS 1-(0,0)  2-(1,0)  3-(0,1)    1..2  ...S
      H(1) = 1. - S - T
      H(2) = S
      H(3) = T
      RETURN
      END
```

(a)

Fig. 6.12—continued

```
      SUBROUTINE  DER3T (S,T,DH)
C     * * * * * * * * * * * * * * * * * * * * * * * * * * * *
C     LOCAL DERIVATIVES OF A THREE NODE UNIT TRIANGLE
C     * * * * * * * * * * * * * * * * * * * * * * * * * * * *
CDP   IMPLICIT REAL*8 (A-H,O-Z)
      DIMENSION  DH(2,3)
C     S,T = LOCAL COORDINATES OF THE POINT
C     DH(1,K) = DH(K)/DS,    DH(2,K)=DH(K)/DT
C     NODAL COORDS ARE :  1-(0,0) 2-(1,0) 3-(0,1)
      DH(1,1) = -1.
      DH(1,2) = 1.
      DH(1,3) = 0.0
      DH(2,1) = -1.
      DH(2,2) = 0.0
      DH(2,3) = 1.
      RETURN
      END
```

(b)

Fig. 6.13 **(a)** Subroutine SHP6T, and **(b)** subroutine DER6T

```
      SUBROUTINE  SHP6T (S,T,H)
C     * * * * * * * * * * * * * * * * * * * * * * * * * * * *
C     LOCAL SHAPE FUNCTIONS FOR A SIX NODE UNIT TRIANGLE
C     * * * * * * * * * * * * * * * * * * * * * * * * * * * *
CDP   IMPLICIT REAL*8 (A-H,O-Z)
      DIMENSION  H(6)
C     S,T = LOCAL COORDINATES                              3
C           OF A POINT IN THE UNIT TRIANGLE
C     H = SIX SHAPE FUNCTIONS FOR A QUADRATIC    T      6 5
C         ELEMENT  WITH SIX NODES                |
C     THE NODAL ORDER IS SHOWN TO THE RIGHT      .-S    1 4 2
C     NODAL COORDS : 1-(0,0)    2-(1,0)    3-(0,1)
C                    4-(0.5,0)  5-(0.5,0.5)  6-(0,0.5)
      H(1) = 1. - 3.*S - 3.*T + 2.*S*S + 4.*S*T + 2.*T*T
      H(2) = 2.*S*S - S
      H(3) = 2.*T*T - T
      H(4) = 4.*(S - S*S - S*T)
      H(5) = 4.*S*T
      H(6) = 4.*(T - S*T - T*T)
      RETURN
      END
```

(a)

```
      SUBROUTINE  DER6T (S,T,DH)
C     * * * * * * * * * * * * * * * * * * * * * * * * * * * *
C     LOCAL DERIVATIVES FOR A SIX NODE UNIT TRIANGLE
C     * * * * * * * * * * * * * * * * * * * * * * * * * * * *
CDP   IMPLICIT REAL*8 (A-H,O-Z)
      DIMENSION  DH(2,6)
C     S,T = LOCAL COORDINATES
C     DH = LOCAL DERIVATIVES OF SHAPE FUNCTIONS
C     DH(1,K) = DH(K)/DS,    DH(2,K)=DH(K)/DT
C     NODAL COORDS : 1-(0,0)     2-(1,0)     3-(0,1)
C                    4-(0.5,0)   5-(0.5,0.5)   6-(0,0.5)
      DH(1,1) = -3. + 4.*S + 4.*T
      DH(1,2) = -1. + 4.*S
      DH(1,3) = 0.0
      DH(1,4) = 4. - 8.*S - 4.*T
      DH(1,5) = 4.*T
      DH(1,6) = -4.*T
      DH(2,1) = -3. + 4.*S + 4.*T
      DH(2,2) = 0.0
      DH(2,3) = -1. + 4.*T
      DH(2,4) = -4.*S
      DH(2,5) = 4.*S
      DH(2,6) = 4. -4.*S - 8.*T
      RETURN
      END
```

(b)

Fig. 6.14 Boundary curves through element nodes

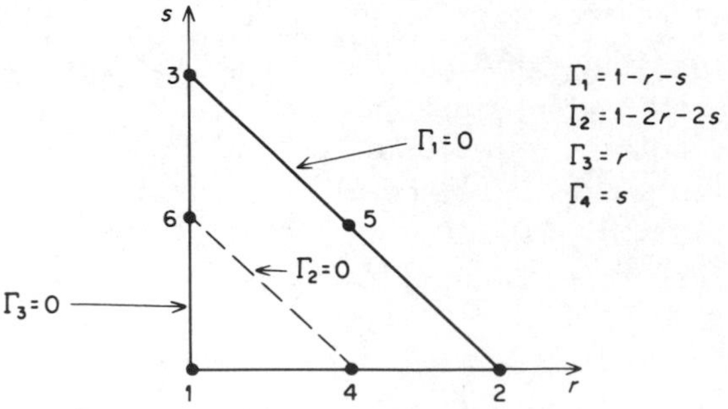

Fig. 6.15 Subroutine SHAPE

```
      SUBROUTINE  SHAPE (QPT,H,N,NSPACE)
C     * * * * * * * * * * * * * * * * * * * * * * * * *
C     EVALUATE C0 ELEMENT INTERPOLATION FUNCTIONS
C     * * * * * * * * * * * * * * * * * * * * * * * * *
CDP   IMPLICIT REAL*8 (A-H,O-Z)
      DIMENSION  H(N), QPT(NSPACE)
C     N = NUMBER OF NODES PER ELEMENT
C     NSPACE = NO OF SPATIAL DIMENSIONS
C     H = ELEMENT INTERPOLATION FUNCTIONS AT QPT
C     QPT = LOCAL COORD OF A POINT
C     BRANCH ON SPACE, THEN NO OF NODES
      GO TO (100,200,300,1), NSPACE
C-->  1-D ELEMENTS
 100  GO TO (1,102,103,1), N
  102 CALL SHP2L (QPT(1),H)
      RETURN
  103 CALL SHP3L (QPT(1),H)
      RETURN
C-->  2-D ELEMENTS
 200  GO TO (1,1,203,204,1,206,1,208,
     1        1,1,1,212,1), N
  203 CALL SHP3T (QPT(1),QPT(2),H)
      RETURN
  204 CALL SHP4Q (QPT(1),QPT(2),H)
      RETURN
  206 CALL SHP6T (QPT(1),QPT(2),H)
      RETURN
  208 CALL SHP8Q (QPT(1),QPT(2),H)
      RETURN
  212 CALL SHP412 (QPT(1),QPT(2),H)
      RETURN
C---> 3-D ELEMENTS
 300  GO TO (1,1,1,1,1,1,1,308,1), N
  308 CALL SHP8H (QPT(1),QPT(2),QPT(3),H)
      RETURN
C     UNSUPPORTED OPTION
 1    WRITE (6,1000)
 1000 FORMAT (' INVALID DATA IN SUBR SHAPE')
      RETURN
      END
```

6.7 Interpolation enhancement for C^0 transition elements

The procedures in the previous sections were designed for elements having the same order of interpolation on each edge or face. If one wishes to go from one order of interpolation to another then it is necessary to supply transition elements between the two groups of elements or to define generalized constraint equations on the boundaries between the two different groups. The former method is not difficult to implement and will be illustrated here.

Several special *transition* elements have been developed by various authors using trial and error methods. Special elements can be generated by enhancing low order elements, usually linear, or by degrading high order elements. Here a typical enhancement procedure will be utilized. Assume that the element to be enriched satisfies the usual conditions

$$u(r, s) = \sum_{i=1}^{n} H_i(r, s)u_i, \tag{6.15}$$

where

$$\sum_{i=1}^{n} H_i(r, s) = 1, \qquad H_i(r_j, s_j) = \delta_{ij}.$$

Here the H_i are the original unenhanced interpolation functions, the u_i are nodal values, δ_{ij} is the Kronecker delta, and n is the original number of nodes. Let the enriching functions, say $M_j(r, s)$, be defined at m additional points. These satisfy the similar conditions that

$$M_i(r_j, s_j) = \delta_{ij}, \quad i > n, \quad 1 \leqslant j \leqslant (m+n). \tag{6.16}$$

Thus the enriched elements have values defined by

$$u(r, s) = \sum_{i=1}^{(m+n)} H_i(r, s)u_i, \tag{6.17}$$

where the enriched interpolation functions, H_i, are given by

$$H_i(r, s) \leftarrow H_i(r, s) - \sum_{j=n+1}^{n+m} H_i(r_j, s_j)M_j(r, s), \quad i \leqslant n \tag{6.18}$$

$$H_i(r, s) \leftarrow M_i(r, s), \quad i > n.$$

where "$\leftarrow$" reads "is replaced by". These functions satisfy analogue relations to the second and third of Eqn. (6.15).

That can be seen by summing the enriched function so that

$$\sum_{i=1}^{m+n} H_i \leftarrow \sum_{i=1}^{n} H_i - \sum_{i=1}^{n} \sum_{j=1+n}^{m+n} H_i(r_j, s_j) M_j + \sum_{j=1+n}^{m+n} M_j$$

$$= 1 - \sum_{j=1+n}^{m+n} M_j(r, s) \left(1 - \sum_{i=1}^{n} H_i(r_j, s_j)\right)$$

$$= 1 - 0 = 1$$

as required. Similarly if $i > n$

$$H_i(r_k, s_k) \leftarrow M_i(r_k, s_k) = \delta_{ik},$$

from Eqns. (6.16) and (6.18) while if $i \leqslant n$

$$H_i(r_k, s_k) \leftarrow H_i(r_k, s_k) - \sum_{j=1+n}^{m+n} H_i(r_j, s_j) M_j(r_k, s_k)$$

$$= H_i(r_k, s_k) - \sum_{j=1+n}^{m+n} H_i(r_j, s_j)\, \delta_{jk}$$

$$= \begin{cases} 0 & \text{if } k > n \\ \delta_{ik} & \text{if } k \leqslant n, \end{cases}$$

thus, as required,

$$H_i(r_k, s_k) = \delta_{ik}.$$

As the simplest example of enhancement, consider a line element to which a centre node is to be added, Fig. 6.16. First, assume that the enriching function is to be quadratic. Then from Eqn. (5.15) the interpolation functions are

$$\begin{aligned} H_1(r) &= 1 - r, \\ H_2(r) &= r, \\ M_3(r) &= 4r - 4r^2. \end{aligned} \tag{6.19}$$

Then the enriched interpolation function for node 1 is

$$\begin{aligned} H_1(r) &= H_1(r) - H_1(r_3) M_3(r) \\ &= [1 - r] - (1/2)\,[4r - 4r^2] \\ &= 1 - 3r + 2r^2. \end{aligned} \tag{6.20}$$

This, of course, is the standard function for a full quadratic line element.

The value of M_2 is obtained in a similar manner.

Figure 6.16(b) shows a standard bilinear quadrilateral to which a fifth node is to be added. Again only interpolation functions for nodes 1 and 2 need be enriched. These functions are

$$\begin{aligned} H_1(r,s) &= 1 - r - s + rs \\ H_2(r,s) &= \quad r \quad - rs. \end{aligned} \tag{6.21}$$

If the enriching function is quadratic then it is

$$M_5(r,s) = 4r(1-r)(1-s),$$

so that

$$\begin{aligned} H_1(r,s) &= H_1(r,s) - H_1(r_5,s_5)M_5(r,s) \\ &= (1-r)(1-s) - (\tfrac{1}{2})[4r(1-r)(1-s)] \\ &= (1-2r)(1-r)(1-s). \end{aligned} \tag{6.22}$$

This corresponds to the linear–quadratic transition quadrilateral given by Bathe and Wilson [14].

One may also wish to have a linear–linear transition element for grid refinement. Then we may take

$$\begin{aligned} M_5(r,s) &= 2r(1-s), & 0 \leqslant r \leqslant 1/2 \\ &= 2(1-r)(1-s), & 1/2 \leqslant r \leqslant 1, \end{aligned} \tag{6.23}$$

and the corresponding modified interpolation functions are

$$H_1 = (1-r)(1-s) - (\tfrac{1}{2})[2r(1-s)],\ r \leqslant \tfrac{1}{2},$$

$$H_1 = (1-r)(1-s) - (\tfrac{1}{2})[2(1-r)(1-s)],\ r > \tfrac{1}{2},$$

which reduce to

$$\begin{aligned} H_1 &= (1-2r)(1-s), & 0 \leqslant r \leqslant \tfrac{1}{2}, \\ &= 0 & \tfrac{1}{2} \leqslant r \leqslant 1. \end{aligned} \tag{6.24}$$

These correspond to the functions given by Whiteman [78] for a linear–linear transition quadrilateral. The same functions were later developed by Gupta [37]. The above elements, and corresponding triangles, are shown in Fig. 6.17.

The enrichment of higher elements proceeds in a similar manner. A linear to cubic enrichment is summarized in Fig. 6.18. The results for a series of quadrilateral elements are shown in Fig. 6.20. The figure presents the interpolation functions for an element which can be linear, quadratic or cubic on any of its four sides. It is generated by enhancing the bilinear procedure. These operations can be programmed in different ways. Figure 6.19 illustrates only one way of generating such an element. Subroutine SHP412 could easily be extended to include one or four interior nodes so as to generate Lagrangian elements. A similar procedure can be used to generate a 3 to 10 node enhanced triangle. Of course, it is also easily extended to three-dimensional elements.

Since such elements allow for local node numbers to be zero it is necessary to avoid zero or negative subscripts. This is easily done with a few IF statements. For example, during assembly if a local node is zero there are no corresponding terms in the element matrices so the zero (or negative) subscript is skipped during assembly. (See subroutine STORCL Fig. 7.2.)

Fig. 6.16 Simple element enhancement

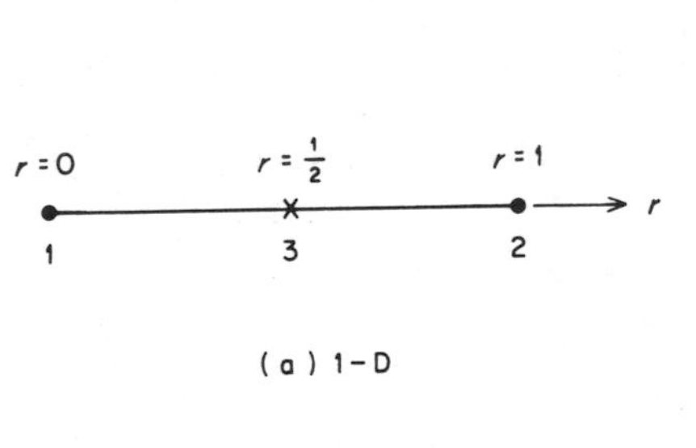

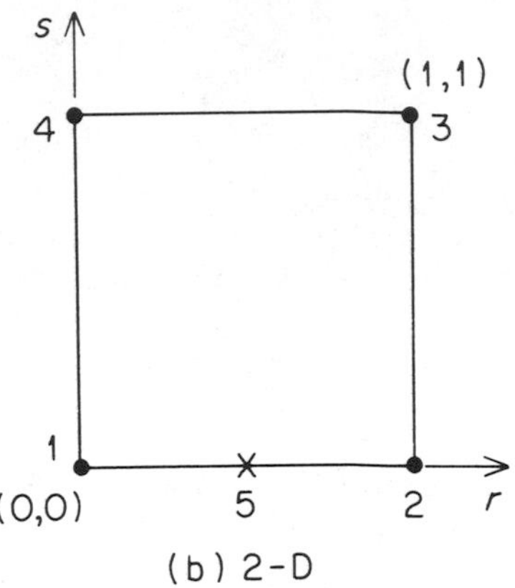

Fig. 6.17 Typical transaction elements

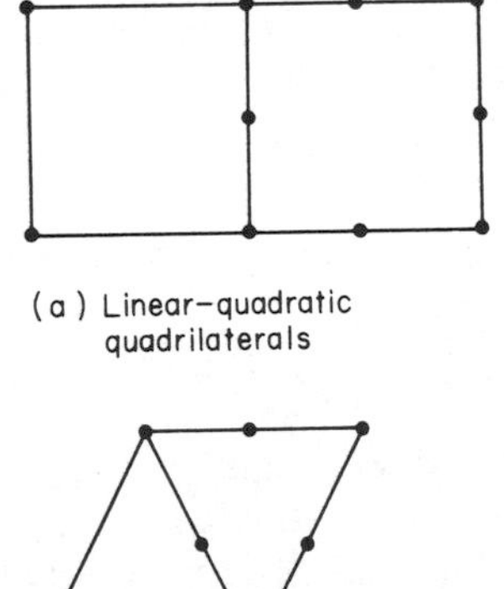

(a) Linear-quadratic quadrilaterals

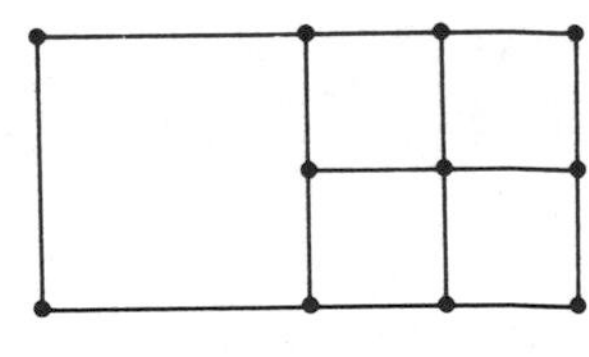

(b) Linear-linear quadrilaterals

(c) Linear-quadratic triangles

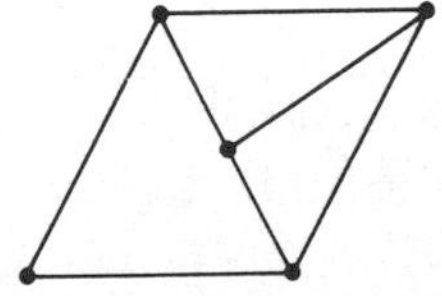

(d) Linear-linear triangles

Fig. 6.18 Enrichment to a cubic element

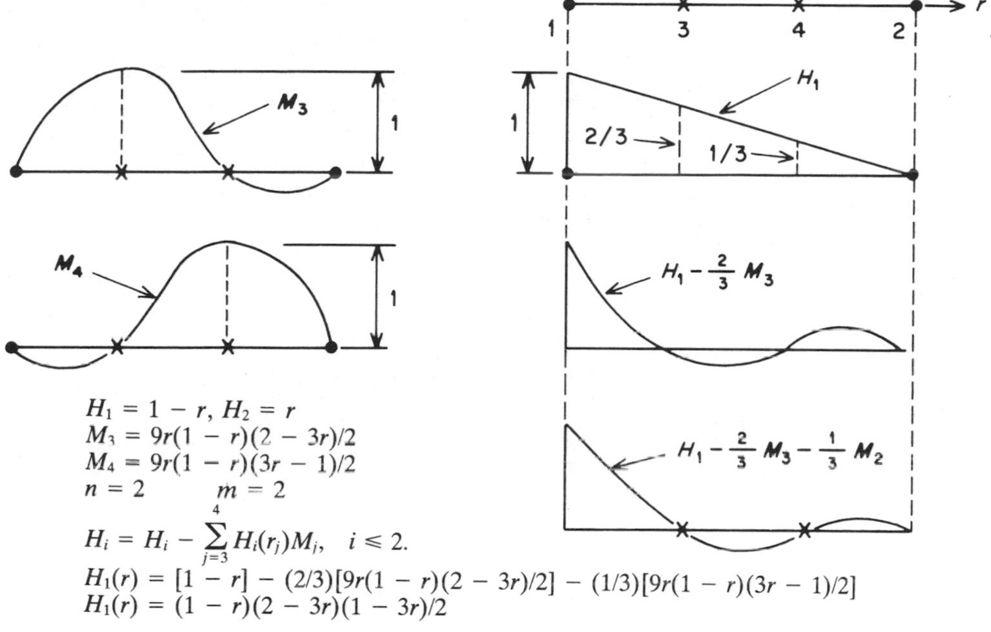

Fig. 6.19 A 4 to 12 node enhanced quadrilateral; subroutine SHP412

```
      SUBROUTINE  SHP412 (R,S,H,LNODE)
C     * * * * * * * * * * * * * * * * * * * * * * * * *
C     SHAPE FUNCTIONS OF 4 TO 12 NODE QUADRILATERAL
C     * * * * * * * * * * * * * * * * * * * * * * * * *
CDP   IMPLICIT REAL*8 (A-H,O-Z)
      DIMENSION  H(12), LNODE(12), IP(4), JP(4), NEXT(4)
      DATA IP,JP,NEXT /1,0,-1,0,   0,1,0,-1,   2,3,4,1/
C     R,S = LOCAL COORDS OF PT         4--11---7---3
C     H = ELEM SHAPE FUNCTIONS         I           I
C     LNODE = ELEM INCIDENCES LIST     8      S    10
C     ELEMENT SKETCH TO RIGHT          I      .R   I
C     LOCAL COORD OF NODES:           12           6
C     1-(-1,-1)  3-(+1,+1)             I           I
C     SIDES OF ORDER 1, 2, OR 3        1---5---9---2
C-->  GENERATE FOUR NODE QUAD
      CALL  SHP4Q (R,S,H)
C-->  LOOP OVER SIDES
      DO 10  I = 1,4
      H(I+4) = 0.0
      H(I+8) = 0.0
C     IS SIDE LINEAR
      IF ( LNODE(I+4).EQ.0 )   GO TO 10
      K = NEXT(I)
C     FIND PT RELATIVE TO SIDE
      P = R*IP(I) + S*JP(I)
      Q =-R*JP(I) + S*IP(I)
      TEMP = (1. - Q)*0.5
C-->  IS SIDE CUBIC
      IF ( LNODE(I+8).EQ.0 )   GO TO 20
      H(I+8) = TEMP*(1. - 3.*P - P*P + 3.*P*P*P)*9./16.
      H(I+4) = TEMP*(1. + 3.*P - P*P - 3.*P*P*P)*9./16.
C     CORRECT CORNER POINTS FOR CUBIC
      H(I) = H(I) - H(I+4)*2./3. - H(I+8)/3.
      H(K) = H(K) - H(I+8)*2./3. - H(I+4)/3.
      GO TO 10
C-->  QUADRATIC SIDE
 20   H(I+4) = TEMP*(1. - P*P)
C     CORRECT CORNER POINTS FOR QUADRATIC
      H(I) = H(I) - H(I+4)*0.5
      H(K) = H(K) - H(I+4)*0.5
 10   CONTINUE
      RETURN
      END
```

Fig. 6.20 Cubic to linear transaction quadrilateral

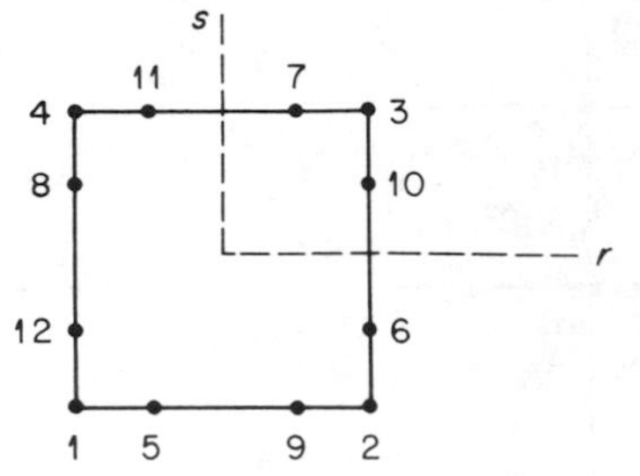

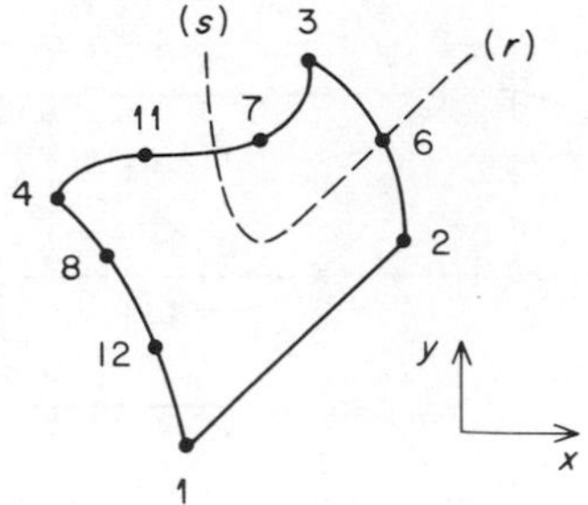

Cubic side:

$$H_i(r, s) = (1 - s^2)(1 + 9ss_i)(1 + rr_i)9/32, \; r_i = \pm 1, s_i = \pm 1/3 \quad (i = 6, 8, 10, 12)$$

$$H_i(r, s) = (1 - r^2)(1 + 9rr_i)(1 + ss_i)9/32, \; r_i = \pm 1/3, s_i = \pm 1 \quad (i = 5, 7, 9, 11)$$

Quadratic side:

$$H_i(r, s) = (1 + rr_i)(1 - s^2)/2, \; r_i = \pm 1, s_i = 0 \quad (i = 6, 8)$$

$$H_i(r, s) = (1 + ss_i)(1 - r^2)/2, \; r_i = 0, s_i = \pm 1 \quad (i = 5, 7)$$

$$H_j = 0, j = i + 4$$

Linear side

$$H_j = H_k = 0, j = i + 4, k = i + 8, (i = 1, 2, 3, 4)$$

Corners:

$$H_i(r, s) = C_i(1 + rr_i)(1 + ss_i)/4, \; r_i = \pm 1, s_i = \pm 1, (i = 1, 2, 3, 4)$$

Corner corrections:

$$C_i = P_r + P_s, (i = 1, 2, 3, 4)$$

Order of side	P_r, $s_i = \pm 1$	P_s, $r_i = \pm 1$
Linear	1/2	1/2
Quadratic	$rr_i - 1/2$	$ss_i - 1/2$
Cubic	$(9r^2 - 5)/8$	$(9s^2 - 5)/8$

Fig. 6.21 Distorted elements for singularities; **(a)** point singularity at P, and **(b)** line singularity on Γ

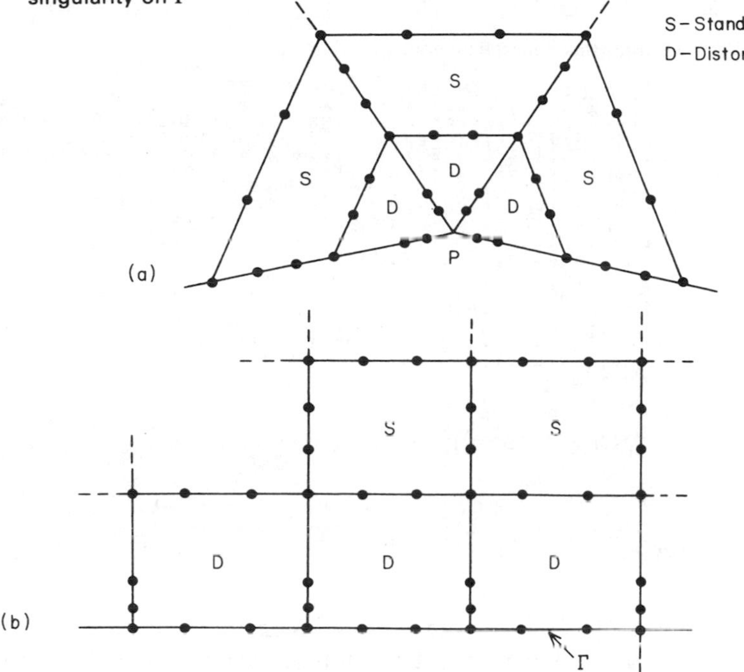

Fig. 6.22 Global node locations $x_i = x(r_i)$ for inerpolations of order $\rho^{1/m}$

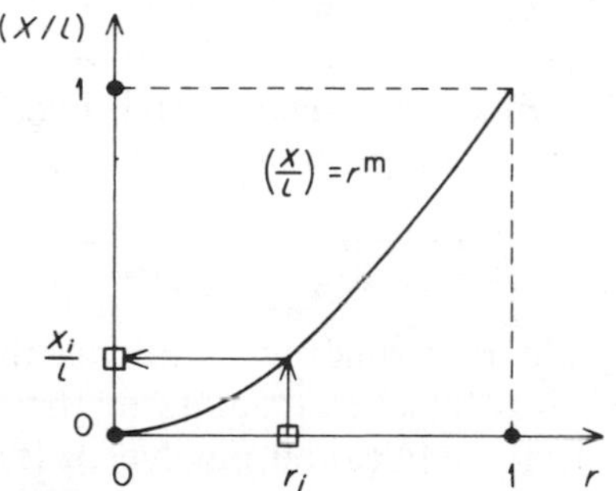

(a) Mapping for Jacobian Manipulation

Number of nodes $n > m$	Standard node r_i $1 \leqslant i \leqslant n$	Mapped node $(x/l)_i = r_i^m$ $1 \leqslant i \leqslant n$	m
2	0 1	0 1	1
3	0 1/2 1	0 1/2 1	1
		0 1/4 1	2
4	0 1/3 2/3 1	0 1/3 2/3 1	1
		0 1/9 4/9 1	2
		0 1/27 8/27 1	3

(b) Exact mapping to physical space (integer m)

Fig. 6.23 Subroutine SINGLR

```
      SUBROUTINE  SINGLR (P,H,DH,N,NSPACE)
C     * * * * * * * * * * * * * * * * * * * * * * * * * * * *
C     CONVERT STANDARD FUNCTIONS TO SINGULAR FUNCTIONS
C     WITH DERIV SINGULARITIES AT NODE 1 OF O(R**(-P))
C     * * * * * * * * * * * * * * * * * * * * * * * * * * * *
CDP   IMPLICIT REAL*8 (A-H,O-Z)
      DIMENSION  H(N), DH(NSPACE,N)
C     H=SHAPE FUNCTION ARRAY, DH=LOCAL DERIVATIVES OF H
C     N=NO OF SHAPE FUNCTIONS, NSPACE=DIMENSION OF SPACE
C     REQUIRES SUM OF H(I) = 1, & CONST JACOBIAN
      IF ( P.EQ.0.0 )  RETURN
      W = 1.0 - H(1)
      R = W**P
      DO 20  I = 1,NSPACE
      DO 10  J = 2,N
   10 DH(I,J) = DH(I,J)/R + P*DH(I,1)*H(J)/R/W
      DH(I,1) = (1.0 - P)*DH(I,1)/R
   20 CONTINUE
      DO 30  J = 2,N
   30 H(J) = H(J)/R
      H(1) = 1.0 - W/R
      RETURN
      END
```

6.8 Special elements

Current applications of the finite element method of analysis show a trend towards the enhancement of solution accuracy through the utilization of elements with *special interpolation* functions. A survey of this type of special element, developed with the aid of physical insight, is presented. This class includes point and line singularity elements, semi-infinite elements, and elements with *upwind* weighting. The relative advantages and disadvantages of these special elements are also discussed.

6.8.1 Singularity elements

There are numerous analysis problems that involve the presence of singular derivatives at known points in the solution domain. The most common singularity application involves a fracture mechanics analysis of the stresses near a crack tip or crack edge. In two dimensions let ρ denote the radial distance from the singular point. In a linear fracture mechanics analysis the displacements and stresses are $0(\rho^{1/2})$ and $0(\rho^{-1/2})$, respectively. Originally these problems were approached by a direct solution that employed standard elements and an extensive grid refinement near the singularity. Engineers expected that if they employed their physical insight and used special elements around the singularity, they would obtain more accurate results. Thus they attempted to modify standard interpolation functions to include the $\rho^{1/2}$ displacement behaviour near the crack tip. Many special two-dimensional elements have been proposed for this type of application.

There are numerous other problems that involve point singularities. For example, in the solution of the Poisson equation in a domain with a re-entrant corner there is a singularity at the corner point. If b denotes the interior obtuse corner angle, the solution near the corner has leading term $0(\rho^{\pi/b})$. Since $\pi < b \leqslant 2\pi$ one can encounter a wide range of singularity strengths. For L-shaped regions $b = 3\pi/2$ so that the solution is $0(\rho^{2/3})$ and its derivatives are singular with order $0(\rho^{-1/3})$. Only if the region contains a slit, $b = 2\pi$, does one encounter the stronger singularity for which many special elements have been proposed by the fracture mechanics' researchers. The above variable singularity could be encountered in such contexts as heat conduction, potential flow, and transverse electromagnetic lines.

Before considering the common special elements we will review the relationship between global and local derivatives. Recall Eq. (5.8)

$$\boldsymbol{\partial}_g u = \mathbf{J}^{-1}\boldsymbol{\partial}_\ell \mathbf{u},$$

$$\mathbf{J} = \boldsymbol{\partial}_\ell \mathbf{H}[\mathbf{X}^e : \mathbf{Y}^e],$$

where

$$\boldsymbol{\partial}_\ell u = \boldsymbol{\partial}_\ell \mathbf{H}\mathbf{U}^e.$$

From the first equation we note that a global derivative singularity in the physical coordinate can be introduced by way of the local derivatives of the interpolation functions or the inverse of the Jacobian.

An easy way of affecting the Jacobian in the second equation is to vary the global coordinates of the nodes, i.e. to *distort* the element. Henshell and Shaw [38] were among the first to note that the point at which the Jacobian goes singular can be controlled through the use of distorted elements. Generally the order of the geometric interpolation functions is quadratic or higher [39]. One can control the location of the singular point by moving nodes in the global coordinates. The singular point usually lies outside the element and one must force it to occur on an element boundary or node. To illustrate this concept consider a one-dimensional quadratic element, $0 \leqslant x \leqslant h$, of length h, see Fig. 6.16(a). The coordinate transformation of the standard unit element, $0 \leqslant \rho \leqslant 1$, with an interior node at ah is

$$x(\rho) = h(4a - 1)\rho + 2h(1 - 2a)\rho^2 \tag{6.25}$$

and the Jacobian is

$$J = \frac{\partial x}{\partial \rho} = h(4a - 1) + 4h(1 - 2a)\rho. \tag{6.26}$$

Thus if one wants a singularity to occur at $\rho = 0$, i.e. node 1, one must set

$a = 1/4$. That means that the global coordinates of node 3 must be specified so that $x_3 = h/4$. Then one has $x = h\rho^2$ so that $\rho = (x/h)^{1/2}$ and the inverse Jacobian becomes

$$J^{-1} = \frac{\partial \rho}{\partial x} = \frac{1}{2}(hx)^{-1/2}$$

which has a singularity of order $0(x^{-1/2})$ at $x = 0$. The derivative singularity can also be seen by noting that for quadratic interpolation on the unit interval

$$u(r) = \alpha_1 + \alpha_2\rho + \alpha_3\rho^2$$

becomes, in physical space,

$$u(x) = \alpha_1 = \beta_2 x^{1/2} + \beta_3 x.$$

Although the above one-dimensional example is simple, the algebra involved in this type of procedure can be rather involved in higher-dimensional elements. In two and three dimensions, respectively, the point or line of singularity is surrounded with the special distorted elements as shown in Fig. 6.21. The locations of nodes on the element sides are usually the same as for the one-dimensional case. Any interior nodes usually lie on lines connecting the shifted side nodes. The above procedure is easily generalized [80, 39] and typical nodal locations for function interpolations of order $0(r^{1/m})$ are shown in Fig. 6.22.

The second alternative, in Eqn. (5.8), of introducing the singularity through the local derivative requires special element interpolation functions. A number of special elements of this type have been developed for the form $u = 0(\rho^{1/2})$ by Blackburn [20]. However, most such elements cannot be generalized to other singularities. The first conforming element having a general form of $u = 0(\rho^{1-a})$ for $0 < a < 1$ was given by Akin [2]. This procedure is valid for, and compatible with, an element that satisfies the condition (5.4) that $\Sigma_i N_i = 1$. This easily programmed procedure is outlined here. If the singularity occurs at node K in an element one defines a function

$$W(p, q) = 1 - H_k(r, s)$$

which is zero at node k and equal to unity at all other nodes. To obtain the power singularity form one can divide by a function $R = W_a$. If N_i denote the standard functions, then the modified singular functions, H_i, are defined as

$$H_i = [1 - 1/R]\delta_{ki} + N_i/R.$$

These also satisfy Eqn. (5.4) and near node K give a solution of the form $\phi(\rho^{1-a})$ so that the first derivative is $O(\rho^{-a})$. Thus since $a > 0$ the first derivative is singular. This procedure has been extended to line singularities by Akin [7]. The programming of these operations for point singularities is given in subroutine SINGLR in Fig. 6.23.

The examples of Section 6.7 illustrated how elements can be enriched to give higher order functions on edges and grid refinement on edges. The method can also be utilized to incorporate known singularities. For example, consider the line element again as in Eqn. (6.19) and let the centre node have a more general enrichment of the form:

$$M_3(r) = K(r^a - r^b), \quad a \neq b \tag{6.27}$$

where $K = 1/[1/2)^a - (1/2)^b]$. This can yield the standard quadratic function, $a = 1$ and $b = 2$, or a singularity function. For example, if one desires to have terms or order $0(r^{1/2})$ enriched into the previous linear element one could set $a = 1$ and $b = 1/2$. Then

$$M_3 = 2K(r - r^{1/2}), \quad K = 1/(1 - \sqrt{2})$$

and the modified functions are

$$H_1(r) = 1 - (1 + K)r + Kr^{1/2}$$

$$H_2(r) = (1 + K)r - Kr^{1/2}.$$

If standard interpolation is utilized for the geometry then this element gives derivative singularities at node 1 of order $0(\rho^{1/2})$ where ρ is the distance from the node. The above special case was used by Stern [67] to define a five node singularity triangle for functions of $0(\rho^b)$. He also introduced an additional M nodes, on the non-singular side, to model the transverse behaviour as an (M + 1) polynomial.

6.8.2 Semi-infinite elements

A number of practical applications involve infinite or semi-infinite domains. They include soil mechanics, external flows, and electromagnetic fields. The most common method of solution has been to extend the *standard elements* out as far as one can afford and thereto approximate the boundary conditions at infinity. Recently, special spatial coordinate interpolation functions have been suggested to map standard elements into a semi-infinite domain. Thus in a typical problem one would surround the region of interest with standard elements. The outer layer of the elements would then be mapped to infinity and the corresponding boundary conditions applied at the points at infinity.

In a standard element one defines the global coordinates in terms of the local coordinates by the standard mapping; for example, for a one-dimensional quadratic element $x(r) = H_1(r)x_1 + H_2(r)x_2 + H_3(r)x_3$. Thus $x_1 \leqslant x(r) \leqslant x_3$ and the element is a finite domain. To generate a semi-infinite domain one could replace the standard interpolation functions, $H_i(r)$, with a special set, say $N_i(r)$, that lead to the proper physical behaviour as $x \rightarrow \infty$ (i.e. $r \rightarrow 1$). Recently, two approaches to this problem have been suggested. Gartling and Becker [36] proposed the use of rational functions for the N_i while Bettess [18] suggested a combined exponential and Lagrangian interpolation for the N_i. Both approaches lead to functions involving a characteristic length, say L, that is related to the input coordinates. The results obtained from their analyses can vary significantly with L.

As an example, consider the mapping which takes the local coordinates, $r = 0$, $r = 1/2$, and $r = 1$, into the semi-infinite domain x_1, x_2, ∞. The lower limit, x_1, is necessary to maintain compatibility with the standard element mesh. One possible mapping is $x(r) = x_1 + Lr/(1 - r)$ where $L = (x_2 - x_1)$. That is, where $N_1(r) = 1 - r/(1 - r)$, $N_2(r) = r/(1 - r)$ and $N_3 = 0$.

Using the standard local interpolation of the dependent variable, $u(r) = H_1(r)u_1 + H_2(r)u_2 + H_3(r)u_3$, together with the spatial mapping, the gradient in the semi-infinite element is $\partial u/\partial x = (\partial r/\partial x)\ (\partial u/\partial r) = ((1 - r)(1 - r)/L)\partial u/\partial p$. Thus as $r \rightarrow 1$, $x \rightarrow \infty$, $u \rightarrow u_3$, and $\partial u/\partial x \rightarrow 0$. The value of L affects the 2nd and 4th of the limits. To find the optimum L one would have to use physical insight. If one arbitrarily selects L but knows a physical condition such as $\partial u/\partial x \rightarrow 1/x^2$ as $x \rightarrow \infty$ then the above form implies a multi-point constraint between u_1 and u_2. The generation of semi-infinite elements and procedures for finding optimum parameters, such as L, still merits attention.

An alternative procedure for treating semi-infinite domains was suggested by Silvester *et al.* [66]. It does not introduce arbitrary constants as most mappings do and is computationally efficient. Consider a finite region R containing a standard finite element mesh. The region has been selected such that there exists at least one point, P, which can see all points on the boundary, Γ. Assume that this region needs to be extended to infinity to satisfy a boundary condition. Define a mapping of Γ onto a series of temporary interface boundaries, γ_k, such that for each pair of points on γ_k and γ_{k+1}:

$$\mathbf{r}_{k+1} = \mathrm{K}\mathbf{r}_k$$

where $\mathbf{r}$ is the position vector from P, $\mathrm{K} > 1$ is a constant and $\gamma_0 \equiv \Gamma$. Thus at the exterior boundary, say γ_n:

$$\mathbf{r}_n = \mathrm{K}^n r_0.$$

Between two interfaces, say γ_k and γ_{k+1}, an angular segment defines an element e_{k+1}. The volumes of the extended elements e_{k+1} and e_k will have a ratio of K^{NSPACE} and the elements are geometrically similar.

Assume the exterior domain $\left(E = \bigcup_{k=1}^{n} e_k\right)$ satisfies a homogeneous differential equation and that the material properties are constant. Then if the square matrix involves an inner product of the first global derivatives the element matrices in e_k and e_{k+1} are given by

$$\mathbf{S}^{k+1} = \lambda \mathbf{S}^k$$

where $\lambda = K^{-NSPACE}$. This observation allows a recursion relation to be developed so that only contributions to nodes on Γ and γ_n need be retained. Since the function of interest on γ_n, say ϕ_n, is usually zero it may only be necessary to consider the net contributions to nodes of Γ. Full details (and minor errors) are given in [66].

6.8.3 Upwind Elements

The use of physical insight to generate an *upwind* difference procedure for the finite difference analysis of flows is known to be very useful. Thus it is logical to develop a similar procedure for finite elements. Zienkiewicz *et al.* have proposed such a special element [86], and evaluated a typical application [22]. For one-dimensional applications the special elements seem quite promising. However, the programming difficulties associated with the extension to higher-dimensional problems may limit the usefulness of the procedure. The upwind procedure recommended by Hughes [45] seems more practical from the computational point of view.

6.9 Exercises

1. Utilize Eqn. (6.9) to verify Eqn. (6.10).
2. Write a subroutine, say SYMTRI, to tabulate the symmetric triangle quadrature rules in Fig. 6.6.
3. Locate and tabulate the symmetric four point rule for triangles that has a negative weight at the centroid. (Some data have typographical errors.)
4. Using Fig. 6.14 as a guide, complete the derivation of the remaining four interpolation functions for the quadratic triangle.
5. Make random checks of the shape functions and local derivatives presented in Section 6.5 using SCHECK and DCHECK from Section 5.2.
6. Write a subroutine for C^0 Lagrangian interpolation on a line element.

7. Write a subroutine to generate Lagrangian quadrilaterals from the tensor product of the one-dimensional Lagrangian interpolations in each local direction.
8. Make the changes in SHAPE to include a two node line element and a twenty node hexahedron.
9. Referring to SHAPE, in Fig. 6.15, write a similar routine, DERIV, to determine the local derivatives.
10. Repeat the triangular integration example of Section 6.4 using an alternate three point rule having the same weights. The coordinates of the quadrature points are (1/6, 1/6), (2/3, 1/6) and (1/6, 2/3). Which of these two rules do you think would be best for axisymmetric problems and why?
11. Develop a table, similar to Fig. 6.19, for a three to six node triangle.
12. Write a subroutine, DER412, to find the local derivatives of the functions in Fig. 6.20.
13. Develop the interpolation functions for an eight to twenty node hexahedron.
14. Evaluate the element in Eqn. (6.25) when $a = 1$ and $b = 2$.
15. Verify that the elements in Fig. 6.22(b) retain the linear interpolation ability.
16. The bilinear quadrilateral elements of Section 5.2 can be degenerated to a triangle. Requiring that $x_3 = x_4$, etc. suggests a modified set of interpolation functions where $H_3 = H_3 + H_4$ (and $N_4 = 0$). Sketch the element in the global space and show the local axes. Determine the Jacobian on line 3–4 in the local space. Is this element identical to the three node simplex element?
17. Write a routine that will take the outputs of SHP4Q and DER4Q and modify them to generate the corresponding data for the degenerate triangle.

7

Assembly of element equations into system equations

7.1 Introduction

An important but often misunderstood topic is the procedure for *assemblying* the system equations from the element equations. Here assemblying is defined as the operation of adding the coefficients of the element equations into the proper locations in the system equations. There are various methods for accomplishing this but, as will be illustrated later, most are numerically inefficient. The numerically efficient *direct assembly* technique will be described here in some detail. First, it is desirable to review the simple but important relationship between a set of local (nodal point, or element) degree of freedom numbers and the corresponding system degree of freedom numbers. These simple calculations were described earlier in the discussions of INDXPT and INDXEL (Figs. 3.10 and 3.11, respectively).

Once the system degree of freedom numbers for the element have been stored in a vector, say INDEX, then the subscripts of a coefficient in the element equation can be directly converted to the subscripts of the corresponding system coefficient to which it is to be added. This correspondence between local and system subscripts is illustrated in Fig. 7.1. The expressions are generally of the form

$$C_I = C_I + C_i^e$$

$$S_{I,J} = S_{I,J} + S_{i,j}^e$$

where i and j are the local subscripts of a coefficient in the element square matrix and I, J are the corresponding subscripts of the system equation coefficient to which the element contribution is to be added. The direct conversions are $I = \text{INDEX}(i)$, $J = \text{INDEX}(j)$, where the INDEX array for element, e, is generated from Eqn. (1.1) by subroutine INDXEL.

Fig. 7.1 Correspondence of local and system degrees of freedom for a typical element, $\mathbf{S}^e \mathbf{D}^e = \mathbf{C}^e$

	Columns							
System	Index (1)	…	Index (i)	…	Index (j)	…	Index (N*NG)
Local	1	…	i	…	j	…	N*NG

$$\begin{bmatrix} S^e_{11} & \cdots & S^e_{1i} & \cdots & S^e_{1j} & \cdots & S^e_{1,N*NG} \\ \vdots & & \vdots & & \vdots & & \vdots \\ S^e_{i1} & \cdots & S^e_{ii} & \cdots & S^e_{1j} & \cdots & S^e_{i,N*NG} \\ \vdots & & \vdots & & \vdots & & \vdots \\ S^e_{j1} & \cdots & S^e_{ji} & \cdots & S^e_{jj} & \cdots & S^e_{j,N*NG} \\ \vdots & & \vdots & & \vdots & & \vdots \\ S^e_{N*NG,1} & \cdots & S^e_{N*NG,i} & \cdots & S^e_{N*NG,j} & \cdots & S^e_{N*NG,N*NG} \end{bmatrix}$$

Local	System
1	Index (1)
⋮	⋮
I	Index (i)
⋮	⋮
J	Index (i)
⋮	⋮
N*NG	Index (N*NG)

$$\begin{Bmatrix} C^e_1 \\ \vdots \\ C^e_i \\ \vdots \\ C^e_j \\ \vdots \\ C^e_{N*NG} \end{Bmatrix}$$

7.2 Assembly programs

Utilizing the vector containing the system degree of freedom numbers for the element, the coefficients of the element column matrix are easily added to the coefficients of the system column matrix. This simple operation is performed by subroutine STORCL, which is illustrated in Fig. 7.2.

The addition of the NELFRE*NELFRE square element matrix to the "square" matrix of the system is accomplished by using the system degree of freedom number array, INDEX, to convert both the element subscripts to system subscripts. This is known as the *direct method* of assemblying the system equations. A simple FORTRAN routine, STRFUL, to accomplish this direct assembly is illustrated in Fig. 7.3. Although this type of routine can be used on small problems, most practical problems have a large number of unknowns and are not economical if the full system equations are retained and solved. Figure 7.4 gives a graphical example of full assembly. The more practical procedure for storing the element equations in the upper half-bandwidths of the system equations is illustrated in subroutine STORSQ, Fig. 7.5.

The procedure utilized in STORSQ is slightly more involved since, due to the banded symmetric nature of the equations, the complete square system matrix is not stored in the computer. As mentioned before, the upper half-bandwidth (including the diagonal) of the system equations is stored as a rectangular array. If the element coefficient corresponds to a system coefficient that is below the main diagonal then it is ignored, but if it corresponds to a coefficient in the upper half-bandwidth then it is stored in the rectangular array.

As illustrated in Fig. 1.6, the main diagonal of the square array becomes the first column of the rectangular array. In general, a row in the square array becomes a row in the rectangular array. A column in the upper half-bandwidth of the square array becomes a diagonal in the rectangular array. Thus, if a coefficient belongs in the Ith row and J^{th} column of the square array ($J \geqslant I$) then that coefficient is stored in the Ith row and the JJ^{th} column of the rectangular array where

$$JJ = J - I + 1 \tag{7.1}$$

The program defines NELFRE*NELFRE storage locations for the element matrix, NDFREE*MAXBAN locations for the coefficient of the system upper half-bandwidth, and NELFRE storage locations for the system degree of freedom numbers associated with the element.

Again referring to Fig. 1.6, one should note for the future reference that, due to symmetry, the coefficients of the lower half-bandwidth portion of the Jth column are also contained in the upper half-bandwidth portion of the Jth row.

As shown in Fig. 2.4, the generation and assembly of the element matrices often takes place in a single element loop. A typical subroutine, ASYMBL, for controlling these operations is shown in Fig. 7.6. It employs GENELM, given in Fig. 4.9, to generate the element data. Subroutine STORCL is used to assembly the force vector, **C**, and STORSQ is used to assembly **S** in a banded storage mode.

Of course, an alternative storage mode like the skyline procedure could have been used to assembly **S** as shown in Fig. 3.17. Both options require the use of a few extra calculations. If the skyline method is used then STORSQ would be replaced by a procedure such as SKYSTR in Fig. 7.7. The goal in both cases is the same. It is to take a coefficient that is to be added to the term $S_{I,J}$ in the full matrix and to determine where it must actually be placed when the sparse matrix storage method is considered. In the skyline mode, Fig. 7.8, the vector position, IDIAG(J), of the diagonal term, $S_{J,J}$, is found and then steps are taken backwards (up the column) to locate where $S_{I,J}$ is stored. In the bandwidth mode, Fig. 7.9, the diagonal term $S_{I,I}$ is found in the first column of row I of the rectangular storage array. Then Eqn. (7.1) is employed to count over the correct number of columns to the location of $S_{I,J}$. In both cases symmetry is assumed. In a non-symmetric problem both the upper and lower band segments are needed. Then the same pointers could be used with two arrays (upper and lower parts of **S**). Then only the diagonal is duplicated in storage and the same routines are still valid.

Fig. 7.2 Subroutine STORCL

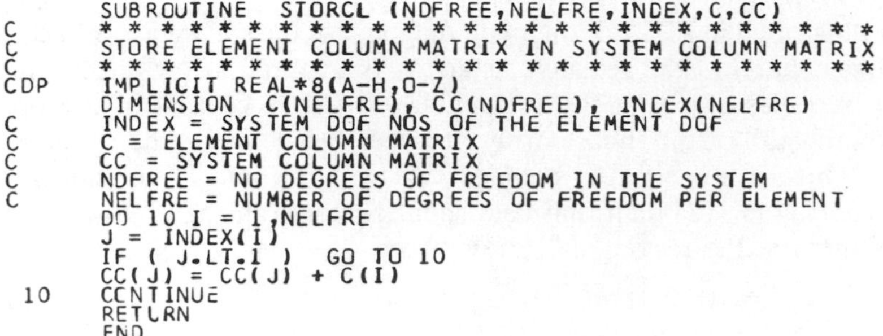

```
      SUBROUTINE  STORCL (NDFREE,NELFRE,INDEX,C,CC)
C     * * * * * * * * * * * * * * * * * * * * * * * * *
C     STORE ELEMENT COLUMN MATRIX IN SYSTEM COLUMN MATRIX
C     * * * * * * * * * * * * * * * * * * * * * * * * *
CDP   IMPLICIT REAL*8(A-H,O-Z)
      DIMENSION  C(NELFRE), CC(NDFREE), INDEX(NELFRE)
C     INDEX = SYSTEM DOF NOS OF THE ELEMENT DOF
C     C = ELEMENT COLUMN MATRIX
C     CC  = SYSTEM COLUMN MATRIX
C     NDFREE = NO DEGREES OF FREEDOM IN THE SYSTEM
C     NELFRE = NUMBER OF DEGREES OF FREEDOM PER ELEMENT
      DO 10 I = 1,NELFRE
      J = INDEX(I)
      IF ( J.LT.1 )  GO TO 10
      CC(J) = CC(J) + C(I)
   10 CONTINUE
      RETURN
      END
```

Fig. 7.4 Graphical assembly for a line element with two parameters per node

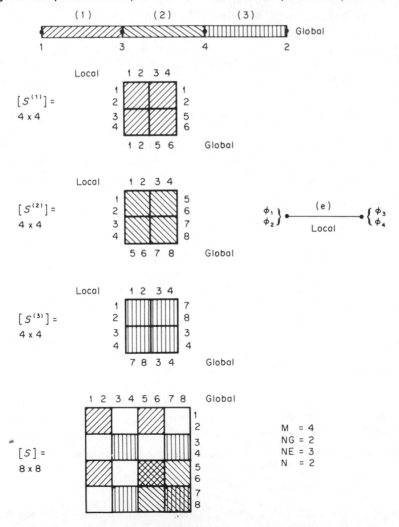

Fig. 7.3 Subroutine STRFUL

```
      SUBROUTINE  STRFUL (NDFREE,NELFRE,S,SS,INDEX)
C     * * * * * * * * * * * * * * * * * * * * * * * * * *
C     STORE ELEMENT SQ MATRIX IN FULL SYSTEM SQ MATRIX
C     * * * * * * * * * * * * * * * * * * * * * * * * * *
CDP   IMPLICIT REAL*8 (A-H,O-Z)
      DIMENSION  S(NELFRE,NELFRE), SS(NDFREE,NDFREE),
     1           INDEX(NELFRE)
C     NELFRE = NO DEGREES OF FREEDOM PER ELEMENT
C     NDFREE = TOTAL NO OF SYSTEM DEGREES OF FREEDOM
C     SS = FULL SYSTEM SQUARE MATRIX
C     S = FULL ELEMENT SQUARE MATRIX
C     INDEX = SYSTEM DOF NOS OF ELEMENT PARAMETERS
      DO 20  I = 1,NELFRE
      II = INDEX(I)
      IF ( II.LT.1 )  GO TO 20
      DO 10  J = 1,NELFRE
      JJ = INDEX(J)
      IF ( JJ.LT.1 )  GO TO 10
      SS(II,JJ) = SS(II,JJ) + S(I,J)
   10 CONTINUE
   20 CONTINUE
      RETURN
      END
```

Fig. 7.5 Subroutine STORSQ

```
      SUBROUTINE  STORSQ (NDFREE,IBW,NELFRE,INDEX,S,SS)
C     * * * * * * * * * * * * * * * * * * * * * * * * * *
C     ADD ELEMENT SQUARE MATRIX TO UPPER HALF BANDWIDTH
C     OF THE SYMMETRIC SYSTEM SQUARE MATRIX
C     * * * * * * * * * * * * * * * * * * * * * * * * * *
CDP   IMPLICIT REAL*8(A-H,O-Z)
      DIMENSION  S(NELFRE,NELFRE), SS(NDFREE,IBW),
     1           INDEX(NELFRE)
C     INDEX = SYSTEM DOF NOS OF THE ELEMENT DOF
C     S = SQUARE ELEMENT MATRIX
C     SS = SQUARE SYSTEM MATRIX
C     NDFREE = DEGREES OF FREEDOM IN THE SYSTEM
C     IBW = HALF BAND WIDTH INCLUDING THE DIAGONAL
C     NELFRE = NUMBER OF PARAMETERS (DOF) PER ELEMENT
C     I AND J ARE THE ROW AND COLUMN POSITIONS IN THE
C        UNPACKED SYSTEM MATRIX
C     JJ = COLUMN POSITION IN PACKED SYSTEM MATRIX
      DO 20 L = 1,NELFRE
      I = INDEX(L)
      IF ( I.LT.1 )  GO TO 20
      DO 10 K = 1,NELFRE
      J = INDEX(K)
      IF ( J.LT.1 .OR. I.GT.J )  GO TO 10
      JJ = J - I + 1
      SS(I,JJ) = SS(I,JJ) + S(L,K)
   10 CONTINUE
   20 CONTINUE
      RETURN
      END
```

Fig. 7.6 Subroutine ASYMBL

```
      SUBROUTINE  ASYMBL (M,N,NE,NG,NSPACE,NELFRE,MAXBAN,
     1            NDFREE,NITER,LPTEST,LHOMC,NNPFLC,IP1,
     2            IP3,IP4,IP5,JP2,KP1,KP2,NTAPE1,NODES,
     3            LNODE,INDEX,LPFIX,LPRCP,X,FLTNP,FLTEL,
     4            FLTMIS,ELPROP,COORD,S,C,SS,CC,D,DDCLD,
     5            PRTLPT,KP3,NULCOL,NTAPE2,NTAPE3,NTAPE4,
     6            NHCMO)
C     * * * * * * * * * * * * * * * * * * * * * * * * * *
C      ASSEMBLE SYSTEM EQUATIONS AND STCRE POST
C      SOLUTION ELEMENT DATA
C     * * * * * * * * * * * * * * * * * * * * * * * * * *
C     GENERATE ELEMENT EQUATIONS & POST SOLUTICN MATRICES
CDP   IMPLICIT REAL*8 (A-H,O-Z)
      DIMENSION   S(NELFRE,NELFRE), C(NELFRE), CC(NDFREE),
     1            SS(NDFREE,NDFREE),NODES(NE,N), LNCDE(N),
     2            INDEX(NELFRE)
C     JUST PASSING THROUGH
      DIMENSION   X(1), COORD(1), FLTNP(1), FLTEL(1),
     1            ELPROP(1), D(1), DDOLD(1), PRTLPT(1),
     2            FLTMIS(1), LPFIX(1), LPROP(1)
C     NG = NO. OF PARAMETERS PER NODE
C     N = NUMBER OF NODES PER ELEMENT
C     NE = NUMBER OF ELEMENTS IN SYSTEM
C     NELFRE = NUMBER OF DEGREES OF FREEDOM PER ELEMENT
C     NDFREE = TOTAL NUMBER OF SYSTEM DEGREES CF FREEDCM
C     NODES = NODAL INCIDENCES OF ALL ELEMENTS
C     LNODE = THE N NODAL INCIDENCES OF THE ELEMENT
C     INDEX = SYSTEM DOF NUMBERS ASSOCIATED WITH ELEMENT
C     LPFIX = SYSTEM ARRAY OF FIXED PT ELEM PRCP
C     SS = SYS EQ SQ MATRIX,  CC = SYS EQ COL MATRIX
C     S = ELEMENT SQ MATRIX, C= ELEMENT COLUM MATRIX
C     MAXBAN = MAX HALF BANDWIDTH OF SYS SQ MATRIX
C     IF NULCOL.NE.0  ELEMENT COLUMN MATRIX IS ALWAYS ZERO
C
      DO 10  IE = 1,NE
C-->   EXTRACT ELEMENT NODE NUMBERS
       CALL  LNODES (IE,NE,N,NODES,LNODE)
C-->   CALCULATE DEGREE OF FREEDOM NUMBERS
       CALL  INDXEL (N,NELFRE,NG,LNODE,INDEX)
C-->   GENERATE ELEMENT PROBLEM DEPENDENT MATRICES
       CALL  GENELM (IE,M,N,NSPACE,NELFRE,
     1       NDFREE,NITER,LPTEST,LHOMO,NNPFLO,IP1,
     2       IP3,IP4,IP5,JP2,KP1,KP2,NTAPE1,
     3       LNODE,INDEX,LPFIX,LPROP,X,FLTNP,FLTEL,
     4       FLTMIS,ELPROP,COORD,S,C,D,DCCLD,NHCMO,
     5       PRTLPT,KP3,NULCOL,NTAPE2,NTAPE3,NTAPE4)
C-->   STORE THE MATRICES IN SYSTEM EQUATIONS
       CALL  STORSQ (NDFREE,MAXBAN,NELFRE,INDEX,S,SS)
       IF ( NULCOL.EQ.0 )  CALL STORCL (NDFREE,NELFRE,INDEX,
     1                                  C,CC)
   10  CONTINUE
C      ASSEMBLY COMPLETED
       RETURN
       END
```

Fig. 7.7 Store element square matrix in skyline

```
      SUBROUTINE  SKYSTR (NOCOEF,NDFREE,NELFRE,INDEX,
     1                    IDIAG,S,SS)
C     * * * * * * * * * * * * * * * * * * * * * * * * *
C     STORE ELEMENT SQUARE MATRIX TO SYSTEM SQUARE
C     MATRIX STIRED IN SYMMETRIC SKYLINE MCDE
C     * * * * * * * * * * * * * * * * * * * * * * * * *
CDP   IMPLICIT REAL*8 (A-H,O-Z)
      DIMENSICN   SS(NOCOEF), S(NELFRE,NELFRE),
     1            INDEX(NELFRE), IDIAG(NCFREE)
C     NOCCEF = NO COEFF IN SYS SQ MATRIX = ICIAG(NDFREE)
C     NDFREE = TOTAL NO OF DOF IN SYSTEM
C     NELFRE = NUMBER OF ELEMENT DEGREES OF FREEDCM
C     INDEX(I) = SYS DOF NO OF ELEMENT DOF I
C     IDIAG(I) = LOCATION OF DIAGONAL OF I-TH EQ
C     S = ELEMENT SQUARE MATRIX
C     SS = SYS SQUARE MATRIX IN SKYLINE VECTOR MCDE
C
```

Fig. 7.7—continued

```
C       LOOP OVER ELEMENT COEFFICIENTS
        DO 20  J = 1,NELFRE
        NDXJ = INDEX(J)
        IF ( NDXJ.LT.1 )  GO TO 20
        JTEMP = IDIAG(NDXJ) - NDXJ
        DO 10  I = 1,NELFRE
        NDXI = INDEX(I)
        IF ( NDXI.GT.NDXJ .OR. NDXI.LT.1 )  GO TO 10
C       FIND SYSTEM COEFF IN VECTOR S
        NDXV = JTEMP + NDXI
C       NDXV = IDIAG(AMAXO(NDXI,NDXJ)) - IABS(NDXJ-NDXI)
        SS(NDXV) = SS(NDXV) + S(I,J)
 10     CONTINUE
 20     CONTINUE
        RETURN
        END
```

Fig. 7.8 **(a)** Subroutine SKYSUB, and **(b)** location of coefficients in a symmetric skyline array

```
        SUBROUTINE  SKYSUB (NDFREE,IDIAG,I,J,IJV)
C       * * * * * * * * * * * * * * * * * * * * * * * * *
C       CONVERT (I,J) FULL SYMMETRIC MATRIX SUBSCRIPTS
C       TO IJV SUBSCRIPT OF VECTOR SKYLINE STORAGE MODE
C       * * * * * * * * * * * * * * * * * * * * * * * * * *
C       ASSUMING SYMM EQS, COLS STORED FROM TOP DOWN
        DIMENSION IDIAG(NDFREE)
C       NDFREE = TOTAL NO OF SYSTEM EQUATIONS
C       IDIAG(I) = LOCATION OF DIAG OF I-TH EQ
        ID = MAXO (I,J)
        IJV = IDIAG(ID) - IABS(I-J)
        RETURN
        END
```

(a)

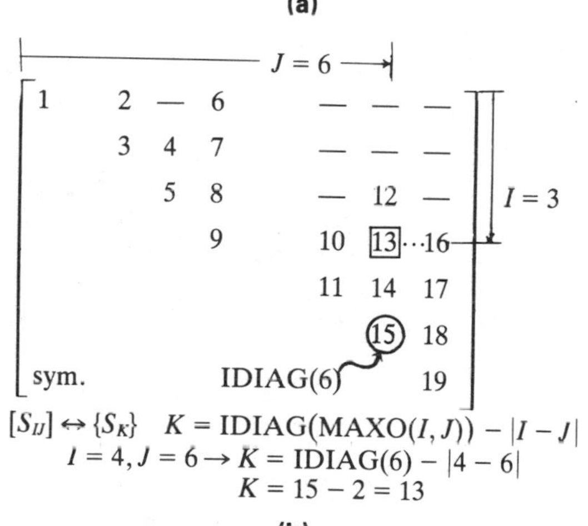

$[S_{IJ}] \leftrightarrow \{S_K\} \quad K = \text{IDIAG}(\text{MAXO}(I, J)) - |I - J|$

$I = 4, J = 6 \rightarrow K = \text{IDIAG}(6) - |4 - 6|$

$K = 15 - 2 = 13$

(b)

Fig. 7.9 **(top)** Subroutine BANSUB, and **(bottom)** location of typical coefficients in a symmetric band array

```
      SUBROUTINE  BANSUB (I,J,K,L)
C     * * * * * * * * * * * * * * * * * * * * * * * * *
C     CONVERT SUBSCRIPTS (I,J) OF SYMMETRIC SQ MATRIX
C     TO SUBSCRIPTS (K,L) IN UPPER HALF BANDWIDTH
C     * * * * * * * * * * * * * * * * * * * * * * * * *
      ITEST = I - J
      IF ( ITEST )  10,20,30
   10 K = I
      L = 1 - ITEST
      RETURN
   20 K = I
      L = 1
      RETURN
   30 K = J
      L = 1 + ITEST
      RETURN
      END
```

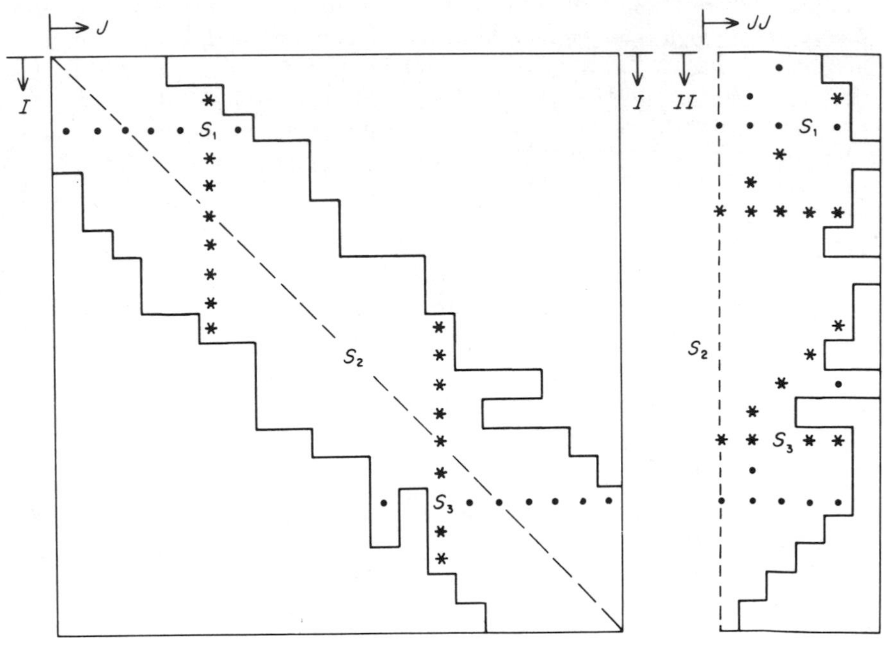

7.3 Example

Consider a two-dimensional problem (NSPACE = 2) involving 4000 nodal points (M = 4000) and 3500 elements (NE = 3500). Assume two parameters per node (NG = 2) and let these parameters represent the horizontal and vertical components of some vector. In a stress analysis problem, the vector could represent the displacement vector of the node, whereas in a fluid flow problem it could represent the velocity vector at the nodal point.

Assume the elements to be triangular with three corner nodes (N = 3). The local numbers of these nodes will be defined in some consistent manner; e.g., by numbering counter-clockwise from one corner. This mesh is illustrated in Fig. 7.10. For simplicity assume that two floating point material properties are associated with each element, and that the elements are homogeneous throughout the domain. (As discussed in Section 2.2, this implies that NLPFLO = 2, LHOMO = 1, NNPFIX = NNPFLO = NLPFIX = MISCFX = MISCFL = NHOMO = 0.) In addition, assume that the nodal parameter boundary conditions specify the following conditions:

(1) both parameters are zero at nodal point 1;
(2) at nodal point 500 the first parameter equals zero; and
(3) at nodal point 4000 the two parameters are related by the equation $\frac{3}{5}P_1 - \frac{4}{5}P_2 = 0$, where P_1 and P_2 denote the first and second parameters, respectively (at that node).

Define the two material properties of the elements to have the values of 30×10^6 and 0.25, respectively. Assume that in the y-direction at node 270 there is a known initial contribution to the load vector (system column matrix). Let this coefficient be associated with the second nodal parameter at the node and have a value of −1000. Finally, assume that there are specified flux components in the y-direction along the two element segments (NSEG = 2 and LBN = 2) between nodes 2 and 3, and 3 and 4. Let the magnitudes of the fluxes be −100, −150, and −25 at nodes 2, 3 and 4, respectively.

The input data to, and the arrays stored by the building block codes for this example, are shown in Tables 7.1 to 7.6. In these tables the input data are enclosed in dashed lines and the resulting arrays stored by the program are enclosed in solid lines.

By utilizing the above control parameters, it is easy to determine the total number of degrees of freedom in the system, NDFREE, and associated with a typical element, LEMFRE:

$$\text{NDFREE} = \text{M} * \text{NG} = 4000 * 2 = 8000$$

$$\text{LEMFRE} = \text{N} * \text{NG} = 3 * 2 = 6.$$

In addition to the total number of degrees of freedom in the system, it is important to be able to identify the system degree of freedom number that is associated with any parameter in the system. Equation (1.1), or subroutine DEGPAR, provides this information. This relation has many practical uses. For example, when one specifies that the first parameter ($J = 1$) at node 500 ($I = 500$) has some given value what one is indirectly saying is that system degree of freedom number NDF = 2 * (500 − 1) + 1 = 999 has a given value.

In a similar manner we often need to identify the system degree of freedom numbers that correspond to the ELMFRE local degrees of freedom of the element. In order to utilize Eqn. (1.1) to do this, one must be able to identify the N node numbers associated with the element of interest. This is relatively easy to accomplish since these data are part of the input data (element incidences). For example, for element number 421 we find the 3 element incidences (by extracting row 421 from array NODES) to be

System	261	270	310	←Array LNODE
Local	1	2	3	(1 * N)

Therefore, by applying Eqn. (1.1), we find the following degree of freedom numbers (for element 421):

Node local i	Number system I	Parameter number J	Degree of freedom System NDF_s	local NDF_L
1	261	1	521	1
		2	522	2
2	270	1	539	3
		2	540	4
3	310	1	619	5
		2	620	6
	↖Array LNODE		↖Array INDEX	

The element array INDEX has many programming uses. Its most important application is to aid in the assembly of the element equations to form the governing system equations.

Recall from Fig. 7.1 that the element equations are expressed in terms of local degree of freedom numbers. In order to add these element coefficients into the system equations one must identify the relation between the local degree of freedom numbers and the corresponding system degree of freedom numbers. Array INDEX provides this information for a specific element.

In practice, the assembly procedure is as follows. First the system matrices are set equal to zero. Then a loop over all the elements is performed. For each element, the element matrices are generated in terms of the local degrees of freedom. The coefficients of the element matrices are added to the corresponding coefficients in the system matrices. Before the addition is carried out, the element array INDEX is used to convert the local subscripts of the coefficient to the system subscripts of the term in the system equations to which the coefficient is to be added. That is,

$$\begin{aligned} S^e_{i,j} &\xrightarrow{+} S_{I,J} \\ C^e_i &\xrightarrow{+} C_I \end{aligned} \tag{7.2}$$

where $I_s = \text{INDEX}\ (i_L)$ and $J_s = \text{INDEX}\ (j_L)$ are the corresponding row and column numbers in the system equations, i_L, j_L are the subscripts of the coefficients in terms of the local degrees of freedom, and $\xrightarrow{+}$ reads "is added to". Considering the coefficients in the element matrices for element 421 in the previous example, one finds

$$\begin{aligned} S^e_{1,1} &\xrightarrow{+} S_{521,521} & \qquad C^e_1 &\xrightarrow{+} C_{521} \\ S^e_{2,3} &\xrightarrow{+} S_{522,539} & C^e_2 &\xrightarrow{+} C_{522} \\ S^e_{3,4} &\xrightarrow{+} S_{549,540} & C^e_3 &\xrightarrow{+} C_{539} \\ S^e_{4,5} &\xrightarrow{+} S_{540,619} & C^e_4 &\xrightarrow{+} C_{540} \\ S^e_{5,6} &\xrightarrow{+} S_{619,620} & C^e_5 &\xrightarrow{+} C_{619} \\ S^e_{1,6} &\xrightarrow{+} S_{521,620} & C^e_6 &\xrightarrow{+} C_{620}. \end{aligned}$$

Let $L_{\min}$ and $L_{\max}$ denote, respectively, the maximum and minimum system degree of freedom numbers associated with a particular element. Then one can make the general observation that the coefficients of the square element matrix will be included in a square subregion of the system matrix. As illustrated in Fig. 3.4, the limits of this square region are defined by $L_{\min}$ and $L_{\max}$. From this figure, one can also note that the half-bandwidth associated with a typical element (including the diagonal) is given by

$$\text{HBW} = L_{max} - L_{min} + 1 \tag{7.3}$$

For example, for element 421, the $\text{HBW}_{421} = 620 - 521 + 1 = 100$.

The equation for the half-bandwidth (including the diagonal) may be expressed in a much more useful form. Note that from Eqn. (1.1)

$$\begin{aligned} L_{max} &= \text{NG} * (I_{max} - 1) + J_{max} \\ L_{min} &= \text{NG} * (I_{min} - 1) + J_{min} \end{aligned} \tag{7.4}$$

where $J_{min} = 1$ and $J_{max} = \text{NG}$ and where I_{max} and I_{min} denote the largest and smallest system node numbers associated with the element, respectively. Substituting these two relations into the above expression for the half-bandwidth yields:

$$\text{HBW} = [\text{NG} * (I_{max} - 1) + \text{NG}] - [\text{NG} * (I_{min} - 1) + 1] + 1$$

or simply (7.5)

$$\text{HBW} = \text{NG} * (I_{max} - I_{min} + 1)$$

which was stated earlier in Eqn. (1.1). As a check, we can return to element 421 of the example to determine that $\text{HBW}_{421} = 2*(310 - 261 + 1) = 100$ which agrees with the previous calculations.

After the equations have been completely assembled, it is necessary to apply the boundary conditions. The types of allowed constraints and the data array IBC were introduced in Chapter 2. Searching array IBC, one determines that there are three Type 1 boundary indicators and two Type 2 indicators, thus there are three Type 1 constraint equations and one (2/2) Type 2 constraint equation. Again Eqn. (1.1) is applied to determine the system degree of freedom that must be modified to include these constraint equations. The programming concepts associated with applying the boundary conditions will be discussed in Section 8.2.

Fig. 7.10 An example mesh

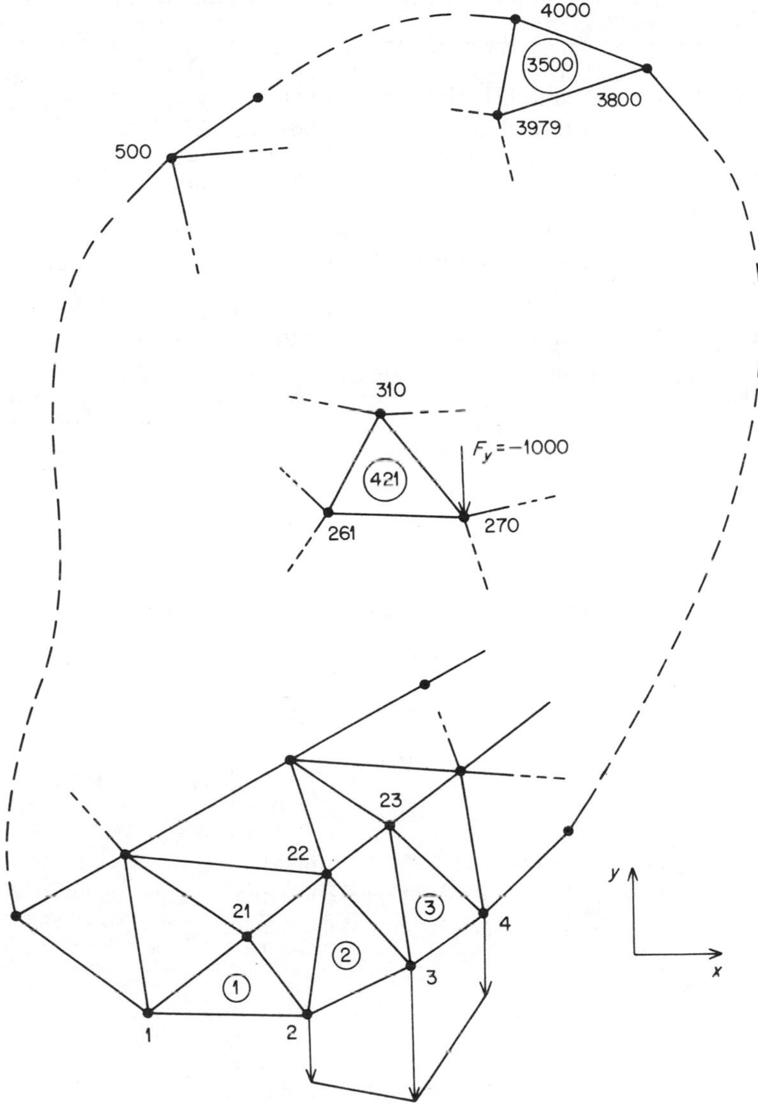

Table 7.1 Nodal Point Data for Input

Node	Boundary condition code	Spatial coordinates	
1	11	0.00	0.00
2	00	0.25	0.00
3	00	0.50	0.1
⋮	⋮	⋮	⋮
261	00	2.25	1.50
⋮	⋮	⋮	⋮
270	00	2.75	1.50
⋮	⋮	⋮	⋮
310	00	2.60	1.90
⋮	⋮	⋮	⋮
500	10	0.00	3.40
⋮	⋮	⋮	⋮
4000	22	9.80	5.60

Array IBC — Array X

Table 7.2 Element Connectivity Data for Input

Element	Element incidences		
1	1	2	21
2	2	3	22
3	3	4	23
⋮	⋮	⋮	⋮
421	261	270	310
⋮	⋮	⋮	⋮
3500	3979	3980	4000

Array NODES

Table 7.4 Problem Properties Data for INPROP

Element properties

Element no.	Properties	
1	30.E6	0.25

Array FLTEL

Table 7.3 Nodal Parameter Boundary Condition Data for INCEQ

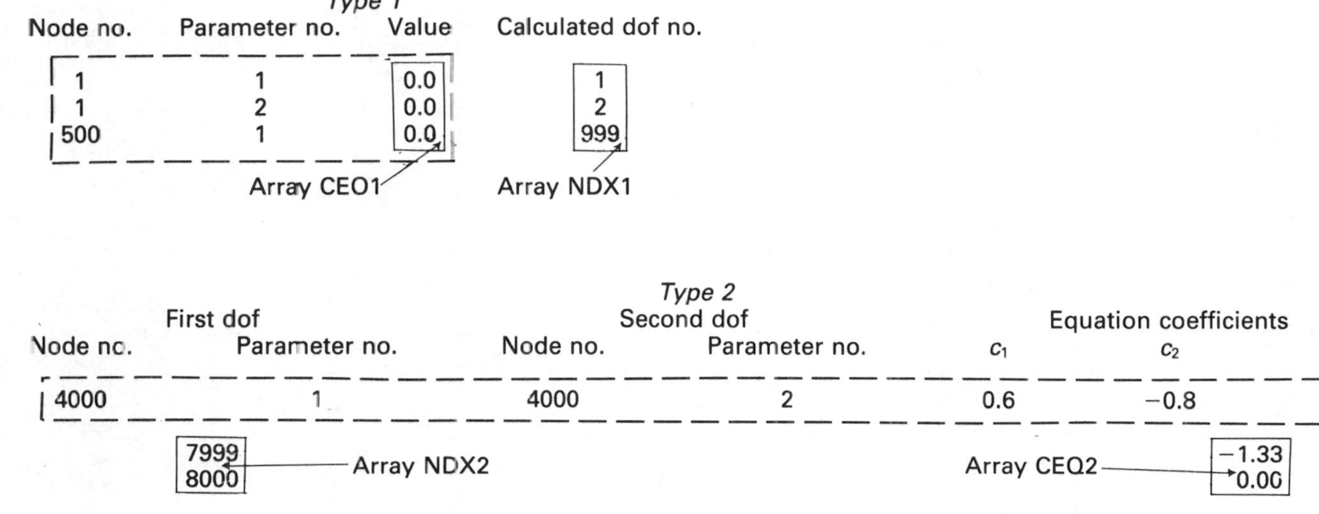

Type 1

Node no.	Parameter no.	Value	Calculated dof no.
1	1	0.0	1
1	2	0.0	2
500	1	0.0	999
		Array CEO1	Array NDX1

Type 2

First dof		Second dof		Equation coefficients		
Node no.	Parameter no.	Node no.	Parameter no.	c_1	c_2	c_3
4000	1	4000	2	0.6	−0.8	0.0

Array NDX2: 7999, 8000

Array CEQ2: −1.33, 0.00

Table 7.5 Initialization Data for INVECT

Node no.	Parameter no.	Value
270	2	−1000
4000	2	0

$J = 2(270 - 1) + 2 = 540$

CC(540) = −1000, plus 7999 zero terms

Table 7.6 Flux Data for INFLUX

Segment 1

Nodes	2	3		
Flux components	0.0	−100	0.0	−150

Segment 2

Nodes	3	4		
Flux components	0.0	−100	0,0	−25

7.4 Symbolic element assembly for quadratic forms

Many finite element problems can be expressed symbolically in terms of a scalar quantity, I, such as

$$\mathrm{I}(\boldsymbol{\Delta}) = \tfrac{1}{2}\boldsymbol{\Delta}^T\mathbf{K}\boldsymbol{\Delta} + \boldsymbol{\Delta}^T\mathbf{P} \tag{7.6}$$

where $\boldsymbol{\Delta}$ is a vector containing the unknown nodal parameters associated with the problem, and **K** and **P** are matrices defined in terms of the element properties and geometry. The above quantity is known as a *quadratic form.* If one uses a variational formulation or a least squares weighted residual formulation then the solution of the finite element problem is usually required to satisfy the following system equations:

$$\frac{\partial \mathrm{I}}{\partial \boldsymbol{\Delta}} = \mathbf{0}. \tag{7.7}$$

In the finite element analysis one assumes that the (scalar) value of I is given by the sum of the element contributions. That is, one assumes

$$\mathrm{I}(\boldsymbol{\Delta}) = \sum_{i=1}^{\mathrm{NE}} j_i(\boldsymbol{\Delta}) \tag{7.8}$$

where j_i is the contribution of element number 'i', and NE represents the total number of elements in the system. Usually one can (but does not in practice) define j_i in terms of $\mathbf{\Delta}$ such that

$$j_i(\mathbf{\Delta}) = \mathbf{\Delta}^T\mathbf{K}_i\mathbf{\Delta} + \mathbf{\Delta}^T\mathbf{P}_i. \tag{7.9}$$

Therefore, Eqn. (7.8) can be expressed as

$$\mathrm{I}(\mathbf{\Delta}) = \tfrac{1}{2}\mathbf{\Delta}^T\left(\sum_{i=1}^{\mathrm{NE}} \mathbf{K}_i\right)\mathbf{\Delta} + \mathbf{\Delta}^T\left(\sum_{i=1}^{\mathrm{NE}} \mathbf{P}_i\right) \tag{7.10}$$

and comparing this with Eqn. (7.6) one can identify the relations

$$\begin{aligned} \mathbf{K} &= \sum_{i=1}^{\mathrm{NE}} \mathbf{K}_i \\ \mathbf{P} &= \sum_{i=1}^{\mathrm{NE}} \mathbf{P}_i \end{aligned} \tag{7.11}$$

If NDFREE represents the total number of unknowns in the system, then the size of these matriccs are NDFREE*NDFREE and NDFREE*1, respectively.

As a result of Eqn. (7.11) one often sees the statement, "the system matrices are simply the sum of the corresponding element matrices." This is true, and indeed the symbolic operations depicted in Eqn. (7.11) are simple but one should ask (while preparing for the ensuing handwaving), "in practice, how are the element matrices obtained and how does one carry out the summations?"

Before attempting to answer this question, it will be useful to backtrack a little. First, recall that it has been assumed that an element's behaviour, and thus its contribution to the problem, depends *only* on those nodal parameters that are associated with the element. In practice, the number of parameters associated with a single element lies between a minimum of two and a maximum of sixty (or ninety-six); with the most common range being from three to eight. By way of comparison, NDFREE can easily reach a value of several thousand. Consider an example of a system where NDFREE = 5000. Let this system consist of one-dimensional elements with two parameters per element. A typical matrix $\mathbf{P}_i$ will contain 5000 terms and all but two of these terms will be identically zero since only those two terms of $\mathbf{\Delta}$, 5000 * 1, associated with element 'i' are of any significance to element 'i'. In a similar manner one concludes that, for the present example, only four of the 25,000 terms of $\mathbf{K}_i$ would not be identically zero. Therefore, it becomes obvious that the symbolic procedure introduced here is not numerically efficient and would not be used in practice.

There are some educational uses of the symbolic procedure that justify pursuing it a little further. Recalling that it is assumed that the element behaviour depends only on those parameters, say $\boldsymbol{\delta}_i$, that are associated with element '*i*', it is logical to assume that

$$j_i = \boldsymbol{\delta}_i^T \mathbf{k}_i \boldsymbol{\delta}_i + \boldsymbol{\delta}_i^T \mathbf{p}_i. \tag{7.12}$$

If NELFRE represents the number of degrees of freedom associated with the element then $\boldsymbol{\delta}_i$ and $\mathbf{p}_i$ are NELFRE*1 in size and the size of $\mathbf{k}_i$ is NELFRE*NELFRE. Note that in practice NELFRE is much less than NDFREE.

The matrices $\mathbf{k}_i$ and $\mathbf{p}_i$ are commonly known as *the* element matrices. For the one-dimensional element discussed in the previous example $\mathbf{k}_i$ and $\mathbf{p}_i$ would be 2*2 and 2*1 in size, respectively, and would represent the only coefficients in $\mathbf{K}_i$ and $\mathbf{P}_i$ that are not identically zero.

All that remains is to relate $\mathbf{k}_i$ to $\mathbf{K}_i$ and $\mathbf{p}_i$ to $\mathbf{P}_i$. Obviously Eqns. (7.9) and (7.12) are equal and are the key to the desired relations. In order to utilize these equations, it is necessary to relate the degrees of freedom of the element. $\boldsymbol{\delta}_i$, to the degrees of freedom of the total system, $\boldsymbol{\Delta}$. This is done symbolically by introducing a NELFRE*NDFREE matrix, $\mathbf{B}_i$ (to be discussed later), such that the following identity is satisfied:

$$\boldsymbol{\delta}_i \equiv \mathbf{B}_i \boldsymbol{\Delta}. \tag{7.13}$$

Substituting this identity, Eqn. (7.12) becomes

$$j_i(\boldsymbol{\Delta}) = \boldsymbol{\Delta}^I \mathbf{B}_i^T \mathbf{k}_i \mathbf{B}_i \boldsymbol{\Delta} + \boldsymbol{\Delta}^T \mathbf{B}_i^T \mathbf{p}_i. \tag{7.14}$$

Comparing this relation with Eqn. (7.9), one can establish the symbolic relationships

$$\begin{aligned} \mathbf{K}_i &\equiv \mathbf{B}_i^T \mathbf{k}_i \mathbf{B}_i, \\ \mathbf{P}_i &\equiv \mathbf{B}_i^T \mathbf{p}_i \end{aligned} \tag{7.15}$$

and

$$\begin{aligned} \mathbf{K} &\equiv \sum_{i=1}^{\mathrm{NE}} \mathbf{B}_i^T \mathbf{k}_i \mathbf{B}_i, \\ \mathbf{P} &\equiv \sum_{i=1}^{\mathrm{NE}} \mathbf{B}_i^T \mathbf{p}_i. \end{aligned} \tag{7.16}$$

Equations (7.16) can be considered as the symbolic definitions of the assembly procedures relating *the* element matrices, $\mathbf{k}_i$ and $\mathbf{p}_i$, to the total system matrices, $\mathbf{K}$ and $\mathbf{P}$. Note that these relations involve the element connectivity (topology), $\mathbf{B}_i$, as well as the element behaviour, $\mathbf{k}_i$ and $\mathbf{p}_i$. Although some programs do use this procedure, it is very inefficient and thus very expensive.

For the sake of completeness the $\mathbf{B}_i$ matrix will be briefly considered. To simplify the discussion, it will be assumed that each nodal point has only a single unknown scalar nodal parameter (degree of freedom). Define a mesh consisting of four triangular elements. Figure 7.11 shows both the system and element degree of freedom numbers. The system degrees of freedom are defined as

$$\mathbf{\Delta}^T = [\Delta_1 \Delta_2 \Delta_3 \Delta_4 \Delta_5 \Delta_6] \tag{7.17}$$

and the degrees of freedom of element 'i' are

$$\boldsymbol{\delta}_i^T = [\delta_1 \delta_2 \delta_3]_i \tag{7.18}$$

For element number four ($i = 4$), these quantities are related by

$$\boldsymbol{\delta}_4 = \begin{Bmatrix} \delta_1 \\ \delta_2 \\ \delta_3 \end{Bmatrix}_4 = \begin{Bmatrix} \Delta_3 \\ \Delta_6 \\ \Delta_5 \end{Bmatrix}$$

which can be expressed as

$$\boldsymbol{\delta}_4 = \begin{Bmatrix} \delta_1 \\ \delta_2 \\ \delta_3 \end{Bmatrix}_4 = \begin{bmatrix} 0 & 0 & 1 & 0 & 0 & 0 \\ 0 & 0 & 0 & 0 & 0 & 1 \\ 0 & 0 & 0 & 0 & 1 & 0 \end{bmatrix} \begin{Bmatrix} \Delta_1 \\ \Delta_2 \\ \Delta_3 \\ \Delta_4 \\ \Delta_5 \\ \Delta_6 \end{Bmatrix}. \tag{7.19}$$

Defining $\boldsymbol{\delta}_4 \equiv \mathbf{B}_4 \mathbf{\Delta}$ yields

$$\mathbf{B}_4 \equiv \begin{bmatrix} 0 & 0 & 1 & 0 & 0 & 0 \\ 0 & 0 & 0 & 0 & 0 & 1 \\ 0 & 0 & 0 & 0 & 1 & 0 \end{bmatrix}. \tag{7.20}$$

The matrices $\mathbf{B}_1$, $\mathbf{B}_2$, and $\mathbf{B}_3$ can be defined in a similar manner.

Since the matrix $\mathbf{B}_i$ contains only ones and zeros, it is called the element Boolean or binary matrix. Note that there is a single unity term in each row and all other coefficients are zero. Therefore, this NELFRE*NDFREE array will contain only NELFRE non-zero (unity) terms, and since NELFRE $\ll$ NDFREE, the matrix multiplications of Eqn. (7.16) are numerically very inefficient.

Although these symbolic relations have certain educational uses their gross inefficiency for practical computer calculations led to the development of the equivalent programming procedure of the "direct method" of assembly that was discussed earlier.

It is useful at times to note that the identity (7.13) leads to the relation

$$\frac{\partial(\,)_i}{\partial \mathbf{\Delta}} = \mathbf{B}_i^T \frac{\partial(\,)_i}{\partial \boldsymbol{\delta}_i}, \tag{7.21}$$

where $(\,)_i$ is some quantity associated with element 'i'.

Fig. 7.11 Relationship between system and element degrees of freedom

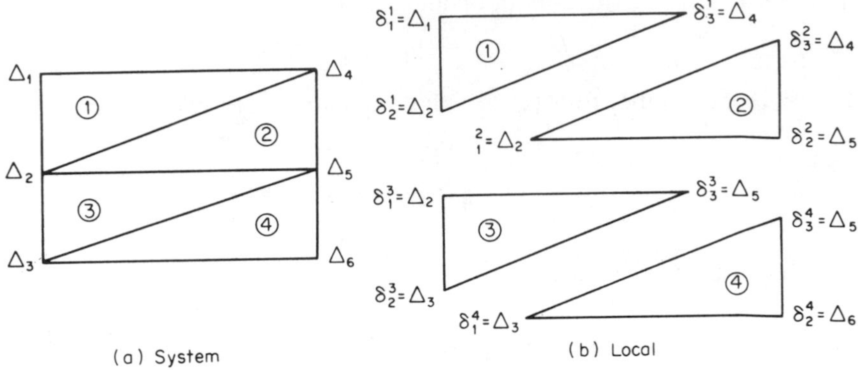

7.5 Frontal assembly and solution procedures

In addition to banded and skyline or profile methods of storage and solution many finite element codes utilize *frontal methods*. Usually an element by element frontal assembly and elimination procedure is employed. Such a procedure can utilize a relatively small memory. Unlike the procedures considered to this point, the efficiency of frontal methods is independent of node numbers and is dependent on the element processing order.

A very detailed description of the frontal method, along with the supporting software, was given in the text by Hinton and Owen [42]. Here only a limited insight into the frontal assembly procedure will be presented. This involves the definition between the element degree of freedom numbers and the current system dof number destination number in memory.

The frontal method stores only equations that are currently active. Equations that have received all of their element (and constraint) contributions are eliminated to provide storage for equations that may become active with the next element. To illustrate these points consider a mesh with eight nodes, fourteen line elements, two nodes per element and one degree of freedom per node (M = 8, NE = 14, N = 2, NG = 1, NDFREE = 8). If assembled in full form before elimination takes place, storage must be provided for the total NDFREE = 8 equations. The frontal method generally requires less storage. This is illustrated in Fig. 7.12 which shows the mesh and the storage location (or destination) of each system equation as the elements are read. Note that after the first four elements are read, node 3 (destination 2) has made its last contribution. Thus equation 3 can be eliminated and its storage locations freed for the next new equation. When the next element, 5, is read, node 6 makes its first appearance and can be stored where equation 3 used to be located. Note that proceeding in this manner requires storage for 7 equations instead of 8. Note that equations 5 and 7 are the last two to be eliminated.

Figure 7.13 shows that if the elements are input in a different order only a maximum of 4 equations must be stored instead of 8. The maximum number of equations required in memory is usually called the front size or wavefront. Algorithms for reordering elements to yield reduced fronts have been given by Akin and Pardue [5].

Fig. 7.12 Frontal storage example

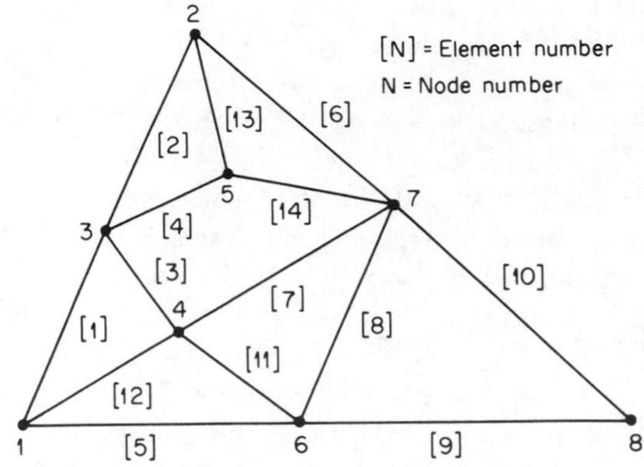

	Destination number						
	1	2	3	4	5	6	7
Element	Node number						
1	1	3	–	–	–	–	–
2	1	3	2	–	–	–	–
3	1	3	2	4	–	–	–
4	1	3	2	4	5	–	–
5	1	6	2	4	5	–	–
6	1	6	2	4	5	7	–
7	1	6	2	4	5	7	–
8	1	6	2	4	5	7	–
9	1	6	2	4	5	7	8
10	1	6	2	4	5	7	8
11	1	6	2	4	5	7	–
12	1	–	2	4	5	7	–
13	–	–	2	–	5	7	–
14	–	–	–	–	5	7	–

Fig. 7.13 Reduced frontal storage

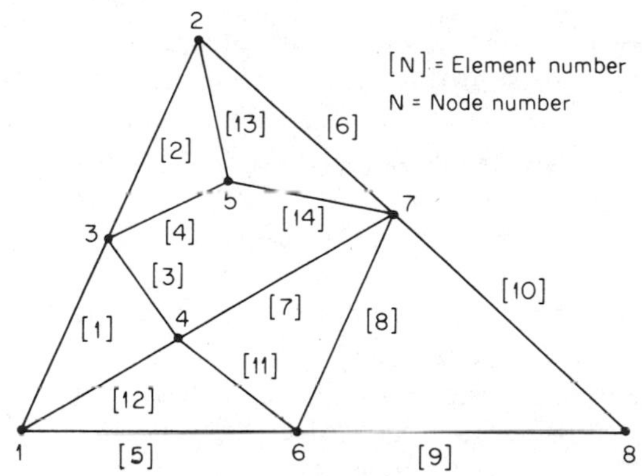

	Destination number			
	1	2	3	4
Revised element order	Node number			
9 → 1	6	8	–	–
10 → 2	6	8	7	–
5 → 3	6	1	7	–
8 → 4	6	1	7	–
11 → 5	6	1	7	4
12 → 6	–	1	7	4
1 → 7	3	1	7	4
7 → 8	3	–	7	4
3 → 9	3	–	7	4
14 → 10	3	5	7	–
6 → 11	3	5	7	2
4 → 12	3	5	–	2
13 → 13	3	5	–	2
2 → 14	3	–	–	2

7.6 Exercises

1. Draw the upper half triangle of the system square matrix for the mesh in Fig. 7.12. Use an x to represent a non-zero contribution from an element. Draw the skyline and maximum band limit. How many storage locations are required for the (a) skyline storage, and (b) band storage (remember the triangle of zeros at the end)?
2. Repeat the above problem using a different node numbering of your choice.
3. Reproduce Fig. 7.13 where the destination numbers are based on an element input order of your choice.
4. The cost of eliminating the equations is proportional to the root mean square, RMS, of the element front sizes. In Fig. 7.13 the RMS = 3.128; what is it in Fig. 7.12?
5. Redraw Fig. 7.4 if there are three dof per node.
6. Redraw Fig. 3.16 for two dof per node.
7. Assume constant element matrices of

$$\mathbf{S}^e = \begin{bmatrix} 3 & 4 \\ 4 & 3 \end{bmatrix}, \quad \mathbf{C}^e = \begin{Bmatrix} 1 \\ 2 \end{Bmatrix}$$

and assemble the system equations for the mesh in Fig. 3.19.

8

Application of nodal parameter boundary constraints

8.1 Introduction

After the system equations have been fully assembled in the form of Eqn. (1.2), it is necessary to apply the boundary conditions before the equations can be solved for the unknown nodal parameters. There are at least two different formal procedures for applying these boundary constraints. In this section, three types of nodal parameter boundary constraints will be defined. The most common boundary conditions (Type 1) are of the form

$$D_j = b, \tag{8.1}$$

where D_j represents system degree of freedom number, j, and b is the specified value assigned to that degree of freedom. Obviously, this boundary condition has many practical applications. For example, in a heat transfer problem this can be used to assign a specific temperature to a particular nodal parameter. Also, this type of condition occurs very frequently in structural analysis problems.

The next boundary constraint (Type 2) to be considered is of the form

$$AD_j + BD_k = C, \tag{8.2}$$

where D_j and D_k represent two system degrees of freedom, and A, B, and C are arbitrary constants ($A \neq 0$). This restraint contains the first condition as a special case. The second type of constraint also has various practical applications. For example, in a two-dimensional inviscid fluid flow problem, this can be used to require that the two velocity components at a node be tangential to a solid boundary at that node. The third and final boundary constraint (Type 3) to be considered is of the form

$$AD_j + BD_k + CD_\ell = E. \tag{8.3}$$

This type of constraint can occur in several three-dimensional problems. For example, in a structural analysis problem one could have a node constrained to displace tangential to an arbitrary plane in space (general roller support). Such a support condition can be described by Eqn. (8.3).

The analyst should observe that if there are m Type 2 constraint equations (and corresponding input data) then there are $2m$ nodal parameters that will have Type 2 boundary code indicators. Similarly, if there are m Type 3 constraint equations then there are $3m$ nodal parameters with Type 3 indicator codes, etc.

These three constraints are probably the most commonly used (nodal parameter) boundary conditions found in practical finite element problems. As mentioned earlier, there are two common methods used to apply the boundary conditions. The first method is to modify the assembled equations to reflect the boundary constraints and then to solve the complete set of equations. The second method is to eliminate the restrained parameters and then to solve the reduced set of equations. The first method has the advantage that it can more easily handle general forms of restraint. However, it requires all the nodal parameters to be calculated; even those with specified values. The second method is often used in structural analysis programs and has the advantage that only the unrestrained parameters have to be calculated. Its major disadvantage is that it is difficult to program for constraints other than those of the first type.

8.2 Matrix manipulations

The first method will be illustrated by outlining the modifications necessary to incorporate a Type 2 constraint. To simplify the procedure, Eqn. (8.2) will be rewritten as $D_j + aD_k = b$ where $a = B/A$ and $b = C/A$. Since the procedure is not restricted to banded matrices, the assumption of a banded matrix will temporarily be lifted. The assumption of symmetry is retained. Figure 8.1 shows the original form of the system equations. Recalling that $D_j = b - aD_k$ one can replace D_j times the jth column by $(b - aD_k)$ times the jth column. This substitution results in the two modifications shown in Fig. 8.2. Note that these changes destroy the symmetry of the equations and result in a determinant of zero. To restore the symmetry subtract 'a' times the jth row from the kth row and then multiply the jth row by zero. This second modification is illustrated in Fig. 8.3. These operations restore the symmetry of the problem but the equations are still singular since the dependence on D_j has been removed. Either the equations must be reduced in size by one or modified still further. To be consistent, the latter approach will be used.

Recall that the identities $D_j + aD_k = b$ and $aD_j - aD_j = 0$ can be expressed in a symmetric matrix form as

$$\begin{bmatrix} 1 & a \\ a & a^2 \end{bmatrix} \begin{Bmatrix} D_j \\ D_k \end{Bmatrix} = \begin{Bmatrix} b \\ ab \end{Bmatrix}. \tag{8.4}$$

Combining these relations with the last modification results in the final form necessary to include the second type of constraint. This final form is illustrated in Fig. 8.4. The solution of this complete set of symmetric equations will exactly satisfy the given constraint equation. It should be noted that the modification necessary to incorporate the first type of boundary constraint is obtained as a special case ($a = 0$) of the above procedure. The final modification for this type of constraint is shown in Fig. 8.5.

Note that if the original matrix is banded then the modification for constraint Types 2 and 3 may increase the bandwidth size. The Type 1 constraint, which is most common in practice, does not increase the bandwidth. In practice, the increase in bandwidth for Types 2 and 3 will be small since the constrained parameters are usually at the same node. The increase in bandwidth can be minimized by making programming changes (see REVISE and CONVRT) that will partition, during assembly, the system equations so that the parameters are separated according to their constraint type number.

Figure 8.6 illustrates a simple FORTRAN subroutine, MODFUL, for applying a Type 1 boundary constraint to a system (or element) square matrix. Again it should be noted that this type of routine would only be practical for problems involving a small number of unknowns.

It was pointed out above that, except for Type 1 constraint equations (which fortunately are the most common), the modification of the system equations may result in an increase in the size of the half-bandwidth of the equations. The increased bandwidth results from the addition of two columns (or rows) that takes place during the modification procedure and the size of the new bandwidth depends on the original size of the bandwidth, due to the element topology, and the difference in the column (or row) numbers that are being added. Once the constraint equation data and the system list of element incidences have been input, it is possible to calculate the increase (if any) in the bandwidth. First, the original half-bandwidth is determined (by subroutine SYSBAN). Next the degree of freedom numbers of the parameters in the constraint equations are determined (by subroutine INDXPT). The differences in these (column and row) numbers are combined with the original bandwidth to determine the new bandwidth.

These calculations (executed by subroutine MODBAN) take place early in the solution process so that adequate storage space can be set aside for the upper half-bandwidth of the system equations. The increased bandwidth (if any) results in many additional zero coefficients being introduced into the equations to be solved. This is serious since the initial zeros become non-zeros during the factorization of the system square matrix. Thus, the exact modifications are undesirable for banded modes with other than Type 1 nodal constraints. The banded storage mode modifications are executed by MODFY1, MODFY2 and MODFY3. Figure 8.7 shows subroutine MODFY1. These modifications would usually be under the control of a routine (such as APLYBC) which recovers the input constraint equation degree of freedom numbers and coefficients. Its flow chart was given in Chapter 2 while the actual details used in MODEL are given in the Appendix.

The second method for applying Type 1 boundary conditions will be outlined at this point. This procedure has been described in detail in several texts. Note that the system equations could be arranged in the following partitioned form

$$\begin{bmatrix} S_{\alpha\alpha} & S_{\alpha\beta} \\ S_{\beta\alpha} & S_{\beta\beta} \end{bmatrix} \begin{Bmatrix} \mathrm{D}_\alpha \\ \mathrm{D}_\beta \end{Bmatrix} = \begin{Bmatrix} \mathrm{C}_\alpha \\ \mathrm{C}_\beta \end{Bmatrix}$$

where $\mathbf{D}_\alpha$ represents the unknown nodal parameters, and $\mathbf{D}_\beta$ represents the boundary values of the prescribed Type 1 nodal parameters. The submatrices $\mathbf{S}_{\alpha\alpha}$ and $\mathbf{S}_{\beta\beta}$ are square, whereas $\mathbf{S}_{\alpha\beta}$ and $\mathbf{S}_{\beta\alpha}$ are rectangular, in general. It has been assumed that the equations have been assembled in a manner that places the prescribed Type 1 parameters at the end of the system equations. The above matrix relations can be rewritten as

$$\begin{aligned} \mathbf{S}_{\alpha\alpha}\mathbf{D}_\alpha + \mathbf{S}_{\alpha\beta}\mathbf{D}_\beta &= \mathbf{C}_\alpha \\ \mathbf{S}_{\beta\alpha}\mathbf{D}_\alpha + \mathbf{S}_{\beta\beta}\mathbf{D}_\beta &= \mathbf{C}_\beta \end{aligned} \tag{8.5}$$

so that the unknown nodal parameters are obtained by inverting the matrix $\mathbf{S}_{\alpha\alpha}$, that is

$$\mathbf{D}_\alpha = \mathbf{S}_{\alpha\alpha}^{-1}(\mathbf{C}_\alpha - \mathbf{S}_{\alpha\beta}\mathbf{D}_\beta). \tag{8.6}$$

Also, if desired, the value of $\mathbf{C}_\beta$ is determined from

$$\mathbf{C}_\beta = \mathbf{S}_{\beta\alpha}\mathbf{S}_{\alpha\alpha}^{-1}\mathbf{C}_\alpha - (\mathbf{S}_{\beta\alpha}\mathbf{S}_{\alpha\alpha}^{-1}\mathbf{S}_{\alpha\beta} - \mathbf{S}_{\beta\beta})\mathbf{D}_\beta. \tag{8.7}$$

The extension of these matrix operations to multiple constraints will not be discussed here.

Fig. 8.1 Original system equations

$$
\begin{array}{c}
\\ 1 \\ \vdots \\ j \\ \vdots \\ k \\ \vdots \\ p
\end{array}
\begin{array}{c}
\begin{array}{ccccccc} 1 & \dots & j^* & \dots & k^* & \dots & p \end{array} \\
\begin{bmatrix}
S_{11} & \dots & S_{1j} & \dots & S_{1k} & \dots & S_{1p} \\
\vdots & & \vdots & & \vdots & & \vdots \\
S_{j1} & \dots & S_{jj} & \dots & S_{jk} & \dots & S_{jp} \\
\vdots & & \vdots & & \vdots & & \vdots \\
S_{k1} & \dots & S_{kj} & \dots & S_{kk} & \dots & S_{kp} \\
\vdots & & \vdots & & \vdots & & \vdots \\
S_{p1} & \dots & S_{pj} & \dots & S_{pk} & \dots & S_{pp}
\end{bmatrix}
\end{array}
\begin{Bmatrix} D_1 \\ \vdots \\ D_j \\ \vdots \\ D_k \\ \vdots \\ D_p \end{Bmatrix}
=
\begin{Bmatrix} C_1 \\ \vdots \\ C_j \\ \vdots \\ C_k \\ \vdots \\ C_p \end{Bmatrix}
$$

*arbitrary integers $1 \leqslant j, k \leqslant p$

Fig. 8.2 First modification

$$
\begin{array}{c}
\\ 1 \\ \vdots \\ j \\ \vdots \\ k \\ \vdots \\ p
\end{array}
\begin{array}{c}
\begin{array}{ccccccc} 1 & \dots & j & \dots & k & \dots & p \end{array} \\
\begin{bmatrix}
S_{11} & \dots & 0 & \dots & (S_{1k} - aS_{1j}) & \dots & S_{1p} \\
\vdots & & \vdots & & \vdots & & \vdots \\
S_{j1} & \dots & 0 & \dots & (S_{jk} - aS_{jj}) & \dots & S_{jp} \\
\vdots & & \vdots & & \vdots & & \vdots \\
S_{k1} & \dots & 0 & \dots & (S_{kk} - aS_{kj}) & \dots & S_{kp} \\
\vdots & & \vdots & & \vdots & & \vdots \\
S_{p1} & \dots & 0 & \dots & (S_{pk} - aS_{pj}) & \dots & S_{pp}
\end{bmatrix}
\end{array}
\begin{Bmatrix} D_1 \\ \vdots \\ D_j \\ \vdots \\ D_k \\ \vdots \\ D_p \end{Bmatrix}
=
\begin{Bmatrix} C_1 - bS_{1j} \\ \vdots \\ C_j - bS_{jj} \\ \vdots \\ C_k - bS_{kj} \\ \vdots \\ C_p - bS_{pj} \end{Bmatrix}
$$

Fig. 8.3 Second modification

$$
\begin{array}{c}
\\ 1 \\ \vdots \\ j \\ \vdots \\ k \\ \vdots \\ p
\end{array}
\begin{array}{c}
\begin{array}{ccccccc} 1 & \dots & j & \dots & k & \dots & p \end{array} \\
\begin{bmatrix}
S_{11} & \dots & 0 & \dots & (S_{1k} - aS_{1j}) & \dots & S_{1p} \\
\vdots & & \vdots & & \vdots & & \vdots \\
0 & \dots & 0 & \dots & 0 & \dots & 0 \\
\vdots & & \vdots & & \vdots & & \vdots \\
(S_{k1} - aS_{j1}) & \dots & 0 & \dots & (S_{kk} - 2aS_{jk} + a2S_{jj}) & \dots & (S_{kp} - aS_{jp}) \\
\vdots & & \vdots & & \vdots & & \vdots \\
S_{p1} & \dots & 0 & \dots & (S_{pk} - aS_{pj}) & \dots & S_{pp}
\end{bmatrix}
\end{array}
\begin{Bmatrix} D_1 \\ \vdots \\ D_j \\ \vdots \\ D_k \\ \vdots \\ D_p \end{Bmatrix}
=
\begin{Bmatrix} C_1 - bS_{1j} \\ \vdots \\ 0 \\ \vdots \\ C_k - aC_j - bS_{kj} + abS_{jj} \\ \vdots \\ C_p - bS_{pj} \end{Bmatrix}
$$

Fig. 8.4 Final modification (Type 2) $D_j + aD_k = b$

$$\begin{array}{c} \\ 1 \\ \vdots \\ j \\ \vdots \\ k \\ \vdots \\ p \end{array}
\begin{array}{c} \begin{array}{ccccc} 1 & \ldots j \ldots & k & \ldots & p \end{array} \\
\left[\begin{array}{ccccc}
S_{11} & \ldots 0 \ldots & (S_{1k} - aS_{1j}) & \ldots & S_{1p} \\
\vdots & \vdots & \vdots & & \vdots \\
0 & \ldots 1 \ldots & +a & \ldots & 0 \\
\vdots & \vdots & \vdots & & \vdots \\
(S_{k1} - aS_{j1}) & \ldots +a \ldots & (S_{kk} - 2aS_{jk} + a^2 S_{jj} + a^2) & \ldots & (S_{kp} - aS_{jp}) \\
\vdots & \vdots & \vdots & & \vdots \\
S_{p1} & \ldots 0 \ldots & (S_{pk} - aS_{pj}) & \ldots & S_{pp}
\end{array}\right] \end{array}
\left\{\begin{array}{c} D_1 \\ \vdots \\ D_j \\ \vdots \\ D_k \\ \vdots \\ D_p \end{array}\right\}
=
\left\{\begin{array}{c} C_1 - bS_{1j} \\ \vdots \\ b \\ \vdots \\ C_k - aC_j - bS_{kj} + abS_{jj} + ab \\ \vdots \\ C_p - bS_{pj} \end{array}\right\}$$

Fig. 8.5 Final modification (type 1) $D_j = b$

$$\begin{array}{c} \\ 1 \\ \vdots \\ j \\ \vdots \\ k \\ \vdots \\ p \end{array}
\begin{array}{c} \begin{array}{ccccccc} 1 & \cdots & j^* & \cdots & k & \cdots & p \end{array} \\
\left[\begin{array}{ccccccc}
S_{11} & \cdots & 0 & \cdots & S_{1k} & \cdots & S_{1p} \\
\vdots & & \vdots & & \vdots & & \vdots \\
0 & \cdots & 1 & \cdots & 0 & \cdots & 0 \\
\vdots & & \vdots & & \vdots & & \vdots \\
S_{k1} & \cdots & 0 & \cdots & S_{kk} & \cdots & S_{kp} \\
\vdots & & \vdots & & \vdots & & \vdots \\
S_{p1} & \cdots & 0 & \cdots & S_{pk} & \cdots & S_{pp}
\end{array}\right] \end{array}
\left\{\begin{array}{c} D_1 \\ \vdots \\ D_j \\ \vdots \\ D_k \\ \vdots \\ D_p \end{array}\right\}
=
\left\{\begin{array}{c} C_1 - bS_{1j} \\ \vdots \\ b \\ \vdots \\ C_k - bS_{kj} \\ \vdots \\ C_p - bS_{pj} \end{array}\right\}$$

$^*1 \leqslant j \leqslant p$

Fig. 8.6 Subroutine MODFUL

```
      SUBROUTINE  MODFUL (NTOTAL,N,VALUE,S,C)
C     * * * * * * * * * * * * * * * * * * * * * * * * * *
C     APPLY TYPE 1 MODIFICATION TO FULL SYMM EQS
C          S*D = C,   D(N) = VALUE
C     * * * * * * * * * * * * * * * * * * * * * * * * * *
CDP   IMPLICIT REAL*8 (A-H,O-Z)
      DIMENSION  S(NTOTAL,NTOTAL), C(NTOTAL)
C     NTOTAL = TOTAL NUMBER OF EQUATIONS
C     N = DOF NUMBER OF CONSTRAINED PARAMETER
C     VALUE = GIVEN VALUE OF DOF NUMBER N
C     S = FULL SQUARE MATRIX, C = FULL COLUMN MATRIX
      DO 10  I = 1,NTOTAL
      C(I) = C(I) - VALUE*S(I,N)
      S(I,N) = 0.D0
   10 S(N,I) = 0.D0
      S(N,N) = 1.D0
      C(N) = VALUE
      RETURN
      END
```

Fig. 8.7 Subroutine MODFYI

```
      SUBROUTINE  MODFY1 (NDFREE,MBW,L1,C1,SS,CC)
C     * * * * * * * * * * * * * * * * * * * * * * * * * *
C     APPLY TYPE 1 CONSTRAINT EQUATION MODIFICATIONS
C           IN UPPER HALF BANDWIDTH MODE
C              SS*DD = CC,   DD(L1) = C1
C     * * * * * * * * * * * * * * * * * * * * * * * * * *
CDP   IMPLICIT REAL*8(A-H,O-Z)
      DIMENSION  SS(NDFREE,MBW), CC(NDFREE)
C     SS = RECT MATRIX WITH UPPER HALF BANDWIDTH OF
C        SYMMETRIC SYSTEM EQUATIONS
C     CC = SYSTEM COLUMN MATRIX
C     L1 = SPECIFIED SYSTEM DEGREE OF FREEDOM NUMBER
C     C1 = SPECIFIED CONSTRAINT EQUATION COEFFICIENT
C     MBW = MAX. HALF BANDWIDTH OF SYSTEM
C     NDFREE = TOTAL DEGREES OF FREEDOM OF SYSTEM
      ZERO = 0.D0
      M1 = MIN0 (L1,MBW) - 1
      IF ( M1.LT.1 )  GO TO 20
      DO 10  I = 1,M1
      IROW = L1 - I
      ICOL = I + 1
      IF ( C1.EQ.ZERO )  GO TO 10
      CC(IROW) = CC(IROW) - C1*SS(IROW,ICOL)
   10 SS(IROW,ICOL) = ZERO
   20 M1 = MIN0 ( (NDFREE + 1 - L1),MBW )
      DO 30  I = 1,M1
      IROW = L1 - 1 + I
      ICOL = I
      IF ( C1.EQ.ZERO )  GO TO 30
      CC(IROW) = CC(IROW) - C1*SS(L1,ICOL)
   30 SS(L1,ICOL) = ZERO
      SS(L1,1) = 1.D0
      CC(L1) = C1
      RETURN
      END
```

8.3 Constraints applied at the element level

The previous sections on the application of boundary nodal constraints have assumed that the constraints would be applied at the system level after the equations have been fully assembled. This is generally the most efficient way of applying the nodal constraints. However, there are situations where it is necessary or useful to apply the nodal constraints at the element level. The above procedures for introducing the nodal constraints can also be applied to the element matrices, $\mathbf{S}^e$, and $\mathbf{C}^e$. This is true so long as all the degrees of freedom given in the linear constraint equation occur at a common node. For example, consider a problem with one parameter per node and assume that the value at a particular node is defined to be V, i.e. a Type 1 constraint. Further, assume that there are g elements connected to that node. Then the matrices of each of the g elements would be modified such that the corresponding square matrix diagonal term and column matrix terms are one and V respectively. When the g elements are assembled, the system matrix has a corresponding diagonal term of g and the column matrix term has a value of g^*V. Thus the final result gives the same as would be obtained if the modifications were carried out at the system level. Since the element matrices usually are stored in their full form, instead of in upper half-bandwidth form, one would need a subroutine such as MODFUL (Fig. 8.6) to apply the Type 1 modification at the element level.

8.4 Penalty modifications for nodal constraints

Nodal parameter constraints can also be accomplished by the use of an artifice such as the following. Let 'i' denote the degree of freedom to be assigned a given value, say V. This trick consists of modifying two terms of the system matrix and column vector, i.e. $S_{i,i}$, and C_i. These terms are redefined to be

$$C_i = \beta V S_{i,i} \qquad (8.8)$$
$$S_{i,i} = \beta S_{i,i}$$

where β is a very large number. This yields the ith system equation

$$S_{i,1}D_1 + \ldots + \beta S_{i,i}D_i + \ldots + S_{i,n}D_n = \beta V S_{i,i},$$

which is a good approximation of the boundary condition $D_i = V$, if β is sufficiently large. This artifice is the fastest method for introducing Type 1 constraints but it can lead to numerical ill-conditioning. There is also a problem of how large a value should be assigned to β. Commonly used values range from 10^{12} to 10^{28}.

The higher order constraints could be introduced in a similar manner. For example, a Type 2 constraint such as the one in Eqn. (8.4) could be also added to the system equations after being multiplied by some penalty weight. Such a procedure has the advantage that it adds very little to skyline or frontal storage procedures.

However, there are numerical difficulties. The multiplication by the weighting factor introduces large numbers into off-diagonal terms and may cause ill-conditioning of the equations. There is also a problem of how to select the penalty weight. If a very large penalty is used, as in the Type 1 case, the system tends to *lock-up*, i.e. yield a zero solution, and the constraints approach a set of Type 1 constraints. To prevent this one must retain some information about element contributions to the corresponding equations. One way of doing this is to find the average value of the original diagonal terms to be modified by the constraint and then the constraint equation penalty could be selected to be 100 or 1000 times larger. This gives reasonable accuracy but the exact matrix manipulation procedure is better.

The symmetric constraints in Eqn. (8.4) can be generalized to any number of terms. Recall that the normalized coefficients, CEQ, and the degree of freedom numbers, NDXC, are stored for each constraint of Type n, where

$$1 \cdot \mathrm{D}_{\mathrm{NDXC}(1)} + \mathrm{CEQ}(1) \cdot \mathrm{D}_{\mathrm{NDXC}(2)} + \ldots + \mathrm{CEQ}(i) \cdot \mathrm{D}_{\mathrm{NDXC}(i+1)} + \ldots + \mathrm{CEQ}(n-1) \cdot \mathrm{D}_{\mathrm{NDXC}(n)} = \mathrm{CEQ}(n) \tag{8.9}$$

Thus a least squares procedure can be used to form the penalty square and column matrices, **SP**, and **CP**. Form a temporary vector, $\mathbf{V}(n+1)$, such that $V(1) = 1$ and $V(i+1) = \mathrm{CEQ}(i)$ for $1 \leqslant i \leqslant n$. Calculate the product $\mathbf{V}^T\mathbf{V}$ and delete the last row. Then the last column contains **CP** while the remainder of the expression is **SP**. Several other types of penalty constraints have been given by Campbell [21].

The matrix manipulation procedure was originally used in MODEL (see Chapter 2) but the penalty constraints are now employed for simplicity. If a skyline solution procedure is utilized then the exact manipulation procedure could be used with little additional cost if high accuracy in multi-point constraints is required. Of course there are other procedures, such as transformation methods [25], that can be employed to enforce constraints. In problems with a high percentage of constrained degrees of freedom one should also consider reducing the rank of the system matrices as an additional cost-saving procedure.

The subroutines for generating and applying the penalty constraints are given in Figs. 8.8 and 8.9. The constraint matrices are generated by PENLTY. The data are extracted by PENMOD which also selects the penalty weight and the additions to the system equations.

The contributions of the constraint equations to the system bandwidth are determined by subroutine CEQBAN illustrated in Fig. 8.10. If a skyline storage mode were utilized it would be necessary to carry out similar calculations to see if the constraints change the system column heights.

Fig. 8.8 Subroutine PENLTY

```
      SUBROUTINE PENLTY (NPFRE,CEQ,CP,SP,WT)
C     * * * * * * * * * * * * * * * * * * * * * * * * *
C     DEFINE CONSTRAINT PENALTY SQ AND COL MATRICES
C     * * * * * * * * * * * * * * * * * * * * * * * * *
CDP   IMPLICIT REAL*8 (A-H,O-Z)
      DIMENSION  CEQ(NPFRE), CP(NPFRE), SP(NPFRE,NPFRE)
C     NPFRE = NO DOF IN CONSTRAINT EQUATION
C     CEQ(I) = CONSTR EQ COEFFICIENT I+1
C     CP = CONSTRAINT COLUMN MATRIX
C     SP = CONSTRAINT SQUARE MATRIX
C     WT = PENALTY WEIGHT FACTOR
C     INITIAL CALCULATIONS
      CP(1) = 1.D0
      TEMP = CEQ(NPFRE)
      IF ( NPFRE.EQ.1 )  GO TO 20
      DO 10  I = 2,NPFRE
   10 CP(I) = CEQ(I-1)
C     CALCULATE LEAST SQ CONSTRAINT FORMS
   20 DO 40  I = 1,NPFRE
      DO 30  J = 1,NPFRE
   30 SP(J,I) = WT*CP(I)*CP(J)
   40 CONTINUE
      DO 50  I = 1,NPFRE
   50 CP(I) = CP(I)*TEMP*WT
      RETURN
      END
```

Fig. 8.9 Subroutine PENMOD

```
      SUBROUTINE  PENMOD (MAXACT,NUMCE,NREQ,NDXC,CEQ,
     1                   CP,SP,CC,SS,NDFREE,MAXBAN)
C     * * * * * * * * * * * * * * * * * * * * * * * * *
C     APPLY CONSTRAINT EQS BY PENALTY MODIFICATIONS
C     * * * * * * * * * * * * * * * * * * * * * * * * *
CDP   IMPLICIT REAL*8 (A-H,O-Z)
      DIMENSION SS(NDFREE,MAXBAN), CC(NDFREE), CP(MAXACT),
     1          SP(MAXACT,MAXACT), CEQ(MAXACT,NUMCE),
     2          NDXC(MAXACT,NUMCE), NREQ(MAXACT)
C     MAXACT = NO ACTIVE CONSTRAINT TYPES
C     NUMCE = NO OF CONSTRAINT EQUATIONS
C     NREQ(I) = NO CONSTR EQS OF TYPE I
C     NDXC(I,J) = DOF NO OF TERM I OF EQ J
C     CEQ(I,J) = COEFF TERM I+1 OF EQ J
C     CP = PENALTY SQUARE MATRIX
C     SP = PENALTY SQUARE MATRIX
C     SS = SYSTEM SQ MATRIX UPPER HALF BANDWIDTH
C     CC = SYSTEM COLUMN MATRIX
C     FACTOR = PENALTY WEIGHT FACTOR
      IEQ = 0
      IF ( NREQ(1).EQ.0 )   GO TO 40
C-->  TYPE 1 CONSTRAINT
      DO 5  J = 1,NREQ(1)
      IEQ = IEQ + 1
```

Fig. 8.9—continued

```
5       CALL MODFY1 (NDFREE,MAXBAN,NDXC(1,IEQ),
     2                CEQ(1,IEQ),SS,CC)
40      IF ( MAXACT.EQ.1 )  RETURN
C-->    OTHER TYPES
        FACTOR = 1.0D3
        DO 30 IC = 2,MAXACT
        NTEST = NREQ(IC)
        IF ( NTEST.EQ.0 )  GO TO 30
C       LOOP OVER NC EQS
        DO 20 J = 1,NTEST
        IEQ = IEQ + 1
C       SELECT AVERAGE WEIGHT
        WT = 0.0
        DO 10 K = 1,IC
10      WT = WT + SS(NDXC(K,IEQ),1)
        WT = WT*FACTOR/IC
C       EXTRACT COEFF AND FORM LEAST SQ MATRICES
        CALL PENLTY (IC,CEQ(1,IEQ),CP,SP,WT)
C       EXTRACT DOF NOS AND ADD PENALTY TO SYS EQS
        CALL STORCL (NDFREE,IC,NDXC(1,IEQ),CP,CC)
        CALL STORSQ (NDFREE,MAXBAN,IC,NDXC(1,IEQ),SP,SS)
20      CONTINUE
30      CONTINUE
        RETURN
        END
```

Fig. 8.10 Penalty bandwidth calculation; subroutine CEQBAN

```
        SUBROUTINE  CEQBAN (JBW,NREQ,MAXACT,NUMCE,NDXC,
     1                      NDFREE)
C       * * * * * * * * * * * * * * * * * * * * * * * * *
C       FIND MAXIMUM HALF BANDWIDTH REQUIRED BY
C       CONSTRAINT EQUATION MODIFICATION PROCEDURES
C       * * * * * * * * * * * * * * * * * * * * * * * * *
        DIMENSION  NDXC(MAXACT,NUMCE), NREQ(MAXACT)
C       JBW = MAX HALF BANDWIDTH THAT WILL RESULT FROM
C           CONSTRAINT EQUATIONS MODIFICATIONS
C       MAXACT = NO ACTIVE CONSTR TYPES
C       NUMCE = TOTAL NO CONSTR EQS
C       NDXC(I,J) = CONSTR DOF NO I OF EQ J
C       NDFREE = TOTAL NO OF SYSTEM DEGREES OF FREEDOM
        JBW = 1
        IEQ = 0
C-->    LOOP OVER NON DIAGONAL CONSTRAINTS
        DO 30  IC = 2,MAXACT
        NTEST = NREQ(IC)
        IF ( NTEST.EQ.0 )  GO TO 30
C-->    LOOP OVER TYPE IC EQUATIONS
        DO 20  J = 1,NTEST
        IEQ = IEQ + 1
        IMIN = NDXC(1,IEQ)
        IMAX = IMIN
C-->    FIND EQUATION BANDWIDTH
        DO 10  I = 1,IC
        INDEX = NDXC(I,IEQ)
        IF ( INDEX.LT.IMIN )  IMIN = INDEX
10      IF ( INDEX.GT.IMAX )  IMAX = INDEX
        LBW = IMAX - IMIN + 1
C       UPDATE MAXIMUM
        IF ( LBW.GT.JBW )  JBW = LBW
20      CONTINUE
30      CONTINUE
        RETURN
        END
```

8.5 Exercises

1. Modify the matrix in Fig. 4.11(a) to include the constraint that $D_2 = 3$. Use both the manipulations of Fig. 8.5 and the penalty method.
2. Write an algorithm to apply the modifications of Fig. 8.4 if the square matrix is stored in a skyline node (see [26]).
3. Discuss the effects of renumbering the equations so that Type 1 constraints are considered on the right-hand side but are not retained in the system equations. How does it effect the storage, output of variables, etc.?
4. Consider a system where $D_4 = 0$ and $D_1 - 0.5\,D_2 = 0.5$. Use the method in Fig. 8.4 to show that the solution is $\mathbf{D}^T = [1 \;\; 3 \;\; 0.5 \;\; 0]$ if $\mathbf{C}^T = [5 \;\; 6.5 \;\; 9 \;\; 10]$ and

$$\mathbf{s} = \begin{bmatrix} 1 & 1 & 2 & 2 \\ & 1.5 & 2 & 2 \\ & & 2 & 4 \\ (\text{sym}) & & & 1 \end{bmatrix}$$

9

Solution and result output

9.1 Economical solution techniques for the system equations

9.1.1 Introduction

Modern computer methods of finite element analysis present us with the problem of having to solve large systems of simultaneous algebraic equations. If the analysis is to be performed economically, it is critical that we give careful consideration to the equation solving process. Present experience indicates that from one-third to one-half of the computer time involved in a linear analysis is associated with solving simultaneous equations. In nonlinear and dynamic analysis, as much as 80% of the computer time is associated with solving the system equations.

Certain properties of the square system matrix in the finite element method allow us to employ a number of techniques that reduce not only the solution time, but the amount of computer storage required to perform the analysis. As mentioned before, the system equations are *symmetric*, *positive definite*, *sparse*, and are often *banded*. Symmetry allows us to economize on computer storage since only the elements on the main diagonal and those above (or below) the diagonal need be stored. Because the matrix is positive definite, the process of *pivoting* is never required in order to ensure a stable solution. This allows us to save on computational steps. Banding, however, depends directly on the way the node numbers are assigned and the analyst's ability to prepare an efficient numbering scheme. Banding provides us with the capability to further save on computer storage and computational steps in that only those elements within the band will be stored and involved in the operations. Figure 1.6 is illustrative of a banded, symmetric matrix.

Since the solution time varies approximately as the square of the bandwidth, the nodal numbering producing the minimum bandwidth results in the most economical solution. By way of comparison it should be mentioned here that the solution time for equation solvers which operate on all elements in a square array varies cubically with the number of equations.

9.1.2 Efficient node numbering

The previous section has again highlighted the importance of the half-bandwidth on the storage requirements and the solution time. There are more advanced programming techniques that can be utilized to reduce the significance of the half-bandwidth but these methods will not be considered here. Since the solution procedures described here are banded techniques, the effects of the nodal numbering scheme will be briefly reviewed in the hope that the user will exercise the proper care when numbering nodes. Recall that the half-bandwidth caused by a typical element is given by Eqn. (7.5):

$$\mathrm{HBW} = \mathrm{NG}^*(I_{\max} - I_{\min} + 1)$$

and is illustrated graphically in Fig. 7.5. Note that the analyst assigns the node numbers. Thus he has control over largest and smallest node numbers ($I_{\max}$ and $I_{\min}$, respectively) that are assigned to any element. He therefore has direct control over the maximum half-bandwidth that occurs in the system of elements. Usually the analyst can number the nodes so that the minimum half-bandwidth is achieved by inspection. However, for three-dimensional problems and multiple connected regions, one may find it necessary to rely on algorithms that automatically renumber the nodes so as to reduce (but not necessarily minimize) the bandwidth.

9.1.3 Equation solving techniques

Before discussing specific equation solving algorithms, it is necessary to discuss briefly those capabilities that are essential for efficient and economical solutions of the system equations. The equation solver should be able to take advantage of the sparse, symmetric, banded nature of the equations. It should also be capable of handling multiple solution vectors economically. In addition, solution algorithms should be compared on the basis of calculation and data handling efficiencies.

There are three principle direct solution algorithms in wide use today. These are (1) Gauss elimination, (2) Choleski factorization, and (3) modified Gauss–Choleski factorization. From the standpoint of calculation efficiency, it has been proven that no algorithm for equation solving can involve fewer calculations than Gauss elimination. The original Choleski algorithm requires only $10N$ (where N is the number of equations) more calculations than Gauss elimination. This results from the square-root operation in the original Choleski algorithm. It has been shown that there is much less error associated with the Gauss process than the original Choleski procedure. This is attributed to the fact that the repeated square-root operation severely degrades the accuracy of the process on some computers. This is the fault of the computer and not the algorithm. Despite the disadvantages of the original Choleski process, it has important data storage advantages. It is possible to retain the best features of both the Gauss and Choleski algorithms within a modified Gauss–Choleski algorithm.

For simplicity we consider first full matrices. Basically the Gauss elimination procedure consists of two parts: a factorization procedure; and a back-substitution process. The effect of the factorization procedure on a full matrix is illustrated in Fig. 9.1. The coefficients below the diagonal are reduced to zero (i.e. *eliminated*) by performing a sequence of row operations. For example, to eliminate the terms below the diagonal in column i, one multiplies row i by $(-k_{ji}/k_{ii})$ and adds the result to row j, where $(i + 1) \leq j \leq n$.

After the factorization is completed as shown in Fig. 9.1, the last row contains one equation with one unknown. At this point the back-substitution process may be started. That is, the last equation, $S_{nn}D_n = C_n$, is solved for the value of D_n. Once D_n is known, it is substituted into the next to the last equation, which is then solved for D_{n-1}. The back-substitution continues in a similar manner until all the D_i are known. Algorithms for the Gaussian elimination technique can be found in many texts on numerical analysis. However, most finite element codes do not use this method since other methods, such as the Choleski method, result in more efficient storage utilization and simpler handling of multiple solution vectors, **C**.

The (original) Choleski factorization method factors the system square matrix **S** into the product of a lower triangular matrix (i.e. coefficients above the diagonal are zero), **L**, and the transpose of this lower triangular matrix. That is,

$$\mathbf{SD} = \mathbf{C} \tag{9.1}$$

is factored into

$$\mathbf{LL}^T\mathbf{D} = \mathbf{C}. \tag{9.2}$$

Then defining

$$\mathbf{G} = \mathbf{L}^T\mathbf{D}, \tag{9.3}$$

the previous equation can be written as

$$\mathbf{LG} = \mathbf{C}. \tag{9.4}$$

Since **L** and **C** are known and **L** is a triangular matrix (with one equation and one unknown in the first row), the intermediate vector, **G**, can be determined by a forward-substitution process. After **G** has been determined from Eqn. (9.4), the unknown nodal parameters, **D**, can be obtained by a back-substitution into Eqn. (9.3).

The attractive features of the Choleski algorithm should now be more apparent. First, note that it is only necessary to store **L** or $\mathbf{L}^T$, whichever is more convenient. Secondly, once the system square matrix, **S**, has been factored the unknown nodal parameters, **D**, can be evaluated by a simple forward- and backward-substitution utilizing the given solution vector(s), **C** (and the intermediate vector, **G**). In most problems the equations only need be solved for one value of **C**. However, in dynamic or transient problems and in structural problems with several load conditions one may require the solutions involving many different values of **C**. With the Choleski scheme the time-consuming operations of assembling and factoring **S** are only performed once in linear problems. For every given **C** only the relatively fast forward- and back-substitutions need be repeated. The difficulty with the original Choleski algorithm was that the calculation of the diagonal terms of **L** required a square-root operation. This operation required an undesirable amount of time and also led to relatively inaccurate results. Thus a modified Gauss–Choleski algorithm was developed.

A modified Gauss–Choleski algorithm retains the better features of both techniques and at the same time eliminates the undesirable square-root operations in the factorization. The algorithm is outlined below in symbolic form. The system equations

$$\mathbf{SD} = \mathbf{C}$$

are factored into

$$(\mathbf{LdL}^T)\mathbf{D} = \mathbf{C} \tag{9.5}$$

in which **L** is a lower triangular matrix with ones on the diagonal matrix, see Fig. 9.2. It is the introduction of the diagonal matrix that eliminates the square-root operations. Proceeding as before, let

$$\mathbf{G} \equiv \mathbf{dL}^T\mathbf{D}, \tag{9.6}$$

then

$$\mathbf{LG} = \mathbf{C}. \tag{9.7}$$

Given the value of **C** one obtains **G** by a forward-substitution into Eqn. (9.7) and then **G** is back-substituted into Eqn. (9.6) to yield the nodal parameters, **D**. The last operation can be divided into two operations; first, a scaling calculation, using the diagonal matrix **d**

$$\mathbf{L}^T\mathbf{D} = \mathbf{d}^{-1}\mathbf{G} \equiv \mathbf{H}$$

and then the back substitution of **H** to find the **D**.

9.1.3 Programming the solution

Clearly the first step in programming the solution of the system equations is to identify the algorithm for calculating the arrays **L** and **d**. To illustrate these operations consider a system with only four equations. Carrying out the products in Fig. 9.2 and equating the lower (or upper) triangle gives the identities of Fig. 9.3(a). As shown in Fig. 9.3(b) the identities can be solved to define the necessary terms in **L** and **d**. The above operations can be generalized for any number of equations by the expressions:

$$\begin{aligned} d_{11} &= S_{11} \\ d_{ii} &= S_{ii} - \sum_{k=1}^{i-1} d_{kk}L_{ik}^2, \quad i > 1, \\ L_{ii} &= 1 \\ L_{ij} &= \left(S_{ij} - \sum_{k=1}^{j-1} d_{kk}L_{ik}L_{jk}\right)/d_{jj}, \quad i > j. \\ L_{ij} &= 0, \quad i < j \end{aligned} \tag{9.8}$$

Of course, one only calculates the L_{ij} terms when $i > j$. The L_{ij} terms are stored in the original locations of the S_{ij}. The d_{ii} are stored on the diagonal in place of the $L_{ii} = 1$ terms.

Since the above factorization operations can represent a significant part of the total computational costs every effort is usually made to program Eqns. (9.8) as efficiently as possible. Thus it is often difficult to compare the software with Eqns. (9.8). This is in part due to the fact that large finite element matrices are not stored in their full mode as was assumed in Eqns. (9.8). If they are banded the summations must be restricted to the bandwidth. Conversely, if a skyline storage is used the summations are carried out only with the stored columns.

Since both **L** and **C** are known, the identity of Eqn. (9.7) gives the forward-substitution algorithm for the determination of **G**. The procedure is shown in Fig. 9.4. The general expression is

$$G_i = C_i - \sum_{k=1}^{i-1} G_k L_{ik}. \tag{9.9}$$

Note that **G** could be stored in either the old **C** locations or the, as yet unneeded, **D** locations. Here we use the latter. The backward-substitution is defined by Eqn. (9.6). The operations are illustrated (in reverse order) in Fig. 9.5 and the resulting expressions are

$$D_i = G_i/d_{ii} - \sum_{k=1}^{n-1} D_{i+k}L_{i+k,i} \tag{9.10}$$

for n equations. Again note that **D** could be stored in **G**, that is in the old **C**. Many codes utilize this storage efficiency. In the actual subroutines in use here, **D** and **C** are given as different arrays. Of course they could be assigned the same storage pointer for added storage efficiency. A sample solution is shown in Fig. 9.6.

The actual subroutines used for solutions, FACTOR and SOLVE, are given in Figs 9.7 and 9.8, respectively; for linear problems FACTOR need be called only once while SOLVE would be called once for each *loading condition.* For a skyline storage mode the programming becomes more involved. Typical programs have been given by Taylor [69] and Bathe and Wilson [14]. Programs SKYFAC and SKYSOL, shown in Figs. 9.9 and 9.10, respectively, illustrate one approach. They employ a function program, DOT, to evaluate the dot (or scalor) product of two vectors. They are compatible with the previously discussed skyline subroutines.

Fig. 9.1 Result of Gauss factorization

$$\begin{bmatrix} S_{11} & S_{12} \cdots S_{1j} & \cdots S_{1n} \\ S_{21} & S_{22} & S_{2n} \\ \vdots & \ddots & \vdots \\ S_{j1} & S_{jj} & S_{jn} \\ \vdots & \ddots & \vdots \\ S_{n1} & \cdots & S_{nn} \end{bmatrix} \begin{Bmatrix} D_1 \\ D_2 \\ \vdots \\ D_j \\ \vdots \\ D_n \end{Bmatrix} = \begin{Bmatrix} C_1 \\ C_2 \\ \vdots \\ C_j \\ \vdots \\ C_n \end{Bmatrix}$$

(a) Original equations

Equations after factorization (b)

$$\begin{bmatrix} S'_{11} & S'_{12} \cdots S'_{1j} \cdots & S'_{1n} \\ & S'_{22} & S'_{2n} \\ & \ddots & \vdots \\ & S'_{jj} & \vdots \\ & & \vdots \\ \text{zero} & & S'_{nn} \end{bmatrix} \begin{Bmatrix} D_1 \\ D_2 \\ \vdots \\ D_j \\ \vdots \\ D_n \end{Bmatrix} = \begin{Bmatrix} C'_1 \\ C'_2 \\ \vdots \\ C'_j \\ \vdots \\ C'_n \end{Bmatrix}$$

Fig. 9.2 A modified Choleski factorization

$$\begin{bmatrix} S_{11} & S_{12} \cdots S_{1n} \\ S_{21} & S_{22} & \vdots \\ \vdots & \ddots & \vdots \\ S_{n1} & \cdots & S_{nn} \end{bmatrix} =$$

$$\begin{bmatrix} 1 & & & \text{zero} \\ L_{21} & 1 & & \\ L_{31} & L_{32} & 1 & \\ \vdots & & \ddots & \\ L_{n1} & L_{n2} & \cdots & 1 \end{bmatrix} \begin{bmatrix} d_{11} & & \text{zero} \\ & d_{22} & \\ & & \ddots \\ \text{zero} & & d_{nn} \end{bmatrix} \begin{bmatrix} 1 & L_{21} & L_{31} \cdots L_{n1} \\ & 1 & L_{32} \quad \vdots \\ & & 1 \quad \vdots \\ \text{zero} & & \ddots \quad 1 \end{bmatrix}$$

$$\mathbf{S} = \mathbf{LdL}^T$$

Fig. 9.3 A factorization of four equations

$$\begin{aligned}
S_{11} &= d_{11} \\
S_{21} &= d_{11}L_{21} \\
S_{22} &= d_{11}L_{21}L_{21} + d_{22} \\
S_{31} &= d_{11}L_{31} \\
S_{32} &= d_{11}L_{31}L_{21} + d_{22}L_{32} \\
S_{33} &= d_{11}L_{31}L_{31} + d_{22}L_{32}L_{32} + d_{33} \\
S_{41} &= d_{11}L_{41} \\
S_{42} &= d_{11}L_{41}L_{21} + d_{22}L_{42} \\
S_{43} &= d_{11}L_{41}L_{31} + d_{22}L_{42}L_{32} + d_{33}L_{43} \\
S_{44} &= d_{11}L_{41}L_{41} + d_{22}L_{42}L_{42} + d_{33}L_{43}L_{43} + d_{44}
\end{aligned}$$

(a) The $\mathbf{S} = \mathbf{LdL}^T$ identity

$$\begin{aligned}
d_{11} &= S_{11} \\
L_{21} &= S_{21}/d_{11} \\
d_{22} &= S_{22} - d_{11}L_{21}L_{21} \\
L_{31} &= S_{31}/d_{11} \\
L_{32} &= (S_{32} - d_{11}L_{31}L_{21})/d_{22} \\
d_{33} &= S_{33} - d_{11}L_{31}L_{31} - d_{22}L_{32}L_{32} \\
L_{41} &= S_{41}/d_{11} \\
L_{42} &= (S_{42} - d_{11}L_{41}L_{21})/d_{22} \\
L_{43} &= (S_{43} - d_{11}L_{41}L_{31} - d_{22}L_{42}L_{32})/d_{33} \\
d_{44} &= S_{44} - d_{11}L_{41}L_{41} - d_{22}L_{42}L_{42} - d_{33}L_{43}L_{43}
\end{aligned}$$

(b) Solution for coefficients of **L** and **d**

Fig. 9.4 Forward substitution for equations

$$G_1 = C_1$$
$$L_{21}G_1 + G_2 = C_2$$
$$L_{31}G_1 + L_{32}G_2 + G_3 = C_3$$
$$L_{41}G_1 + L_{42}G_2 + L_{43}G_3 = C_4$$

(a) The $\mathbf{L}\,\mathbf{G} = \mathbf{C}$ identity

$$G_1 = C_1$$
$$G_2 = C_2 - L_{21}G_1$$
$$G_3 = C_3 - L_{31}G_1 - L_{32}G_2$$
$$G_4 = C_4 - L_{41}G_1 - L_{42}G_2 - L_{43}G_3$$

(b) Solution for $\mathbf{G}$

Fig. 9.5 Back-substitution for four equations

$$G_4 = d_{44}D_4$$
$$G_3 = d_{33}(D_3 + L_{43}D_4)$$
$$G_2 = d_{22}(D_2 + L_{32}D_3 + L_{42}D_4)$$
$$G_1 = d_{22}(D_1 + L_{21}D_2 + L_{31}D_3 + L_{41}D_4)$$

(a) The (reverse) $\mathbf{G} = \mathbf{d}\mathbf{L}^T\mathbf{D}$ identity

$$D_4 = G_4/d_{44}$$
$$D_3 = G_3/d_{33} - L_{43}D_4$$
$$D_2 = G_2/d_{22} - L_{32}D_3 - L_{42}D_4$$
$$D_1 = G_1/d_{11} - L_{21}D_2 - L_{31}D_3 - L_{41}D_4$$

(b) Solution for the coefficients of $\mathbf{D}$

Fig. 9.6 Sample factorization and solution

(a) Original system, $\mathbf{SD} = \mathbf{C}$, $\begin{bmatrix} 1 & -1 & 1 & 2 \\ & 5 & -3 & 0 \\ & & 3 & 0 \\ & \text{sym} & & 7 \end{bmatrix} \begin{Bmatrix} D_1 \\ D_2 \\ D_3 \\ D_4 \end{Bmatrix} = \begin{Bmatrix} 2 \\ -4 \\ 4 \\ 1 \end{Bmatrix}$

(b) Factorization, $\mathbf{S} = \mathbf{L}\mathbf{d}\mathbf{L}^T$, $\mathbf{L}^T = \begin{bmatrix} 1 & -1 & 1 & 2 \\ 0 & 1 & -1/2 & 1/2 \\ 0 & 0 & 1 & -1 \\ 0 & 0 & 0 & 1 \end{bmatrix}$, $\mathbf{d} = \begin{bmatrix} 1 & & & \\ & 4 & & \\ & & 1 & \\ & & & 1 \end{bmatrix}$

(c) Forward-substitution, $\mathbf{LG} = \mathbf{C}$, $\mathbf{G}^T = [2 \quad -2 \quad 1 \quad -1]$

(d) Scaling, $\mathbf{dH} = \mathbf{G}$, $\mathbf{H}^T = [2 \quad -1/2 \quad 1 \quad -1]$

(e) Back-substitution, $\mathbf{L}^T\mathbf{D} = \mathbf{H}$, $\mathbf{D}^T = [4 \quad 0 \quad 0 \quad -1]$

Fig. 9.7 Subroutine FACTOR

```
      SUBRCUTINE  FACTOR (NDFREE,IBW,S)
C     * * * * * * * * * * * * * * * * * * * * * * * * * * * * * * *
C     LDLT FACTOR OF BANDED SYMMETRIC SQUARE MATRIX
C     * * * * * * * * * * * * * * * * * * * * * * * * * * * * * * *
CDP   IMPLICIT REAL*8(A-H,O-Z)
      DIMENSICN  S(NDFREE,IBW)
C     NDFREE= MAX. DEGREES OF FREEDOM OF SYSTEM
C     IBW = MAXIMUM HALF BANDWIDTH OF SYSTEM EQS
C     S = RECT MATRIX WITH UPPER HALF BAND CF SYS EQS
      TEMP = 1.DO/S(1,1)
      DO 10 J = 2,IBW
   10 S(1,J) = S(1,J)*TEMP
      DO 70  I = 2,NDFREE
      LL = I - 1
      NN = NDFREE - LL
   20 IF (NN .GT. IBW) NN = IBW
      DO 60 J = 1,NN
      L = IBW - J
      SUM = 0.DO
      IF( L .EQ. 0 ) GO TO 50
      IF( LL .LT. L ) L = LL
   30 DO 40 K = 1,L
      K1 = I - K
      K2 = 1 + K
      K3 = J + K
   40 SUM = SUM + S(K1,K2)*S(K1,K3)*S(K1,1)
   50 S(I,J) = S(I,J) - SUM
      IF( J .EQ. 1 ) GO TO 60
      S(I,J) = S(I,J) / S(I,1)
   60 CCNTINUE
   70 CONTINUE
      RETURN
      END
```

Fig. 9.8 Subroutine SOLVE

```
      SUBROUTINE  SOLVE (NDFREE,IBW,S,P,D)
C     * * * * * * * * * * * * * * * * * * * * * * * * * * * * * * *
C     FOWARD AND BACK SUBSTITUTION OF SYSTEM ECUATIONS
C     PART TWO OF CHOLESKY-GAUSSIAN SOLUTICN
C     * * * * * * * * * * * * * * * * * * * * * * * * * * * * * * *
CDP   IMPLICIT REAL*8(A-H,O-Z)
      DIMENSION  S(NDFREE,IBW), P(NDFREE), D(NDFREE)
C     NDFREE = MAX. DEGREES OF FREEDOM IN SYSTEM
C     IBW = MAXIMUM HALF BANDWIDTH OF THE SYSTEM
C     S = FACTORED SYS SQ MATRIX FROM SUBR FACTOR
C     D = SYSTEM DEGREES OF FREEDOM TO BE DETERMINED
C     P = SYSTEM COLUMN MATRIX (KNOWN)
      D(1) = P(1) / S(1,1)
C-->  FOWARD SUBSTITUTION
      DO 20  I = 2,NDFREE
      II = I + 1
      J = II - IBW
      IF(II .LE. IBW ) J = 1
      IK = I - 1
      SUM = 0.DO
      DO 10 K = J,IK
      KK = II - K
   10 SUM = SUM + S(K,KK)*S(K,1)*D(K)
   20 D(I) = ( P(I) - SUM ) / S(I,1)
C-->  BACK SUBSTITUTION
      DO 40  NN = 2,NDFREE
      I = NDFREE + 1 - NN
      LL = I - 1
      J = LL + IBW
      IF ( J.GT.NDFREE )  J = NDFREE
      L = I + 1
      SUM = 0.DO
      DO 30 K = L,J
      KK = K - LL
   30 SUM = SUM + S(I,KK)*D(K)
   40 D(I) = D(I) - SUM
      RETURN
      END
```

Fig. 9.9 Subroutine SKYFAC

```
      SUBROUTINE SKYFAC (NDFREE,NOCOEF,IDIAG,S)
C     * * * * * * * * * * * * * * * * * * * * * * * * * *
C     L*D*LT FACTORIZATION OF SYSTEM SQUARE MATRIX S
C     STORED IN SYMMETRIC SKYLINE VECTOR MODE
C     * * * * * * * * * * * * * * * * * * * * * * * * * *
CDP   IMPLICIT REAL*8 (A-H,O-Z)
      DIMENSION  S(NOCOEF), IDIAG(NDFREE)
C     NDFREE = TOTAL NUMBER OF SYSTEM DOF
C     NOCOEF = IDIAG(NDFREE) = NO OF COEFFS IN S
C     IDIAG(I) = LOCATION OF DIAGONAL OF EQ I
C     S = SYSTEM SQUARE MATRIX
      ZERO = 0.D0
C     FACTOR OFF DIAGONAL TERMS
      DO 300  J = 2,NDFREE
      JOFF = IDIAG(J) - IDIAG(J-1) - 1
      JTOP = J - JOFF
      ISTART = JTOP + 1
      ISTOP = J - 1
      JD = IDIAG(ISTOP) - JTOP + 1
      IF ( ISTART.GT.ISTOP )  GO TO 110
      DO 100  I = ISTART,ISTOP
      IOFF = IDIAG(I) - IDIAG(I-1) - 1
      ITOP = I - IOFF
      NUM = MAX0 (ITOP,JTOP)
      NEND = I - NUM
      IJV = JD + I
      IL = IDIAG(I-1) + NUM - ITOP + 1
      JG = JD + NUM
  100 S(IJV) = S(IJV) - DOT(NEND,S(IL),S(JG))
C     FACTOR DIAGONAL
  110 ISTART = JTOP
      IF ( ISTART.GT.ISTOP )  GO TO 300
      SUM = ZERO
      DO 200  I = ISTART,ISTOP
      IF ( S(IDIAG(I)).LE.ZERO )  GO TO 200
      D =  S(I+JD)/S(IDIAG(I))
      SUM = SUM + S(I+JD)*D
      S(I+JD) = D
  200 CONTINUE
      S(IDIAG(J)) = S(IDIAG(J)) - SUM
  300 CONTINUE
      RETURN
      END

      FUNCTION  DOT (N,A,B)
C     * * * * * * * * * * * * * * * * * * * * * * * * * *
C     DOT PRODUCT OF VECTORS A(N)*B(N)
C     * * * * * * * * * * * * * * * * * * * * * * * * * *
CDP   IMPLICIT REAL*8 (A-H,O-Z)
      DIMENSION A(N), B(N)
      DOT = 0.D0
      DO 10  I = 1,N
   10 DOT = DOT + A(I)*B(I)
      RETURN
      END
```

Fig. 9.10 Subroutine SKYSOL

```
      SUBROUTINE  SKYSOL (NDFREE,NOCOEF,IDIAG,S,C,D)
C     * * * * * * * * * * * * * * * * * * * * * * * * * *
C     FOWARD AND BACK SUBSTITUTION OF L*D*LT
C     FACTORIZATION OF SYSTEM EQS  S*D=C
C     * * * * * * * * * * * * * * * * * * * * * * * * * *
C     S IN SYMMERIC SKYLINE STORAGE VECTOR
CDP   IMPLICIT REAL*8 (A-H,O-Z)
      DIMENSION S(NOCOEF), C(NDFREE), D(NDFREE),
     1          IDIAG(NDFREE)
C     NDFREE = TOTAL NUMBER OF SYSTEM DOF
C     NOCOEF = IDIAG(NDFREE) = NO OF COEFFS IN MATRIX S
C     IDIAG(I) = LOCATION OF DIAGONAL OF EQ I
C     S = FACTOR OF SYS SQ MATRIX, FROM SKYFAC
C     C = SYSTEM COLUMN MATRIX
C     D = SYSTEM DEGREES OF FREEDOM (RETURNED)
C     DOT = DOT PROD OF 2 VECTORS, FUNCTION PROG
      ZERO = 0.D0
```

Fig. 9.10—continued

```
C       FORWARD SUBSTITUTION
        D(1) = C(1)
        DO 100  I = 2,NDFREE
        IOFF = IDIAG(I) - IDIAG(I-1) - 1
        IF ( IOFF.LT.1 )  GO TO 100
        ITOP = I - IOFF
        IS = IDIAG(I-1) + 1
        C(I) = C(I) - DOT(IOFF,S(IS),D(ITOP))
  100   CONTINUE
C       BACK SUBSTITUTION
        DO 200  I = 1,NDFREE
        IF ( S(IDIAG(I)).NE.ZERO )  D(I)=D(I)/S(IDIAG(I))
  200   CONTINUE
        NDFM1 = NDFREE - 1
        DO 300  K = 1,NDFM1
        I = NDFREE - K + 1
        IOFF = IDIAG(I) - IDIAG(I-1) - 1
        ITOP = I - IOFF
        JSTART = ITOP
        JSTOP = I - 1
        IF ( JSTART.GT.JSTOP )  GO TO 300
        JD = IDIAG(I-1) - ITOP + 1
        DI = D(I)
        IF ( DI.EQ.ZERO )  GO TO 250
        DO 250  J = JSTART,JSTOP
        D(J) = D(J) - DI*S(J+JD)
  250   CONTINUE
  300   CONTINUE
        RETURN
        END
```

9.2 Output of results

Once the nodal parameters of the system have been calculated it is necessary to output these quantities in a practical form. The number of nodal parameters (degrees of freedom) can be quite large. Common problems often involve 2000 parameters but some special problems have required as many as 50,000 (or more) parameters. Also the reader should recall that these nodal parameters are often utilized to calculate an equally large set of secondary quantities. Obviously, a simple printed list of these nodal parameters is not the optimum form of output for the analyst. Data presented in a graphical form are probably the most practical. Several computer graphics programs have been developed for use with certain specialized finite element codes. However, these graphics packages are often expensive and usually highly hardware-dependent. Thus it is not feasible here to delve into the subject of computer graphics as a mode of data presentation. Instead the following four sections will discuss the utilization of printed output.

The most straightforward way to list the calculated nodal parameters is to list each of the M nodal points in the system and the values of the NG parameters that correspond to each of these nodal points. It is common practice to also list the NSPACE spatial coordinates of the nodal points so as to save the analyst the inconvenience of referring back to the input data to determine the location of the point. The above format for printing the calculated nodal parameters is utilized by subroutine WRTPT. It determines the system degree of freedom to be printed by substituting the system node number (I) of the point and the local parameter (J) into Eqn. (1.1). Subroutine WRTPT is shown in Fig. 9.11.

It is desirable at times to know the values of the calculated parameters that are associated with some particular element. This can be accomplished by listing the data of interest associated with each of the NE elements in the system. These data usually include a list of all nodal points (element incidences) associated with the element, and a corresponding list of the NG nodal parameters that occur at each of these nodes. When using this approach it is obvious that several of the nodal parameters will be listed more than once since most of the nodal points are shared by more than one element. The element incidences are available as input data and thus can be used with Eqn. (1.1) to identify which system degrees of freedom should be printed with a particular element. These elementary operations are executed by subroutine WRTELM, which is shown in Fig. 9.12.

Once the nodal parameters, D, have been calculated, the analyst may desire to know the extreme values of each of the NG nodal parameters and the node numbers where each extreme value occurred. These simple operations are performed by subroutine MAXMIN, which is shown in Fig. 9.13. This returns two arrays, RANGE and NRANGE. Both arrays are NG* 2 in size. The first column of RANGE contains the NG maximum values of D and the second column contains the corresponding minimum values. Array NRANGE contains the node numbers where the corresponding extreme values in RANGE were encountered.

It is desirable to also have information on the contours of the nodal dof. One would prefer to have the computer plot the contour lines illustrating the spatial variation of each of the NG types of nodal parameters. Lacking this ability one would like the computer to supply data that would make plotting by hand as simple as possible. As discussed in the previous section, the extreme values of any particular type of parameter (there are NG) are readily determined. The procedure utilized here for each parameter type is to print the spatial coordinates of points lying on specified contour curves. The analyst specifies the number of contours (>2) to range from five percent above the minimum value to five percent below the maximum value for each parameter. The contour curves can be generated in several ways. One simple procedure (not the most efficient) is as follows. Examine, in order, each element in the system to ascertain if the contour passes through that element. That is, determine if the specified contour value is bounded by the nodal parameters associated with that element. If so, assume that the contour passes through the element only once and calculate the spatial coordinates of the points where it pierces the boundary of the element. For simplicity, the present program, subroutine CONTUR, utilizes linear interpolation in these calculations. This subroutine is shown in Fig. 9.14.

Once the intersection points have been located their NSPACE spatial coordinates are printed along with an identifying integer point number. For printed output this procedure is inefficient since for NSPACE > 1 most points will be listed (and numbered) twice. (This repetition could be reduced by calculating, printing, and numbering only the first piercing point.) Nevertheless, this procedure is much more efficient than doing the calculations by hand. If the analyst has access to a plotter, the above procedure is easily modified to yield plotted contours (for NSPACE ⩾ 2). Once the two intersection points for an element have been identified, one simply makes the plotter draw a straight line through the two points. Of course, it would be desirable to have a number or symbol plotted that would clearly identify the contour curve to which the segment belongs. After all the elements have been surveyed, the completed contour, consisting of many straight line segments, will have been plotted.

Subroutines MAXMIN and CONTUR have other possible applications. For example, if properties have been defined at each nodal point (in array FLTNP) then these properties values could also be contoured, if NNPFLO ⩽ NG. This would clearly identify regions of different materials as a visual data check. However, since array FLTNP is doubly subscripted and array D in CONTUR has only a single subscript, a few minor changes would be required in subroutine CONTUR.

Fig. 9.11 Subroutine WRTPT

```
      SUBROUTINE  WRTPT ( M,NG,NDF,NSPACE,X,DD,INDEX)
C     * * * * * * * * * * * * * * * * * * * * * * * * * *
C     OUTPUT, BY NODES, OF CALCULATED DEGREES OF FREEDOM
C     * * * * * * * * * * * * * * * * * * * * * * * * * *
CDP   IMPLICIT REAL*8(A-H,O-Z)
      DIMENSION  X(M,NSPACE), DD(NDF), INDEX(NG)
C     M = NUMBER OF NODES IN SYSTEM
C     NG = NUMBER OF PARAMETERS (DOF) PER NODE
C     NDF = NUMBER OF DOF IN THE SYSTEM
C     NSPACE = DIMENSION OF SPACE
C     X =  SYSTEM COORDINATES OF ALL NODES
C     DD = CALCULATED NODAL PARAMETERS
C     INDEX = SYSTEM DOF NOS OF PARAMETERS ON A NODE
      WRITE (6,5000) NSPACE, NG
 5000 FORMAT('1***  OUTPUT OF RESULTS  ***',/,
     1 ' NODE, ',I2,' COORDINATES, ',I2,' PARAMETERS.')
      DO 10  I = 1,M
      CALL  INDXPT (I,NG,INDEX)
      WRITE (6,5010) I,(X(I,L),L=1,NSPACE),
     1                (DD(INDEX(K)),K=1,NG)
 5010 FORMAT ( I5, (7(3X,E15.8)) )
 10   CONTINUE
      RETURN
      END
```

Fig. 9.12 Subroutine WRTELM

```
      SUBROUTINE  WRTELM (NE,N,NG,NDFREE,NELFRE,DD,
     1                    INDEX,NODES,LNODE)
C     * * * * * * * * * * * * * * * * * * * * * * * * * *
C     OUTPUT, BY ELEMS, OF CALCULATED DEGREES OF FREEDOM
C     * * * * * * * * * * * * * * * * * * * * * * * * * *
CDP   IMPLICIT REAL*8(A-H,O-Z)
      DIMENSION  DD(NDFREE), INDEX(NELFRE), NODES(NE,N),
     1           LNODE(N)
C     NE = NO OF ELEMENTS IN SYSTEM
C     N = NO NODES PER ELEMENT
C     NG = NO OF PARAMETERS (DOF) PER NODE
C     NDFREE = NO DEGREES OF FREEDOM IN SYSTEM
C     NELFRE = NO DEGREES OF FREEDOM PER ELEMENT
C     DD = CALCULATED NODAL PARAMETERS (DOF)
C     NODES = NODAL INCIDENCES OF ALL ELEMENTS
C     LNODE = NODAL INCIDENCES OF AN ELEMENT
C     INDEX = SYSTEM DOF NOS FOR ELEMENT PARAMETERS
      WRITE (6,5000)  NG
 5000 FORMAT ('1***  OUTPUT OF RESULTS  ***',/,
     1' ELEMENT NO., NODE NO.,'I3,' PARAMETERS',/)
      DO 20  IE = 1,NE
      CALL LNODES (IE,NE,N,NODES,LNODE)
      DO 10  K = 1,N
      NODE = LNODE(K)
      IF ( NODE.LT.1 )  GO TO 10
      CALL INDXPT (NODE,NG,INDEX)
      WRITE (6,5010) IE, NODE, (DD(INDEX(L)),L=1,NG)
 5010 FORMAT (I5,I8,(6(3X,E15.8)) )
 10   CONTINUE
 20   CONTINUE
      RETURN
      END
```

Fig. 9.13 Subroutine MAXMIN

```
      SUBROUTINE  MAXMIN (M,NG,NDFREE,IPRINT,RANGE,DD,
     1                    INDEX,NRANGE)
C     * * * * * * * * * * * * * * * * * * * * * * * * * *
C     FIND EXTREME RANGE OF VALUES OF THE NG NODAL DOF
C     * * * * * * * * * * * * * * * * * * * * * * * * * *
CDP   IMPLICIT REAL*8(A-H,O-Z)
      DIMENSION  DD(NDFREE), RANGE(NG,2), INDEX(NG),
     1           NRANGE(NG,2)
C     M = NUMBER OF NODES IN SYSTEM
C     NG = NUMBER OF PARAMETERS (DOF) PER NODE
C     NDFREE = TOTAL NUMBER OF SYSTEM DEGREES OF FREEDOM
C     IF IPRINT.NE.0 PRINT RANGE OF VALUES
C     RANGE: 1-MAXIMUM VALUE, 2-MINIMUM VALUE
C     DD = ARRAY OF SYSTEM DEGREES OF FREEDOM
C     INDEX = LIST OF SYSTEM DOF NOS FOR DOF AT NODE
C     NRANGE = ARRAY OF NODE NOS OF EXTREME VALUE POINTS
      DO 10  J = 1,NG
      NRANGE(J,1) = 0
      NRANGE(J,2) = 0
      RANGE(J,1) = DD(J)
 10   RANGE(J,2) = DD(J)
      DO 40  I = 1,M
      CALL  INDXPT (I,NG,INDEX)
      DO 30  J = 1,NG
      DDTEST = DD(INDEX(J))
      IF ( DDTEST.LT.RANGE(J,1) )  GO TO 20
      RANGE(J,1) = DDTEST
      NRANGE(J,1) = I
 20   IF ( DDTEST.GT.RANGE(J,2) )  GO TO 30
      RANGE(J,2) = DDTEST
      NRANGE(J,2) = I
 30   CONTINUE
 40   CONTINUE
      IF ( IPRINT.EQ.0 )  RETURN
C     PRINT RANGE OF VALUES
      WRITE (6,5000)
 5000 FORMAT ('1*** EXTREME VALUES OF THE NODAL ',
     1'PARAMETERS ***',//,
     2' PARAMETER NO.        MAX. VALUE,NODE        ',
     3'MIN. VALUE,NODE')
      DO 50  J = 1,NG
 50   WRITE (6,5010) J,RANGE(J,1),NRANGE(J,1),RANGE(J,2),
     1                 NRANGE(J,2)
 5010 FORMAT (I10,2X,E15.8,' , ',I5,2X,E15.8,' , ',I5)
      RETURN
      END
```

Fig. 9.14 Subroutine CONTUR

```
      SUBROUTINE  CONTUR (M,NE,N,NG,NSPACE,NDFREE,NELFRE,
     1                    NCURVE,X,COORD,XPT,DD,D,RANGE,
     2                    NODES,LNODE,INDEX)
C     * * * * * * * * * * * * * * * * * * * * * * * * *
C     LIST COORDINATES OF POINTS ON CONTOURS BETWEEN THE
C             5TH AND 95TH PERCENTILE CURVES
C     * * * * * * * * * * * * * * * * * * * * * * * * *
CDP   IMPLICIT REAL*8(A-H,O-Z)
      DIMENSION X(M,NSPACE), COORD(N,NSPACE), DD(NDFREE),
     1          D(NELFRE), RANGE(NG,2), XPT(NSPACE),
     2          NODES(NE,N), LNODE(N), INDEX(NELFRE)
C     M = NUMBER OF SYSTEM NODES
C     NG = NUMBER OF PARAMETERS (DOF) PER NODE
C     NDFREE = TOTAL NO. OF SYSTEM DEGREES OF FREEDOM
C     NELFRE = NO. OF ELEMENT DEGREES OF FREEDOM
C     N = NUMBER OF NODES PER ELEMENT
C     NE = NUMBER OF ELEMENTS IN SYSTEM
C     NSPACE = DIMENSION OF SPACE
C     X = COORDINATES OF SYSTEM NODES
C     COORD = COORDINATES OF ELEMENT NODES
C     XPT = SPATIAL COORDINATES OF A CONTOUR POINT
C     DD = SYSTEM LIST OF NODAL PARAMETERS
C     D = NODAL PARAMETERS ASSOCIATED WITH ELEMENT
C     RANGE 1-MAX VALUE 2-MIN VALUE OF NODAL DOF
C     NODES = SYSTEM ARRAY OF ELEMENT INCIDENCES
C     LNODE = ELEMENT INCIDENCES ARRAY OF THE ELEMENT
C     INDEX = SYS. DOF NOS FOR (N*NG) ELEMENT DOF
      WRITE (6,5000)
 5000 FORMAT ('1*** CALCULATION OF CONTOUR CURVES ***',/)
      IF ( NCURVE.LT.2 )  NCURVE = 2
C-->   LOOP OVER PARAMETERS
      DO 100  IPAR = 1,NG
      DIFF = RANGE(IPAR,1) - RANGE(IPAR,2)
      V1 = 0.05*DIFF + RANGE(IPAR,2)
      VNCUR = 0.95*DIFF + RANGE(IPAR,2)
      WRITE (6,5010)  IPAR, V1, VNCUR
 5010 FORMAT ('0PARAMETER NUMBER =',I3,/,
     1 '   5TH PERCENTILE VALUE = ',E15.8,/,
     2 '  95TH PERCENTILE VALUE = ',E15.8,/)
      CL1 = NCURVE - 1
      DO 90  ICUR = 1,NCURVE
      RATIO = ICUR - 1
      RATIO = RATIO/CL1
      VALUE = V1 + RATIO*(VNCUR - V1)
      WRITE (6,5020)  ICUR, VALUE, NSPACE
 5020 FORMAT ('0CONTOUR NUMBER = ',I5,/,
     1 ' VALUE OF CONTOUR = ',E15.8,/,
     2 ' POINT NO.,',I3,' COORDINATES')
      KPOINT = 0
C-->   LOOP OVER ELEMENTS
      DO 80  IE = 1,NE
      KALL = 0
C     EXTRACT LIST OF NODES ON THE ELEMENT
      CALL  LNODES (IE,NE,N,NODES,LNODE)
C     DETERMINE CORRESPONDING SYSTEM DOF NUMBERS
      CALL  INDXEL (N,NELFRE,NG,LNODE,INDEX)
      CALL  ELFRE (NDFREE,NELFRE,D,DD,INDEX)
C     FIND SPATIAL COORDINATES OF NODES
C     CALL  ELCORD (M,N,NSPACE,X,COORD,LNODE)
C-->   TEST SIDES OF ELEMENT
      DO 70  IPT = 1,N
      IFIRST = IPT
      ILAST = IPT + 1
      IF ( IPT.LT.N )  GO TO 10
      IF ( N.EQ.2 )  GO TO 80
      ILAST = 1
   10 KPAR1 = NG*(IFIRST-1) + IPAR
      KPAR2 = NG*(ILAST-1) + IPAR
      TEST1 = D(KPAR1)
      TEST2 = D(KPAR2)
      IF ( (VALUE-TEST1)*(TEST2-VALUE).LT.0.0 )  GO TO 70
C-->   FOUND A POINT
      KPOINT = KPOINT + 1
      RATIO = TEST2 - TEST1
      IF ( RATIO )  20,30,20
C     CALCULATE COORDINATES OF THE CONTOUR POINT
   20 RATIO = ( VALUE-TEST1 )/RATIO
   30 IF ( KALL )  50,40,50
   40 CALL  ELCORD (M,N,NSPACE,X,COORD,LNODE)
      KALL = 1
```

Fig. 9.14—continued

```
50    DO 60  ISPACE = 1,NSPACE
      C1 = COORD(IFIRST,ISPACE)
      C2 = COORD(ILAST,ISPACE)
      DIFF = C2 - C1
60    XPT(ISPACE) = C1 + RATIO*DIFF
      WRITE (6,5030) KPOINT,(XPT(L),L=1,NSPACE)
5030  FORMAT ( I8, (5E18.8,/) )
70    CONTINUE
80    CONTINUE
90    CONTINUE
100   CONTINUE
      RETURN
      END
```

9.3 Post-solution calculations within the elements

As stated earlier in Section 4.3, if NTAPE1 is greater than zero this implies that problem-dependent calculations are to be performed in each element after the nodal parameters have been calculated. These post-solution calculations are under the control of subroutine POST, which is shown in Fig. 9.15. Its argument includes the following quantities:

NTAPE1	= unit on which ELPOST stored the post solution element data;
NE	= total number of elements;
N	= number of nodes per element;
NELFRE	= number of element degrees of freedom;
NDFREE	= total number of system degrees of freedom;
NODES(NE, N)	= element incidences lists for all elements;
DD(NDFREE)	= calculated array of nodal parameters; and
NTAPE2–4	= optical scratch units (if > 0).

For every element in the system, subroutine POST extracts the list of nodes on that element, and the nodal parameters, D, associated with the element. It then calls the problem-dependent subroutine POSTEL. The latter routine is furnished the element number, NTAPE1, NELFRE and the nodal parameters, D(NELFRE), associated with the element. Subroutine POSTEL reads the necessary problem-dependent data (generated by ELPOST) from NTAPE1, combines these with the nodal parameters, D and calculates additional quantities of interest within the element. These problem-dependent quantities are usually printed along with the identifying element number, IE.

Fig. 9.15 Subroutine POST

```
      SUBROUTINE  POST (NTAPE1,NE,N,NG,NELFRE,NDFREE,
     1                 NODES,LNODE,INDEX,DD,D,LPROP,
     2                 NTAPE2,NTAPE3,NTAPE4,IT,NITER,M)
C     * * * * * * * * * * * * * * * * * * * * * * * * * * *
C        ELEMENT LOOP FOR POST SOLUTION CALCULATIONS
C     * * * * * * * * * * * * * * * * * * * * * * * * * * *
CDP   IMPLICIT REAL*8 (A-H,O-Z)
      DIMENSION  DD(NDFREE), D(NELFRE), NODES(NE,N),
     1         LNODE(N), INDEX(NELFRE), LPROP(1)
C     NTAPE1 = UNIT FOR POST SOL MATRICES STORAGE
C     NE = NUMBER OF ELEMENTS IN SYSTEM
C     N = NUMBER OF NODES PER ELEMENT
C     NG = NUMBERS PARAMETERS PER NODE
C     NELFRE = NUMBER OF DEGREES OF FREEDOM PER ELEMENT
C     NDFREE = TOTAL NUMBER OF SYSTEM DEGREES OF FREEDOM
C     NODES = ELEMENT INCIDENCES OF ALL ELEMENTS
C     LNODE = THE N ELEMENT INCIDENCES OF THE ELEMENT
C     INDEX = SYSTEM DOF NOS ASSOCIATED WITH ELEMENT
C     DD = ARRAY OF SYSTEM DEGREES OF FREEDOM
C     D = ARRAY OF ELEMENT DEGREES OF FREEDOM
C     NTAPE2,3,4 = OPTIONAL UNITS FOR USER (WHEN > 0)
C     IT = CURRENT ITERATION NUMBER, NITER = MAX IT
C     M = TOTAL NUMBER OF NODES
      REWIND  NTAPE1
C-->  LOOP OVER ELEMENTS
      DO 10  IE = 1,NE
C     EXTRACT ELEMENT'S NODES
      CALL LNODES (IE,NE,N,NODES,LNODE)
C     EXTRACT NODAL PARAMETERS OF THE ELEMENT
      CALL INDXEL (N,NELFRE,NG,LNODE,INDEX)
      CALL ELFRE  (NDFREE,NELFRE,D,DD,INDEX)
C-->  PERFORM PROBLEM DEPENDENT CALCULATIONS AND OUTPUT
      CALL POSTEL (NTAPE1,NELFRE,D,IE,NTAPE2,NTAPE3,
     1             NTAPE4,IT,NITER,NE,M)
   10 CONTINUE
      RETURN
      END
```

9.4 Exercises

1. Modify the storage array pointers so the system unknowns, **D**, are stored in the old forcing function locations, **C**.
2. Use Fig. 9.2 to verify the indentities of Figs. 9.3(a), 9.4(a), and 9.5(a).
3. Program Eqns. (9.8), (9.9), and (9.10) for a matrix stored in standard full symmetric form.
4. Solve the equations in Fig. 9.6(a) by Gauss elimination.
5. Resolve the system in Fig. 9.6 if $\mathbf{C}^T = [1\ 0\ 4\ 3]$.

10

One-dimensional applications

10.1 Introduction

The simplest applications of the finite element method involve problems which are independent of time. The general programming considerations of this class of problems were presented in detail in the first four chapters. At this point one is ready to consider the programming of the problem dependent subroutines associated with the application of interest. The following sections will present several representative examples of the application of the finite element method to steady state problems.

Generally an element which can be defined by a single spatial variable in the local coordinates is relatively simple to implement. Thus, the typical illustrative applications begin with this class of element. This, of course, does not mean that the global coordinates must also be one-dimensional. For example, a set of one-dimensional *truss* elements can be utilized to construct a three-dimensional structure.

10.2 Conductive and convective heat transfer

The text by Myers [56] presents a detailed study of heat transfer analysis from both the analytic and numerical points of view. Several examples are solved by finite difference and finite element methods as well as by analytic techniques. All three methods were applied to the problem to be illustrated in this section. Consider the heat transfer in a slender rod, illustrated in Fig. 10.1, that is insulated at $x = L$, has a specified temperature, t_0, at $x = 0$, and is surrounded by a medium with a temperature of $t_\infty = 0$. The temperature, $t(x)$, in this one-dimensional problem is governed by the differential equation

$$KA\frac{d^2t}{dx^2} - hPt = 0 \tag{10.1}$$

and the boundary conditions

$$t(0) = t_0$$

$$\frac{dt}{dx}(L) = 0$$

where

A = cross-sectional area of rod,
h = convective transfer coefficient,
K = thermal conductivity,
L = length of rod,
P = perimeter of cross-section,
t = temperature, and
x = spatial coordinate.

The exact solution for this problem is $t(x) = \cosh[m(L - x)]/\cosh(mL)$ where $m^2 = hP/KA$. Myers shows that an integral formulation can be obtained by rendering stationary the functional

$$I = \frac{1}{2}\int_0^L \left[KA\left(\frac{dt}{dx}\right)^2 + hPt^2\right] dx, \tag{10.2}$$

subject to the boundary condition $t(0) = t_0$. The boundary condition imposed at $x = 0$ is an *essential* boundary condition. The second boundary condition is automatically satisfied, and is known as a *natural boundary condition.* Of course, if a second temperature had been specified at $x = L$, it would have taken precedence over the natural boundary condition at that end. The reader may wish to verify that applying the weighted residual method of Galerkin and integrating the first term by parts yields an identical formulation. Assuming a linear element, as shown in Fig. 10.2, in any element the interpolation functions are known to be

$$t^e(x) = \mathbf{M}(x)\,\mathbf{D}^e\mathbf{T}^e$$

where

$$\mathbf{M}(x) = [1 \quad x],$$

$$\mathbf{D}^e = \frac{1}{l}\begin{bmatrix} x_j & -x_i \\ -1 & 1 \end{bmatrix},$$

$$\mathbf{T}^{eT} = [T_i \quad T_j]$$

and $l = (x_j - x_i)$ denotes the length of the element. That is, $t^e(x)$ is a function of an assumed spatial variation, $\mathbf{M}(x)$, the element geometry, $\mathbf{D}^e$, and the element's nodal temperatures, $\mathbf{T}^e$. This can be expressed in a simpler form (see Fig. 1.4) here as

$$t^e(x) = \mathbf{H}^e(x)\,\mathbf{T}^e, \tag{10.3}$$

where

$$\mathbf{H}^e(x) = \mathbf{M}(x)\,\mathbf{D}^e = [(x_j - x)/l \quad (x - x_i)/l]$$

is called *the* element interpolation matrix. Myers shows that the element square matrix for this element, $\mathbf{S}^e$, can be defined as

$$\mathbf{S}^e = \int_{x_i}^{x_j} \left(K^e A^e \frac{\partial \mathbf{H}^{eT}}{\partial x} \frac{\partial \mathbf{H}^e}{\partial x} + h^e P^e \mathbf{H}^{eT} \mathbf{H}^e \right) \mathrm{d}x \tag{10.4}$$

which reduces to

$$\mathbf{S}^e = \mathbf{D}^{eT} \mathcal{S}^e \mathbf{D}^e,$$

where

$$\mathcal{S}^e = \begin{bmatrix} h^e P^e l & \tfrac{1}{2} h^e P^e (x_j^2 - x_i^2) \\ \text{sym.} & (\tfrac{1}{3} h^e P^e (x_j^2 - x_i^2) + K^e A^e l) \end{bmatrix}$$

if the properties are assumed to be constant within the element. In this problem there is no element column matrix. The above expressions could easily be programmed to generate $\mathbf{S}^e$. However, physical intuition suggests that the conduction terms involving K should be inversely proportional to l and the convective part directly proportional to l. For this simple problem the above product can be evaluated in closed form and it reduces to

$$\mathbf{S}^e = \begin{bmatrix} (\tfrac{1}{3} P^e h^e l + A^e K^e / l) & (\tfrac{1}{6} P^e h^e l - A^e K^e / l) \\ \text{sym.} & (\tfrac{1}{3} P^e h^e l + A^e K^e / l) \end{bmatrix} \tag{10.5}$$

which is the form that is utilized in subroutine ELSQ which is shown in Fig. 10.3. This simple expression is all that is required to generate the equations necessary to solve for the nodal temperatures.

Generally, one is also interested in other quantities that can be evaluated once the nodal temperatures have been calculated. For example, one may need to perform a post-solution calculation of the heat lost through convection along the length of the bar. Myers shows that the convective heat loss, $\mathrm{d}q$, from a segment of length $\mathrm{d}x$ is $\mathrm{d}q = HPt\,\mathrm{d}x$. Thus, the loss from a typical element (with constant properties) is

$$\Delta Q^e = \int_{x_i}^{x_j} \mathrm{d}q = h^e P^e \int_{x_i}^{x_j} t^e \,\mathrm{d}x$$

or simply

$$\Delta Q^e = \mathbf{QC}^e\mathbf{T}^e \tag{10.6}$$

where

$$\mathbf{QC}^e = \tfrac{1}{2}h^e P^e l[1 \quad 1].$$

The simple matrix, $\mathbf{QC}^e$, can be generated at the same time as $\mathbf{S}^e$ and stored on auxiliary storage for use after the nodal temperatures, $\mathbf{T}^e$, are known. The matrix $\mathbf{QC}^e$ is generated and stored by subroutine ELPOST. It is retrieved and multiplied by $\mathbf{T}^e$ in subroutine POSTEL which calculates and prints ΔQ^e for each element. The latter two problem-dependent routines are shown in Figs. 10.4 and 10.5, respectively.

The above equations were derived by assuming constant properties within each element. In some practical applications, one or more of the properties might vary along the length of the bar. The above element could still be utilized to obtain a good approximation by assuming average properties over the element. This could be accomplished by defining the variable properties at each node and inputting them as nodal properties. For each element the program would then automatically extract the properties associated with the element's nodal points (array PRTLPT) and pass those data to subroutine ELSQ. One could then use the average of the two nodal properties, e.g. instead of A = ELPROP(1) set A = 0.5(PRTLPT(1,1) + PRTLPT(2,1)).

A more accurate approach would be to integrate Eqn. (10.2) numerically with due consideration for the variable properties. The variation of the properties could be approximated by assuming, for example, that $\mathbf{VALUES}(x) = \mathbf{H}^e(x)\,\mathbf{PRTLPT}$, where **VALUES** denotes the four properties of interest within the element. Recall that the latter operation is carried out by subroutine CALPRT, which was presented in Section 5.3.

As a numerical example consider a problem where $A = 0.01389\ \text{ft}^2$, $h = 2.0$ BTU/h ft^2 °F, $K = 120$ BTU/h ft °F, $L = 4$ ft, $P = 0.5$ ft and $t_0 =$ 10°F. Select a mesh with eight nodes (M = 8), seven elements (NE = 7), one temperature per node (NG = 1), two nodes per element (N = 2), and four floating point properties per element (NLPFLO = 4). This one-dimensional problem, shown in Fig. 10.6, involves homogeneous element properties and a null element column matrix (NSPACE = 1, LHOMO = 1, NULCOL = 1). Since data are to be stored by subroutine ELPOST on NTAPE1, assign unit eight for that purpose (NTAPE1 = 8). The above four properties could also have been stored as miscellaneous system properties. Since node 1 is at $x = 0$, it is necessary to define the first (and only) parameter at node 1 to have a value of 10.0. The input data for this problem listed in Fig. 10.7 should be compared with the formats given in the

Appendix. The output data from the MODEL code are shown in Fig. 10.8. The calculated and exact values are compared in Fig. 10.6. Table 10.1 lists the configuration options of the MODEL code used in the present problem.

Fig. 10.1 Heat transfer in a slender rod

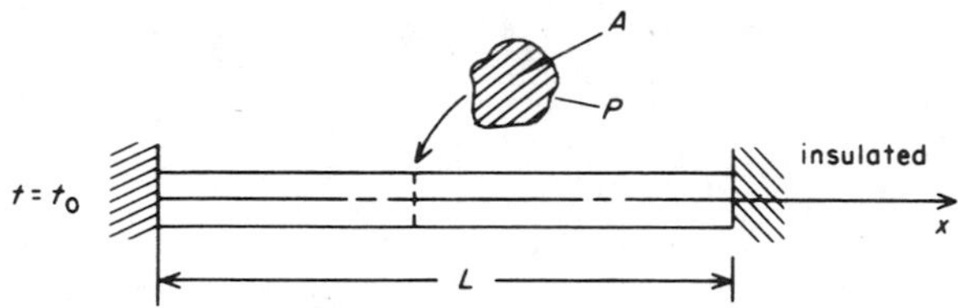

Fig. 10.2 A linear element model

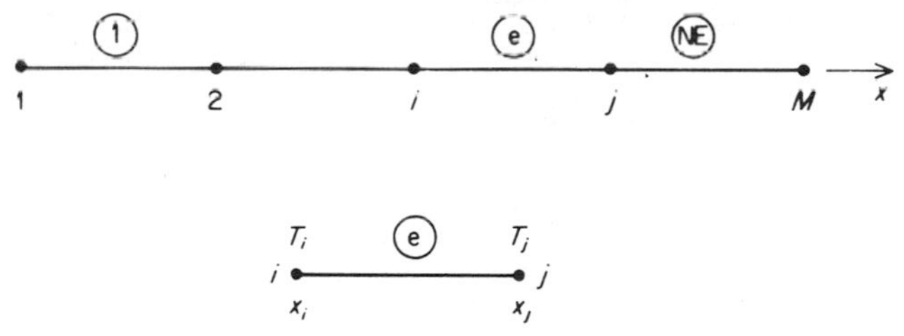

(a) A typical linear element

(b) Element interpolation functions

Fig. 10.3 Subroutine ELSQ

```
      SUBROUTINE  ELSQ  (N,NELFRE,NSPACE,IP5,JP2,KP1,KP2,
     1                KP3,COORD,D,ELPROP,LPROP,PRTLPT,FLTMIS,
     2                S,NTAPE1,NTAPE2,NTAPE3,NTAPE4)
C     * * * * * * * * * * * * * * * * * * * * * * * * * * *
C               GENERATE ELEMENT SQUARE MATRIX
C     * * * * * * * * * * * * * * * * * * * * * * * * * * *
CDP   IMPLICIT REAL*8(A-H,O-Z)
      DIMENSION COORD(N,NSPACE), D(NELFRE), ELPROP(KP2),
     1          LPROP(JP2), S(NELFRE,NELFRE),
     2          PRTLPT(IP5,KP1), FLTMIS(KP3)
C     N = NUMBER OF NODES PER ELEMENT
C     NELFRE = NUMBER OF DEGREES OF FREEDOM PER ELEMENT
C     NSPACE = DIMENSION OF SPACE
C     IP5 = NUMBER OF ROWS IN ARRAY PRTLPT
C     JP2 = NUMBER OF COLUMNS IN ARRAYS LPFIX AND LPROP
C     KP1 = NUMBER OF COLUMNS IN FLTNP & PRTLPT & PTPROP
C     KP2 = NUMBER OF COLUMNS IN ARRAYS FLTEL AND ELPROP
C     KP3 = NUMBER OF COLUMNS IN ARRAY FLTMIS
C     COORD = SPATIAL COORDINATES OF ELEMENT'S NODES
C     D = NODAL PARAMETERS ASSOCIATED WITH AN ELEMENT
C     ELPROP = ELEMENT ARRAY OF FLOATING PT PROPERTIES
C     LPROP = ARRAY OF FIXED POINT ELEMENT PROPERTIES
C     PRTLPT = FLOATING POINT PROP FOR ELEMENT'S NODES
C     FLTMIS = SYSTEM STORAGE OF FLOATING PT MISC PROP
C     S = ELEMENT SQUARE MATRIX
C     NTAPE1 = UNIT FOR POST SOLUTION MATRICES STORAGE
C     NTAPE2,3,4 = OPTIONAL UNITS FOR USER (USED WHEN > 0)

C     .........................................................
C      *** ELSQ PROBLEM DEPENDENT STATEMENTS FOLLOW ***
C     .........................................................
C     APPLICATION: ONE-DIM. STEADY STATE HEAT TRANSFER
C     REFER: MYERS,  ANAL. METH. IN COND. HEAT TRANSFER
C     NSPACE=1, N=2, NG=1, NELFRE=2
      COMMON /ELARG/  PH, XIJ, TOTAL, Q(1,2), QC(1)
C     ELPROP(1)= CROSS-SECTIONAL AREA, ELPROP(2)= PERIMETER
C     ELPROP(3) = THERMAL CONDUCTIVITY
C     ELPROP(4) = CONVECTIVE LOSS COEFFICIENT
      AK = ELPROP(1)*ELPROP(3)
      PH = ELPROP(2)*ELPROP(4)
      XIJ = COORD(2,1) - COORD(1,1)
      S(1,1) = AK/XIJ + PH*XIJ/3.
      S(1,2) = -AK/XIJ + PH*XIJ/6.
      S(2,1) = S(1,2)
      S(2,2) = S(1,1)
      RETURN
      END
```

Fig. 10.4 Subroutine ELPOST

```
      SUBROUTINE ELPOST (NTAPE1,NTAPE2,NTAPE3,NTAPE4)
C     * * * * * * * * * * * * * * * * * * * * * * * * *
C          GENERATE DATA FOR POST SOLUTION CALCULATIONS
C     * * * * * * * * * * * * * * * * * * * * * * * * *
CDP   IMPLICIT REAL*8(A-H,O-Z)
C     NTAPE1 = UNIT FOR POST SOLUTION MATRICES STORAGE
C     NTAPE2,3,4 = OPTIONAL UNITS FOR USER (USED WHEN > 0)

C     .........................................................
C      *** ELPOST PROBLEM DEPENDENT STATEMENTS FOLLOW ***
C     .........................................................
C-->  GENERATE CONVECTIVE HEAT LOSS DATA
C     APPLICATION: ONE-DIM. STEADY STATE HEAT TRANSFER
C     REFER: MYERS,  ANAL. METH. IN COND. HEAT TRANSFER
      COMMON /ELARG/  PH, XIJ, TOTAL, Q(1,2), QC(1)
      Q(1,1) = 0.5*PH*XIJ
      Q(1,2) = 0.5*PH*XIJ
      WRITE (NTAPE1)  Q
      RETURN
      END
```

Fig. 10.5 Subroutine POSTEL

```
      SUBROUTINE POSTEL (NTAPE1,NELFRE,D,IE,NTAPE2,
     1                    NTAPE3,NTAPE4,IT,NITER,NE,M)
C     * * * * * * * * * * * * * * * * * * * * * * * * * * * * *
C           ELEMENT LEVEL POST SOLUTION CALCULATIONS
C     * * * * * * * * * * * * * * * * * * * * * * * * * * * * *
C     NTAPE1 = UNIT FOR POST SOLUTION MATRICES STORAGE
C     NELFRE = NUMBER OF DEGREES OF FREEDOM PER ELEMENT
C     D = NODAL PARAMETERS ASSOCIATED WITH THE ELEMENT
C     IE = ELEMENT NUMBER
C     NTAPE2,3,4 = OPTIONAL UNITS FOR USER (USED WHEN > 0)
C     IT = CURRENT ITERATION NUMBER
C     NITER = MAXIMUM NUMBER OF ITERATIONS
C     NE = TOTAL NUMBER OF ELEMENTS
C     M = TOTAL NUMBER OF NODES
CDP   IMPLICIT REAL*8 (A-H,O-Z)
      DIMENSION  D(NELFRE)

C     ..................................................
C      *** POSTEL PROBLEM DEPENDENT STATEMENTS FOLLOW ***
C     ..................................................
C      APPLICATION: ONE-DIM. STEADY STATE HEAT TRANSFER
C      CALCULATE CONVECTIVE HEAT LOSS BY ELEMENT
C      REFER: MYERS,  ANAL. METH. IN COND. HEAT TRANSFER
       COMMON /ELARG/  PH, XIJ, TOTAL, Q(1,2), QC(1)
       DATA KALL /1/
       IF ( KALL )  10,20,10
C--> SET CONST, PRINT TITLES ON FIRST CALL
 10    WRITE (6,5000)
 5000  FORMAT ('0** CONVECTIVE HEAT LOSS**',//,
     1' ELEMENT            LOSS              TOTAL')
       KALL = 0
       TOTAL = 0.0
 20    CONTINUE
C-->   FIND ELEMENT LOSS
       READ (NTAPE1) Q
       CALL  MMULT (Q,D,QC,1,2,1)
       TOTAL = TOTAL + QC(1)
       WRITE (6,5010)  IE, QC, TOTAL
 5010  FORMAT ( I7, 2E16.8)
       RETURN
       END
```

Fig. 10.6 Results for the slender rod; **(a)** temperature distribution, and **(b)** convective heat loss

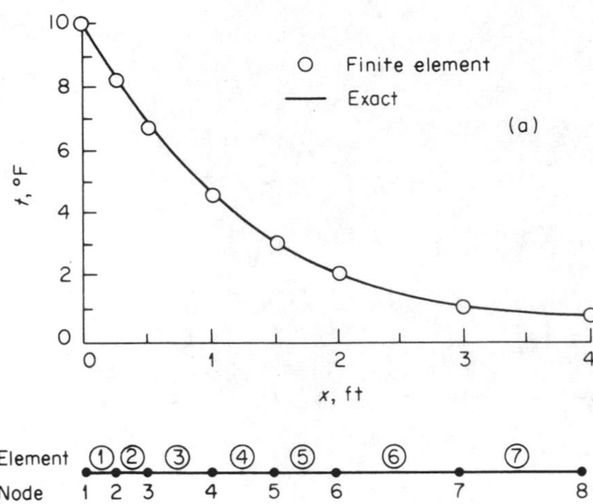

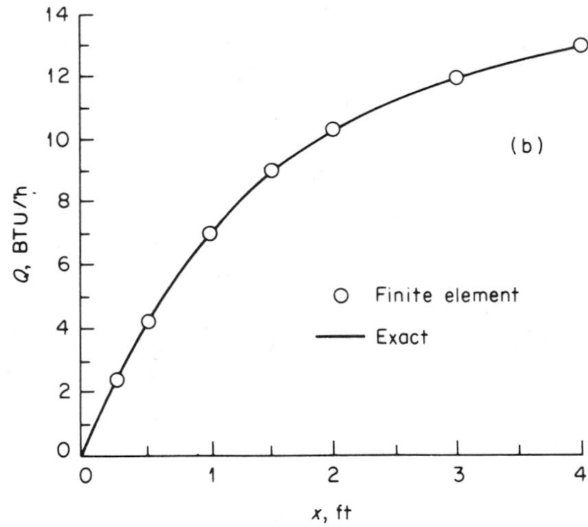

Fig. 10.7 Input data, one-dimensional heat transfer in a slender rod

```
HEAT TRANSFER IN A SLENDER ROD (,=4 SPACES)
,8,7,1,2,1,0,0,0,0,0,0
,0,0,0,4,0,0,0,1,0,0,8,0,0,0
           1                 1  0.0
           2                 0  0.25
           3                 0  0.5
           4                 0  1.0
           5                 0  1.5
           6                 0  2.0
           7                 0  3.0
           8                 0  4.0
     1     1      2
     2     2      3
     3     3      4
     4     4      5
     5     5      6
     6     6      7
     7     7      8
     1     1    10.0
,1 0.01389     0.5,    120.,   2.0
```

Fig. 10.8 Output data, one-dimensional heat transfer in a slender rod

```
*****  PROBLEM PARAMETERS   *****
NUMBER OF NODAL POINTS IN SYSTEM =..........   8
NUMBER OF ELEMENTS IN SYSTEM =..............   7
NUMBER OF NODES PER ELEMENT =...............   2
NUMBER OF PARAMETERS PER NODE =.............   1
DIMENSION OF SPACE =........................   1
NUMBER OF BOUNDARIES WITH GIVEN FLUX=.......   0
NUMBER OF NODES ON BOUNDARY SEGMENT=........   0
NUMBER OF FIXED PT PROP PER NODE =..........   0
NUMBER OF FLOATING PT PROP PER NODE =.......   0
NUMBER OF FIXED PT PROP PER ELEMENT =.......   0
NUMBER OF FLOATING PT PROP PER ELEMENT =....   4
NUMBER OF FIXED PT MISC PROP =..............   0
NUMBER OF FLOATING PT MISC PROP =...........   0

ELEMENT PROPERTIES ARE HOMOGENEOUS.
OPTIONAL UNIT NUMBERS (UTILIZED IF > 0)
NTAPE1=8 NTAPE2=0 NTAPE3=0 NTAPE4=0
NODAL PARAMETERS TO BE LISTED BY NODES
NODAL PARAMETERS TO BE LISTED BY ELEMENTS

*** NODAL POINT DATA ***
NODE, CONSTRAINT INDICATOR, 1 COORDINATES
        1             1       0.0000
        2             0       0.2500
        3             0       0.5000
        4             0       1.0000
        5             0       1.5000
        6             0       2.0000
        7             0       3.0000
        8             0       4.0000

*** ELEMENT CONNECTIVITY DATA ***
ELEMENT NO.,  2 NODAL INCIDENCES.
  1     1     2
  2     2     3
  3     3     4
  4     4     5
  5     5     6
  6     6     7
  7     7     8

***  NODAL PARAMETER CONSTRAINT LIST   ***
CONSTRAINT          NUMBER OF
   TYPE             EQUATIONS
    1                   1
    2                   0
    3                   0

*** CONSTRAINT EQUATION DATA ***
```

Fig. 10.8—continued

```
CONSTRAINT TYPE ONE
EQ. NO.    NODE1    PAR1              A1
      1        1       1    .10000000E+02

***  ELEMENT  PROPERTIES   ***
ELEMENT NO.  PROPERTY NO.        VALUE
      1             1       0.13890000E-01
      1             2       0.50000000E+00
      1             3       0.12000000E+03
      1             4       0.20000000E+01
END OF FLOATING PT PROPERTIES OF ELEMENTS

EQUATION HALF BANDWIDTH =.........       2
AND OCCURS IN ELEMENT NUMBER       1.
CONSTRAINT HALF BANDWIDTH =..............  1
MAXIMUM HALF BANDWIDTH OF SYSTEM =.......  2
TOTAL NUMBER OF SYSTEM EQUATIONS =.......  8
NUMBER OF ELEMENT DEGREES OF FREEDOM =...  2

***  OUTPUT OF RESULTS   ***
NODE,  1 COORDINATES,  1 PARAMETERS.
   1    0.00000000E+00    0.10000000E+02
   2    0.25000000E+00    0.82384766E+01
   3    0.50000000E+00    0.67878143E+01
   4    0.10000000E+01    0.46138443E+01
   5    0.15000000E+01    0.31496384E+01
   6    0.20000000E+01    0.21699524E+01
   7    0.30000000E+01    0.11321939E+01
   8    0.40000000E+01    0.84916429E+00

***  OUTPUT OF RESULTS   ***
ELEMENT NO., NODE NO.,   1 PARAMETERS
   1        1      0.10000000E+02
   1        2      0.82384766E+01
   2        2      0.82384766E+01
   2        3      0.67878143E+01
   3        3      0.67878143E+01
   3        4      0.46138443E+01
   4        4      0.46138443E+01
   4        5      0.31496384E+01
   5        5      0.31496384E+01
   5        6      0.21699524E+01
   6        6      0.21699524E+01
   6        7      0.11321939E+01
   7        7      0.11321939E+01
   7        8      0.84916429E+00

** CONVECTIVE HEAT LOSS**
ELEMENT          LOSS             TOTAL
      1   0.22798096E+01   0.22798096E+01
      2   0.18782864E+01   0.41580960E+01
      3   0.28504147E+01   0.70085106E+01
      4   0.19408707E+01   0.89493814E+01
      5   0.13298977E+01   0.10279279E+02
      6   0.16510732E+01   0.11930352E+02
      7   0.99067911E+00   0.12921031E+02
NORMAL ENDING OF MODEL PROGRAM.
```

Table 10.1 Parameter Definitions for One-dimensional Heat Transfer

CONTROL:
NSPACE = 1   NLPFLO = 4
N = 2   NULCOL = 1
NG = 1   NTAPE1 = 8

DEPENDENT VARIABLES:
1 = temperature, *t*

PROPERTIES:
Element level:
Floating point:
1 = cross-sectional area, *A*
2 = perimeter of cross-section, *P*
3 = thermal conductivity, *K*
4 = convection coefficient ($t_\infty \equiv 0$), *h*

CONSTRAINTS:
Type 1: specified nodal temperatures

POST-SOLUTION CALCULATIONS:
1 = element convective heat loss, *q*

10.3 Plane truss structures

A truss structure is an assemblage of bar members connected at their end points by smooth pins. Most texts on structural analysis present a detailed analysis of the simple truss element. The element equations presented here were taken from the first finite element text by Zienkiewicz [82] and will be checked against a simple example presented in the text by Meek [52]. The truss element is a one-dimensional element that can be utilized in a one-dimensional global space to solve problems involving collinear force systems. However, this is of little practical interest and one is usually more interested in two- and three-dimensional structures.

Consider a two-dimensional truss constructed from the one-dimensional bar elements. Let the bar go from point i to point j and thus have direction angles from the x- and y-axes of θ_x and θ_y, respectively. Let the bar, of length L, have a cross-sectional area, A, with a moment of intertia, I, and a depth of $2d$. The bar has a modulus of elasticity, E, and a coefficient of thermal expansion, α. The displacement at each node has x- and y-components of u and v, respectively so that the element has four degrees of freedom.

If one orders the four element degrees of freedom as $\delta^{e^T} = [u_1 v_1 u_2 v_2]$,

Zienkiewicz [82] shows that the element square matrix (or *stiffness matrix*) is

$$\mathbf{S}^e = \frac{EA}{L}\left[\begin{array}{cc:cc} cxx & cxy & -cxx & -cxy \\ cxy & cyy & -cxy & -cyy \\ \hdashline -cxx & -cxy & cxx & cxy \\ -cxy & -cyy & cxy & cyy \end{array}\right] \tag{10.7}$$

where $cxx = \cos^2\theta_x$, $cxy = \cos\theta_x\cos\theta_y$, and $cyy = \cos^2\theta_y$. The element column matrix (or *load vector*) can have several contributions. If the element is restrained and undergoes a temperature rise ΔT then

$$\mathbf{C}^e = EA\alpha\Delta T\begin{Bmatrix} -cx \\ -cy \\ cx \\ cy \end{Bmatrix}, \tag{10.8}$$

where $cx = \cos\theta_x$, and $cy = \cos\theta_y$. Similarly, if there is a uniformly applied line load, p, acting to the right as one moves from point i to point j, then the statically equivalent nodal loads are

$$\mathbf{C}^e = \frac{pL}{2}\begin{Bmatrix} cy \\ -cx \\ cy \\ -cx \end{Bmatrix}. \tag{10.9}$$

Loads externally applied at the node points would be considered as initial contributions to the system column matrix. If only external loads are in effect, then the element column matrix is null. In summary, the problem under consideration is two-dimensional (NSPACE = 2) and involves an element with two nodes (N = 2), two displacements per node (NG = 2) and seven (A, E, α, ΔT, p, I, d) possible element properties (NLPFLO = 7). A typical truss element is shown in Fig. 10.9.

After the displacements, $\boldsymbol{\delta}^e$, have been calculated, the post-solution stress calculations can be performed. The stresses at the mid-section on the right (σ_1) and left (σ_2) sides (going from i to j) are

$$\begin{Bmatrix} \sigma_1 \\ \sigma_2 \end{Bmatrix} = \frac{E}{L}\begin{bmatrix} -cx & -cy & -cx & cy \\ -cx & -cy & cx & cy \end{bmatrix}\boldsymbol{\delta}^e + \frac{pL^2d}{8I}\begin{Bmatrix} 1 \\ -1 \end{Bmatrix} - E\alpha\Delta T\begin{Bmatrix} 1 \\ 1 \end{Bmatrix}, \tag{10.10}$$

where a negative stress denotes compression. A word of caution should be mentioned in the presentation of this element. A derivation of the truss element considers only the axial effects of applied loads. The effects of the transverse line load, p, are handled in a special way. First, the force effects are simply lumped at the nodes by inspection. Secondly, the calculated bending stresses are obtained from elementary beam theory and not from finite element theory. These bending effects are available in finite element models which utilize six degree of freedom frame elements (see Section 10.6).

As a specific example, consider the problem in Fig. 10.10 which was solved by Meek [52]. It has six nodes (M = 6), eleven elements (NE = 11) with homogeneous element properties (LHOMO = 1), and no element loads due to pressures or temperature changes (NULCOL = 1). At node 1 both displacements are zero and at node 6 the y-displacement is zero. There is a negative externally applied y-load of 10 kips at node 3 (thus INSOLV = 1). The properties are assumed to be $A = 1$ in^2, $E = 30{,}000$ ksi, $\alpha = \Delta T = p = 0.0$, $I = \frac{1}{12}$ in^4 and $d = 0.5$ in. The problem dependent subroutines are shown in Figs. 10.11 to 10.14. The input data are shown in Fig. 10.15 and the MODEL output in Fig. 10.16. The calculated displacements and stresses are much more accurate than the solution given by Meek.

If one wanted to check joint force equilibrium and determine the support reactions, one would have to multiply the original system stiffness matrix by the calculated nodal displacements. To accomplish this, it would be necessary to store the original system square matrix on auxiliary storage and recall it after the displacements are known. A subroutine, BANMLT, for executing the multiplication to find the resultant joint forces is shown in Fig. 10.17. One could easily obtain the member forces in subroutine POSTEL by multiplying the area by the stress resulting from the thermal effects and deformation of the member. Referring to Table 10.2, one finds the list of options used for this element.

As a second simple example, consider a single element with the same properties except that $p = 0.005$ kip/in. As shown in Fig. 10.18, the bar is pinned at the left end (i.e. $u_1 = v_1 = 0$) and has an inclined roller at the right end. The latter condition represents a Type 2 constraint of the form $3u_2 - 4v_2 = 0$, which requires the right end to move tangentially to the inclined plane of support. The data and output are shown in Fig. 10.19 and 10.20. The calculated stresses agree with the results from elementary beam theory and the displacements at node 2 satisfy the specified constraint. In this case, element loads are present (NULCOL = 0) and no external nodal loads are specified (ISOLV = 0).

The Type 2 constraints could have been handled differently. Since both parameters occurred at a common node, the constraint could have been applied at the element level by subroutine MOD2FL. However, programming changes would have been required; for example, one could have assigned an integer property to each node that would indicate when a constraint, that can be applied at the element level, occurs at that node.

Fig. 10.9 A typical truss element

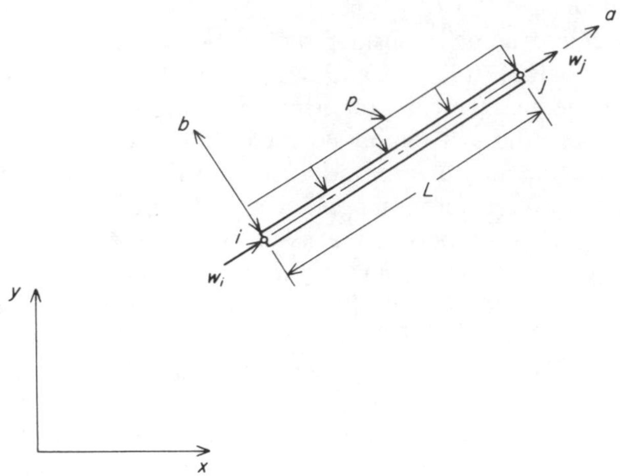

Fig. 10.10 Sample truss problem; **(a)** mesh and **(b)** calculated member forces

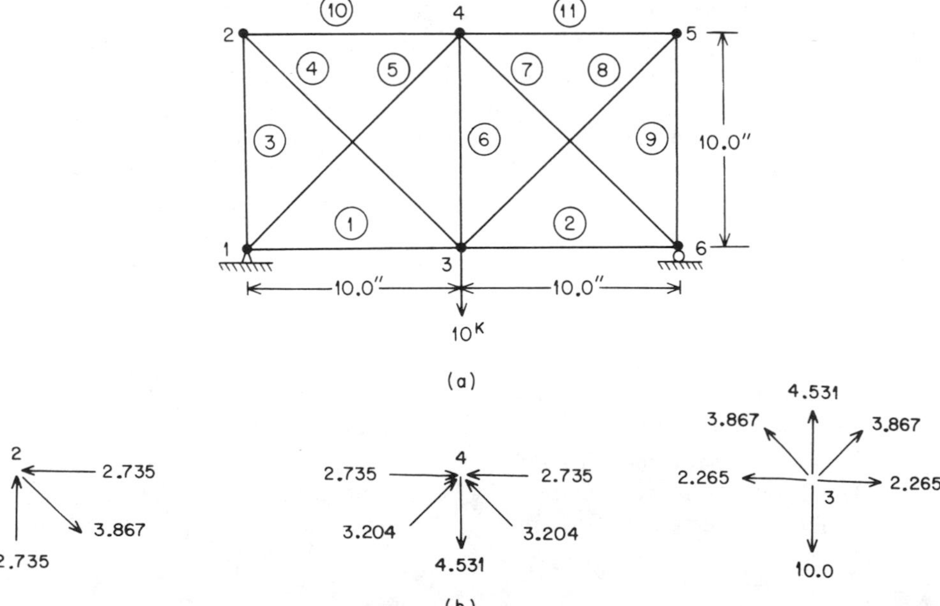

Fig. 10.11 Subroutine ELSQ

```
      SUBROUTINE  ELSQ  (N,NELFRE,NSPACE,IP5,JP2,KP1,KP2,
     1                 KP3,COORD,D,ELPROP,LPROP,PRTLPT,FLTMIS,
     2                 S,NTAPE1,NTAPE2,NTAPE3,NTAPE4)
C     * * * * * * * * * * * * * * * * * * * * * * * * * * *
C                  GENERATE ELEMENT SQUARE MATRIX
C     * * * * * * * * * * * * * * * * * * * * * * * * * * *
CDP   IMPLICIT REAL*8(A-H,O-Z)
      DIMENSION COORD(N,NSPACE), D(NELFRE), ELPROP(KP2),
     1          LPROP(JP2), S(NELFRE,NELFRE),
     2          PRTLPT(IP5,KP1), FLTMIS(KP3)
C     N = NUMBER OF NODES PER ELEMENT
C     NELFRE = NUMBER OF DEGREES OF FREEDOM PER ELEMENT
C     NSPACE = DIMENSION OF SPACE
C     IP5 = NUMBER OF ROWS IN ARRAY PRTLPT
C     JP2 = NUMBER OF COLUMNS IN ARRAYS LPFIX AND LPROP
C     KP1 = NUMBER OF COLUMNS IN FLTNP & PRTLPT & PTPROP
C     KP2 = NUMBER OF COLUMNS IN ARRAYS FLTEL AND ELPROP
C     KP3 = NUMBER OF COLUMNS IN ARRAY FLTMIS
C     COORD = SPATIAL COORDINATES OF ELEMENT'S NODES
C     D = NODAL PARAMETERS ASSOCIATED WITH AN ELEMENT
C     ELPROP = ELEMENT ARRAY OF FLOATING PT PROPERTIES
C     LPROP = ARRAY OF FIXED POINT ELEMENT PROPERTIES
C     PRTLPT = FLOATING POINT PROP FOR ELEMENT'S NODES
C     FLTMIS = SYSTEM STORAGE OF FLOATING PT MISC PROP
C     S = ELEMENT SQUARE MATRIX
C     NTAPE1 = UNIT FOR POST SOLUTION MATRICES STORAGE
C     NTAPE2,3,4 = OPTIONAL UNITS FOR USER (USED WHEN > 0)
C     .....................................................
C      *** ELSQ PROBLEM DEPENDENT STATEMENTS FOLLOW ***
C     .....................................................
C     APPLICATION:  ANALYSIS OF A TWO-DIMENSIONAL TRUSS
C     ZIENKIEWICZ, F.E.M. IN STRUCTURAL & CONTINUUM MECH.
C       NG = 2, N = 2, NSPACE = 2, NLPFLC = 7
C     ELPROP(1) = AREA, ELPROP(2) = MODULUS OF ELASTICITY
C     ELPROP(3) = TEMP RISE, ELPROP(4) = COEF THERMAL EXP
C     ELPROP(5) = LINE LOAD, ELPROP(6) = MOMENT OF INERTIA
C     ELPROP(7) = HALF DEPTH OF BAR
      COMMON /ELARG/  CX,CY,A,E,T,AL,P,BI,BD,BARL
CDP   SQRT(Z) = DSQRT(Z)
      XI = COORD(1,1)
      XJ = COORD(2,1)
      YI = COORD(1,2)
      YJ = COORD(2,2)
      A = ELPROP(1)
      E = ELPROP(2)
      T = ELPROP(3)
      AL = ELPROP(4)
      P = ELPROP(5)
      BI = ELPROP(6)
      IF ( BI.LE.0.0 )  BI = 1.0
      BD = ELPROP(7)
      IF ( BD.LE.0.0 )  BD = 1.0
C-->  FIND BAR LENGTH AND DIRECTION COSINES
      DX = XJ-XI
      DY = YJ-YI
      BARL = SQRT( DX*DX + DY*DY )
      CX = DX/BARL
      CY = DY/BARL
C-->  FIND 1-D STIFFNESS, K=E*A/L
      STIF = E*A/BARL
C-->  TRANSFORM TO 2-D STIFFNESS
      CXX = CX*CX
      CXY = CX*CY
      CYY = CY*CY
      S(1,1) =  STIF*CXX
      S(1,2) =  STIF*CXY
      S(1,3) = -STIF*CXX
      S(1,4) = -STIF*CXY
      S(2,1) =  STIF*CXY
      S(2,2) =  STIF*CYY
      S(2,3) = -STIF*CXY
      S(2,4) = -STIF*CYY
      S(3,1) = -STIF*CXX
      S(3,2) = -STIF*CXY
      S(3,3) =  STIF*CXX
      S(3,4) =  STIF*CXY
      S(4,1) = -STIF*CXY
      S(4,2) =  STIF*CYY
      S(4,3) =  STIF*CXY
      S(4,4) =  STIF*CYY
      RETURN
      END
```

Fig. 10.12 Subroutine ELCOL

```
      SUBROUTINE  ELCOL (N,NELFRE,NSPACE,IP5,JP2,KP1,KP2,
     1              KP3,COORD,D,ELPROP,LPROP,PRTLPT,FLTMIS,
     2              C,NTAPE1,NTAPE2,NTAPE3,NTAPE4)
C     * * * * * * * * * * * * * * * * * * * * * * * * * * *
C             GENERATE ELEMENT COLUMN MATRIX
C     * * * * * * * * * * * * * * * * * * * * * * * * * * *
CDP   IMPLICIT REAL*8(A-H,O-Z)
      DIMENSION COORD(N,NSPACE), D(NELFRE), ELPROP(KP2),
     1            LPROP(JP2), C(NELFRE), PRTLPT(IP5,KP1),
     2            FLTMIS(KP3)
C     N = NUMBER OF NODES PER ELEMENT
C     NELFRE = NUMBER OF DEGREES OF FREEDOM PER ELEMENT
C     NSPACE = DIMENSION OF SPACE
C     IP5 = NUMBER OF ROWS IN ARRAY PRTLPT
C     JP2 = NUMBER OF COLUMNS IN ARRAYS LPFIX AND LPROP
C     KP1 = NUMBER OF COLUMNS IN  FLTNP & PRTLPT & PTPROP
C     KP2 = NUMBER OF COLUMNS IN ARRAYS FLTEL AND ELPROP
C     KP3 = NUMBER OF COLUMNS IN ARRAY FLTMIS
C     COORD = SPATIAL COORDINATES OF ELEMENT'S NODES
C     D = NODAL PARAMETERS ASSOCIATED WITH A ELEMENT
C     ELPROP = ELEMENT ARRAY OF FLOATING POINT PROPERTIES
C     LPROP = ARRAY OF FIXED POINT ELEMENT PROPERTIES
C     PRTLPT = FLOATING POINT PROP OF ELEMENT'S NODES
C     FLTMIS = SYSTEM STORAGE OF FLOATING POINT MISC. PROP
C     C = ELEMENT COLUMN MATRIX
C     NTAPE1 = UNIT FOR POST SOLUTION MATRICES STORAGE
C     NTAPE2,3,4 = OPTIONAL UNITS FOR USER  (USED WHEN > 0)

C     ..........................................................
C      ***  ELCOL PROBLEM DEPENDENT STATEMENTS FOLLOW ***
C     ..........................................................
C     APPLICATION:  ANALYSIS OF A TWO-DIMENSIONAL TRUSS
C     ZIENKIEWICZ, F.E.M. IN STRUCTURAL & CONTINUUM MECH.
C       NG = 2, N = 2, NSPACE = 2, NLPFLO = 7
C     ELPROP(1) = AREA, ELPROP(2) = MODULUS OF ELASTICITY
C     ELPROP(3) = TEMP RISE, ELPROP(4) = COEF THERMAL EXP
C     ELPROP(5) = LINE LOAD, ELPROP(6) = MOMENT OF INERTIA
C     ELPROP(7) = HALF DEPTH OF BAR
      COMMON  /ELARG/  CX,CY,A,E,T,AL,P,BI,BD,BARL
      DO 10  I = 1,4
   10 C(I) = 0.0
C     CHECK FOR P=0 AND DT=0  (NO ELEMENT LOADS)
      IF ( P.EQ.0.0 .AND. T.EQ.0.0 )  RETURN
C-->  INITIAL STRAIN EFFECTS   F = E*A*ALPHA*DT
      IF ( T.EQ.0.0 )  GO TO 20
      F = E*A*AL*T
      C(1) = -CX*F
      C(2) = -CY*F
      C(3) = CX*F
      C(4) = CY*F
   20 CONTINUE
C-->  LINE LOAD EFFECTS  F = 0.5*P*L (POS TO R , I TO J )
      IF ( P.EQ.0.0 )  RETURN
      F = 0.5*P*BARL
      C(1) = C(1)  + CY*F
      C(2) = C(2)  - CX*F
      C(3) = C(3)  + CY*F
      C(4) = C(4)  - CX*F
      RETURN
      END
```

Fig. 10.13 Subroutine ELPOST

```
      SUBROUTINE ELPOST (NTAPE1,NTAPE2,NTAPE3,NTAPE4)
C     * * * * * * * * * * * * * * * * * * * * * * * * * * *
C         GENERATE DATA FOR POST SOLUTION CALCULATIONS
C     * * * * * * * * * * * * * * * * * * * * * * * * * * *
CDP   IMPLICIT REAL*8(A-H,O-Z)
C     NTAPE1 = UNIT FOR POST SOLUTION MATRICES STORAGE
C     NTAPE2,3,4 = OPTIONAL UNITS FOR USER (USED WHEN > 0)

C     ..........................................................
C     *** ELPOST PROBLEM DEPENDENT STATEMENTS FOLLOW ***
C     ..........................................................
C     APPLICATION:  ANALYSIS OF A TWO-DIMENSIONAL TRUSS
C     ZIENKIEWICZ, F.E.M. IN STRUCTURAL & CONTINUUM MECH.
C       NG = 2, N = 2, NSPACE = 2, NLPFLO = 7
```

Fig. 10.13—continued

```
C       ELPROP(1) = AREA, ELPROP(2) = MODULUS OF ELASTICITY
C       ELPROP(3) = TEMP RISE, ELPROP(4) = COEF THERMAL EXP
C       ELPROP(5) = LINE LOAD, ELPROP(6) = MOMENT OF INERTIA
C       ELPROP(7) = HALF DEPTH OF BAR
        COMMON  /ELARG/   CX,CY,A,E,T,AL,P,BI,BD,BARL
        COMMON /ELARG2/   SIGD(2,4),SIGP(2),SIGT(2),SIG(2)
C-->     FORM STRESS-DISPLACEMENT MATRIX
        EDL = E/BARL
        SIGD(1,1) = -EDL*CX
        SIGD(1,2) = -EDL*CY
        SIGD(1,3) =  EDL*CX
        SIGD(1,4) =  EDL*CY
        SIGD(2,1) = -EDL*CX
        SIGD(2,2) = -EDL*CY
        SIGD(2,3) =  EDL*CX
        SIGD(2,4) =  EDL*CY
C-->     BENDING EFFECTS    = P*L*L*D/8*I
        SIGP(1) = 0.125*P*BARL*BARL*BD/BI
        SIGP(2) = -SIGP(1)
C-->     FORM THERMAL EFFECT = -E*ALPHA*DT
        SIGT(1) = -E*AL*T
        SIGT(2) = SIGT(1)
C-->     STORE DATA ON NTAPE1 FOR USE BY POSTEL
        WRITE (NTAPE1)  SIGD,SIGP,SIGT
        RETURN
        END
```

Fig. 10.14 Subroutine POSTEL

```
        SUBROUTINE POSTEL (NTAPE1,NELFRE,D,IE,NTAPE2,
       1                   NTAPE3,NTAPE4,IT,NITER,NE,M)
C       * * * * * * * * * * * * * * * * * * * * * * * * * *
C            ELEMENT LEVEL POST SOLUTION CALCULATIONS
C       * * * * * * * * * * * * * * * * * * * * * * * * * *
C       NTAPE1 = UNIT FOR POST SOLUTION MATRICES STORAGE
C       NELFRE = NUMBER OF DEGREES OF FREEDOM PER ELEMENT
C       D = NODAL PARAMETERS ASSOCIATED WITH THE ELEMENT
C       IE = ELEMENT NUMBER
C       NTAPE2,3,4 = OPTIONAL UNITS FOR USER (USED WHEN > 0)
C       IT = CURRENT ITERATION NUMBER
C       NITER = MAXIMUM NUMBER OF ITERATIONS
C       NE = TOTAL NUMBER OF ELEMENTS
C       M = TOTAL NUMBER OF NODES
CDP     IMPLICIT REAL*8 (A-H,O-Z)
        DIMENSION  D(NELFRE)

C       ..............................................
C        *** POSTEL PROBLEM DEPENDENT STATEMENTS FOLLOW ***
C       ..............................................
C       APPLICATION:  ANALYSIS OF A TWO-DIMENSIONAL TRUSS
C       ZIENKIEWICZ, F.E.M. IN STRUCTURAL & CONTINUUM MECH.
C         NG = 2, N = 2, NSPACE = 2, NLPFLO = 7
C       ELPROP(1) = AREA, ELPROP(2) = MODULUS OF ELASTICITY
C       ELPROP(3) = TEMP RISE, ELPROP(4) = COEF THERMAL EXP
C       ELPROP(5) = LINE LOAD, ELPROP(6) = MOMENT OF INERTIA
C       ELPROP(7) = HALF DEPTH OF BAR
        COMMON  /ELARG/   CX,CY,A,E,T,AL,P,BI,BD,BARL
        COMMON /ELARG2/   SIGD(2,4),SIGP(2),SIGT(2),STRESS(2)
        DATA KODE/1/
        IF ( KODE.EQ.0 )  GO TO 10
        KODE = 0
C-->     WRITE STRESS HEADINGS
        WRITE (6,1000)
 1000   FORMAT (1H1,8X,'E L E M E N T     S T R E S S E S',//,
       A' ELEMENT         MID SECTION STRESS AT:',/,
       B' NUMBER        RIGHT                    LEFT',/)
   10   CONTINUE
C-->     READ STRESS DATA OFF NTAPE1 FROM ELPOST
        READ (NTAPE1)  SIGD,SIGP,SIGT
C--->    CALCULATE STRESS, STRESS=SIGD*D
        CALL  MMULT  (SIGD,D,STRESS,2,4,1)
C       ADD THERMAL AND PRESSURE EFFECTS
        STRESS(1) = STRESS(1) + SIGT(1) + SIGP(1)
        STRESS(2) = STRESS(2) + SIGT(2) + SIGP(2)
        WRITE (6,1010)  IE,STRESS
 1010   FORMAT (I5,7X,2E20.7)
        RETURN
        END
```

Fig. 10.15 Input data, truss analysis problem

```
  TRUSS ANALYSIS PROBLEM  (,=4 SPACES)
,6    11,2,2,2,0,0,0,0,1,0
,0,0,0,7,0,0,0,1,0,C,8,0,0,0
          1          11  0.0         0.0
          2          00  0.0         10.0
          3          0C  10.0        0.0
          4          00  10.0        10.0
          5          C0  20.0        10.0
          6          01  20.0        0.0
    1     1     3
    2     3     6
    3     1     2
    4     2     3
    5     1     4
    6     3     4
    7     4     6
    8     3     5
    9     5     6
   10     2     4
   11     4     5
    1     1   0.C
    1     2   0.C
    6     2   0.C
,1  1.0,   30000., 0.,,0.,,0.0,    0.083333    0.5
    3     2   -1C.
    6     2   0.0
```

Fig. 10.16 Output data, truss analysis problem

```
*****  PROBLEM PARAMETERS  *****
NUMBER OF NODAL POINTS IN SYSTEM =..........  6
NUMBER OF ELEMENTS IN SYSTEM =..............  11
NUMBER OF NODES PER ELEMENT =...............  2
NUMBER OF PARAMETERS PER NODE =.............  2
DIMENSION OF SPACE =........................  2
NUMBER OF BOUNDARIES WITH GIVEN FLUX=.......  0
NUMBER OF NODES ON BOUNDARY SEGMENT=........  0
INITIAL FORCING VECTOR TO BE INPUT
NUMBER OF FIXED PT PROP PER NODE =..........  0
NUMBER OF FLOATING PT PROP PER NODE =.......  0
NUMBER OF FIXED PT PROP PER ELEMENT =.......  0
NUMBER OF FLOATING PT PROP PER ELEMENT =....  7
NUMBER OF FIXED PT MISC PROP =..............  0
NUMBER OF FLOATING PT MISC PROP =...........  0

ELEMENT PROPERTIES ARE HOMOGENEOUS.
OPTIONAL UNIT NUMBERS (UTILIZED IF > 0)
NTAPE1=8 NTAPE2=0 NTAPE3=0 NTAPE4=0
NODAL PARAMETERS TO BE LISTED BY NODES
NODAL PARAMETERS TO BE LISTED BY ELEMENTS

*** NODAL POINT DATA ***
NODE, CONSTRAINT INDICATOR, 2 COORDINATES
           1          11      0.0000      0.0000
           2           0      0.0000     10.0000
           3           0     10.0000      0.0000
           4           0     10.0000     10.0000
           5           0     20.0000     10.0000
           6           1     20.0000      0.0000

*** ELEMENT CONNECTIVITY DATA ***
ELEMENT NO.,  2 NODAL INCIDENCES.
    1      1      3
    2      3      6
    3      1      2
    4      2      3
    5      1      4
    6      3      4
    7      4      6
    8      3      5
    9      5      6
   10      2      4
   11      4      5

***  NODAL PARAMETER CONSTRAINT LIST  ***
CONSTRAINT          NUMBER OF
   TYPE             EQUATIONS
     1                  3
     2                  0
     3                  0
```

Fig. 10.16—continued

```
*** CONSTRAINT EQUATION DATA ***
CONSTRAINT TYPE ONE
EQ. NO.    NODE1    PAR1          A1
      1        1       1    .00000000E+00
      2        1       2    .00000000E+00
      3        6       2    .00000000E+00

***  ELEMENT  PROPERTIES   ***
ELEMENT NO.  PROPERTY NO.        VALUE

      1            1          0.10000000E+01
      1            2          0.30000000E+05
      1            3          0.00000000E+00
      1            4          0.00000000E+00
      1            5          0.00000000E+00
      1            6          0.83333000E-01
      1            7          0.50000000E+00

END OF FLOATING PT PROPERTIES OF ELEMENTS
EQUATION HALF BANDWIDTH =........      8
AND OCCURS IN ELEMENT NUMBER     2.
CONSTRAINT HALF BANDWIDTH =.............   1
MAXIMUM HALF BANDWIDTH OF SYSTEM =.......  8
TOTAL NUMBER OF SYSTEM EQUATIONS =....... 12
NUMBER OF ELEMENT DEGREES OF FREEDOM =...  4

*** INITIAL FORCING VECTOR DATA ***
  NODE   PARAMETER         VALUE         DOF
     3       2     -0.10000000E+02         6
     6       2      0.00000000E+00        12

***  OUTPUT OF RESULTS  ***
NODE,   2 COORDINATES,   2 PARAMETERS.
 1  0.00E+00  0.00E+00  0.0000E+00 -0.1231E-14
 2  0.00E+00  0.10E+02  0.1666E-02 -0.9115E-03
 3  0.10E+02  0.00E+00  0.7551E-03 -0.4401E-02
 4  0.10E+02  0.10E+02  0.7551E-03 -0.2890E-02
 5  0.20E+02  0.10E+02 -0.1563E-03 -0.9115E-03
 6  0.20E+02  0.00E+00  0.1510E-02 -0.1231E-14

***  OUTPUT OF RESULTS   ***
ELEMENT NO., NODE NO.,   2 PARAMETERS
    1         1      0.00000000E+00   -0.12313269E-14
    1         3      0.75513640E-03   -0.44012575E-02
    2         3      0.75513640E-03   -0.44012575E-02
    2         6      0.15102728E-02   -0.12313269E-14
    3         1      0.00000000E+00   -0.12313269E-14
    3         2      0.16666667E-02   -0.91153028E-03
    4         2      0.16666667E-02   -0.91153028E-03
    4         3      0.75513640E-03   -0.44012575E-02
    5         1      0.00000000E+00   -0.12313269E-14
    5         4      0.75513642E-03   -0.28909847E-02
    6         3      0.75513640E-03   -0.44012575E-02
    6         4      0.75513642E-03   -0.28909847E-02
    7         4      0.75513642E-03   -0.28909847E-02
    7         6      0.15102728E-02   -0.12313269E-14
    8         3      0.75513640E-03   -0.44012575E-02
    8         5     -0.15639386E-03   -0.91153029E-03
    9         5     -0.15639386E-03   -0.91153029E-03
    9         6      0.15102728E-02   -0.12313269E-14
   10         2      0.16666667E-02   -0.91153028E-03
   10         4      0.75513642E-03   -0.28909847E-02
   11         4      0.75513642E-03   -0.28909847E-02
   11         5     -0.15639386E-03   -0.91153029E-03

      E L E M E N T    S T R E S S E S
ELEMENT         MID SECTION STRESS AT:
NUMBER       RIGHT                      LEFT
   1       0.2265409E+01            0.2265409E+01
   2       0.2265409E+01            0.2265409E+01
   3      -0.2734591E+01           -0.2734591E+01
   4       0.3867295E+01            0.3867295E+01
   5      -0.3203772E+01           -0.3203772E+01
   6       0.4530818E+01            0.4530818E+01
   7      -0.3203772E+01           -0.3203772E+01
   8       0.3867295E+01            0.3867295E+01
   9      -0.2734591E+01           -0.2734591E+01
  10      -0.2734591E+01           -0.2734591E+01
  11      -0.2734591E+01           -0.2734591E+01
NORMAL ENDING OF MODEL PROGRAM.
```

Fig. 10.17 Subroutine BANMLT

```
      SUBROUTINE  BANMLT (NDFREE,MAXBAN,SS,DD,CC,IOPT)
C     * * * * * * * * * * * * * * * * * * * * * * * * * *
C     MULTIPLY PACKED SQUARE MATRIX, SS, BY MATRIX DD
C        IF IOPT = 0 STORE RESULT IN MATRIX CC
C        OTHERWISE ADD RESULT TO MATRIX CC
C     * * * * * * * * * * * * * * * * * * * * * * * * * *
CDP   IMPLICIT REAL*8 (A-H,O-Z)
      DIMENSION  SS(NDFREE,MAXBAN), DD(NDFREE), CC(NDFREE)
C     NDFREE = TOTAL NUMBER OF SYSTEM DEGREES OF FREEDOM
C     MAXBAN = SYSTEM HALF BANDWIDTH
      MBM1 = MAXBAN - 1
      DO 70  I = 1,NDFREE
      SUM = 0.D0
      J1 = I - MBM1
      J2 = I + MBM1
      J1 = MAX0 (J1,1)
      J2 = MIN0(J2,NDFREE)
      DO 50  J = J1,J2
      IF ( J - I )  10,20,30
   10 JJ = I-J+1
      II = J
      GO TO 40
   20 JJ = 1
      II = I
      GO TO 40
   30 JJ = J-I+1
      II = I
   40 SUM = SUM + SS(II,JJ)*DD(J)
   50 CONTINUE
      IF ( IOPT.EQ.0 )  GO TO 60
      CC(I) = CC(I) + SUM
      GO TO 70
   60 CC(I) = SUM
   70 CONTINUE
      RETURN
      END
```

Fig. 10.18 Bar element on an inclined plane

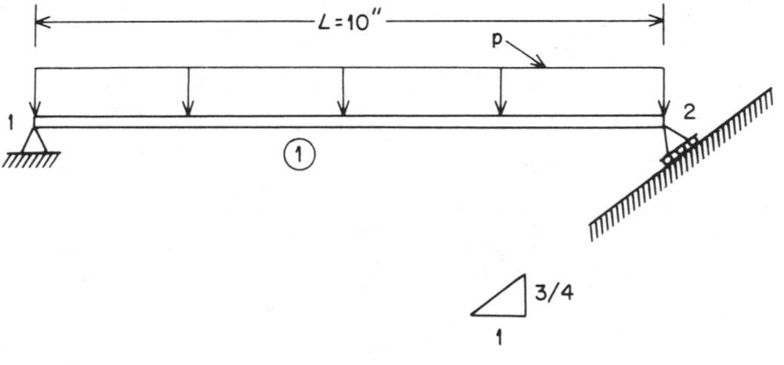

(a) Original state

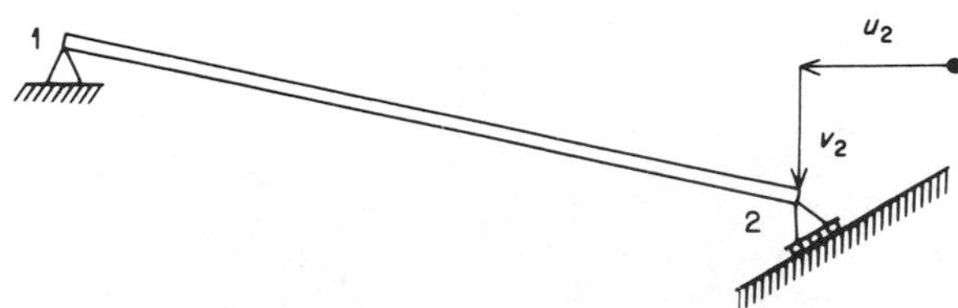

(b) Final position

Fig. 10.19 Bar with uniform pressure and included roller; test problem data

```
 BAR WITH PRESSURE & INCLINED ROLLER (,=4 SPACES)
,2,1,2,2,2,C,0,0,0,0,0
,0,0,0,7,0,0,0,1,0,0,8,0,0,0
           1          11   0.0         0.0
           2          22  10.0         0.0
   1   1    2
   1   1    C.0
   1   2    C.0
   2   1    2     2     3.        -4.        0.
,1  1.0,   3000C., 0.,,0.,,0.01, 0.08333,0.5
```

Fig. 10.20 Bar with uniform pressure and inclined roller support; output data

```
*****  PROBLEM PARAMETERS   *****
NUMBER OF NODAL POINTS IN SYSTEM =.......... 2
NUMBER OF ELEMENTS IN SYSTEM =.............. 1
NUMBER OF NODES PER ELEMENT =............... 2
NUMBER OF PARAMETERS PER NODE =............. 2
DIMENSION OF SPACE =........................ 2
NUMBER OF BOUNDARIES WITH GIVEN FLUX=....... 0
NUMBER OF NODES ON BOUNDARY SEGMENT=........ 0
NUMBER OF FIXED PT PROP PER NODE =.......... 0
NUMBER OF FLOATING PT PROP PER NODE =....... 0
NUMBER OF FIXED PT PROP PER ELEMENT =....... 0
NUMBER OF FLOATING PT PROP PER ELEMENT =.... 7
NUMBER OF FIXED PT MISC PROP =.............. 0
NUMBER OF FLOATING PT MISC PROP =........... 0

ELEMENT PROPERTIES ARE HOMOGENEOUS.
OPTIONAL UNIT NUMBERS (UTILIZED IF > 0)
NTAPE1= 8 NTAPE2= 0 NTAPE3= 0 NTAPE4= 0
NODAL PARAMETERS TO BE LISTED BY ELEMENTS

*** NODAL POINT DATA ***
NODE, CONSTRAINT INDICATOR, 2 COORDINATES
       1           11     0.0000      0.0000
       2           22    10.0000      0.0000

*** ELEMENT CONNECTIVITY DATA ***
ELEMENT NO.,  2 NODAL INCIDENCES.
   1    1    2

***  NODAL PARAMETER CONSTRAINT LIST   ***
CONSTRAINT          NUMBER OF
   TYPE             EQUATIONS
     1                  2
     2                  1
     3                  0

*** CONSTRAINT EQUATION DATA ***
CONSTRAINT TYPE ONE
EQ. NO.    NODE1     PAR1            A1
   1         1          1    .00000000E+00
   2         1          2    .00000000E+00
CONSTRAINT TYPE TWO
EQ,NOD1,PAR1,NOD2,PAR2,    A1,    A2,    A3
 3  2   1   2   2   .300E+01   -.400E+01   .000E+00

***   ELEMENT   PROPERTIES        ***
ELEMENT NO.   PROPERTY NO.          VALUE
     1              1          0.10000000E+01
     1              2          0.30000000E+05
     1              3          0.00000000E+00
     1              4          0.00000000E+00
     1              5          0.10000000E-01
     1              6          0.83333000E-01
     1              7          0.50000000E+00
END OF FLOATING PT PROPERTIES OF ELEMENTS
```

Fig. 10.20—continued

```
EQUATION HALF BANDWIDTH =........      4
AND OCCURS IN ELEMENT NUMBER      1.
CONSTRAINT HALF BANDWIDTH =............      2
MAXIMUM HALF BANDWIDTH OF SYSTEM =.......      4
TOTAL NUMBER OF SYSTEM EQUATIONS =.......      4
NUMBER OF ELEMENT DEGREES OF FREEDOM =...      4

***  OUTPUT OF RESULTS  ***
ELEMENT NO., NODE NO.,  2 PARAMETERS
    1       1     0.00000E+00  0.00000E+00
    1       2    -0.62500E-05 -0.46875E-05

     E L E M E N T   S T R E S S E S
 ELEMENT  MID SECTION STRESSES AT:
 NUMBER  RIGHT                 LEFT
    1     0.731253E+00    -0.768753E+00
NORMAL ENDING OF MODEL PROGRAM.
```

Table 10.2 Parameter Definitions for A Plane Truss

CONTROL:
- NSPACE = 2
- N = 2
- NG = 2
- NLPFLO = 7
- ISOLVT = 0*, or 1**

DEPENDENT VARIABLES:
- 1 = *x*-component of displacement, *u*
- 2 = *y*-component of displacement, *v*

PROPERTIES:
- Element level:
  - Floating point:
    - 1 = cross-sectional area, *A*
    - 2 = modulus of elasticity, *E*
    - 3 = temperature rise from unstressed state, Δt
    - 4 = coefficient of thermal expansion, α
    - 5 = uniform line load (to the right when moving from node 1 to node 2 of the element)
    - 6 = moment of inertia, *I*
    - 7 = half depth of the bar, *d*

CONSTRAINTS:
- Type 1: specified nodal displacement components
- Type 2: inclined roller supports, rigid members, etc.

INITIAL VALUES OF COLUMN VECTOR:
- Specified external nodal forces**

POST-SOLUTION CALCULATIONS:
- 1 = principal element stresses

* Default value
** Optional

10.4 Slider bearing lubrication

Several references are available on the application of the method to lubrication problems. These include the early work of Reddi [63], a detailed analysis and computer program for the three node triangle by Allan [9], and a presentation of higher order elements by Wada and Hayashi [73]. The most extensive discussion is probably found in the text by Huebner [43]. These formulations are based on the Reynolds equation of lubrication. For simplicity a one-dimensional formulation will be presented here.

Consider the slider bearing shown in Fig. 10.21 which is assumed to extend to infinity out of the plane of the figure. It consists of a rigid bearing and a slider moving relative to the bearing with a velocity of U. The extremely thin gap between the bearing and the slider is filled with an incompressible lubricant having a viscosity of ν. For the one-dimensional case the governing Reynolds equation reduces to

$$\frac{\mathrm{d}}{\mathrm{d}x}\left(\frac{h^3\,\mathrm{d}P}{6\nu\,\mathrm{d}x}\right)=\frac{\mathrm{d}}{\mathrm{d}x}(Uh), \tag{10.11}$$

where $P(x)$ denotes the pressure and $h(x)$ denotes the thickness of the gap. The boundary conditions are that P must equal the known external pressures (usually zero) at the two ends of the bearing. It can be shown that the variational equivalent of the one-dimensional Reynolds equation requires the minimization of the functional

$$I=\int_0^L\left[\frac{h^3}{12\nu}\left(\frac{\mathrm{d}P}{\mathrm{d}x}\right)^2+\mathrm{h}U\left(\frac{\mathrm{d}P}{\mathrm{d}x}\right)\right]\mathrm{d}x. \tag{10.12}$$

As a word of warning, it should be noted that, while the pressure P is continuous, the film thickness h is often discontinuous at one or more points on the bearing. Another related quantity of interest is the load capacity of the bearing. From statics one finds the resultant normal force per unit length in the z-direction, F_y, is $F_y=\int_0^L P\,\mathrm{d}x$. This is a quantity which would be included in a typical set of post-solution calculations.

As a specific example of a finite element formulation, consider a linear element with two nodes (N = 2) and one pressure per node (NG = 1). Thus $P(x)=\mathbf{H}^e(x)\,\mathbf{P}^e$ where as before $\mathbf{P}^{eT}=[P_i\quad P_j]$ and the interpolation functions are $\mathbf{H}^e=[(x_j-x)\ (x-x_i)]/l$, where $l=(x_j-x_i)$ is the length of the element. Minimizing the above functional defines the element square and column matrices as

$$\mathbf{S}^e=\frac{1}{6\nu}\int_{x_i}^{x_j}h^3\mathbf{H}^{eT}_{,x}\mathbf{H}^e_{,x}\,\mathrm{d}x,$$

and

$$\mathbf{C}^e = -U \int_{x_i}^{x_j} h\mathbf{H}^{eT}_{,x}\, dx.$$

For the element under consideration $\mathbf{H}^e$ is linear in x so that its first derivative will be constant. That is, $\mathbf{H}^e_{,x} = [-1 \quad 1]/l$ so that the element matrices simplify to

$$\mathbf{S}^e = \frac{1}{6\nu l^2}\begin{bmatrix} 1 & -1 \\ -1 & 1 \end{bmatrix} \int_{x_i}^{x_j} h^3\, dx,$$

and (10.13)

$$\mathbf{C}^e = \frac{U}{l}\begin{Bmatrix} 1 \\ -1 \end{Bmatrix} \int_{x_i}^{x_j} h\, dx.$$

Thus, it is clear that the assumed thickness variation within the element has an important effect on the complexity of the element matrices. It should also be clear that the nodal points of the mesh must be located such that any discontinuity in h occurs at a node. The simplest assumption is that h is constant over the length of the element. In this case the latter two integrals reduce to $h^3 l$ and hl, respectively. One may wish to utilize this element to approximate a varying distribution of h by a series of constant steps. In this case, one could use an average thickness of $h = (h_i + h_j)/2$, where h_i and h_j denote the thickness at the nodal points of the element. Subroutines ELSQ and ELCOL for this element are shown in Figs. 10.22 and 10.23. Note that these subroutines allow for two methods of defining the film thickness, h, in each element. In the default option (NLPFLO = 1, NNPFLO = 0) the value of h is input as a floating point element property, i.e. H = ELPROP(1). In the second option (NLPFLO = 0, NNPFLO = 1) the thickness is specific at each node as a floating point property. Note that for the latter option NLPFLO = 0, which always causes ELPROP(1) = 0.0 and thus $h = 0$. The program checks for this occurrence and then skips to the second definition of h.

The programs for performing the post-solution calculations, ELPOST and POSTEL, are shown in Figs. 10.24 and 10.25. These two routines evaluate the force, F^e_y, carried by each element. The load on a typical element is $F^e_y = \mathbf{Q}^e \mathbf{P}^e$ where $\mathbf{Q}^e = [l/2 \quad l/2]$. Subroutine ELPOST generates and stores $\mathbf{Q}^e$ for each element. Subroutine POSTEL carries out the multiplication once the nodal pressures, $\mathbf{P}^e$, are known. In addition, it sums the force on each element to obtain the total load capacity of the bearing. It prints the element number and its load and the total load on the bearing. With the addition of a few extra post-solution calculations, one could also output the location of the resultant bearing force. In closing, recall that both U and ν are constant along the entire length of the bearing. Thus, they are simply defined as floating point miscellaneous system properties (MISCFL = 2).

As a numerical example consider the step bearing shown in Fig. 10.26 which has two constants gaps of different thicknesses but equal lengths. Select a mesh with three nodes (M = 3) and two elements (NE = 2). Let $l_1 = l_2 = 0.125$ ft, $U = 20$ ft/s, $\nu = 0.002$ lb s/ft^2, $h_1 = 0.025$ ft, and $h_2 = 0.036$ ft. The two boundary conditions are $P_1 = P_3 = 0$ and we desire to calculate the pressure, P_2, at the step. The calculated pressure is $P_2 = 5.299$ psf, which is the exact value, and the total force on the bearing is $F_y = 0.66$ ppf. The accuracy is not surprising since the exact solution for this problem gives a linear pressure variation over each of the two segments of the bearing. Indeed if one solves the above two element problem in closed form one obtains the exact algebraic expression for P_2. The input data are shown in Fig. 10.27 and typical output data are shown in Fig. 10.28. Table 10.3 summarizes this element.

Fig. 10.21 Infinite slider bearing

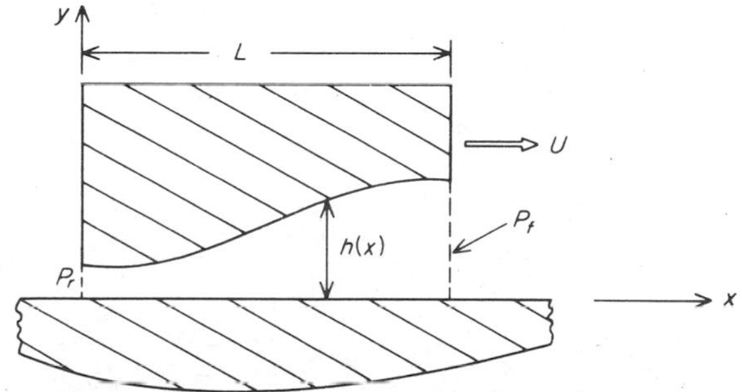

Fig. 10.22 Subroutine ELSQ

```
      SUBROUTINE  ELSQ  (N,NELFRE,NSPACE,IP5,JP2,KP1,KP2,
     1                   KP3,COORD,D,ELPROP,LPROP,PRTLPT,FLTMIS,
     2                   S,NTAPE1,NTAPE2,NTAPE3,NTAPE4)
C     * * * * * * * * * * * * * * * * * * * * * * * * * * * * *
C                   GENERATE ELEMENT SQUARE MATRIX
C     * * * * * * * * * * * * * * * * * * * * * * * * * * * * *
CDP   IMPLICIT REAL*8(A-H,O-Z)
      DIMENSION COORD(N,NSPACE), D(NELFRE), ELPRCP(KP2),
     1          LPROP(JP2), S(NELFRE,NELFRE),
     2          PRTLPT(IP5,KP1), FLTMIS(KP3)
C     N = NUMBER OF NODES PER ELEMENT
C     NELFRE = NUMBER OF DEGREES OF FREEDOM PER ELEMENT
C     NSPACE = DIMENSION OF SPACE
C     IP5 = NUMBER OF ROWS IN ARRAY PRTLPT
C     JP2 = NUMBER OF COLUMNS IN ARRAYS LPFIX AND LPROP
C     KP1 = NUMBER OF COLUMNS IN FLTNP & PRTLPT & PTPROP
C     KP2 = NUMBER OF COLUMNS IN ARRAYS FLTEL AND ELPROP
C     KP3 = NUMBER OF COLUMNS IN ARRAY FLTMIS
C     COORD = SPATIAL COORDINATES OF ELEMENT'S NODES
C     D = NODAL PARAMETERS ASSOCIATED WITH AN ELEMENT
```

Fig. 10.22—continued

```
C       ELPROP = ELEMENT ARRAY OF FLOATING PT PROPERTIES
C       LPROP = ARRAY OF FIXED POINT ELEMENT PROPERTIES
C       PRTLPT = FLOATING POINT PROP FOR ELEMENT'S NODES
C       FLTMIS = SYSTEM STORAGE OF FLOATING PT MISC PROP
C       S = ELEMENT SQUARE MATRIX
C       NTAPE1 = UNIT FOR POST SOLUTION MATRICES STORAGE
C       NTAPE2,3,4 = OPTIONAL UNITS FOR USER (USED WHEN > 0)
C
C       ..............................................................
C        ***  ELSQ PROBLEM DEPENDENT STATEMENTS FOLLOW ***
C       ..............................................................
C       APPLICATION: LINEAR SLIDER BEARING
C       NSPACE = 1, N = 2, NG = 1, NELFRE = 2, MISCFL = 2
C       FLTMIS(1) = VISCOSITY, FLTMIS(2) = VELOCITY
C       ELPROP(1) OR PRTLPT(K,1) = FILM THICKNESS
        COMMON /ELARG/  SP(2,2),CP(2),VIS,VEL,DL,Q(2),H
        DATA KALL /1/
        IF ( KALL.EQ.0 )  GO TO 10
C-->     ON FIRST CALL GENERATE CONSTANT MATRICES
        KALL = 0
        VIS = FLTMIS(1)
        VEL = FLTMIS(2)
        V6I = 1./(6.*VIS)
        SP(1,1) = V6I
        SP(2,2) = V6I
        SP(1,2) = -V6I
        SP(2,1) = -V6I
        CP(1) = VEL
        CP(2) = -VEL
   10   CONTINUE
C-->     DEFINE ELEMENT PROPERTIES
        DL = COORD(2,1) - COORD(1,1)
        H = ELPROP(1)
C       CHECK FOR ALTERNATE DEFINITION
        IF ( H.EQ.0.0 )  H = 0.5*(PRTLPT(1,1) + PRTLPT(2,1))
        IF ( H.EQ.0.0 )  H = 1.0
        H3L = H*H*H/DL
C-->     GENERATE ELEMENT SQUARE MATRIX
        S(1,1) = H3L*SP(1,1)
        S(1,2) = H3L*SP(1,2)
        S(2,2) = H3L*SP(2,2)
        S(2,1) = H3L*SP(2,1)
        RETURN
        END
```

Fig. 10.23 Subroutine ELCOL

```
        SUBROUTINE  ELCOL (N,NELFRE,NSPACE,IP5,JP2,KP1,KP2,
       1            KP3,COORD,D,ELPROP,LPROP,PRTLPT,FLTMIS,
       2            C,NTAPE1,NTAPE2,NTAPE3,NTAPE4)
C       * * * * * * * * * * * * * * * * * * * * * * * * * * * *
C                 GENERATE ELEMENT COLUMN MATRIX
C       * * * * * * * * * * * * * * * * * * * * * * * * * * * *
CDP     IMPLICIT REAL*8(A-H,O-Z)
        DIMENSION COORD(N,NSPACE), D(NELFRE), ELPROP(KP2),
       1          LPROP(JP2), C(NELFRE), PRTLPT(IP5,KP1),
       2          FLTMIS(KP3)
C       N = NUMBER OF NODES PER ELEMENT
C       NELFRE = NUMBER OF DEGREES OF FREEDOM PER ELEMENT
C       NSPACE = DIMENSION OF SPACE
C       IP5 = NUMBER OF ROWS IN ARRAY PRTLPT
C       JP2 = NUMBER OF COLUMNS IN ARRAYS LPFIX AND LPROP
C       KP1 = NUMBER OF COLUMNS IN  FLTNP & PRTLPT & PTPROP
C       KP2 = NUMBER OF COLUMNS IN ARRAYS FLTEL AND ELPROP
C       KP3 = NUMBER OF COLUMNS IN ARRAY FLTMIS
C       COORD = SPATIAL COORDINATES OF ELEMENT'S NODES
C       D = NODAL PARAMETERS ASSOCIATED WITH A ELEMENT
C       ELPROP = ELEMENT ARRAY OF FLOATING POINT PROPERTIES
C       LPROP = ARRAY OF FIXED POINT ELEMENT PROPERTIES
C       PRTLPT = FLOATING POINT PROP OF ELEMENT'S NODES
C       FLTMIS = SYSTEM STORAGE OF FLOATING POINT MISC. PROP
C       C = ELEMENT COLUMN MATRIX
C       NTAPE1 = UNIT FOR POST SOLUTION MATRICES STORAGE
C       NTAPE2,3,4 = OPTIONAL UNITS FOR USER  (USED WHEN > 0)
```

Fig. 10.23—continued

```
C       ....................................................
C        ***  ELCOL PROBLEM DEPENDENT STATEMENTS FCLLOW ***
C       ....................................................
C       APPLICATION: LINEAR SLIDER BEARING
C       NSPACE = 1, N = 2, NG = 1, NELFRE = 2, MISCFL = 2
C       FLTMIS(1) = VISCOSITY, FLTMIS(2) = VELOCITY
C       ELPROP(1) OR PRTLPT(K,1) = FILM THICKNESS
        COMMCN /ELARG/  SP(2,2),CP(2),VIS,VEL,DL,Q(2),H
C -->   FORM ELEMENT COLUMN MATRIX
        C(1) = H*CP(1)
        C(2) = H*CP(2)
        RETURN
        END
```

Fig. 10.24 Subroutine ELPOST

```
        SUBROUTINE ELPOST (NTAPE1,NTAPE2,NTAPE3,NTAPE4)
C       * * * * * * * * * * * * * * * * * * * * * * * * * * *
C           GENERATE DATA FOR POST SOLUTICN CALCULATICNS
C       * * * * * * * * * * * * * * * * * * * * * * * * * * *
CDP     IMPLICIT REAL*8(A-H,O-Z)
C       NTAPE1 = UNIT FOR POST SOLUTION MATRICES STORAGE
C       NTAPE2,3,4 = OPTIONAL UNITS FOR USER (USED WHEN > 0)
C
C       ....................................................
C        *** ELPOST PROBLEM DEPENDENT STATEMENTS FOLLOW ***
C       ....................................................
C       APPLICATION: LINEAR SLIDER BEARING
C       NSPACE = 1, N = 2, NG = 1, NELFRE = 2, MISCFL = 2
        COMMCN /ELARG/  SP(2,2),CP(2),VIS,VEL,DL,Q(2),H
C-->     GENERATE DATA FOR LOAD CALCULATIONS AND STORE
        Q(1) = 0.5*DL
        Q(2) = 0.5*DL
        WRITE (NTAPE1)  Q
        RETURN
        END
```

Fig. 10.25 Subroutine POSTEL

```
        SUBRCUTINE POSTEL (NTAPE1,NELFRE,D,IE,NTAPE2,
      1                  NTAPE3,NTAPE4,IT,NITER,NE,M)
C       * * * * * * * * * * * * * * * * * * * * * * * * * * *
C            ELEMENT LEVEL POST SOLUTION CALCULATICNS
C       * * * * * * * * * * * * * * * * * * * * * * * * * * *
C       NTAPE1 = UNIT FOR POST SOLUTION MATRICES STCRAGE
C       NELFRE = NUMBER OF DEGREES OF FREEDOM PER ELEMENT
C       D = NODAL PARAMETERS ASSOCIATED WITH THE ELEMENT
C       IE = ELEMENT NUMBER
C       NTAPE2,3,4 = OPTIONAL UNITS FOR USER (USED WHEN > 0)
C       IT = CURRENT ITERATION NUMBER
C       NITER = MAXIMUM NUMBER OF ITERATICNS
C       NE = TOTAL NUMBER OF ELEMENTS
C       M = TOTAL NUMBER OF NODES
CDP     IMPLICIT REAL*8 (A-H,O-Z)
        DIMENSION  D(NELFRE)
C
C       ....................................................
C        *** POSTEL PROBLEM DEPENDENT STATEMENTS FCLLOW ***
C       ....................................................
C       APPLICATION: LINEAR SLIDER BEARING
C       NSPACE = 1, N = 2, NG = 1, NELFRE = 2, MISCFL = 2
        COMMCN /ELARG/  SP(2,2),CP(2),TOT,VEL,DL,Q(2),H
        DATA KALL/1/
        IF ( KALL.EQ.0 )  GO TO 10
C-->     PRINT TITLES ON THE FIRST CALL
        KALL = 0
        WRITE (6,5000)
 5000 FORMAT ('1***     E L E M E N T   L O A D S   ***',//,
      1 ' ELEMENT            LOAD            TOTAL')
        TOT = 0.D0
   10 CONTINUE
C-->     CALCULATE LOADS ON THE ELEMENTS, F=Q*D
        READ  (NTAPE1)  Q
        CALL  MMULT (Q,D,F,1,2,1)
        TOT = TOT + F
        WRITE (6,5010)  IE, F, TOT
 5010 FORMAT (I5,E16.5,3X,E16.5)
        RETURN
        END
```

Fig. 10.26 Linear slider bearing

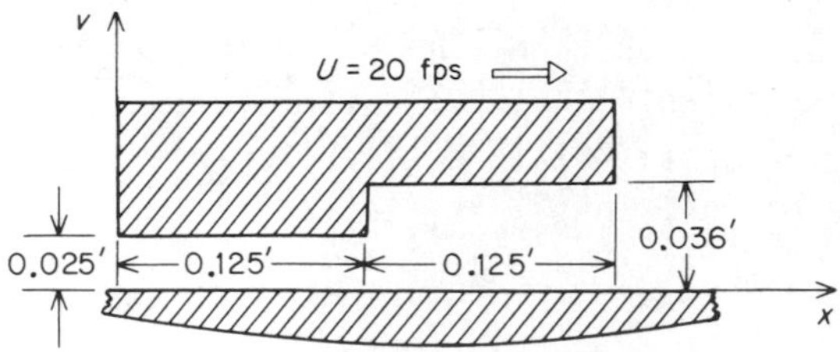

Fig. 10.27 Input data, linear slider bearing

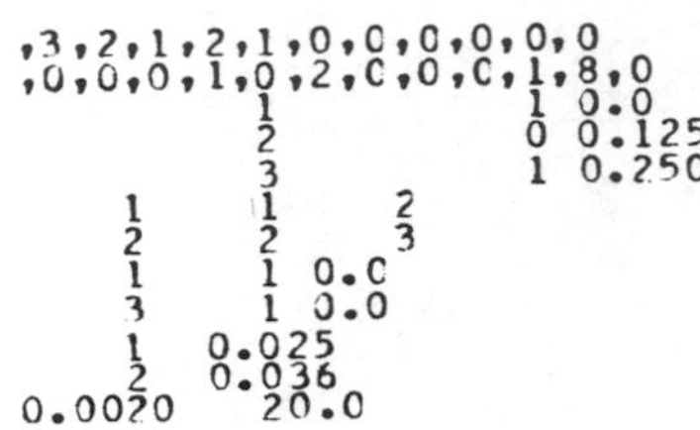

```
,3,2,1,2,1,0,0,0,0,0,0
,0,0,0,1,0,2,C,0,C,1,8,0
          1                1 0.0
          2                0 0.125
          3                1 0.250
     1    1      2
     2    2      3
     1    1  0.C
     3    1  0.0
     1   0.025
     2   0.036
0.0020     20.0
```

Fig. 10.28 Output data, linear slider bearing

```
*****   PROBLEM PARAMETERS   *****
NUMBER OF NODAL POINTS IN SYSTEM =..........     3
NUMBER OF ELEMENTS IN SYSTEM =..............     2
NUMBER OF NODES PER ELEMENT =...............     2
NUMBER OF PARAMETERS PER NODE =.............     1
DIMENSION OF SPACE =........................     1
NUMBER OF BOUNDARIES WITH GIVEN FLUX=.......     0
NUMBER OF NODES ON BOUNDARY SEGMENT=........     0
NUMBER OF FIXED PT PROP PER NODE =..........     0
NUMBER OF FLOATING PT PROP PER NODE =.......     0
NUMBER OF FIXED PT PROP PER ELEMENT =.......     0
NUMBER OF FLOATING PT PROP PER ELEMENT =....     1
NUMBER OF FIXED PT MISC PROP =..............     0
NUMBER OF FLOATING PT MISC PROP =...........     2

ELEMENT PROPERTIES ARE HOMOGENEOUS.
OPTIONAL UNIT NUMBERS (UTILIZED IF > 0)
NTAPE1=8 NTAPE2=0 NTAPE3=0 NTAPE4=0
NODAL PARAMETERS TO BE LISTED BY NODES

*** NODAL POINT DATA ***
NODE, CONSTRAINT INDICATOR, 1 COORDINATES
        1           1       0.0000
        2           0       0.1250
        3           0       0.2500

*** ELEMENT CONNECTIVITY DATA ***
 ELEMENT NO.,  2 NODAL INCIDENCES.
    1     1     2
    2     2     3

*** CONSTRAINT EQUATION DATA ***
CONSTRAINT TYPE ONE
EQ. NO.    NODE1     PAR1          A1
    1        1         1     .00000000E+00
    2        3         1     .00000000E+00

***   ELEMENT   PROPERTIES    ***
ELEMENT NO.   PROPERTY NO.       VALUE
     1             1          0.25000000E-01
     2             1          0.36000000E-01
END OF FLOATING PT PROPERTIES OF ELEMENTS
```

Fig. 10.28—continued

```
***  MISCELLANEOUS SYSTEM PROPERTIES   ***
PROPERTY NO.      VALUE
        1     0.20000000E-02
        2     0.20000000E+02
END OF FLOATING PT PROPERTIES OF SYSTEM

EQUATION HALF BANDWIDTH =.........       2
AND OCCURS IN ELEMENT NUMBER       1.
CONSTRAINT HALF BANDWIDTH =.............. 1
MAXIMUM HALF BANDWIDTH OF SYSTEM =....... 2
TOTAL NUMBER OF SYSTEM EQUATIONS =....... 8
NUMBER OF ELEMENT DEGREES OF FREEDOM =... 2

***  OUTPUT OF RESULTS   ***
NODE,   1 COORDINATES,    1 PARAMETERS.
   1     0.00000000E+00      0.52985662E-11
   2     0.12500000E+00      0.52985662E+01
   3     0.25000000E+00      0.52985662E-11

***  E L E M E N T   L O A D S   ***
ELEMENT          LOAD            TOTAL
   1        0.33116E+00         0.33116E+00
   2        0.33116E+00         0.66232E+00
NORMAL ENDING OF MODEL PROGRAM.
```

Table 10.3 Parameter Definitions for a Slider Bearing

CONTROL:

NSPACE = 1 — NNPFLO = 0*, or 1

N = 2 — NLPFLO = 1*, or 0

NG = 1 — MISCFL = 2

DEPENDENT VARIABLES:

1 = pressure, p

PROPERTIES:

Nodal point level:

Floating point:

1 = none,* or film thickness at the node, h

Element level:

Floating point:

1 = film thickness of the element,* h, or none

MISCELLANEOUS:

Floating point:

1 = viscosity, ν

2 = velocity, U

CONSTRAINTS:

Type 1: specified nodal pressures

POST-SOLUTION CALCULATIONS:

1 = resultant force on the element

* Default option

10.5 Ordinary differential equations

10.5.1 Linear example

There are several weighted residual methods available for formulating finite element solutions of differential equations. As an example, consider the ordinary differential equation $y' + ay = b$, which can be written symbolically in operator form as $L(y) = Q(x)$. Akin [1] and Lynn and Arya [48] have presented a least squares finite element solution of such problems. Let the value of y within an element be $y = \mathbf{H}^e(x)\ \mathbf{y}^e$, where $\mathbf{H}$ is the element interpolation functions matrix, and $\mathbf{y}^e$ is the array of nodal values for the element. These authors show that the corresponding element matrices are

$$\mathbf{S}^e = \int_{l^e} \mathbf{G}^e(x)^T \mathbf{G}^e(x)\, \mathrm{d}x,$$

and (10.14)

$$\mathbf{C}^e = \int_{l^e} \mathbf{G}^e(x)^T Q(x)\, \mathrm{d}x,$$

where $\mathbf{G}^e$ is the matrix obtained from the differential operator acting on the element interpolation functions. That is, $\mathbf{G}^e = \mathrm{L}(\mathbf{H}^e)$. For the equation given above $\mathrm{L} = (\mathrm{d}/\mathrm{d}x) + a$ and $\mathrm{Q} = b$. Since the operator is of first order the above integral definitions of the element matrices will only require C^0 continuity of the approximate finite element solution. Thus, one can utilize a linear element with a node at each end. As stated earlier, the interpolation functions for such an element are

$$\mathbf{H}^e = [(x_j - x) \quad (x - x_i)]/l,$$

where $l = (x_j - x_i)$ is the length of the element. Substituting yields

$$\mathbf{G}^e = [(a(x_j - x) - 1) \quad (a(x - x_i) + 1)]/l$$

so that the resulting element matrices are

$$\mathbf{C}^e = \frac{b}{2}\begin{Bmatrix} al - 2 \\ al + 2 \end{Bmatrix}$$

and (10.15)

$$\mathbf{S}^e = \frac{1}{3l}\begin{bmatrix} (3 - 3al + a^2l^2) & (a^2l^2 - 6)/2 \\ \text{sym.} & (3 + 3al + a^2l^2) \end{bmatrix}.$$

Figures 10.29 and 10.30 show the problem-dependent subroutines for calculating the above element matrices. Note that since a and b are constant over the whole solution domain they are treated as miscellaneous system properties (MISCFL = 2). As a specific example consider the solution of the problem $y' + 2y = +10$, where $y(0) = 0$. Let the domain of interest, $0 \leq x \leq 0.5$, be divided into five elements of equal length (NE = 5, M = 6, NSPACE = 1, N = 2, NG = 1). The input data are shown in Fig. 10.31 and the exact solution, $y(x) = 5(1 - e^{-2x})$, is compared with the MODEL results in Fig. 10.32. The program output is summarized in Fig. 10.33. Table 10.4 gives the problem summary.

Fig. 10.29 Subroutine ELSQ

```
      SUBROUTINE  ELSQ  (N,NELFRE,NSPACE,IP5,JP2,KP1,KP2,
     1                   KP3,COORD,D,ELPROP,LPROP,PRTLPT,FLTMIS,
     2                   S,NTAPE1,NTAPE2,NTAPE3,NTAPE4)
C     * * * * * * * * * * * * * * * * * * * * * * * * *
C                GENERATE ELEMENT SQUARE MATRIX
C     * * * * * * * * * * * * * * * * * * * * * * * * *
CDP   IMPLICIT REAL*8(A-H,O-Z)
      DIMENSION COORD(N,NSPACE), D(NELFRE), ELPROP(KP2),
     1          LPROP(JP2), S(NELFRE,NELFRE),
     2          PRTLPT(IP5,KP1),  FLTMIS(KP3)
C     N = NUMBER OF NODES PER ELEMENT
C     NELFRE = NUMBER OF DEGREES OF FREEDOM PER ELEMENT
C     NSPACE = DIMENSION OF SPACE
C     IP5 = NUMBER OF ROWS IN ARRAY PRTLPT
C     JP2 = NUMBER OF COLUMNS IN ARRAYS LPFIX AND LPROP
C     KP1 = NUMBER OF COLUMNS IN FLTNP & PRTLPT & PTPROP
C     KP2 = NUMBER OF COLUMNS IN ARRAYS FLTEL AND ELPROP
C     KP3 = NUMBER OF COLUMNS IN ARRAY FLTMIS
C     COORD = SPATIAL COORDINATES OF ELEMENT'S NODES
C     D = NODAL PARAMETERS ASSOCIATED WITH AN ELEMENT
C     ELPROP = ELEMENT ARRAY OF FLOATING PT PROPERTIES
C     LPROP = ARRAY OF FIXED POINT ELEMENT PROPERTIES
C     PRTLPT = FLOATING POINT PROP FOR ELEMENT'S NODES
C     FLTMIS = SYSTEM STORAGE OF FLOATING PT MISC PROP
C     S = ELEMENT SQUARE MATRIX
C     NTAPE1 = UNIT FOR POST SOLUTION MATRICES STORAGE
C     NTAPE2,3,4 = OPTIONAL UNITS FOR USER (USED WHEN > 0)
C
C     ..................................................
C      ***  ELSQ PROBLEM DEPENDENT STATEMENTS FOLLOW ***
C     ..................................................
C     APPLICATION:  LEAST SQUARES SOLUTION OF Y'+A*Y=B
C     NSPACE = 1, N = 2, NG = 1
C     DL = LENGTH OF ELEMENT
C     ARRAY FLTMIS CONTAINS PROBLEM COEFFICIENTS A AND B
      COMMON /ELARG/  A,B,DL
      DL = COORD(2,1) - COORD(1,1)
      A = FLTMIS( 1)
      B = FLTMIS( 2)
      S(1,1) = (3.-3.*A*DL+A*A*DL*DL)/3./DL
      S(2,2) = (3.+3.*A*DL+A*A*DL*DL)/3./DL
      S(1,2) = (A*A*DL*DL-6.)/6./DL
      S(2,1) = S( 1,2)
      RETURN
      END
```

Fig. 10.30 Subroutine ELCOL

```
      SUBROUTINE  ELCOL (N,NELFRE,NSPACE,IP5,JP2,KP1,KP2,
     1              KP3,COORD,D,ELPROP,LPROP,PRTLPT,FLTMIS,
     2              C,NTAPE1,NTAPE2,NTAPE3,NTAPE4)
C     * * * * * * * * * * * * * * * * * * * * * * * * *
C              GENERATE ELEMENT COLUMN MATRIX
C     * * * * * * * * * * * * * * * * * * * * * * * * *
CDP   IMPLICIT REAL*8(A-H,O-Z)
      DIMENSION COORD(N,NSPACE), D(NELFRE), ELPROP(KP2),
     1          LPROP(JP2), C(NELFRE), PRTLPT(IP5,KP1),
     2          FLTMIS(KP3)
C     N = NUMBER OF NODES PER ELEMENT
C     NELFRE = NUMBER OF DEGREES OF FREEDOM PER ELEMENT
C     NSPACE = DIMENSION OF SPACE
C     IP5 = NUMBER OF ROWS IN ARRAY PRTLPT
C     JP2 = NUMBER OF COLUMNS IN ARRAYS LPFIX AND LPROP
C     KP1 = NUMBER OF COLUMNS IN  FLTNP & PRTLPT & PTPROP
C     KP2 = NUMBER OF COLUMNS IN ARRAYS FLTEL AND ELPROP
C     KP3 = NUMBER OF COLUMNS IN ARRAY FLTMIS
C     COORD = SPATIAL COORDINATES OF ELEMENT'S NODES
C     D = NODAL PARAMETERS ASSOCIATED WITH A ELEMENT
C     ELPROP = ELEMENT ARRAY OF FLOATING POINT PROPERTIES
C     LPROP = ARRAY OF FIXED POINT ELEMENT PROPERTIES
C     PRTLPT = FLOATING POINT PROP OF ELEMENT'S NODES
C     FLTMIS = SYSTEM STORAGE OF FLOATING POINT MISC. PROP
C     C = ELEMENT COLUMN MATRIX
C     NTAPE1 = UNIT FOR POST SOLUTION MATRICES STORAGE
C     NTAPE2,3,4 = OPTIONAL UNITS FOR USER  (USED WHEN > 0)
C     ...........................................................
C      ***  ELCOL PROBLEM DEPENDENT STATEMENTS FOLLOW ***
C     ...........................................................
C     APPLICATION:  LEAST SQUARES SOLUTION OF Y'+A*Y=B
C     NSPACE = 1, N = 2, NG = 1
      COMMON /ELARG/  A,B,DL
      C(1) = 0.5*B*(A*DL-2.)
      C(2) = 0.5*B*(A*DL+2.)
      RETURN
      END
```

Fig. 10.31 Least squares solution of $y' + 2y = 10$, $y(0) = 0$

```
LEAST SQUARES SOLUTION OF Y'+2Y=10 (,=4 SPACES)
,6,5,1,2,1,0,0,0,0,0,0
,0,0,0,0,0,2,0,0,0,1,0,0,0,0
              1             1  0.0
              2             0  0.1
              3             0  0.2
              4             0  0.3
              5             0  0.4
              6             0  0.5
        1     1      2
        2     2      3
        3     3      4
        4     4      5
        5     5      6
        1     1 0.0
2.0            10.0
```

Fig. 10.32 Mesh and results for a weighted residual procedure

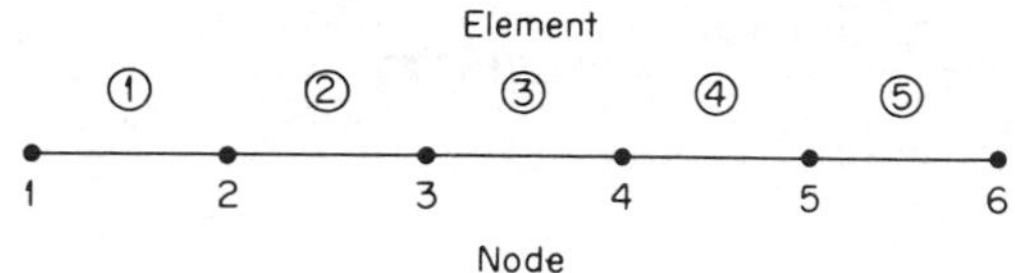

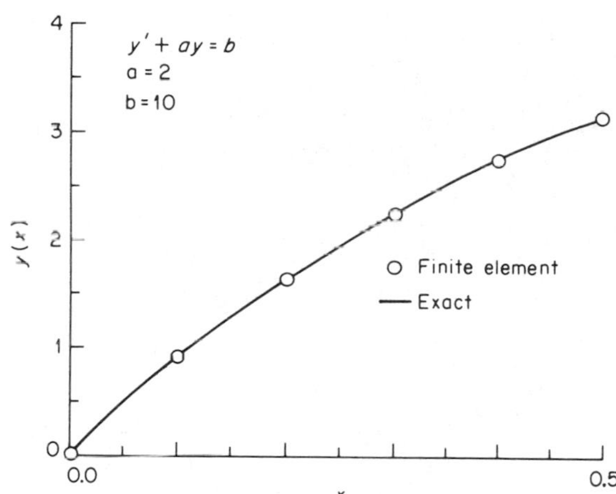

Fig. 10.33 Least squares solution of $y' + 2y = 10$, $y(0) = 0$

```
*****   PROBLEM PARAMETERS   *****
NUMBER OF NODAL POINTS IN SYSTEM =..........      6
NUMBER OF ELEMENTS IN SYSTEM =.............       5
NUMBER OF NODES PER ELEMENT =..............       2
NUMBER OF PARAMETERS PER NODE =............       1
DIMENSION OF SPACE =.......................       1
NUMBER OF BOUNDARIES WITH GIVEN FLUX=......       0
NUMBER OF NODES ON BOUNDARY SEGMENT=.......       0
NUMBER OF FIXED PT PROP PER NODE =.........       0
NUMBER OF FLOATING PT PROP PER NODE =......       0
NUMBER OF FIXED PT PROP PER ELEMENT =......       0
NUMBER OF FLOATING PT PROP PER ELEMENT =...       0
NUMBER OF FIXED PT MISC PROP =.............       0
NUMBER OF FLOATING PT MISC PROP =..........       2

NODAL PARAMETERS TO BE LISTED BY NODES

*** NODAL POINT DATA ***
NODE, CONSTRAINT INDICATOR, 1 COORDINATES
        1              1      0.0000
        2              0      0.1000
        3              0      0.2000
        4              0      0.3000
        5              0      0.4000
        6              0      0.5000
```

Fig. 10.33—continued

```
*** ELEMENT CONNECTIVITY DATA ***
ELEMENT NO.,  2 NODAL INCIDENCES.
    1     1     2
    2     2     3
    3     3     4
    4     4     5
    5     5     6

***  NODAL PARAMETER CONSTRAINT LIST   ***
CONSTRAINT        NUMBER OF
   TYPE           EQUATIONS
     1                1
     2                0
     3                0

*** CONSTRAINT EQUATION DATA ***
CONSTRAINT TYPE ONE
EQ. NO.     NODE1    PAR1           A1
     1         1        1    .00000000E+00

***  MISCELLANEOUS SYSTEM PROPERTIES   ***
PROPERTY NO.      VALUE
           1   0.20000000E+01
           2   0.10000000E+02
END OF FLOATING PT PROPERTIES OF SYSTEM

EQUATION HALF BANDWIDTH =.........      2
AND OCCURS IN ELEMENT NUMBER      1.
CONSTRAINT HALF BANDWIDTH =..............   1
MAXIMUM HALF BANDWIDTH OF SYSTEM =.......   2
TOTAL NUMBER OF SYSTEM EQUATIONS =.......   6
NUMBER OF ELEMENT DEGREES OF FREEDOM =...   2

***  OUTPUT OF RESULTS   ***
NODE,  1 COORDINATES,  1 PARAMETERS.
   1     0.00000000E+00      0.17709900E-14
   2     0.10000000E+00      0.90749034E+00
   3     0.20000000E+00      0.16501816E+01
   4     0.30000000E+00      0.22579809E+01
   5     0.40000000E+00      0.27553633E+01
   6     0.50000000E+00      0.31623579E+01
NORMAL ENDING OF MODEL PROGRAM.
```

Table 10.4 Parameter Definitions for Least Squares Solution of $y' + ay = b$

CONTROL:
- NSPACE = 1
- N = 2
- NG = 1

DEPENDENT VARIABLES:
- 1 = *y*

PROPERTIES:
- MISCELLANEOUS:
  - 1 = coefficient *a*
  - 2 = coefficient *b*

CONSTRAINTS:
- Type 1: specified *y-value*

10.5.2 Nonlinear example

The above procedure can be extended to solve nonlinear operators as well. Akin [1] has shown that if one utilizes an iterative solution of a differential operator of the form $L(y) + N(y) = Q(x)$, where L and N denote linear and nonlinear differential operators, then the element matrices for iteration n become

$$\mathbf{S}_n^e = \int_{l^e} (\mathbf{E}^{eT}\mathbf{E} + \mathbf{E}^{eT}\mathbf{F}_{n-1}^e + \mathbf{F}_{n-1}^{e\ T}\mathbf{E}^e + \mathbf{F}_{n-1}^{eT}\mathbf{F}_{n-1}^e)\, ds,$$

$$\mathbf{C}_n^e = \int_l Q(x)(\mathbf{E}^{eT} + \mathbf{F}_{n-1}^{eT}) ds, \qquad (10.16)$$

where **E** and **F** denote the matrices resulting from the element interpolation functions being acted upon by the operators L and N, respectively. As an example, Akin presents the solution of $2yy'' - (y')^2 + 4y^2 = 0$ subject to the two boundary conditions of $y(\pi/6) = 1/4$ and $y(\pi/2) = 1$. Since this operator is of the second order, the above least square integral definitions will require a finite element model with C^1 continuity; that is, both y and y' must be continuous between elements. Thus, select the one-dimensional element shown in Fig. 10.34 which has two nodes (N = 2) and utilizes both y and y' as nodal parameters (NG = 2). It has been shown that the element interpolation functions for this cubic element are the first order Hermite polynomials. That is,

$$y(s) = \mathbf{H}(s)^e\, \mathbf{Y}^e,$$

where (10.17)

$$\mathbf{Y}^{eT} = [y_i \quad y_i' \quad y_j \quad y_j']$$

and

$$\mathbf{H}(s)^e = [(1 - 3\beta^2 + 2\beta^3) \quad l(\beta - 2\beta^2 + \beta^3) \quad (3\beta^2 - 2\beta^3) \quad l(\beta^3 - \beta^2)]$$

and where $\beta = s/l$, $l = x_j - x_i$, and $s = x - x_i$. These are programmed in Fig. 10.35.

The above definitions of $\mathbf{S}^e$ and $\mathbf{C}^e$ are clearly too complicated to be integrated in closed form so the calculations in subroutine ELSQ, shown in Fig. 10.36, are based on a four point Gaussian quadrature. The initial estimates of $y(x)$ and $y'(x)$ for the first iteration are calculated by the problem-dependent subroutine START which is shown in Fig. 10.37. These starting values simply represent a linear interpolation between the two boundary conditions. Select a mesh with eleven equally spaced nodes (M = 11), and ten elements (NE = 10). In Fig. 10.38 the results of a three iteration solution (NITER = 3) are compared with an exact solution of $y(x) = \sin^2(x)$. Note that the iterative approach rapidly approaches the exact values when given a reasonably good starting estimate. The element summary is given in Table 10.5. The example input and output data are given in Figs. 10.39 and 10.40, respectively. The MODEL program uses routines CHANGE and CORECT, Fig. 10.41, to control the over-relaxation iteration procedure. Clearly, other procedures such as the Newton–Raphson method are better suited to the solution of nonlinear problems.

Fig. 10.34 A one-dimensional cubic element

$$y(x) = \mathbf{H}^e(x)\,\mathbf{Y}^e$$

$$\mathbf{Y}^{eT} = [y_i \quad y_i' \quad y_j \quad y_j']$$

$$\mathbf{H}^e(x) = [(1 - 3\beta^2 + 2\beta^3) \quad l(\beta - 2\beta^2 + \beta^3) \quad (3\beta^2 - 2\beta^3) \quad l(\beta^3 - \beta^2)]$$

$$\mathbf{H}' = [(6\beta^2 - 6\beta)/l \quad (1 - 4\beta + 3\beta^2) \quad (6\beta - 6\beta^2)/l \quad (3\beta^2 - 2\beta)]$$

$$\mathbf{H}'' = [(12\beta - 6i/l^2) \quad (6\beta - 4)/l \quad (6 - 12\beta)/l^2/ \quad (6\beta - 2)/l]$$

$$\beta = s/l, \quad s = x - x_i, \quad \text{and} \quad l = x_j - x_i$$

Fig. 10.35 Cubic element routines

```
      SUBROUTINE  SHPCU (B,A,H)
C     * * * * * * * * * * * * * * * * * * * * * * * * *
C     SHAPE FUNCTIONS FOR A CUBIC ONE DIMENSIONAL ELEMENT
C     * * * * * * * * * * * * * * * * * * * * * * * * *
CDP   IMPLICIT REAL*8 (A-H,O-Z)
      DIMENSION  H(4)
C     A = LENGTH OF ELEMENT        1----------2 -> B
C     B = COORDINATE OF POINT  B=0             B=1
C     H = SHAPE FUNCTIONS ARRAY
      H(1) = 1.-3.*B*B + 2.*B*B*B
      H(2) = (B - 2.*B*B + B*B*B)*A
      H(3) =  3.*B*B - 2.*B*B*B
      H(4) = (B*B*B - B*B)*A
      RETURN
      END
      SUBROUTINE  DERCU (B,A,DH)
C     * * * * * * * * * * * * * * * * * * * * * * * * *
C     FIRST DERIVATIVES OF SHAPE FUNCTIONS FOR 1-D
C     CUBIC ELEMENT    (A C1 ELEMENT)
C     * * * * * * * * * * * * * * * * * * * * * * * * *
CDP   IMPLICIT REAL*8 (A-H,O-Z)
      DIMENSION  DH(4)
C     A = LENGTH OF ELEMENT (SEE SUBR SHPCU)
C     B = COORDINATE OF POINT
C     DH = FIRST DERIVATIVES OF H
      DH(1) = 6.*(B*B - B)/A
      DH(2) = 1. - 4.*B + 3.*B*B
      DH(3) = 6.*(B - B*B)/A
      DH(4) = 3.*B*B - 2.*B
      RETURN
      END
```

Fig. 10.35—continued

```
      SUBROUTINE  DER2CU (B,A,D2H)
C     * * * * * * * * * * * * * * * * * * * * * * * * *
C     SECOND DERIVATIVES OF SHAPE FUNCTIONS FOR 1-D
C      CUBIC ELEMENT    (A C1 ELEMENT)
C     * * * * * * * * * * * * * * * * * * * * * * * * *
CDP   IMPLICIT REAL*8 (A-H,O-Z)
      DIMENSION  D2H(4)
C     A = LENGTH OF ELEMENT (SEE SUBR SHPCU)
C     B = COORDINATE OF POINT
C     D2H = SECOND DERIVATIVES OF H
      D2H(1) = (12.*B - 6.)/A/A
      D2H(2) = (6.*B - 4.)/A
      D2H(3) = (6. - 12.*B)/A/A
      D2H(4) = (6.*B - 2.)/A
      RETURN
      END
```

Fig. 10.36 Subroutine ELSQ

```
      SUBROUTINE  ELSQ  (N,NELFRE,NSPACE,IP5,JP2,KP1,KP2,
     1               KP3,COORD,D,ELPROP,LPROP,PRTLPT,FLTMIS,
     2               S,NTAPE1,NTAPE2,NTAPE3,NTAPE4)
C     * * * * * * * * * * * * * * * * * * * * * * * * *
C              GENERATE ELEMENT SQUARE MATRIX
C     * * * * * * * * * * * * * * * * * * * * * * * * *
CDP   IMPLICIT REAL*8(A-H,O-Z)
      DIMENSION COORD(N,NSPACE), D(NELFRE), ELPROP(KP2),
     1          LPROP(JP2), S(NELFRE,NELFRE),
     2          PRTLPT(IP5,KP1), FLTMIS(KP3)
C     N = NUMBER OF NODES PER ELEMENT
C     NELFRE = NUMBER OF DEGREES OF FREEDOM PER ELEMENT
C     NSPACE = DIMENSION OF SPACE
C     IP5 = NUMBER OF ROWS IN ARRAY PRTLPT
C     JP2 = NUMBER OF COLUMNS IN ARRAYS LPFIX AND LPROP
C     KP1 = NUMBER OF COLUMNS IN FLTNP & PRTLPT & PTPRCP
C     KP2 = NUMBER OF COLUMNS IN ARRAYS FLTEL AND ELPROP
C     KP3 = NUMBER OF COLUMNS IN ARRAY FLTMIS
C     COORD = SPATIAL COORDINATES OF ELEMENT'S NODES
C     D = NODAL PARAMETERS ASSOCIATED WITH AN ELEMENT
C     ELPROP = ELEMENT ARRAY OF FLOATING PT PROPERTIES
C     LPROP = ARRAY OF FIXED POINT ELEMENT PROPERTIES
C     PRTLPT = FLOATING POINT PROP FOR ELEMENT'S NODES
C     FLTMIS = SYSTEM STORAGE OF FLOATING PT MISC PROP
C     S = ELEMENT SQUARE MATRIX
C     NTAPE1 = UNIT FOR POST SOLUTION MATRICES STORAGE
C     NTAPE2,3,4 = OPTIONAL UNITS FOR USER (USED WHEN > 0)
C
C     ....................................................
C      *** ELSQ PROBLEM DEPENDENT STATEMENTS FOLLOW ***
C     ....................................................
C     APPLICATION: LEAST SQ. SOL. OF 2YY''-Y'Y'+4YY=0
      COMMON /ELARG/  GP(4),GWT(4),H(4),DH(4),D2H(4),F(4),
     1                NQP,LSQ
      DATA KALL/1/
C-->  EXTRACT QUADRATURE DATA ON THE FIRST CALL
      IF ( KALL.EQ.0 )  GO TO 10
      KALL = 0
      NQP = 4
      CALL  GAUSCO (NQP,GP,GWT)
      LSQ = NELFRE*NELFRE
   10 XI = COORD(1,1)
      XJ = COORD(2,1)
      DL = XJ - XI
C-->  ZERO ELEMENT SQUARE MATRIX
      CALL ZEROA (LSQ,S)
C     NUMERICAL INTEGRATION LOOP
      DO 50  IP = 1,NQP
C-->  FIND LOCAL COORDINATES OF INTEGRATION POINT
      BETA = 0.5*(1.+GP(IP))
C-->  EVALUATE SHAPE FUNCTIONS AND DERIVATIVES
      CALL  SHPCU (BETA,DL,H)
      CALL  DERCU (BETA,DL,DH)
      CALL  DER2CU (BETA,DL,D2H)
C-->  FORM NONLINEAR MATRIX F USING PREVIOUS D
      CALL  MMULT (H,D,YOLD,1,4,1)
```

Fig. 10.36—continued

```
      CALL  MMULT (DH,D,DYOLD,1,4,1)
      DO 20  I = 1,4
   20 F(I) = 2.*YOLD*D2H(I) - DYOLD*DH(I) + 4.*YOLD*H(I)
C     COMPLETE THE SQUARE MATRIX
C        DX = DX/D BETA * D BETA/DU * DU = 2.*L*DU
      WT = GWT(IP)*2.*DL
      DO 40  I = 1,4
      DO 30  J = 1,4
   30 S(I,J) = S(I,J) + WT*F(I)*F(J)
   40 CONTINUE
   50 CONTINUE
C     INTEGRATION COMPLETED
      RETURN
      END
```

Fig. 10.37 Function START

```
      FUNCTION  START (IG,NSPACE,COORD)
C     * * * * * * * * * * * * * * * * * * * * * * * * * *
C     DEFINE STARTING VALUE OF PARAMETER IG IN TERMS OF
C     COORDINATES OF THE NODE  (FOR ITERATIVE SOLUTIONS)
C     * * * * * * * * * * * * * * * * * * * * * * * * * *
C        A PROBLEM DEPENDENT ROUTINE
CDP   IMPLICIT REAL*8(A-H,O-Z)
      DIMENSION  COORD(1,NSPACE)
C     NSPACE = DIMENSION OF SPACE
C     COORD = SPATIAL COORDINATE ARRAY OF NODE
C     ..................................................
C     ** PROBLEM DEPENDENT START STATEMENTS FOLLOW **
C     ..................................................
C     APPLICATION: LEAST SQ. SOL. OF 2YY''-Y'Y'+4YY=0
C-->  STRAIGHT LINE FIT THROUGH TWO BOUNDARY VALUES
      X = COORD(1,1)
      START = 0.7162D0*X - 0.125D0
      IF ( IG.EQ.2 )  START = 0.7162D0
      RETURN
      END
```

Fig. 10.39 Input data, least squates solution of $2yy'' - y'y' + 4yy = 0$, $y(\pi/6) = 1/4$, $y(\pi/2) = 1$

```
        11    10,2,2,1,0,C,3,0,0,0
   ,0,0,0,0,0,0,0,0,0,1,0,0,0,0
                 1           1  0.5236
                 2           0  0.6283
                 3           0  0.7330
                 4           0  0.8378
                 5           0  0.9425
                 6           0  1.0472
                 7           0  1.1519
                 8           0  1.2566
                 9           0  1.3614
                10           0  1.4661
                11           1  1.5708
         1       1     2
         2       2     3
         3       3     4
         4       4     5
         5       5     6
         6       6     7
         7       7     8
         8       8     9
         9       9    10
        10      10    11
         1       1  0.25
        11       1  1.0
```

Fig. 10.38 Results for the nonlinear example

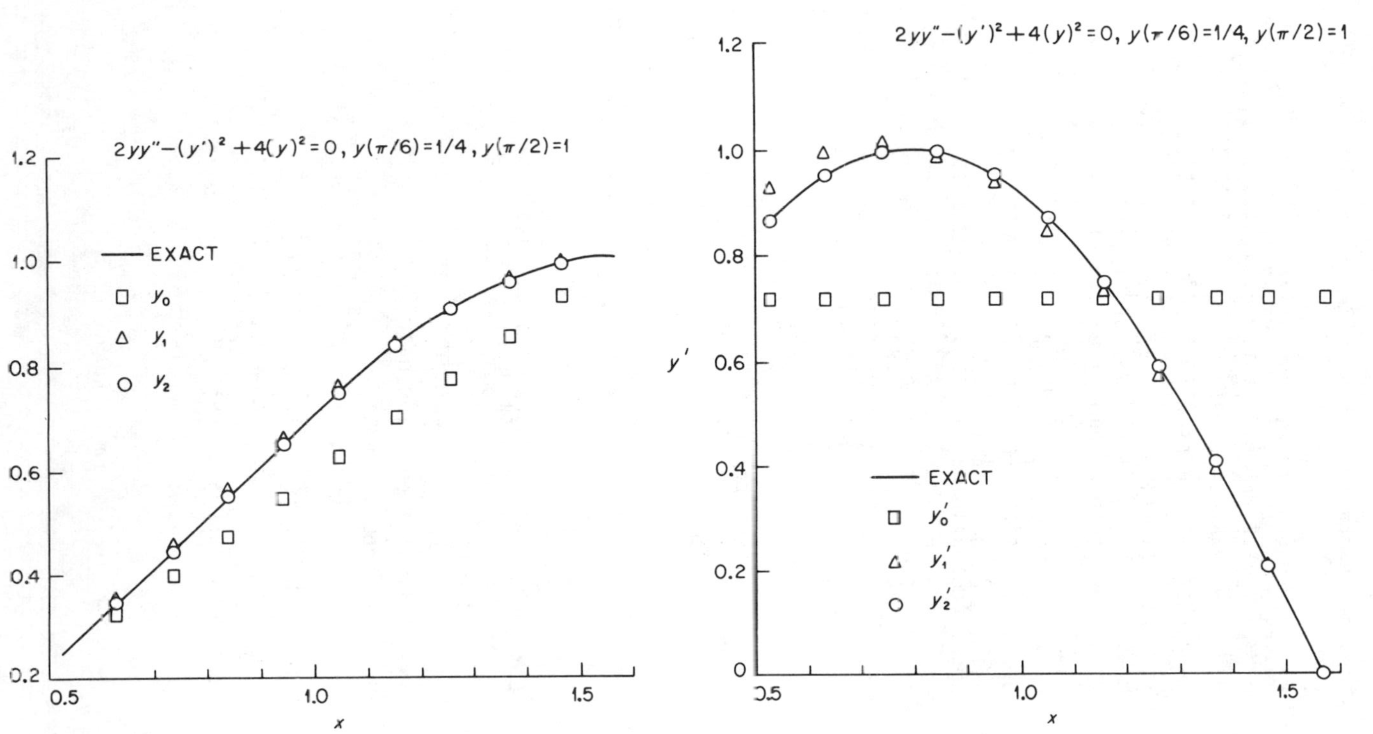

Fig. 10.40 Output data, least squares solution of $2yy'' - y'y' + 4yy = 0$, $y(\pi/6) = 1/4$, $y(\pi/2) = 1$

```
*****  PROBLEM PARAMETERS  *****
NUMBER OF NODAL POINTS IN SYSTEM =..........  11
NUMBER OF ELEMENTS IN SYSTEM =..............  10
NUMBER OF NODES PER ELEMENT =...............   2
NUMBER OF PARAMETERS PER NODE =.............   2
DIMENSION OF SPACE =........................   1
NUMBER OF BOUNDARIES WITH GIVEN FLUX=.......   0
NUMBER OF NODES CN BOUNDARY SEGMENT=........   0
NUMBER OF ITERATIONS TO BE RUN=.............   3
NUMBER OF FIXED PT PROP PER NODE =..........   0
NUMBER OF FLOATING PT PROP PER NODE =.......   0
NUMBER OF FIXED PT PROP PER ELEMENT =.......   0
NUMBER OF FLOATING PT PROP PER ELEMENT =....   0
NUMBER OF FIXED PT MISC PROP =..............   0
NUMBER OF FLOATING PT MISC PROP =...........   0

NODAL PARAMETERS TO BE LISTED BY NODES

*** NODAL POINT DATA ***
NODE, CONSTRAINT INDICATOR, 1 COORDINATES
         1          1       0.5236
         2          0       0.6283
         3          0       0.7330
         4          0       0.8378
         5          0       0.9425
         6          0       1.0472
         7          0       1.1519
         8          0       1.2566
         9          0       1.3614
        10          0       1.4661
        11          1       1.5708

*** ELEMENT CONNECTIVITY DATA ***
ELEMENT NO.,  2 NODAL INCIDENCES.
   1    1    2
   2    2    3
   3    3    4
   4    4    5
   5    5    6
   6    6    7
   7    7    8
   8    8    9
   9    9   1C
  10   10   11

***  NODAL PARAMETER CONSTRAINT LIST   ***
CONSTRAINT        NUMBER OF
   TYPE           EQUATIONS
    1                 2
    2                 0
    3                 0

*** CONSTRAINT EQUATION DATA ***
CONSTRAINT TYPE ONE
EQ. NO.    NODE1    PAR1          A1
   1          1        1    .25000000E+00
   2         11        1    .10000000E+01

EQUATION HALF BANDWIDTH =.........     4
AND OCCURS IN ELEMENT NUMBER      1.
CONSTRAINT HALF BANDWIDTH =..............   1
MAXIMUM HALF BANDWIDTH OF SYSTEM =.......   4
TOTAL NUMBER OF SYSTEM EQUATIONS =.......  22
NUMBER OF ELEMENT DEGREES OF FREEDOM =...   4
```

Fig. 10.40—continued

```
* STARTING VALUES FOR ITERATIVE SOLUTION *
 NODE      PAR. NO.          VALUE
   1           1      0.25000232E+00
   1           2      0.71620000E+00
   2           1      0.32498846E+00
   2           2      0.71620000E+00
   3           1      0.39997460E+00
   3           2      0.71620000E+00
   4           1      0.47503236E+00
   4           2      0.71620000E+00
   5           1      0.55001850E+00
   5           2      0.71620000E+00
   6           1      0.62500463E+00
   6           2      0.71620000E+00
   7           1      0.69999077E+00
   7           2      0.71620000E+00
   8           1      0.77497692E+00
   8           2      0.71620000E+00
   9           1      0.85003468E+00
   9           2      0.71620000E+00
  10           1      0.92502083E+00
  10           2      0.71620000E+00
  11           1      0.10000070E+01
  11           2      0.71620000E+00

***  OUTPUT OF RESULTS   ***
NODE,   1 COORDINATES,   2 PARAMETERS.
 1  0.52360E+00   0.25000000E+00   0.92922297E+00
 2  0.62830E+00   0.35101636E+00   0.99277624E+00
 3  0.73300E+00   0.45636595E+00   0.10126992E+01
 4  0.83780E+00   0.56177522E+00   0.99248312E+00
 5  0.94250E+00   0.66299362E+00   0.93504947E+00
 6  0.10472E+01   0.75637733E+00   0.84333742E+00
 7  0.11519E+01   0.03050601E+00   0.72063925E+00
 8  0.12566E+01   0.90632776E+00   0.57070988E+00
 9  0.13614E+01   0.95724897E+00   0.39761322E+00
10  0.14661E+01   0.98900533E+00   0.20636936E+00
11  0.15708E+01   0.10000000E+01   0.18909482E-02

ITERATION NUMBER =    1
 *** NODAL DOF DATA FOR ITERATIONS ***
ROOT MEAN SQ CF DIFFERENCES . . =  0.11561E+01
ROOT MEAN SQ OF PREVIOUS VALUES =  0.32493E+01
RATIO OF ABOVE QUANTITIES . . . =  0.35578E+00

***  OUTPUT OF RESULTS   ***
NODE,   1 COORDINATES,   2 PARAMETERS.
 1  0.52360E+00   0.25000000E+00   0.84732959E+00
 2  0.62830E+00   0.34428931E+00   0.94596480E+00
 3  0.73300E+00   0.44642277E+00   0.99702212E+00
 4  0.83780E+00   0.55149881E+00   0.10003553E+01
 5  0.94250E+00   0.65437908E+00   0.95737617E+00
 6  0.10472E+01   0.75045903E+00   0.87107845E+00
 7  0.11519E+01   0.83542564E+00   0.74596089E+00
 8  0.12566E+01   0.90551379E+00   0.58797231E+00
 9  0.13614E+01   0.95769181E+00   0.40418970E+00
10  0.14661E+01   0.98960788E+00   0.20336319E+00
11  0.15708E+01   0.10000000E+01  -0.53468031E-02

ITERATION NUMBER =  2
 *** NODAL DOF DATA FOR ITERATIONS ***
ROOT MEAN SQ CF DIFFERENCES . . =  0.32932E+00
ROOT MEAN SQ OF PREVIOUS VALUES =  0.36833E+01
RATIO OF ABOVE QUANTITIES . . . =  0.89407E-01

***  OUTPUT OF RESULTS   ***

NODE,   1 COORDINATES,   2 PARAMETERS.
 1  0.52360E+00   0.25000000E+00   0.87112817E+00
 2  0.62830E+00   0.34566671E+00   0.95050485E+00
 3  0.73300E+00   0.44771570E+00   0.99212629E+00
 4  0.83780E+00   0.55206010E+00   0.99205408E+00
 5  0.94250E+00   0.65405827E+00   0.94927157E+00
 6  0.10472E+01   0.74939343E+00   0.86519349E+00
 7  0.11519E+01   0.83390693E+00   0.74329105E+00
 8  0.12566E+01   0.90389954E+00   0.58882204E+00
 9  0.13614E+01   0.95634352E+00   0.40833761E+00
10  0.14661E+01   0.98883598E+00   0.21004948E+00
11  0.15708E+01   0.10000000E+01   0.23847056E-02

ITERATION NUMBER =    3
 *** NODAL DOF UATA FOR ITERATIONS ***
ROOT MEAN SQ CF DIFFERENCES . . =  0.87376E-01
ROOT MEAN SQ OF PREVIOUS VALUES =  0.34989E+01
RATIO OF ABOVE CUANTITIES . . . =  0.24972E-01
NORMAL ENDING OF MODEL PROGRAM.
```

Fig. 10.41 **(a)** Subroutine CHANGE, and **(b)** subroutine CORECT

```
      SUBROUTINE CHANGE (NDFREE,DD,DDOLD,TOTAL,DIFF,
     1                   RATIO,IPRINT)
C     * * * * * * * * * * * * * * * * * * * * * * * * * *
C     CALCULATE THE MEAN CHANGE IN NODAL PARAMETERS FROM
C     LAST ITERATION
C     * * * * * * * * * * * * * * * * * * * * * * * * * *
CDP   IMPLICIT REAL*8 (A-H,O-Z)
      DIMENSION  DD(NDFREE), DDOLD(NDFREE)
C     * CHANGE SHOULD BE CALLED BEFORE CORECT *
C     RATIO = DIFF/TOTAL
C     DIFF = SQRT(SUM OF (DD(I)-DDOLD(I))**2)
C     TOTAL = SQRT(SUM OF DDOLD(I)**2)
C     DDOLD = NODAL PARAMETER LIST FROM LAST ITERATION
C     DD = NODAL PARAMETERS FROM CURRENT ITERATION
C     NDFREE = TOTAL NO OF DEGREES OF FREEDOM IN SYS
C     IF IPRINT.NE.0  PRINT DIFF, TOTAL, AND RATIO
CDP   SQRT(Z) = DSQRT(Z)
      DIFF = 0.0
      TOTAL = 0.1E-10
      DO 10  I = 1,NDFREE
      TOTAL = TOTAL + DDOLD(I)*DDOLD(I)
   10 DIFF = DIFF + (DD(I)-DDOLD(I))**2
      TOTAL = SQRT(TOTAL)
      DIFF = SQRT(DIFF)
      RATIO = DIFF/TOTAL
      IF ( IPRINT.EQ.0 )  RETURN
      WRITE (6,5000)  DIFF, TOTAL, RATIO
 5000 FORMAT ('0 *** NODAL DOF DATA FOR CURRENT ',
     1  'AND PREVIOUS ITERATIONS ***',//,
     2' ROOT MEAN SQ OF DIFFERENCES . . . . . ',E16.5,/,
     3' ROOT MEAN SQ OF PREVIOUS VALUES . . . ',E16.5,/,
     4' RATIO OF ABOVE QUANTITIES . . . . . . ',E16.5)
      RETURN
      END
```

(a)

```
      SUBROUTINE  CORECT (NDFREE,DD,DDOLD)
C     * * * * * * * * * * * * * * * * * * * * * * * * * *
C     CALCULATE NEW STARTING VALUES FOR NEXT ITERATION
C     * * * * * * * * * * * * * * * * * * * * * * * * * *
C        OVER RELAXATION METHOD
CDP   IMPLICIT REAL*8(A-H,O-Z)
      DIMENSION  DD(NDFREE),   DDOLD(NDFREE)
      DATA  OMEGA/1.25/
C     DD = CALCULATED DOF FROM LAST ITERATION
C     DDOLD = DOF TO BE USED TO START NEXT ITERATION
C     NDFREE = TOTAL NO OF SYS DEGREES OF FREEDOM
      DO 10  I = 1,NDFREE
   10 DDOLD(I) = DDOLD(I) + OMEGA*(DD(I)-DDOLD(I))
      RETURN
      END
```

(b)

Table 10.5 Parameter Definitions for Least Square Solution of $L(y) + N(y) = Q(x)$

CONTROL:
- NSPACE = 1
- N = 2
- NG = 2

DEPENDENT VARIABLES:
- $1 = y$
- $2 = y' = dy/dx$

CONSTRAINTS:
- Type 1: specified values of y and/or y'

10.6 Plane frame structures

Weaver [74] has considered the detailed programming aspects of several one-dimensional structural elements. Among them is the plane frame element which has two displacements and one rotation per node. A typical element is shown in Fig. 10.42 where XM, YM are the member axes and x, y are the global axes. A frame element with an area of A, having a moment of inertia I, a length L, and made of a material having a modulus of E has an element square matrix (stiffness matrix) of

$$\mathbf{S}_m^e = \frac{E}{L}\begin{bmatrix} A & 0 & 0 & -A & 0 & 0 \\ & 12\beta/L & 6\beta & 0 & -12\beta/L & 6\beta \\ & & 4I & 0 & -6\beta & 2I \\ & & & A & 0 & 0 \\ & & & & 12\beta/L & -6\beta \\ \text{sym.} & & & & & 4I \end{bmatrix} \tag{10.18}$$

where $\beta = I/L$ and where the element displacements have been ordered as $\boldsymbol{\delta}^{eT} = [u_1\ v_1\ \theta_1\ u_2\ v_2\ \theta_2]$. Weaver shows that to convert this expression to one valid in the global (x, y) coordinates one must consider how the nodal degrees of freedom transform from the element axes to the global axes. This relation is shown to be

$$\mathbf{S}^e = \mathbf{T}^T\mathbf{S}_m^e\mathbf{T}, \tag{10.19}$$

where the transformation matrix is

$$\mathbf{T} = \begin{bmatrix} R & 0 \\ 0 & R \end{bmatrix}$$

and

$$\mathbf{R} = \begin{bmatrix} cx & cy & 0 \\ -cy & cx & 0 \\ 0 & 0 & 1 \end{bmatrix},$$

where $cx = \cos\theta_x$ and $cy = \cos\theta_y$ are the member direction cosines. If one is only interested in the deflections resulting from joint forces and moments these are the only equations that must be programmed. They are presented in subroutine ELSQ in Fig. 10.43. As an example application consider example structure three presented by Weaver. The structure and its loads are shown in Fig. 10.44. The homogeneous element properties are $E =$ 10,000 ksi, $A = 10\ \text{in}^2$, and $I = 1{,}000\ \text{in}^4$. Nodes 2 and 3 are completely restrained from translating or rotating. The displacements and rotation at node 1 given by Weaver are −0.0202608 in, −0.0993600 in, and −0.00179756 radians, which are in good agreement with the MODEL results.

This problem involves three nodes (M = 3) with three degrees of freedom per node (NG = 3), two elements (NE = 2) with two nodes per element (N = 2), and three (A, I, E) homogeneous properties per element (NLPFLO = 3, LHOMO = 1). There are no element column matrix contributions (NULCOL = 1) and the applied loads at point 1 must be considered as initial contributions to the system matrix (ISOLVT = 1). The input data are shown in Fig. 10.45 and the MODEL output is in Fig. 10.46. The element is summarized in Table 10.6.

Fig. 10.42 A plane frame element

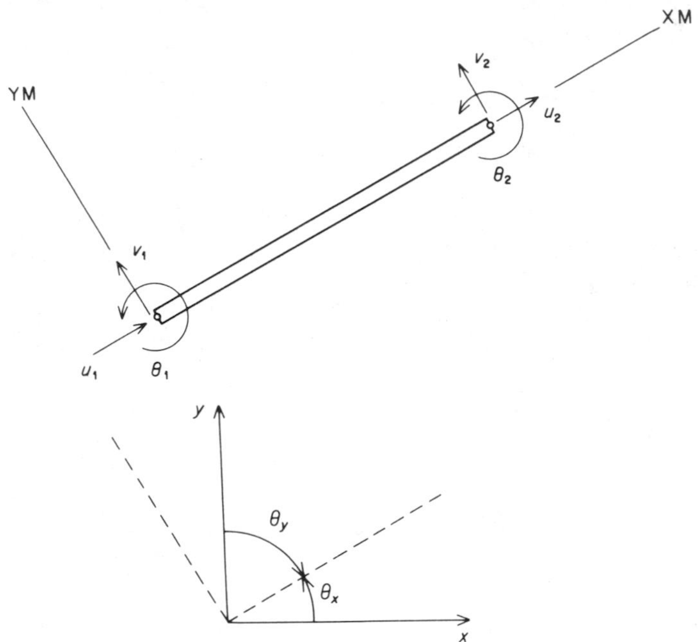

Fig. 10.43 Subroutine ELSQ

```
      SUBROUTINE  ELSQ  (N,NELFRE,NSPACE,IP5,JP2,KP1,KP2,
     1                   KP3,COORD,D,ELPROP,LPROP,PRTLPT,FLTMIS,
     2                   S,NTAPE1,NTAPE2,NTAPE3,NTAPE4)
C     * * * * * * * * * * * * * * * * * * * * * * * * *
C                 GENERATE ELEMENT SQUARE MATRIX
C     * * * * * * * * * * * * * * * * * * * * * * * * *
CDP   IMPLICIT REAL*8(A-H,O-Z)
      DIMENSION COORD(N,NSPACE), D(NELFRE), ELPROP(KP2),
     1          LPROP(JP2), S(NELFRE,NELFRE),
     2          PRTLPT(IP5,KP1), FLTMIS(KP3)
C     N = NUMBER OF NODES PER ELEMENT
```

Fig. 10.43—continued

```
C      NELFRE = NUMBER OF DEGREES OF FREEDOM PER ELEMENT
C      NSPACE = DIMENSION OF SPACE
C      IP5 = NUMBER OF ROWS IN ARRAY PRTLPT
C      JP2 = NUMBER OF COLUMNS IN ARRAYS LPFIX AND LPROP
C      KP1 = NUMBER OF COLUMNS IN FLTNP & PRTLPT & PTPROP
C      KP2 = NUMBER OF COLUMNS IN ARRAYS FLTEL AND ELPROP
C      KP3 = NUMBER OF COLUMNS IN ARRAY FLTMIS
C      COORD = SPATIAL COORDINATES OF ELEMENT'S NODES
C      D = NODAL PARAMETERS ASSOCIATED WITH AN ELEMENT
C      ELPROP = ELEMENT ARRAY OF FLOATING PT PROPERTIES
C      LPROP = ARRAY OF FIXED POINT ELEMENT PROPERTIES
C      PRTLPT = FLOATING POINT PROP FOR ELEMENT'S NODES
C      FLTMIS = SYSTEM STORAGE OF FLOATING PT MISC PROP
C      S = ELEMENT SQUARE MATRIX
C      NTAPE1 = UNIT FOR POST SOLUTION MATRICES STORAGE
C      NTAPE2,3,4 = OPTIONAL UNITS FOR USER (USED WHEN > 0)
C
C      ..................................................
C      *** ELSQ PROBLEM DEPENDENT STATEMENTS FOLLOW ***
C      ..................................................
C      APPLICATION:  PLANE FRAME ANALYSIS
C      NSPACE = 2, N = 2, NG = 3, NLPFLO = 3, NELFRE = 6
C      ELPROP(1) = AREA, ELPROP(2) = MOMENT OF INERTIA
C      ELPROP(3) = MODULUS OF ELASTICITY
       DIMENSION  SM(6,6), T(6,6), DUMMY(6,6)
C      INITIALIZE ZERO TERMS IN SM AND T
       DATA SM,T/36*0.0,36*0.0/
CDP    SQRT(Z) = DSQRT(Z)
C      DEFINE ELEMENT PROPERTIES
       A = ELPROP(1)
       ZI = ELPROP(2)
       E = ELPROP(3)
C      FIND MEMBER LENGTH AND DIRECTION COSINES
       DX = COORD(2,1) - COORD(1,1)
       DY = COORD(2,2) - COORD(1,2)
       FL = SQRT (DX*DX + DY*DY)
       CX = DX/FL
       CY = DY/FL
C      DEFINE NON-ZERO TERMS IN MEMBER STIFFNESS MATRIX
       EDL = E/FL
       BETA = ZI/FL
       SM(1,1) = A*EDL
       SM(2,2) = 12.*EDL*BETA/FL
       SM(2,3) = 6.*EDL*BETA
       SM(3,2) = SM(2,3)
       SM(3,3) = 4.*EDL*ZI
       SM(1,4) = -A*EDL
       SM(4,1) = SM(1,4)
       SM(4,4) = A*EDL
       SM(2,5) =  -SM(2,2)
       SM(5,2) = SM(2,5)
       SM(3,5) = -SM(2,3)
       SM(5,3) = SM(3,5)
       SM(5,5) = SM(2,2)
       SM(2,6) = SM(2,3)
       SM(6,2) = SM(2,6)
       SM(3,6) = 2.*EDL*ZI
       SM(6,3) = SM(3,6)
       SM(5,6) = -SM(2,3)
       SM(6,5) = SM(5,6)
       SM(6,6) = 4.*EDL*ZI
C      DEFINE NON-ZERO TERMS IN THE TRANSFORMATION MATRIX
       T(1,1) = CX
       T(2,2) = CX
       T(3,3) = 1.
       T(4,4) = CX
       T(5,5) = CX
       T(6,6) = 1.
       T(1,2) =  CY
       T(2,1) = -CY
       T(4,5) =  CY
       T(5,4) = -CY
C      TRANSFORM STIFFNESS MATRIX TO GLOBAL COORDINATES
       CALL  MMULT (SM,T,DUMMY,6,6,6)
       CALL  MTMULT (T,DUMMY,S,6,6,6)
       RETURN
       END
```

Fig. 10.44 Plane frame structure

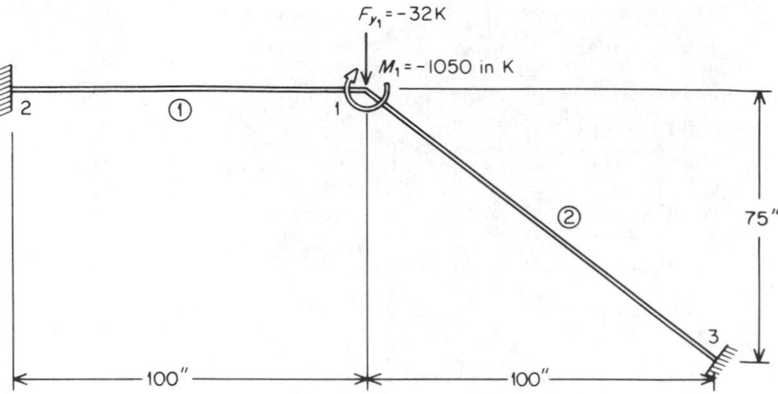

Fig. 10.45 Plane frame analysis, test problem data

```
,3,2,3,2,2
,0,0,0,3,0,0,0,1,0,1,0,0,0,0,1
            1        000  100.        75.
            2        111  0.0         75.
            3        111  200.        0.0
     1      2     1
     2      1     3
     2      1   0.0
     2      2   0.0
     2      3   0.0
     3      1   0.0
     3      2   0.0
     3      3   0.0
     1     10.          1000.       10000.
     1      2 -32.
     1      3 -1050.
     3      3 0.0
```

Fig. 10.46 Plane frame analysis (see Weaver's example structure three), test problem output

```
*****  PROBLEM PARAMETERS  *****
NUMBER OF NODAL POINTS IN SYSTEM =..........  3
NUMBER OF ELEMENTS IN SYSTEM =..............  2
NUMBER OF NODES PER ELEMENT =...............  2
NUMBER OF PARAMETERS PER NODE =.............  3
DIMENSION OF SPACE =........................  2
NUMBER OF BOUNDARIES WITH GIVEN FLUX=.......  0
NUMBER OF NODES ON BOUNDARY SEGMENT=........  0
INITIAL FORCING VECTOR TO BE INPUT
NUMBER OF FIXED PT PROP PER NODE =..........  0
NUMBER OF FLOATING PT PROP PER NODE =.......  0
NUMBER OF FIXED PT PROP PER ELEMENT =.......  0
NUMBER OF FLOATING PT PROP PER ELEMENT =....  3
NUMBER OF FIXED PT MISC PROP =..............  0
NUMBER OF FLOATING PT MISC PROP =...........  0

ELEMENT PROPERTIES ARE HOMOGENEOUS.
NODAL PARAMETERS TO BE LISTED BY NODES

*** NODAL POINT DATA ***
NODE, CONSTRAINT INDICATOR, 2 COORDINATES
      1          0   100.0000    75.0000
      2        111     0.0000    75.0000
      3        111   200.0000     0.0000
```

Fig. 10.46—continued

```
*** ELEMENT CONNECTIVITY DATA ***
ELEMENT NO., 2 NODAL INCIDENCES.
   1     2     1
   2     1     3

*** CONSTRAINT EQUATION DATA ***

CONSTRAINT TYPE ONE
EQ. NO.    NODE1    PAR1           A1
   1         2        1     .00000000E+00
   2         2        2     .00000000E+00
   3         2        3     .00000000E+00
   4         3        1     .00000000E+00
   5         3        2     .00000000E+00
   6         3        3     .00000000E+00

***  ELEMENT  PROPERTIES    ***
ELEMENT NO.  PROPERTY NO.      VALUE
     1             1        0.10000000E+02
     1             2        0.10000000E+04
     1             3        0.10000000E+05
END OF FLOATING PT PROPERTIES OF ELEMENTS

   *** INITIAL FORCING VECTOR DATA ***
   NODE    PARAMETER         VALUE                DOF
     1         2     -0.32000000E+02          2
     1         3     -0.10500000E+04          3
     3         3      0.00000000E+00          9

***  OUTPUT OF RESULTS  ***
NODE,  2 COORDINATES,   3 PARAMETERS.
1 1.00E2  7.5E1   -2.03E-02   -9.94E-02    -1.80E-03
2 0.00E0  7.5E1   -2.03E-14   -9.48E-15    -5.92E-16
3 2.00E2  0.0E0    3.79E-14   -9.43E-14     2.00E-15
NORMAL ENDING OF MODEL PROGRAM.
```

Table 10.6 Parameter Definitions for a Plane Frame

CONTROL:

NSPACE = 2	NLPFLO = 3
N = 2	NULCOL = 1
NG = 3	ISOLUT = 1

DEPENDENT VARIABLES:
- 1 = x-component of displacement, u
- 2 = y-component of displacement, v
- 3 = z-rotation, θ

PROPERTIES:
Element level:
- 1 = cross-sectional area, A
- 2 = moment of inertia, I
- 3 = modulus of elasticity, E

CONSTRAINTS:
Type 1: specified nodal displacements and/or rotations
Type 2: inclined rollers, or rigid members, etc.

INITIAL VALUES OF COLUMN VECTOR:
Specified external nodal forces or moments

11

Two-dimensional applications

11.1 Introduction

From the very beginning finite element methods have been applied to two-dimensional problems. The three node (linear) triangle has been utilized in the derivation of the element matrices for several engineering applications. Most of the current texts present the results for such applications as stress analysis, heat conduction and ground water seepage. The four node rectangular (bilinear) element has also been presented in some detail but it is not often used due to the difficulty of fitting complex engineering shapes with rectangles. Meek [53] made a useful contribution in presenting the closed form element matrices for a six node (quadratic) triangle applied to the Poisson equation. The equations given by Meek are quite useful and should be of interest to a beginner. However, one soon learns that it is easier to program such elements when numerical integration is utilized.

A significant feature of this section is the detailed formulation of isoparametric elements (see Section 11.3). This section, combined with Section 5.3, should supply the reader with the ability to master this very useful concept. The extensions of the isoparametric element to a three-dimensional application will be considered in Section 12.2 to illustrate how easy it is to change from one element type to another.

This chapter will also begin to introduce the reader to some of the useful capabilities of the MODEL program that may not be immediately apparent. For example, it will be shown how one can reduce the data requirements for non-homogeneous problems by assigning to each element a material number and a corresponding array of floating point properties.

11.2 Plane stress analysis

The application of the finite element method to the plane stress analysis of solids has been considered in detail in several texts[14, 25, 29, 34]. The element matrices for the three node linear displacement (constant strain) triangle have been presented by Ural [7], Gallagher [34], and others. The nodal parameters are u and v, the nodal displacement components in the x- and y-directions, respectively. If one orders the element degrees of freedom as $\delta^{eT} = [u_1\, v_1\, u_2\, v_2\, u_3\, v_3]$ it has been shown that the element load vector due to distributed body forces per unit volume are

$$\mathbf{C}^e_{6\times 1} = t\Delta/3 \begin{Bmatrix} X \\ Y \\ X \\ Y \\ X \\ Y \end{Bmatrix} \tag{11.1}$$

where Δ is the area of the triangle and where X and Y denote the components of the distributed body force. The element square matrix (or *stiffness matrix*) for an element of constant thickness, t, is

$$\mathbf{S}^e_{6\times 6} = \mathbf{B}^{eT}\mathbf{D}^e\mathbf{B}^e t\Delta,$$

where

$$\mathbf{B}^e_{6\times 3} = \frac{1}{2\Delta}\begin{bmatrix} b_i & 0 & b_j & 0 & b_m & 0 \\ 0 & c_i & 0 & c_j & 0 & c_m \\ c_i & b_i & c_j & b_j & c_m & b_m \end{bmatrix} \tag{11.2}$$

denotes the *strain-displacement matrix* (i.e. $\boldsymbol{\varepsilon}^e = \mathbf{B}^e\boldsymbol{\delta}^e$), and where the material constitutive matrix for an isotropic material is

$$\mathbf{D}^e_{3\times 3} = \begin{bmatrix} \beta & \nu\beta & 0 \\ & \beta & 0 \\ \text{sym.} & & G \end{bmatrix} \tag{11.3}$$

with $\beta \equiv E/(1 - \nu^2)$ and $G = E/2(1 + \nu)$. In the latter equation, E equals the modulus of elasticity, ν is Poisson's ratio, and G is the shear modulus. Although G, in theory, is not an independent variable, one often encounters applications where it is desirable to treat G as a third material property.

Thus, the MODEL code requires the user to define E, ν, and G in the input data.

The constants in the **B** matrix are defined in terms of the spatial coordinates of the element's nodes. For example, $b_i = y_j - y_m$ and $c_i = x_m - x_i$, and the other terms are obtained by a cyclic permutation of the subscripts. The strains and stresses under consideration are $\boldsymbol{\varepsilon}^T = [\varepsilon_x \quad \varepsilon_y \quad \gamma_{xy}]$ and $\boldsymbol{\sigma}^T = [\sigma_x \quad \sigma_y \quad \tau_{xy}]$ where $\boldsymbol{\sigma}^e = \mathbf{D}^e \boldsymbol{\varepsilon}^e$. If one has applied surface traction components on an element boundary segment then the contributions to the boundary column matrix are as shown in Fig. 4.1.

Thus far, the problem under consideration is two-dimensional (NSPACE = 2), involves triangular elements (N = 3), with six [E, ν, G, t, X, Y) floating point properties (NLPFLO = 6), and two nodes per typical boundary segments (LBN = 2). In addition, each node has two nodal parameters (NG = 2) so that the element has a total of six degrees of freedom. The above discussion provides enough data to calculate the nodal displacements of the mesh. Generally, the post-solution calculation of stresses and strains is important to the analyst. Since these quantities are constant in this type of element, they are usually listed as occurring at the centroidal coordinates of the element. The constant stress–displacement and strain–displacement matrices are generated by subroutine ELSQ, Fig. 11.1. These quantities are passed through the named common ELARG to subroutine ELPOST along with the global coordinates of the element centroid. If NTAPE1 > 0, subroutine ELPOST simply writes these data on unit NTAPE1 for later use by subroutine POSTEL. Similarly, if NTAPE1 > 0, subroutine POSTEL is called after the nodal displacements have been calculated. It reads the above data, extracts the element's nodal displacements and multiplies them by the strain–displacement and stress–displacement matrices to obtain the element strains and stresses, respectively. Then the element's number, centroidal coordinates, stresses and strains are printed. After the post-solution calculations have been completed, one may elect to request the calculation of the displacement contours. This element is summarized in Table 11.1.

The problem-dependent portions of the element subroutines are shown in Figs. 11.1 to 11.4. As a numerical example consider the problem presented in detail by Ural [72]. As shown in Fig. 11.5, the problem has two elements (NE = 2) and four nodes (M = 4). The two nodes (1 and 2) on the left are completely fixed and at node 3 there is a 25 k = 25,000 lb load in the positive x-direction. No other loads are present ($X = Y = 0$) and the material has properties of $E = 30{,}000$ ksi and $\nu = 0.25$ so that $G = 12{,}000$ ksi. The properties are homogeneous (LHOMO = 1) and each element has a thickness of $t = 2$ in. The input data are shown in Fig. 11.6 and the output is presented in Fig. 11.7. The results agree with these given by Ural but the present solution is more accurate since only slide-rule accuracy was utilized in the former solution.

Many stress analysis problems involve non-homogeneous properties. It might prove inconvenient to specify all these properties for each element. Thus, several finite element codes allow a user to assign a single material number code to each element. This code is utilized along with a table of material properties to define the properties of each element. This same concept is easily implemented in the MODEL code. Subroutine MATPRT, shown in Fig. 11.8, can be used for this purpose. The list of material properties for each material number is stored in order of material number in the miscellaneous floating point system array FLTMIS. That is, if there is a maximum of NMAX materials each having NLPFLT floating point properties then array FLTMIS contains NMAX sub-arrays each having NLPFLT floating point numbers (MISCFL = NMAX * NLPFLT). Given the material number, NUM, this subroutine extracts the appropriate properties list from FLTMIS and stores it in a material array called PRTMAT. Of course, the latter array must be dimensioned in one of the problem-dependent subroutines. The integer material number of each element is input by the user in array LPFIX (i.e. NLPFIX = 1) and it exists at the element level in LPROP(1). Recall that array LPFIX is initially zeroed; thus, one only has to input element material numbers that are greater than 1 if one defines LPROP(1) = 0 or =1 to both imply a material number of unity in the problem-dependent subroutines that call MATPRT. The typical changes in the previous program to utilize this concept are shown in later applications.

To illustrate the effects of a surface traction load, consider the problem shown in Fig. 11.9. The problem dependent subroutine BFLUX for this loading condition is shown in Fig. 11.10. This problem was worked out in great detail, with slide-rule accuracy, by Ural in Example 5.2 of his second text [72]. The problem involves five nodes (M = 5), and four elements (NE = 4, N = 3, NG = 2). The plate has a thickness of $t = 1.5$ in., a Young's modulus of $E = 29{,}000$ ksi, and a Poisson's ratio of $\nu = 0.3$. There are no body force loads ($X = Y = 0$) and no externally applied nodal loads (ISOLVT = 0).

Nodes 1 and 2 are completely fixed and a surface traction is applied to edge 5–4 (NSEG = 1, LBN = 2). The traction consists of constant x- and y-components of 1/3 k/in ($T_{x4} = T_{x5} = 0.333$ k/in, $T_{y4} = T_{y5} = 0.0$). The calculated displacements and stresses agree with the results of Ural as expected. The input and output data for this example are in Figs. 11.11 and 11.12, respectively.

In general one would want to define the surface tractions in normal and tangential components. The additional computations in this approach involve finding the direction cosines of the boundary segment and using these data to calculate the corresponding resultant x- and y-components.

Note that the stiffness matrix involved a product of the form $\mathbf{B}^T\mathbf{D}\mathbf{B}$ where the matrix $\mathbf{D}$ is symmetric. This form will occur in most of the linear finite element problems. Thus it is desirable to calculate this product in an efficient manner. Subroutine BTDB, shown in Fig. 11.13(a), is designed for that purpose. Since $\mathbf{D}$ is often a diagonal matrix, a special version, BTDIAB, is given in Fig. 11.13(b).

Fig. 11.1 Subroutine ELSQ

```
      SUBROUTINE  ELSQ  (N,NELFRE,NSPACE,IP5,JP2,KP1,KP2,
     1                  KP3,COORD,D,ELPROP,LPROP,PRTLPT,FLTMIS,
     2                  S,NTAPE1,NTAPE2,NTAPE3,NTAPE4)
C     * * * * * * * * * * * * * * * * * * * * * * * * *
C                GENERATE ELEMENT SQUARE MATRIX
C     * * * * * * * * * * * * * * * * * * * * * * * * *
CDP   IMPLICIT REAL*8(A-H,O-Z)
      DIMENSION COORD(N,NSPACE), D(NELFRE), ELPRCP(KP2),
     1             LPROP(JP2), S(NELFRE,NELFRE),
     2             PRTLPT(IP5,KP1), FLTMIS(KP3)
C     N = NUMBER OF NODES PER ELEMENT
C     NELFRE = NUMBER OF DEGREES OF FREEDOM PER ELEMENT
C     NSPACE = DIMENSION OF SPACE
C     IP5 = NUMBER OF ROWS IN ARRAY PRTLPT
C     JP2 = NUMBER OF COLUMNS IN ARRAYS LPFIX AND LPROP
C     KP1 = NUMBER OF COLUMNS IN FLTNP & PRTLPT & PTPROP
C     KP2 = NUMBER OF COLUMNS IN ARRAYS FLTEL AND ELPROP
C     KP3 = NUMBER OF COLUMNS IN ARRAY FLTMIS
C     COORD = SPATIAL COORDINATES OF ELEMENT'S NCDES
C     D = NODAL PARAMETERS ASSOCIATED WITH AN ELEMENT
C     ELPRCP = ELEMENT ARRAY OF FLOATING PT PROPERTIES
C     LPROP = ARRAY OF FIXED POINT ELEMENT PROPERTIES
C     PRTLPT = FLOATING POINT PROP FOR ELEMENT'S NCDES
C     FLTMIS = SYSTEM STORAGE OF FLOATING PT MISC PRCP
C     S = ELEMENT SQUARE MATRIX
C     NTAPE1 = UNIT FOR POST SOLUTION MATRICES STORAGE
C     NTAPE2,3,4 = OPTIONAL UNITS FOR USER (USED WHEN > 0)
C     ...........................................................
C      ***  ELSQ PROBLEM DEPENDENT STATEMENTS FOLLCW ***
C     ...........................................................
C     APPLICATION: PLANE STRESS, ISOTRCPIC MATERIAL, CST
C     NSPACE = 2, N= 3, NG = 2, NELFRE = 6, NLPFLC = 4
C     ELPROP(1) = ELASTIC MODULUS ELPROP(2) = POISSCN RATIO
C     ELPRCP(3) = SHEAR MODULUS    ELPROP(4) = THICKNESS
C     ELPRCP(5) = X-BODY FORCE      ELPRCP(6) = Y-BODY FORCE
C     B = ELEMENT STRAIN-DISPLACEMENT MATRIX
C     DM = MATERIAL CONSTITUTIVE MATRIX
C     STRESS = ELEMENT STRESS-DISPLACEMENT MATRIX
      COMMON /ELARG/ XB,YB,E,V,G,T,FX,FY,B(3,6),
     1                  STRESS(3,6),DM(3,3),AREA
      CATA KALL/1/
C-->  DEFINE NODAL COORDINATES
      XI = COORD(1,1)
      XJ = COORD(2,1)
      XK = COORD(3,1)
      YI = COORD(1,2)
      YJ = COORD(2,2)
      YK = COORD(3,2)
C-->  DEFINE PROPERTIES
      E = ELPROP(1)
      V = ELPROP(2)
      G = ELPROP(3)
      T = ELPROP(4)
      FX = ELPROP(5)
      FY = ELPROP(6)
C     DEFINE CENTROID COORDINATES
      XB = (XI + XJ + XK)/3.
      YB = (YI + YJ + YK)/3.
```

Fig. 11.1—continued

```
C-->  DEFINE GEOMETRIC PARAMETERS
      AI = XJ*YK-XK*YJ
      AJ = XK*YI-XI*YK
      AK = XI*YJ-XJ*YI
      BI = YJ-YK
      BJ = YK-YI
      BK = YI-YJ
      CI = XK-XJ
      CJ = XI-XK
      CK = XJ-XI
C     CALCULATE ELEMENT AREA
      TWOA = AI + AJ + AK
      AREA = TWOA*0.5
      TA = T*AREA
C-->  ON THE FIRST CALL DEFINE ZERO TERMS IN DM AND B
      IF ( KALL.EQ.0 )  GO TO 10
      KALL = 0
      DM(1,3) = 0.0
      DM(3,1) = 0.0
      DM(2,3) = 0.0
      DM(3,2) = 0.0
      B(1,2) = 0.0
      B(1,4) = 0.0
      B(1,6) = 0.0
      B(2,1) = 0.0
      B(2,3) = 0.0
      B(2,5) = 0.0
C-->  DEFINE VARIABLE TERMS IN DM AND B
   10 DM(1,1) = E/(1.-V*V)
      DM(2,2) = E/(1.-V*V)
      DM(3,3) = G
      DM(1,2) = V*E/(1.-V*V)
      DM(2,1) = V*E/(1.-V*V)
      B(1,1) = BI/TWOA
      B(1,3) = BJ/TWOA
      B(1,5) = BK/TWOA
      B(2,2) = CI/TWOA
      B(2,4) = CJ/TWOA
      B(2,6) = CK/TWOA
      B(3,1) = CI/TWOA
      B(3,3) = CJ/TWOA
      B(3,5) = CK/TWOA
      B(3,2) = BI/TWOA
      B(3,4) = BJ/TWOA
      B(3,6) = BK/TWOA
C     FORM STRESS-DISPLACEMENT MATRIX
      CALL  MMULT  (DM,B,STRESS,3,3,6)
C-->  COMPLETE STIFFNESS MATRIX     ,K= BT*DM*B
      CALL  MTMULT (B,STRESS,S,3,6,6)
      CALL  MSMULT (TA,S,6,6)
      RETURN
      END
```

Fig. 11.2 Subroutine ELCOL

```
      SUBROUTINE  ELCOL (N,NELFRE,NSPACE,IP5,JP2,KP1,KP2,
     1                   KP3,COORD,D,ELPROP,LPROP,PRTLPT,FLTMIS,
     2                   C,NTAPE1,NTAPE2,NTAPE3,NTAPE4)
C     * * * * * * * * * * * * * * * * * * * * * * * * * * *
C                 GENERATE ELEMENT COLUMN MATRIX
C     * * * * * * * * * * * * * * * * * * * * * * * * * * *
CDP   IMPLICIT REAL*8(A-H,O-Z)
      DIMENSION COORD(N,NSPACE), D(NELFRE), ELPRCP(KP2),
     1          LPROP(JP2), C(NELFRE), PRTLPT(IP5,KP1),
     2            FLTMIS(KP3)
C     N = NUMBER OF NODES PER ELEMENT
C     NELFRE = NUMBER OF DEGREES OF FREEDOM PER ELEMENT
C     NSPACE = DIMENSION OF SPACE
C     IP5 = NUMBER OF ROWS IN ARRAY PRTLPT
C     JP2 = NUMBER OF COLUMNS IN ARRAYS LPFIX AND LPROP
C     KP1 = NUMBER OF COLUMNS IN  FLTNP & PRTLPT & PTPROP
C     KP2 = NUMBER OF COLUMNS IN ARRAYS FLTEL AND ELPROP
C     KP3 = NUMBER OF COLUMNS IN ARRAY FLTMIS
C     COORD = SPATIAL COORDINATES OF ELEMENT'S NODES
C     D = NODAL PARAMETERS ASSOCIATED WITH A ELEMENT
C     ELPROP = ELEMENT ARRAY OF FLOATING POINT PROPERTIES
C     LPROP = ARRAY OF FIXED POINT ELEMENT PROPERTIES
C     PRTLPT = FLOATING POINT PROP OF ELEMENT'S NODES
C     FLTMIS = SYSTEM STORAGE OF FLOATING POINT MISC. PROP
C     C = ELEMENT COLUMN MATRIX
C     NTAPE1 = UNIT FOR POST SOLUTION MATRICES STORAGE
C     NTAPE2,3,4 = OPTIONAL UNITS FOR USER  (USED WHEN > 0)
C
C    ..........................................................
C     *** ELCOL PROBLEM DEPENDENT STATEMENTS FOLLOW ***
C    ..........................................................
C    APPLICATION: PLANE STRESS, ISOTROPIC MATERIAL, CST
C    NSPACE = 2, N= 3, NG = 2, NELFRE = 6, NLPFLO = 4
C    ELPROP(1) = ELASTIC MODULUS ELPROP(2) = POISSON RATIO
C    ELPROP(3) = SHEAR MODULUS   ELPROP(4) = THICKNESS
C    ELPROP(5) = X-BODY FORCE      ELPROP(6) = Y-BODY FORCE
     COMMON /ELARG/ XB,YB,E,V,G,T,FX,FY,B(3,6),
    1               STRESS(3,6),DM(3,3),AREA
     TA = T*AREA
     CX = TA*FX/3.
     CY = TA*FY/3.
C---> BODY FORCE LOADS
     C(1) = CX
     C(2) = CY
     C(3) = CX
     C(4) = CY
     C(5) = CX
     C(6) = CY
     RETURN
     END
```

Fig. 11.3 Subroutine ELPOST

```
      SUBROUTINE ELPOST (NTAPE1,NTAPE2,NTAPE3,NTAPE4)
C     * * * * * * * * * * * * * * * * * * * * * * * * * *
C          GENERATE DATA FOR POST SOLUTION CALCULATIONS
C     * * * * * * * * * * * * * * * * * * * * * * * * * *
CDP   IMPLICIT REAL*8(A-H,O-Z)
C     NTAPE1 = UNIT FOR POST SOLUTION MATRICES STORAGE
C     NTAPE2,3,4 = OPTIONAL UNITS FOR USER (USED WHEN > 0)
C
C    ..........................................................
C     *** ELPOST PROBLEM DEPENDENT STATEMENTS FOLLOW ***
C    ..........................................................
C     APPLICATION: PLANE STRESS, ISOTROPIC MATERIAL, CST
C-->  STORE DATA FOR STRESS AND STRAIN CALCULATIONS
C     B = ELEMENT STRAIN-DISPLACEMENT MATRIX
C     STRESS = ELEMENT STRESS-DISPLACEMENT MATRIX
      COMMON /ELARG/ XB,YB,E,V,G,T,FX,FY,B(3,6),
     1               STRESS(3,6),DM(3,3),AREA
      WRITE (NTAPE1)  XB,YB,B,STRESS
      RETURN
      END
```

Fig. 11.4 Subroutine POSTEL

```
      SUBROUTINE POSTEL (NTAPE1,NELFRE,D,IE,NTAPE2,
     1                  NTAPE3,NTAPE4,IT,NITER,NE,M)
C     * * * * * * * * * * * * * * * * * * * * * * * * * * *
C          ELEMENT LEVEL POST SOLUTION CALCULATIONS
C     * * * * * * * * * * * * * * * * * * * * * * * * * * *
C     NTAPE1 = UNIT FOR POST SOLUTION MATRICES STORAGE
C     NELFRE = NUMBER OF DEGREES OF FREEDOM PER ELEMENT
C     D = NODAL PARAMETERS ASSOCIATED WITH THE ELEMENT
C     IE = ELEMENT NUMBER
C     NTAPE2,3,4 = OPTIONAL UNITS FOR USER (USED WHEN > 0)
C     IT = CURRENT ITERATION NUMBER
C     NITER = MAXIMUM NUMBER OF ITERATIONS
C     NE = TOTAL NUMBER OF ELEMENTS
C     M = TOTAL NUMBER OF NODES
CDP   IMPLICIT REAL*8 (A-H,O-Z)
      DIMENSION  D(NELFRE)
C
C     ..........................................................
C      *** POSTEL PROBLEM DEPENDENT STATEMENTS FOLLOW ***
C     ..........................................................
C     APPLICATION: PLANE STRESS, ISOTROPIC MATERIAL, CST
C     NSPACE = 2, N= 3, NG = 2, NELFRE = 6, NLPFLC = 4
C-->  CALCULATE STRESSES AND STRAINS AT ELEMENT CENTROID
      COMMON /ELARG/ XB,YB,EE,V,G,T,FX,FY,B(3,6),
     1               STRESS(3,6),E(3),SIG(3),DUM(3),AREA
C     E = ELEMENT CENTROIDAL STRAINS
C     SIG = ELEMENT CENTROIDAL STRESSES
      DATA KODE/1/
C-->  PRINT HEADINGS ON FIRST CALL
      IF ( KODE.EQ.0 )  GO TO 10
      WRITE (6,5000)
 5000 FORMAT ('1* * * ELEMENT STRESSES * * *',/,
     1' ELEMENT',5X,'X',10X,'Y',14X,'STRAIN',33X,'STRESS',/,
     2 26X,'XX',9X,'YY',9X,'XY',9X,'XX',9X,'YY',9X,'XY',//)
      KODE = 0
   10 READ ( NTAPE1)   XB,YB,B,STRESS
C-->  CALCULATE STRAINS
      CALL   MMULT  (B,D,E,3,6,1)
C-->  CALCULATE STRESSES
      CALL   MMULT  (STRESS,D,SIG,3,6,1)
C     PRINT RESULTS
      WRITE (6,5010)  IE,XB,YB,(E(K),K=1,3),(SIG(L),L=1,3)
 5010 FORMAT (I5,2F10.2,6E11.3)
      RETURN
      END
```

Fig. 11.5 Plane stress problem with nodal forces

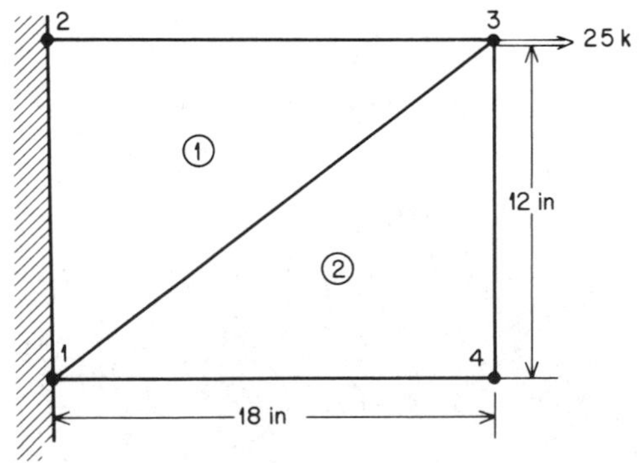

Fig. 11.6 Plane stress analysis (Ural's example); test problem data

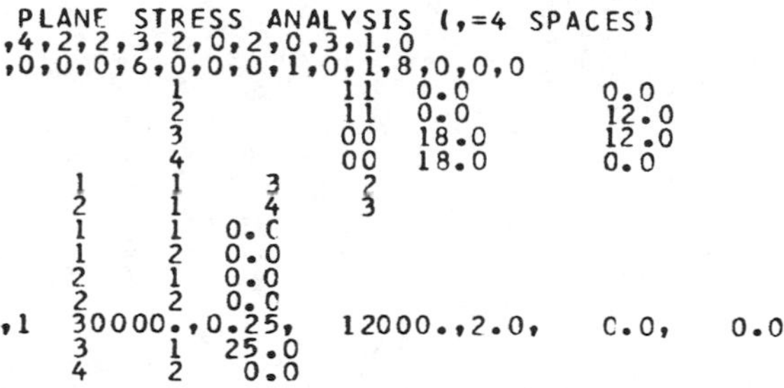

```
PLANE STRESS ANALYSIS (,=4 SPACES)
,4,2,2,3,2,0,2,0,3,1,0
,0,0,0,6,0,0,0,1,0,1,8,0,0,0
            1             11  0.0          0.0
            2             11  0.0         12.0
            3             00  18.0        12.0
            4             00  18.0         0.0
     1      1      3       2
     2      1      4       3
     1      1    0.C
     1      2    0.0
     2      1    0.0
     2      2    0.C
  ,1   30000.,0.25,    12000.,2.0,     C.0,     0.0
     3      1    25.0
     4      2     0.0
```

Fig. 11.7 Plane stress analysis (Ural's example); test problem output

```
*****  PROBLEM PARAMETERS   *****
NUMBER OF NODAL POINTS IN SYSTEM =..........     4
NUMBER OF ELEMENTS IN SYSTEM =.............      2
NUMBER OF NODES PER ELEMENT =..............      3
NUMBER OF PARAMETERS PER NODE =............      2
DIMENSION OF SPACE =.......................      2
NUMBER OF BOUNDARIES WITH GIVEN FLUX=.......     0
NUMBER OF NODES CN BOUNDARY SEGMENT=........     2
NUMBER OF CCNTOUR CURVES BETWEEN
  5TH & 95TH PERCENTILE OF EACH PARAMETER =      3
INITIAL FORCING VECTOR TO BE INPUT
NUMBER OF FIXED PT PROP PER NODE =..........     0
NUMBER OF FLOATING PT PROP PER NODE =.......     0
NUMBER OF FIXED PT PROP PER ELEMENT =.......     0
NUMBER OF FLOATING PT PROP PER ELEMENT =....     6
NUMBER OF FIXED PT MISC PROP =..............     0
NUMBER OF FLOATING PT MISC PROP =...........     0

ELEMENT PROPERTIES ARE HOMOGENEOUS.
OPTIONAL UNIT NUMBERS (UTILIZED IF > 0)
NTAPE1=8 NTAPE2=0 NTAPE3=0 NTAPE4=0
NODAL PARAMETERS TO BE LISTED BY NODES

*** NODAL POINT DATA ***
NODE, CONSTRAINT INDICATOR, 2 COORDINATES
         1            11       0.0000      0.0000
         2            11       0.0000     12.0000
         3             0      18.0000     12.0000
         4             0      18.0000      0.0000

*** ELEMENT CONNECTIVITY DATA ***
 ELEMENT NO.,  3 NODAL INCIDENCES.
    1      1      3      2
    2      1      4      3

 *** CONSTRAINT EQUATION DATA ***
OCONSTRAINT TYPE ONE
 EQ. NO.      NODE1      PAR1            A1
     1          1          1     .00000E+00
     2          1          2     .00000E+00
     3          2          1     .00000E+00
     4          2          2     .00000E+00

***   ELEMENT    PROPERTIES      ***
 ELEMENT NO.     PROPERTY NO.          VALUE
     1                1          0.30000E+05
     1                2          0.25000E+00
     1                3          0.12000E+05
     1                4          0.20000E+01
     1                5          0.00000E+00
     1                6          0.00000E+00
 END OF FLOATING PT PROPERTIES OF ELEMENTS
```

Fig. 11.7—continued

```
*** INITIAL FCRCING VECTOR DATA ***
  NODE    PARAMETER          VALUE          DOF
     3         1      0.25000000E+02         5
     4         2      0.00000000E+C0         8

***  OUTPUT OF RESULTS  ***
NODE,   2 COORCINATES,   2 PARAMETERS.
1   0.00E+00   0.COE+00   -0.1654E-23    0.1969F-15
2   0.00E+00   C.12E+02    0.6355E-15   -0.1969E-15
3   0.18E+02   0.12E+02    0.9061F-03   -0.4723E-03
4   0.18E+02   C.COE+00    0.2519E-03   -0.5090E-03

* * * ELEMENT STRESSES * * *
ELEM   X      Y       STRAIN: XX,    YY,     XY
ELEM                  STRESS: XX,    YY,     XY
 1    6.0    8.0   0.503E-04  -0.328E-16  -0.262E-04
 1                 0.161E+01   0.403F+00  -0.315E+00
 2   12.0    4.0   0.140E-04   0.306E-05   0.262E-04
 2                 0.472E+00   0.210E+00   0.315E+00

*** EXTREME VALUES OF THE NODAL PARAMETERS ***
PARAMETER NO.   MAX. VALUE,NODE    MIN. VALUE,NODE
   1     0.90618587E-03 ,  3   -0.16543612E-23 ,  1
   2     0.19693173E-15 ,  1   -0.50907353E-03 ,  4
NORMAL ENDING OF MODEL PROGRAM.
```

Fig. 11.8 **(a)** Subroutine MATPRT, and **(b)** subroutine MATWRT

```
      SUBROUTINE MATPRT (NUM,NLPFLO,MISCFL,FLTMIS,PRTMAT)
C     * * * * * * * * * * * * * * * * * * * * * * * * *
C     EXTRACT FLOATING PT MATERIAL PROPERTIES CF MATERIAL
C     NUM FROM MISCELLANEOUS FLOATING PT SYSTEM PROPERTIES
C     * * * * * * * * * * * * * * * * * * * * * * * * *
CDP   IMPLICIT REAL*8 (A-H,O-Z)
      DIMENSION  FLTMIS(MISCFL), PRTMAT(NLPFLO)
C     NUM = MATERIAL NUMBER
C     NMAX = MAXIMUM ALLOWABLE MATERIAL NUMBER
C     NLPFLO = NUMBER OF FLOATING POINT ELEMENT PRCPERTIES
C     MISCFL = NO OF MISC FLOATING POINT SYSTEM PRCPERTIES
C     FLTMIS = SYSTEM STORAGE FOR MISC FLOATING POINT PROP
C     PROPERTIES ARE STORED IN FLTMIS IN CRDER OF MAT NO
C     PRTMAT = FLOATING POINT PROP. ARRAY FOR MATERIAL NUM
      NMAX = MISCFL/NLPFLO
      IF ( NUM.GT.NMAX )  WRITE (6,5000)
 5000 FORMAT (' DIMENSIONS EXCEEDED IN SUBROUTINE MATPRT')
      ISTART = NLPFLO*(NUM - 1)
      DO 10  I = 1,NLPFLO
   10 PRTMAT(I) = FLTMIS(ISTART+I)
      RETURN
      END
```

(a)

```
      SUBROUTINE  MATWRT (NLPFLO,MISCFL,FLTMIS,PRTMAT)
C     * * * * * * * * * * * * * * * * * * * * * * * * *
C     LIST FLOATING POINT PROPERTIES BY MATERIAL NUMBER
C     * * * * * * * * * * * * * * * * * * * * * * * * *
CDP   IMPLICIT REAL*8 (A-H,O-Z)
      DIMENSION  FLTMIS(MISCFL), PRTMAT(NLPFLO)
C     NLPFLO = NUMBER OF FLOATING POINT ELEMENT PROP
C     MISCFL = NUMBER OF MISC FLOATING PCINT SYSTEM PRCP
C     FLTMIS = SYSTEM STORAGE OF MISC FLOATING POINT PROP
C     PRTMAT = FLOATING POINT PROP ARRAY FCR MATERIAL NUM
C     NMAX = MAXIMUM ALLOWABLE MATERIAL NUMBER
      NMAX = MISCFL/NLPFLO
      WRITE (6,5000)  NLPFLO
 5000 FORMAT ('1 *** LIST OF FLOATING PCINT ',
     1'MATERIAL PROPERTIES ***',//,
     2' MATERIAL NUMBER,',I8,' FLOATING PCINT PRCPERTIES')
      DO 10  I = 1,NMAX
      CALL  MATPRT (I,NLPFLO,MISCFL,FLTMIS,PRTMAT)
      WRITE (6,5010) I,(PRTMAT(J),J=1,NLPFLO)
 5010 FORMAT (I10,(10E12.5))
   10 CONTINUE
      RETURN
      END
```

(b)

Fig. 11.9 Plane stress with surface traction

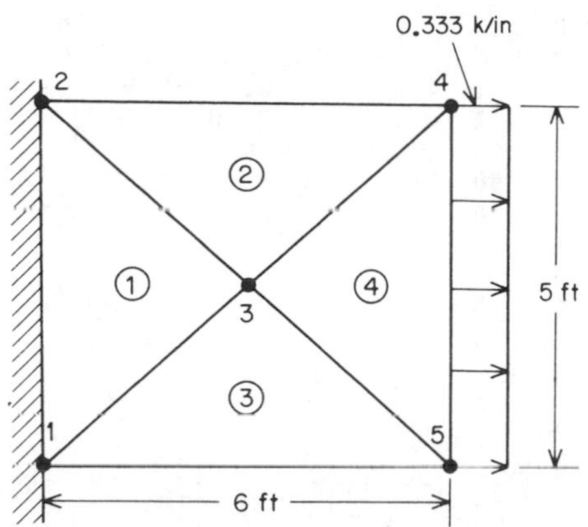

Fig. 11.10 Subroutine BFLUX

```
      SUBROUTINE  BFLUX (FLUX,COORD,LBN,NSPACE,NFLUX,
     1                   NG,C,S,IOPT)
C     * * * * * * * * * * * * * * * * * * * * * * * * * * *
C          PROBLEM DEPENDENT BOUNDARY FLUX CONTRIBUTIONS
C     * * * * * * * * * * * * * * * * * * * * * * * * * * *
CDP   IMPLICIT REAL*8 (A-H,O-Z)
      DIMENSION COORD(LBN,NSPACE), FLUX(LBN,NG), C(NFLUX),
     1          S(NFLUX,NFLUX)
C     FLUX = SPECIFIED BOUNDARY FLUX COMPONENTS
C     COORD = SPATIAL COORDINATES OF SEGMENT NODES
C     LBN = NO. OF NODES ON AN ELEMENT BOUNDARY SEGMENT
C     NSPACE = DIMENSION OF SOLUTION SPACE
C     NFLUX = LBN*NG = MAXIMUM NUMBER OF FLUX CONTRIBUTIONS
C     C = BOUNDARY FLUX COLUMN MATRIX CONTRIBUTIONS
C     S = BOUNDARY FLUX SQUARE MATRIX
C     NG = NUMBER OF PARAMETERS PER NODE POINT
C     IOPT = PROBLEM MATRIX REQUIREMENTS
C          = 1, CALCULATE C ONLY
C          = 2, CALCULATE S ONLY
C          = 3, CALCULATE BOTH C AND S
C     ...........................................................
C       ** BFLUX PROBLEM DEPENDENT STATEMENTS FOLLOW **
C     ...........................................................
C     LINEAR VARIATION OF 2 COMPONENTS ON STRAIGHT LINE
CDP   SQRT(Z) = DSQRT(Z)
      IOPT = 1
      DX  = COORD(2,1) - COORD(1,1)
      DY  = COORD(2,2) - COORD(1,2)
      DL  = SQRT (DX*DX + DY*DY)
      DLD6 = DL/6.
C     ODD = X FORCE, EVEN = Y FORCE
      C(1) = DLD6*( 2.*FLUX(1,1) + FLUX(2,1) )
      C(2) = DLD6*( 2.*FLUX(1,2) + FLUX(2,2) )
      C(3) = DLD6*( FLUX(1,1) + 2.*FLUX(2,1) )
      C(4) = DLD6*( FLUX(1,2) + 2.*FLUX(2,2) )
      RETURN
      END
```

Fig. 11.11 Plate with surface fraction (Ural's example 5.2); test problem data

```
5    4    2    3    2    1    2    0    0    0    0
0    0    0    6    0    0    0    1    0    1    8    0
     1         11  0.0       0.0
     2         11  0.0       60.0
     3         00  36.       30.
     4         00  72.       60.
     5         00  72.       0.0
1    1    3    2
2    2    3    4
3    1    5    3
4    3    5    4
1    1  0.0
1    2  0.0
2    1  0.0
2    2  0.0
1 29000.    0.3       11154.    1.5       0.0       0.0
5    4
0.3333333 0.0       0.3333333 0.0
```

Fig. 11.12 Plate with surface traction (Ural's example 5.2); test problem output

```
*****  PROBLEM PARAMETERS  *****
NUMBER OF NODAL POINTS IN SYSTEM =..........   5
NUMBER OF ELEMENTS IN SYSTEM =.............   4
NUMBER OF NODES PER ELEMENT =..............   3
NUMBER OF PARAMETERS PER NODE =............   2
DIMENSION OF SPACE =.......................   2
NUMBER OF BOUNDARIES WITH GIVEN FLUX=......   1
NUMBER OF NODES ON BOUNDARY SEGMENT=.......   2
NUMBER OF CONTOUR CURVES BETWEEN
  5TH & 95TH PERCENTILE OF EACH PARAMETER =   0
INITIAL FORCING VECTOR TO BE INPUT
NUMBER OF FIXED PT PROP PER NODE =.........   0
NUMBER OF FLOATING PT PROP PER NODE =......   0
NUMBER OF FIXED PT PROP PER ELEMENT =......   0
NUMBER OF FLOATING PT PROP PER ELEMENT =...   6
NUMBER OF FIXED PT MISC PROP =.............   0
NUMBER OF FLOATING PT MISC PROP =..........   0

ELEMENT PROPERTIES ARE HOMOGENEOUS.
OPTIONAL UNIT NUMBERS (UTILIZED IF > 0)
NTAPE1=8 NTAPE2=0 NTAPE3=0 NTAPE4=0
NODAL PARAMETERS TO BE LISTED BY NODES

*** NODAL POINT DATA ***
NODE, CONSTRAINT INDICATOR, 2 COORDINATES
         1          11     0.0000      0.0000
         2          11     0.0000     60.0000
         3           0    36.0000     30.0000
         4           0    72.0000     60.0000
         5           0    72.0000      0.0000

*** ELEMENT CONNECTIVITY DATA ***
ELEMENT NO., 3 NODAL INCIDENCES.
    1    1    3    2
    2    2    3    4
    3    1    5    3
    4    3    5    4

*** CONSTRAINT EQUATION DATA ***
CONSTRAINT TYPE ONE
EQ. NO.    NODE1    PAR1          A1
     1        1        1    .00000000E+00
     2        1        2    .00000000E+00
     3        2        1    .00000000E+00
     4        2        2    .00000000E+00
```

Fig. 11.12—continued

```
*** ELEMENT PROPERTIES ***
ELEMENT NO.  PROPERTY NO.     VALUE
    1            1        0.29000000E+05
    1            2        0.30000000E+00
    1            3        0.11154000E+05
    1            4        0.15000000E+01
    1            5        0.00000000E+00
    1            6        0.00000000E+00
 END OF FLOATING PT PROPERTIES OF ELEMENTS

*** ELEMENT BOUNDARY FLUXES ***
SEGMENT                   2 NODES,/,
                          2 FLUX COMPONENTS PER NODE
     1         5           4
       0.3333             0.0
       0.3333             0.0

 ***  OUTPUT OF RESULTS  ***
 NODE,  2 COORDINATES,   2 PARAMETERS.
 1  0.00E+00  0.00E+00  0.00000E+00  0.00000E+00
 2  0.00E+00  0.00E+00  0.00000E+00  0.00000E+00
 3  0.36E+02  0.60E+02  0.24387E-03  0.30493E-19
 4  0.72E+02  0.60E+00  0.53455E-03 -0.90230E-04
 5  0.72E+02  0.00E+00  0.53455E-03  0.90230E-04

 * * * ELEMENT STRESSES * * *
 ELEM   X     Y           STRAIN: XX,    YY,    XY
 ELEM                     STRESS: XX,    YY,    XY
 1  12.0  30.0  0.677E-05  0.000E+00  0.847E-21
 1              0.216E+00  0.648E-01  0.945E-17
 2  36.0  50.0  0.742E-05 -0.150E-05 -0.473E-06
 2              0.222E+00  0.231E-01 -0.528E-02
 3  36.0  10.0  0.742E-05 -0.150E-05  0.473E-06
 3              0.222E+00  0.231E-01  0.528E-02
 4  60.0  30.0  0.807E-05 -0.301E-05 -0.212E-21
 4              0.229E+00 -0.187E-01  0.260E-17
 NORMAL ENDING OF MODEL PROGRAM
```

Fig. 11.13 (**a**) Subroutine BTDB, and (**b**) subroutine BTDIAB

```
      SUBROUTINE BTDB (D,B,C,M,N,IOPT,S)
C     * * * * * * * * * * * * * * * * * * * * * * * *
C      SPECIAL MATRIX MULTIPLICATION OPERATION
C              IF IOPT=0,   C = (B)T*D*B
C              IF IOPT=1,   C = (B)T*D*B*S + C
C     * * * * * * * * * * * * * * * * * * * * * * * *
CDP   IMPLICIT REAL*8 (A-H,O-Z)
      DIMENSION D(M,M), B(M,N), C(N,N)
C     D(M,M) = SYMMETRIC SQUARE MATRIX
C     B(M,N) = RECTANGLUAR ARRAY
C     C(N,N) = RETURNED SYMMETRIC SQ MATRIX
C     S = SCALAR COEFFICIENT
      DO 50  L = 1,N
      DO 40  K = 1,L
      SUM = 0.D0
      DO 20  I = 1,M
      DBIK = 0.D0
      DO 10  J = 1,M
C     USE SYMMETRY OF D
   10 DBIK = DBIK + D(J,I)*B(J,K)
      SUM = SUM + B(I,L)*DBIK
   20 CONTINUE
      IF ( IOPT.EQ.0 )  GO TO 30
      C(L,K) = C(L,K) + SUM*S
      GO TO 40
   30 C(L,K) = SUM
   40 C(K,L) = C(L,K)
   50 CONTINUE
      RETURN
      END
```

(**a**)

Fig. 11.13—continued

```
      SUBROUTINE BTDIAB (D,B,C,M,N,IOPT,S)
C     * * * * * * * * * * * * * * * * * * * * * * * * * * *
C      SPECIAL MATRIX MULTIPLICATION OPERATION
C              IF IOPT=0,  C = (B)T*D*B
C              IF IOPT=1,  C = (B)T*D*B*S + C
C     * * * * * * * * * * * * * * * * * * * * * * * * * * *
CDP   IMPLICIT REAL*8 (A-H,O-Z)
      DIMENSION D(M), B(M,N), C(N,N)
C     D(M,M) = DIAGCNAL MATRIX
C     B(M,N) = RECTANGLUAR ARRAY
C     C(N,N) = RETURNED SYMMETRIC SQ MATRIX
C     S = SCALAR COEFFICIENT
      DO 50  L = 1,N
      DO 40  K = 1,L
      SUM = 0.D0
      DO 20  I = 1,M
   20 SUM = SUM + B(I,L)*D(I)*B(I,K)
      IF ( IOPT.EQ.0 )  GO TO 30
      C(L,K) = C(L,K) + SUM*S
      GO TC 40
   30 C(L,K) = SUM
   40 C(K,L) = C(L,K)
   50 CONTINUE
      RETURN
      END
```

(b)

Table 11.1 Parameter Definitions for Plane Stress

CONTROL:
- NSPACE = 2
- N = 3
- NG = 2
- NLPFLO = 6
- ISOLVT = 0*, or 1**
- NTAPE1 = 8
- LBN = 0*, or 2**

DEPENDENT VARIABLES:
- 1 = *x*-component of displacement, u
- 2 = *y*-component of displacement, v

BOUNDARY FLUX COMPONENTS AT NODES**:
- 1 = *x*-component of surface traction, T_x
- 2 = *y*-component of surface traction, T_y

PROPERTIES:
- Element level:
  - Floating point:
    - 1 = modulus of elasticity, E
    - 2 = Poisson's ratio, ν
    - 3 = shear modulus, G
    - 4 = element thickness, t
    - 5 = *x*-body force component, X
    - 6 = *y*-body force component, Y

CONSTRAINTS:
- Type 1: specified nodal displacements
- Type 2: inclined roller supports

Table 11.1—continued

INITIAL VALUES OF COLUMN VECTOR**:
Specified external nodal forces

POST-SOLUTION CALCULATIONS:
1 = centroidal coordinates
2 = centroidal strains
3 = centroidal stresses

* default value
** Optional

11.3 Heat conduction

Several authors including Zienkiewicz [87], Desai and Abel [29] and Huebner [43] have considered two-dimensional heat conduction. The problem is governed by the two-dimensional Poisson equation

$$\frac{\partial}{\partial x}\left(K_x \frac{\partial T}{\partial x}\right) + \frac{\partial}{\partial y}\left(K_y \frac{\partial T}{\partial y}\right) + g = 0 \tag{11.4}$$

where $T(x, y)$ denotes the temperature, K_x and K_y are the material conductivities in the x- and y-directions, respectively, and $g(x, y)$ is the heat generation per unit area. Usually one encounters boundary conditions on T on some segment of the boundary and specifies heat fluxes, q_n, on the remaining segments. The latter quantity is positive in the direction of the outward normal vector. It has been shown [43] that the equivalent functional to be minimized is

$$I = \frac{1}{2}\int_A \left[K_x \left(\frac{\partial T}{\partial x}\right)^2 + K_y \left(\frac{\partial T}{\partial y}\right)^2\right] \mathrm{d}a - \int_A gT \,\mathrm{d}a - \int_{\Gamma_n} q_n T \,\mathrm{d}s \tag{11.5}$$

where the last boundary integral is evaluated over the boundary segments on which q_n is specified. Other types of boundary conditions, such as convective heat losses, can be accounted for in the functional. The first term in the functional leads to the definition of the element square matrix, the second contributes to the element column matrix, and the third contributes to a boundary segment column matrix.

A four node bilinear isoparametric element is used and r and s are taken to be the local element coordinates. The element square matrix and column matrix are

$$\mathbf{S}^{e}{}_{4\times 4} = \int_{-1}^{1}\int_{-1}^{1} [K_x^e \mathbf{d}_x^{eT} \mathbf{d}_x^e + K_y^e \mathbf{d}_y^{eT} \mathbf{d}_y^e]\, |J^e|\, dr\, ds$$

and (11.6)

$$\mathbf{C}^{e}{}_{4\times 1} = \int_{-1}^{1}\int_{-1}^{1} g^e \mathbf{H}^{eT}\, |J^e|\, dr\, ds$$

where $\mathbf{H}^e(r, s)$ is the element interpolation matrix, and $\mathbf{d}_x^e$ and $\mathbf{d}_y^e$ denote the global derivatives of $\mathbf{H}^e$; that is, $\mathbf{d}_x^e \equiv \partial/\partial x\, \mathbf{H}^e$, etc., as defined in Chapter 5. The column matrix associated with a specified flux on an exterior boundary segment of an element is

$$\mathbf{C}^{b}{}_{2\times 1} = \int_{-1}^{1} q_n^b \mathbf{H}^{bT}\, |J^b|\, d\Gamma. \tag{11.7}$$

So far, we have defined a two-dimensional element (NSPACE = 2), with four nodes (N = 4), one parameter, t, per node (NG = 1), and three floating point properties, g, K_x, and K_y per element (NLPFLO = 3). In addition, a typical element boundary segment has two nodes (LBN = 2). These three matrices are evaluated in subroutines ELSQ, ELCOL, and BFLUX, respectively. These subroutines are shown in Fig. 11.14 to 11.16. The first two matrices are evaluated by Gaussian quadratures and the last is evaluated by using the closed form expression shown in Fig. 4.1. It would be more economical to form both $\mathbf{S}^e$ and $\mathbf{C}^e$ in the same integration loop but since this is not a production code the alternative approach has been selected for clarity. Since distorted (non-rectangular) elements may require higher order integration rules the user is required to input the desired integration order for each element; that is, there is one fixed point property per element (NLPFIX = 1). If one wishes to utilize the default value (2) of the integration order then one simply sets NLPFIX = 0 which always causes LPROP(1) = 0. Subroutine ELSQ tests for this option. Note from Fig. 11.14 that subroutine ELSQ calls subroutines GAUS2D, SHP4Q, DER4Q, JACOB, I2BY2, and GDERIV. All of these subroutines have been presented in the previous chapters.

In this case no specific post-solution calculations are required (NTAPE1 = 0). However, if one desired to calculate the x- and y-thermal gradients at each integration point within the element it could be easily accomplished. Since the global derivative matrices $\mathbf{d}_x^e$ and $\mathbf{d}_y^e$ must be generated at each integration point they could be written on NTAPE1 from subroutine ELSQ for later use in subroutine POSTEL. Similarly, since $\mathbf{H}^e$ must be evaluated at each integration point in ELCOL, it is simple to find the global coordinates of the integration points, i.e. $\mathbf{x}:\mathbf{y} = \mathbf{H}$ **COORD**, and write them on NTAPE1 for later use by subroutine POSTEL.

The non-zero value of NTAPE1 would cause subroutine ELPOST to be called, but it would consist of only a single RETURN statement. Once the nodal temperatures had been calculated, subroutine POSTEL would be called for each element and then three operations would be performed for each integration point in the element. First, $\mathbf{d}_x$ and $\mathbf{d}_y$ would be read from NTAPE1 and multiplied by the nodal temperatures to obtain the thermal gradients at the point, e.g. $\partial t/\partial x = \mathbf{d}_x^e\mathbf{t}^e$. Next the global coordinates of the point would be read from NTAPE1. Finally, the element number, integration point number, point coordinates, and thermal gradients would be printed by subroutine POSTEL.

It would be desirable to obtain contours of constant temperatures. Thus, we request five contours (NCURVE = 5) between the minimum and maximum temperatures calculated by the program. In closing, note that this formulation is summarized in Table 11.2.

As the first example of two-dimensional heat conduction, consider the problem of uniform heat generation in a unit square plate. In non-dimensional form, let the governing equation be $T_{,xx} + T_{,yy} + 8 = 0$, i.e. $K_x = K_y = 1$ and $g = 8$, and let $T = 0$ around the perimeter of the square. Myers [56] presents the closed form solution (see problem 3.39) and shows that the steady state temperature at the centre is 0.5894. The problem is shown in Fig. 11.17 along with the finite element mesh. Since the problem has symmetric geometry, properties, and boundary conditions only one quarter of the region was utilized (although one-eighth could have been used). Along the two lines of symmetry one has a new boundary condition that the heat flux, q_n, must be zero. Since this leads to null contributions on those boundary lines, these conditions require no special consideration. The problem under consideration is two-dimensional (NSPACE = 2), involves twenty-five nodes (M = 25) with one temperature per node (NG = 1). There are sixteen quadrilateral elements (NE = 16) with four nodes each (N = 4) and three floating point properties that are homogeneous throughout the mesh (NLPFLO = 3, LHOMO = 1). The internal heat generation will define a non-zero element column matrix (NULCOL = 0). There are no boundary segments with a specified non-zero normal heat flux (NSEG = 0).

The input data for this problem are shown in Fig. 11.18 and the output of the model code is shown in Fig. 11.19. Note that the centre temperature at node 1 differs by only $1\frac{1}{4}\%$ from the exact value stated by Myers. The calculated contour data points from subroutine CONTUR are illustrated in Fig. 11.17. The contour curves exhibit the required symmetry along the diagonal line.

Before leaving this application it is desirable to consider the programming changes necessary to utilize a higher order element. For example, assume that one wishes to utilize an eight node quadratic element instead of the present linear element. Then it would be necessary to set N = 8, and replace the calls to shape function and local derivative subroutines, SHP4Q and DER4Q, with an equivalent set of subroutines, say SHP8Q and DER8Q, for the quadratic element. With a higher order element one would usually need to use a higher order quadrature rule in integrating the element matrices. The simplicity of introducing higher elements is a major practical advantage of programming with numerical integration.

The changes necessary to convert to a three-dimensional analysis are almost as simple. If one is utilizing an eight node hexahedron (or *brick* element) one would again supply shape function and local derivative routines, say SHP8H and DER8H. Since NSPACE = 3, the Jacobian would be a 3×3 matrix and to obtain its inverse we would replace the call to I2BY2 with a similar subroutine, say I3BY3, to invert the matrix. Finally the local coordinate definition of the quadrature points would be obtained by an alternative routine, say GAUS3D. The three-dimensional formulation will be illustrated in a later section.

Fig. 11.14 Subroutine ELSQ

```
      SUBROUTINE  ELSQ  (N,NELFRE,NSPACE,IP5,JP2,KP1,KP2,
     1                KP3,COORD,D,ELPROP,LPROP,PRTLPT,FLTMIS,
     2                S,NTAPE1,NTAPE2,NTAPE3,NTAPE4)
C     * * * * * * * * * * * * * * * * * * * * * * * * * * *
C                GENERATE ELEMENT SQUARE MATRIX
C     * * * * * * * * * * * * * * * * * * * * * * * * * * *
CDP   IMPLICIT REAL*8(A-H,O-Z)
      DIMENSION COORD(N,NSPACE), D(NELFRE), ELPROP(KP2),
     1          LPROP(JP2), S(NELFRE,NELFRE),
     2          PRTLPT(IP5,KP1), FLTMIS(KP3)
C     N = NUMBER OF NODES PER ELEMENT
C     NELFRE = NUMBER OF DEGREES OF FREEDOM PER ELEMENT
C     NSPACE = DIMENSION OF SPACE
C     IP5 = NUMBER OF ROWS IN ARRAY PRTLPT
C     JP2 = NUMBER OF COLUMNS IN ARRAYS LPFIX AND LPROP
C     KP1 = NUMBER OF COLUMNS IN FLTNP & PRTLPT & PTPROP
C     KP2 = NUMBER OF COLUMNS IN ARRAYS FLTEL AND ELPROP
C     KP3 = NUMBER OF COLUMNS IN ARRAY FLTMIS
C     COORD = SPATIAL COORDINATES OF ELEMENT'S NODES
C     D = NODAL PARAMETERS ASSOCIATED WITH AN ELEMENT
C     ELPROP = ELEMENT ARRAY OF FLOATING PT PROPERTIES
C     LPROP = ARRAY OF FIXED POINT ELEMENT PROPERTIES
C     PRTLPT = FLOATING POINT PROP FOR ELEMENT'S NODES
C     FLTMIS = SYSTEM STORAGE OF FLOATING PT MISC PROP
C     S = ELEMENT SQUARE MATRIX
C     NTAPE1 = UNIT FOR POST SOLUTION MATRICES STORAGE
C     NTAPE2,3,4 = OPTIONAL UNITS FOR USER (USED WHEN > 0)

C     ..........................................................
C     *** ELSQ PROBLEM DEPENDENT STATEMENTS FOLLOW ***
C     ..........................................................
C     SOLUTION OF POISSON EQUATION IN TWO DIMENSIONS
C     USING A FOUR NODE ISOPARAMETRIC QUADRILATERIAL ELEMENT
C     NSPACE = 2, N = 4, NG = 1, NELFRE = 4
C     ELPROP(1) = CONDUCTIVITY IN X-DIRECTION
C     ELPROP(2) = CONDUCTIVITY IN Y-DIRECTION
C     ELPROP(3) = HEAT GENERATION PER UNIT AREA
C     LPROP(1) = GAUSSIAN INTEGRATION ORDER IN EACH DIMEN
      COMMON /ELARG/ GPT(4),GWT(4),PT(2,16),WT(16),NIP
      COMMON /ELARG2/ XK, YK, G, DELTA(2,4), AJ(2,2),
     1               AJINV(2,2), GLOBAL(2,4), H(4), LSQ
      COMMON /ELARG3/ COL(4), QE(4), TOTAL
      DATA NQPOLD /0/
C-->  DEFINE PROPERTIES
      XK = ELPROP(1)
      YK = ELPROP(2)
      G  = ELPROP(3)
      NQP = LPROP(1)
```

Fig. 11.14—continued

```
C      DEFAULT NQP = 2 SINCE LPROP(1) = 0 IF NLPFIX = 0
       IF ( NQP.LT.2 )  NQP = 2
       IF ( NQP.GT.4 )  NQP = 4
C-->   CALCULATE QUADRATURE DATA ( IF REQUIRED )
       IF ( NQPOLD.EQ.NQP )  GO TO 30
       NQPOLD = NQP
       CALL GAUS2D (NQP,GPT,GWT,NIP,PT,WT)
       LSQ = NELFRE*NELFRE
   30  CONTINUE
C-->   ZERO CALCULATED ARRAYS OF ELEMENT
       CALL ZEROA (LSQ,S)
       CALL ZEROA (NELFRE,COL)
       CALL ZEROA (NELFRE,QE)
C-->   NUMERICAL INTEGRATION LOOP
       DO 60  IP = 1,NIP
C      FORM INTERPOLATION FUNCTIONS
       CALL SHP4Q (PT(1,IP),PT(2,IP),H)
C      FORM LOCAL DERIVATIVES
       CALL  DER4Q (PT(1,IP),PT(2,IP),DELTA)
C      FORM JACOBIAN MATRIX
       CALL   JACOB (N,NSPACE,DELTA,COORD,AJ)
C      FORM DETERMINATE AND INVERSE OF JACOBIAN
       CALL   I2BY2 (AJ,AJINV,DET)
       DETWT = DET*WT(IP)
C      FORM GLOBAL DERIVATIVES
       CALL   GDERIV (NSPACE,N,AJINV,DELTA,GLOBAL)
C      EVALUATE PRODUCTS
       DO 50  J = 1,NELFRE
C      COLUMN MATRIX
       COL(J) = COL(J) + DETWT*G*H(J)
C      AREA INTEGRAL
       QE(J) = QE(J) + DETWT*H(J)
C      SQUARE MATRIX
       DO 40  I = 1,NELFRE
   40  S(I,J) = S(I,J) + DETWT*( XK*GLOBAL(1,I)*GLOBAL(1,J)
     1                         + YK*GLOBAL(2,I)*GLOBAL(2,J) )
   50  CONTINUE
   60  CONTINUE
C      NUMERICAL INTEGRATION COMPLETED
       RETURN
       END
```

Fig. 11.15 Subroutine ELCOL

```
       SUBROUTINE  ELCOL (N,NELFRE,NSPACE,IP5,JP2,KP1,KP2,
     1             KP3,COORD,D,ELPROP,LPROP,PRTLPT,FLTMIS,
     2             C,NTAPE1,NTAPE2,NTAPE3,NTAPE4)
C      * * * * * * * * * * * * * * * * * * * * * * * * * * *
C                  GENERATE ELEMENT COLUMN MATRIX
C      * * * * * * * * * * * * * * * * * * * * * * * * * * *
CDP    IMPLICIT REAL*8(A-H,O-Z)
       DIMENSION COORD(N,NSPACE), D(NELFRE), ELPROP(KP2),
     1           LPROP(JP2), C(NELFRE), PRTLPT(IP5,KP1),
     2           FLTMIS(KP3)
C      N = NUMBER OF NODES PER ELEMENT
C      NELFRE = NUMBER OF DEGREES OF FREEDOM PER ELEMENT
C      NSPACE = DIMENSION OF SPACE
C      IP5 = NUMBER OF ROWS IN ARRAY PRTLPT
C      JP2 = NUMBER OF COLUMNS IN ARRAYS LPFIX AND LPROP
C      KP1 = NUMBER OF COLUMNS IN  FLTNP & PRTLPT & PTPROP
C      KP2 = NUMBER OF COLUMNS IN ARRAYS FLTEL AND ELPROP
C      KP3 = NUMBER OF COLUMNS IN ARRAY FLTMIS
C      COORD = SPATIAL COORDINATES OF ELEMENT'S NODES
C      D = NODAL PARAMETERS ASSOCIATED WITH A ELEMENT
C      ELPROP = ELEMENT ARRAY OF FLOATING POINT PROPERTIES
C      LPROP = ARRAY OF FIXED POINT ELEMENT PROPERTIES
C      PRTLPT = FLOATING POINT PROP OF ELEMENT'S NODES
C      FLTMIS = SYSTEM STORAGE OF FLOATING POINT MISC. PROP
C      C = ELEMENT COLUMN MATRIX
C      NTAPE1 = UNIT FOR POST SOLUTION MATRICES STORAGE
C      NTAPE2,3,4 = OPTIONAL UNITS FOR USER  (USED WHEN > 0)

C      ..................................................
C       ***  ELCOL PROBLEM DEPENDENT STATEMENTS FOLLOW ***
C      ..................................................
C      SOLUTION OF POISSON EQUATION IN TWO DIMENSIONS
C      USING A FOUR NODE ISOPARAMETRIC QUADRILATERIAL ELEMENT
C      NSPACE = 2, N = 4, NG = 1, NELFRE = 4
       COMMON /ELARG3/ COL(4), QE(4), TOTAL
C-->   RECOVER COLUMN MATRIX
       DO 10  I = 1,NELFRE
 10    C(I) = COL(I)
       RETURN
       END
```

Fig. 11.16 Subroutine BFLUX

```
      SUBROUTINE  BFLUX (FLUX,COORD,LBN,NSPACE,NFLUX,
     1                   NG,C,S,IOPT)
C     * * * * * * * * * * * * * * * * * * * * * * * * *
C         PROBLEM DEPENDENT BOUNDARY FLUX CONTRIBUTIONS
C     * * * * * * * * * * * * * * * * * * * * * * * * *
CDP   IMPLICIT REAL*8 (A-H,O-Z)
      DIMENSION COORD(LBN,NSPACE), FLUX(LBN,NG), C(NFLUX),
     1          S(NFLUX,NFLUX)
C     FLUX = SPECIFIED BOUNDARY FLUX COMPONENTS
C     COORD = SPATIAL COORDINATES OF SEGMENT NODES
C     LBN = NO. OF NODES ON AN ELEMENT BOUNDARY SEGMENT
C     NSPACE = DIMENSION OF SOLUTION SPACE
C     NFLUX = LBN*NG = MAXIMUM NUMBER OF FLUX CONTRIBUTIONS
C     C = BOUNDARY FLUX COLUMN MATRIX CONTRIBUTIONS
C     S = BOUNDARY FLUX SQUARE MATRIX
C     NG = NUMBER OF PARAMETERS PER NODE POINT
C     IOPT = PROBLEM MATRIX REQUIREMENTS
C          = 1, CALCULATE C ONLY
C          = 2, CALCULATE S ONLY
C          = 3, CALCULATE BOTH C AND S
C
C     ..........................................................
C        ** BFLUX PROBLEM DEPENDENT STATEMENTS FOLLOW **
C     ..........................................................
C     SOLUTION OF POISSON EQUATION IN TWO DIMENSIONS
C     USING A FOUR NODE ISOPARAMETRIC QUADRILATERAL ELEMENT
C     LINEAR NORMAL FLUX VARIATION ALONG SEGMENT
C     LBN = 2, NSPACE = 2, NG = 1, NFLUX = 2
C     FLUX(K,1) = OUTWARD NORMAL HEAT FLUX AT NODE K
C     BL = BOUNDARY SEGMENT LENGTH
CDP   SQRT(Z) = DSQRT(Z)
      IOPT = 1
      DX = COORD(2,1) - COORD(1,1)
      DY = COORD(2,2) - COORD(1,2)
      BL = SQRT(DX*DX + DY*DY)
      C(1) = BL*(FLUX(1,1)*2. + FLUX(2,1))/6.
      C(2) = BL*(FLUX(1,1) + FLUX(2,1)*2.)/6.
      RETURN
      END
```

Fig. 11.17 Isothermal curves

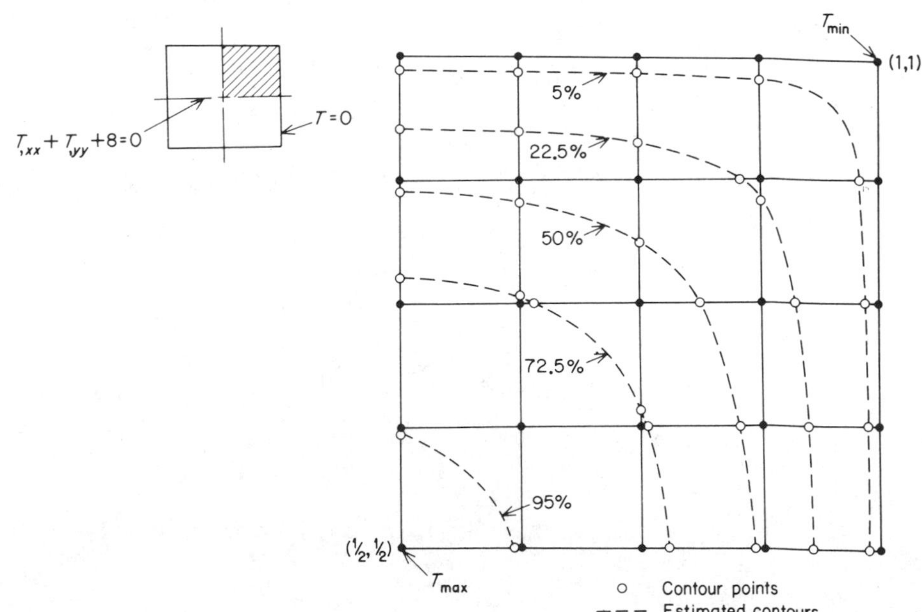

Fig. 11.18 Conduction in a square with uniform source (Myers 3.39); test problem data

```
CONDUCTION IN A SQ. WITH UNIFORM SOURCE (,=4 SPACES)
 25    16,1,4,2,0,0,0,5,0,0
,0,0,1,3,0,0,0,1,0,1,8,0,0,0
            1            C    0.5          0.5
            2            0    0.625        0.5
            3            0    0.75         0.5
            4            0    0.875        0.5
            5            1    1.0          0.5
            6            0    0.5          0.625
            7            0    0.625        0.625
            8            0    0.75         0.625
            9            0    0.875        0.625
           10            1    1.0          0.625
           11            0    0.5          0.75
           12            0    0.625        0.75
           13            0    0.75         0.75
           14            0    0.875        0.75
           15            1    1.0          0.75
           16            0    0.5          0.875
           17            0    0.625        0.875
           18            C    0.75         0.875
           19            0    0.875        0.875
           20            1    1.0          0.875
           21            1    0.5          1.0
           22            1    0.625        1.0
           23            1    0.75         1.0
           24            1    0.875        1.0
           25            1    1.0          1.0
  1     1      2     7     6
  2     2      3     8     7
  3     3      4     9     8
  4     4      5    10     9
  5     6      7    12    11
  6     7      8    13    12
  7     8      9    14    13
  8     9     10    15    14
  9    11     12    17    16
 10    12     13    18    17
 11    13     14    19    18
 12    14     15    20    19
 13    16     17    22    21
 14    17     18    23    22
 15    18     19    24    23
 16    19     20    25    24
  5     1   0.0
 10     1   0.0
 15     1   0.0
 20     1   0.C
 21     1   0.C
 22     1   0.0
 23     1   0.C
 24     1   0.C
 25     1   0.0
  1            3
  1   1.0          1.0          8.0
```

Fig. 11.19 Conduction in a square with uniform source (Myers 3.39); test problem output

```
*****   PROBLEM PARAMETERS   *****
NUMBER CF NODAL POINTS IN SYSTEM =..........  25
NUMBER CF ELEMENTS IN SYSTEM =..............  16
NUMBER CF NODES PER ELEMENT =...............   4
NUMBER CF PARAMETERS PER NODE =.............   1
DIMENSICN OF SPACE =........................   2
NUMBER CF BOUNDARIES WITH GIVEN FLUX=.......   0
NUMBER CF NODES CN BOUNDARY SEGMENT=........   0
NUMBER CF CONTOUR CURVES BETWEEN
   5TH & 95TH PERCENTILE CF EACH PARAMETER =   5
NUMBER CF FIXED PT PROP PER NODE =..........   0
NUMBER CF FLOATING PT PRCP PER NODE =.......   0
NUMBER CF FIXED PT PROP PER ELEMENT =.......   1
NUMBER CF FLOATING PT PRCP PER ELEMENT =....   3
NUMBER OF FIXED PT MISC PROP =..............   0
NUMBER CF FLOATING PT MISC PROP =...........   0
```

Fig. 11.19—continued

```
ELEMENT PROPERTIES ARE HCMCGENECUS.
OPTIONAL UNIT NUMBERS (UTILIZED IF > 0)
NTAPE1=8 NTAPE2=0 NTAPE3=0 NTAPE4=0
NODAL PARAMETERS TO BE LISTED BY NODES

*** NODAL POINT DATA ***
NODE, CONSTRAINT INDICATOR, 2 COORDINATES
      1      0    0.5000     0.5000
      2      0    0.6250     0.5000
      3      0    0.7500     0.5000
      4      0    0.8750     0.5000
      5      1    1.0000     0.5000
      6      0    0.5000     0.6250
      7      0    0.6250     0.6250
      8      0    0.7500     0.6250
      9      0    0.8750     0.6250
     10      1    1.0000     0.6250
     11      0    0.5000     0.7500
     12      0    0.6250     0.7500
     13      0    0.7500     0.7500
     14      0    0.8750     0.7500
     15      1    1.0000     0.7500
     16      0    0.5000     0.8750
     17      0    0.6250     0.8750
     18      0    0.7500     0.8750
     19      0    0.8750     0.8750
     20      1    1.0000     0.8750
     21      1    0.5000     1.0000
     22      1    0.6250     1.0000
     23      1    0.7500     1.0000
     24      1    0.8750     1.0000
     25      1    1.0000     1.0000

*** ELEMENT CONNECTIVITY DATA ***
 ELEMENT NO.,   4 NODAL INCIDENCES.
   1     1    2    7    6
   2     2    3    8    7
   3     3    4    9    8
   4     4    5   10    9
   5     6    7   12   11
   6     7    8   13   12
   7     8    9   14   13
   8     9   10   15   14
   9    11   12   17   16
  10    12   13   18   17
  11    13   14   19   18
  12    14   15   20   19
  13    16   17   22   21
  14    17   18   23   22
  15    18   19   24   23
  16    19   20   25   24

 ***  NODAL PARAMETER CONSTRAINT LIST   ***
CONSTRAINT         NUMBER OF
   TYPE            EQUATIONS
     1                 9
     2                 0
     3                 0

*** CONSTRAINT EQUATION DATA ***
CONSTRAINT TYPE ONE
EQ. NO.    NODE1    PAR1          A1
    1        5        1    .00000000E+00
    2       10        1    .00000000E+00
    3       15        1    .00000000E+00
    4       20        1    .00000000E+00
    5       21        1    .00000000E+00
    6       22        1    .00000000E+00
    7       23        1    .00000000E+00
    8       24        1    .00000000E+00
    9       25        1    .00000000E+00
```

Fig. 11.19—continued

```
***   ELEMENT   PROPERTIES     ***
ELEMENT NO.  PROPERTY NC.      VALUE
      1                1          2
END OF FIXED POINT PROPERTIES OF ELEMENTS
      1                1        0.10000000E+01
      1                2        0.10000000E+01
      1                3        0.80000000E+01
END OF FLOATING PT PROPERTIES OF ELEMENTS

***   OUTPUT OF RESULTS   ***
NODE,   2 COORDINATES,   1 PARAMETERS.
   1  0.500E+00   0.500E+00   0.59678641E+00
   2  0.625E+00   0.500E+00   0.56478982E+00
   3  0.750E+00   0.500E+00   0.46439090E+00
   4  0.875E+00   0.500E+00   0.28274026E+00
   5  0.100E+01   0.500E+00   0.25256895E-12
   6  0.500E+00   0.625E+00   0.56478982E+00
   7  0.625E+00   0.625E+00   0.53503298E+00
   8  0.750E+00   0.625E+00   0.44124783E+00
   9  0.875E+00   0.625E+00   0.27001777E+00
  10  0.100E+01   0.625E+00   0.24235019E-12
  11  0.500E+00   0.750E+00   0.46439090E+00
  12  0.625E+00   0.750E+00   0.44124782E+00
  13  0.750E+00   0.750E+00   0.36762036E+00
  14  0.875E+00   0.750E+00   0.22914274E+00
  15  0.100E+01   0.750E+00   0.20919344E-12
  16  0.500E+00   0.875E+00   0.28274025E+00
  17  0.625E+00   0.875E+00   0.27001776E+00
  18  0.750E+00   0.875E+00   0.22914275E+00
  19  0.875E+00   0.875E+00   0.15011323E+00
  20  0.100E+01   0.875E+00   0.1416890E-12
  21  0.500E+00   0.100E+01   0.25256893E-12
  22  0.625E+00   0.100E+01   0.24235019E-12
  23  0.750E+00   0.100E+01   0.20919343E-12
  24  0.875E+00   0.100E+01   0.14168899E-12
  25  0.100E+01   0.100E+01   0.12193162E-12

AREA INTEGRAL OF DEPENDENT VARIABLE
ELEMENT          INTEGRAL
      1          0.88335898E-02
      2          0.78338342E-02
      3          0.56968624E-02
      4          0.21592110E-02
      5          0.78338340E-02
      6          0.69732384E-02
      7          0.51094872E-02
      8          0.19498458E-02
      9          0.56968622E-02
     10          0.51094872E-02
     11          0.38125746E-02
     12          0.14814687E-02
     13          0.21592110E-02
     14          0.19498458E-02
     15          0.14814687E-02
     16          0.58637981E-03
TOTAL AREA INTEGRAL =    0.68667202E-01

*** EXTREME VALUES OF THE NODAL PARAMETERS ***
PARAMETER NO.   MAX.VALUE,NODE    MIN.VALUE,NODE
 1    0.59678641E+00 ,  1    0.12193162E-12 , 25

*** CALCULATION OF CONTOUR CURVES ***
PARAMETER NUMBER =  1
 5TH PERCENTILE VALUE =  0.29839320E-01
95TH PERCENTILE VALUE =  0.56694709E+00
```

Fig. 11.19—continued

```
CONTOUR NUMBER =     1
VALUE OF CONTOUR =  0.29839320E-01
POINT NO.,  2 COORDINATES
    1      0.98680798E+00      0.50000000E+00
    2      0.98618641E+00      0.62500000E+00
    3      0.98618641E+00      0.62500000E+00
    4      0.98372231E+00      0.75000000E+00
    5      0.98372231E+00      0.75000000E+00
    6      0.97515266E+00      0.87500000E+00
    7      0.62500000E+00      0.98618641E+00
    8      0.50000000E+00      0.98680798E+00
    9      0.75000000E+00      0.98372231E+00
   10      0.62500000E+00      0.98618641E+00
   11      0.87500000E+00      0.97515266E+00
   12      0.75000000E+00      0.98372231E+00
   13      0.97515266E+00      0.87500000E+00
   14      0.87500000E+00      0.97515266E+00
CONTOUR NUMBER =  2
VALUE OF CONTOUR =  0.16411626E+00
POINT NO.,  2 COORDINATES
    1      0.92744389E+00      0.50000000E+00
    2      0.92402525E+00      0.62500000E+00
    3      0.92402525E+00      0.62500000E+00
    4      0.91047269E+00      0.75000000E+00
    5      0.87500000E+00      0.85285158E+00
    6      0.85285158E+00      0.87500000E+00
    7      0.91047269E+00      0.75000000E+00
    8      0.87500000E+00      0.85285158E+00
    9      0.62500000E+00      0.92402525E+00
   10      0.50000000E+00      0.92744389E+00
   11      0.75000000E+00      0.91047269E+00
   12      0.62500000E+00      0.92402525E+00
   13      0.85285158E+00      0.87500000E+00
   14      0.75000000E+00      0.91047269E+00
CONTOUR NUMBER =  3
VALUE OF CONTOUR =  0.29839320E+00
POINT NO.,  2 COORDINATES
    1      0.86422867E+00      0.50000000E+00
    2      0.85428559E+00      0.62500000E+00
    3      0.85428559E+00      0.62500000E+00
    4      0.81248948E+00      0.75000000E+00
    5      0.62500000E+00      0.85428558E+00
    6      0.50000000E+00      0.86422857E+00
    7      0.75000000E+00      0.81248949E+00
    8      0.62500000E+00      0.85428558E+00
    9      0.81248948E+00      0.75000000E+00
   10      0.75000000E+00      0.81248949E+00
CONTOUR NUMBER =  4
VALUE OF CONTOUR =  0.43267015E+00
POINT NO.,  2 COORDINATES
    1      0.77182814E+00      0.50000000E+00
    2      0.75626181E+00      0.62500000E+00
    3      0.75000000E+00      0.63956264E+00
    4      0.63956264E+00      0.75000000E+00
    5      0.75626181E+00      0.62500000E+00
    6      0.75000000E+00      0.63956264E+00
    7      0.62500000E+00      0.75626181E+00
    8      0.50000000E+00      0.77182813E+00
    9      0.63956264E+00      0.75000000E+00
   10      0.62500000E+00      0.75626181E+00
CONTOUR NUMBER =  5
VALUE OF CONTOUR =  0.56694709E+00
POINT NO.,  2 COORDINATES
    1      0.61657229E+00      0.50000000E+00
    2      0.50000000E+00      0.61657229E+00
NORMAL ENDING OF MODEL PROGRAM.
```

Table 11.2 Parameter Definitions for Two-dimensional Heat Conduction

CONTROL:
NSPACE = 2 NLPFIX = 1
N = 4 NLPFLO = 3
NG = 1 LBN = 0*, or 2**

DEPENDENT VARIABLES:
1 = temperature, T

BOUNDARY FLUX COMPONENTS AT NODES**:
1 = specified normal outward heat flux per unit length, q_n

PROPERTIES:
Element level:
Fixed point:
1 = Gaussian quadrature order in each dimension, (2*)
Floating point:
1 = thermal conductivity in the x-direction, K_x
2 = thermal conductivity in the y-direction, K_y
3 = heat generation per unit area, g

CONSTRAINTS:
Type 1: specified nodal temperatures

* Default value
** Optional

11.4 Viscous flow in straight ducts

Consider the flow of a viscous fluid in a straight duct with an arbitrary cross-section. The transverse velocity component, w, is of course zero on the perimeter of the cross-section. The distribution of the velocity $w(x, y)$ over the area has been investigated by Schechter [65]. He shows that the distribution of the transverse velocity is governed by the equation

$$\mu \frac{\partial^2 w}{\partial x^2} + \mu \frac{\partial^2 w}{\partial y^2} - \frac{\partial P}{\partial z} = 0 \tag{11.8}$$

where μ denotes the viscosity of the fluid, and $\partial P/\partial Z$ is the pressure gradient in the direction of flow (z-direction). This is another example of the application of the Poisson equation. The previous section presented the isoparametric element matrices for the general Poisson equation

$$\frac{\partial}{\partial x}\left(K_x \frac{\partial w}{\partial x}\right) + \frac{\partial}{\partial y}\left(K_y \frac{\partial w}{\partial y}\right) + G = 0.$$

Clearly, the flow problem is a special case of the above equation where $K_x = K_y = \mu =$ constant and $G = -\partial P/\partial z$. For the three node triangle, Zienkiewicz [82] shows that the element column matrix is

$$\mathbf{C}^e = -\frac{G\Delta}{3}\begin{Bmatrix}1\\1\\1\end{Bmatrix}, \tag{11.9}$$

where Δ is the area of the triangle. Similarly, the element square matrix is

$$\mathbf{S}^e = \frac{K_x}{4\Delta}\begin{bmatrix}b_ib_i & b_ib_j & b_ib_m\\ & b_jb_j & b_jb_m\\ \text{sym.} & & b_mb_m\end{bmatrix} + \frac{K_y}{4\Delta}\begin{bmatrix}c_ic_i & c_ic_j & c_ic_m\\ & c_jc_j & c_jc_m\\ \text{sym.} & & c_mc_m\end{bmatrix}, \tag{11.10}$$

where i, j, and m denote the three nodes of the triangle and where the constants b_i, and c_i are defined in terms of the coordinates of the nodes as

$$b_i = y_j - y_m \qquad c_i = x_m - x_j$$

$$b_j = y_m - y_i \qquad c_j = x_i - x_m$$

$$b_m = y_i - y_m \qquad c_m = x_j - x_i.$$

In this particular application one might be interested in a post-solution calculation of the flow rate, Q, which is defined as $Q = \int w\,da$. In this approximation, w is defined within a typical element by the interpolation functions, i.e. $w(x, y) = \mathbf{H}^e(x, y)\,\mathbf{W}^e$. Thus, the contribution of a typical flow rate is

$$Q^e = \int_{\Delta^e} w^e\,da = \left(\int_{\Delta^e} \mathbf{H}^{eT}\,da\right)\mathbf{W}^e \equiv \mathbf{q}^e\mathbf{W}^e \tag{11.11}$$

where

$$\mathbf{q}^e = \Delta[(a_i + b_i\bar{x} + c_i\bar{y})(a_j + b_j\bar{x} + c_j\bar{y})(a_m + b_m\bar{x} + c_m\bar{y})] \tag{11.12}$$

$$\bar{x} = (x_i + x_i + x_m)/3$$

$$\bar{y} = (y_i + y_j + y_m)/3$$

$$a_i = x_jy_m - x_my_j$$

$$a_j = x_my_i - x_iy_m$$

$$a_m = x_iy_j - x_jy_i$$

$$2\Delta = a_i + a_j + a_m$$

The equation for $\mathbf{q}^e$ could be calculated by subroutine ELPOST and stored for later use. After the nodal velocities have been calculated subroutine POSTEL could read $\mathbf{q}^e$ from NTAPE1, calculate $Q^e = \mathbf{q}^e\mathbf{W}^e$, print the element number and Q^e, and end with a summary total of the total flow rate.

Rather than change element types at this point the previous isoparametric element will be used. Only the post-solution calculations need to be added. That is, we need to generate, store and recover, $\mathbf{Q}^e$. The calculation of $\mathbf{Q}^e$ has already been included in Fig. 11.14. Thus it could be stored there or in ELPOST. The latter option is used here. After the nodal values have been calculated POSTEL recovers $\mathbf{Q}^e$, carries out the multiplication in Eqn. (11.11) and prints the results for each element. These two additional subroutines are illustrated in Fig. 11.20. The previous mesh was used to consider flow in a square duct. Figure 11.21 shows the comparison with the results of Schechter. The element area integrals are included in Fig. 11.22.

Fig. 11.20 **(a)** Subroutine ELPOST, and **(b)** subroutine POSTEL

(a)

```
      SUBROUTINE ELPOST (NTAPE1,NTAPE2,NTAPE3,NTAPE4)
C     * * * * * * * * * * * * * * * * * * * * * * * * * * *
C          GENERATE DATA FOR POST SOLUTICN CALCULATICNS
C     * * * * * * * * * * * * * * * * * * * * * * * * * * *
CDP   IMPLICIT REAL*8(A-H,O-Z)
C     NTAPE1 = UNIT FOR POST SOLUTION MATRICES STORAGE
C     NTAPE2,3,4 = OPTIONAL UNITS FOR USER (USED WHEN > 0)
C
C     .........................................................
C      *** ELPOST PROBLEM DEPENDENT STATEMENTS FCLLCW ***
C     .........................................................
C     QE = AREA INTEGRAL OF ELEMENT SHAPE FUNCTICNS
      COMMCN /ELARG3/ COL(4), QE(4), TOTAL
      WRITE (NTAPE1)  QE
      RETURN
      END
```

(b)

```
      SUBROUTINE POSTEL (NTAPE1,NELFRE,D,IE,NTAPE2,
     1                   NTAPE3,NTAPE4,IT,NITER,NE,M)
C     * * * * * * * * * * * * * * * * * * * * * * * * * * *
C          ELEMENT LEVEL POST SOLUTION CALCULATICNS
C     * * * * * * * * * * * * * * * * * * * * * * * * * * *
C     NTAPE1 = UNIT FOR POST SOLUTION MATRICES STCRAGE
C     NELFRE = NUMBER OF DEGREES OF FREEDOM PER ELEMENT
C     D = NODAL PARAMETERS ASSOCIATED WITH THE ELEMENT
C     IE - ELEMENT NUMBER
C     NTAPE2,3,4 = OPTIONAL UNITS FOR USER (USED WHEN > 0)
C     IT = CURRENT ITERATION NUMBER
C     NITER = MAXIMUM NUMBER OF ITERATICNS
C     NE = TOTAL NUMBER OF ELEMENTS
C     M = TOTAL NUMBER OF NODES
CDP   IMPLICIT REAL*8 (A-H,O-Z)
      DIMENSION  D(NELFRE)
C
C     .........................................................
C      *** POSTEL PROBLEM DEPENDENT STATEMENTS FOLLCW ***
C     .........................................................
C-->  EVALUATE AREA INTEGRAL OF DEPENDENT VARIABLE
C     QE = AREA INTEGRAL OF ELEMENT SHAPE FUNCTICNS
      CCMMCN /ELARG3/ COL(4), QE(4), TOTAL
      DATA KALL/1/
      IF ( KALL.EQ.0 )  GO TO 10
C     PRINT TITLES CN THE FIRST CALL
      KALL = 0
      TOTAL = 0.0
      WRITE (6,1000)
```

Fig. 11.10—continued

```
 1000 FORMAT ('1AREA INTEGRAL OF DEPENDENT VARIABLE',//,
     1 ' ELEMENT         INTEGRAL')
   10 READ (NTAPE1)   QE
C-->  CALCULATE ELEMENT CONTRIBUTION
      CALL  MMULT (QE,D,VALUE,1,NELFRE,1)
      TOTAL = TOTAL + VALUE
      WRITE (6,1010)   IE, VALUE
 1010 FORMAT (I8,E20.8)
      IF ( IE.EQ.NE )  WRITE (6,1020)   TOTAL
 1020 FORMAT ('0TOTAL AREA INTEGRAL = ',E16.8)
      RETURN
      END
```

Fig. 11.21 Velocity estimates in duct

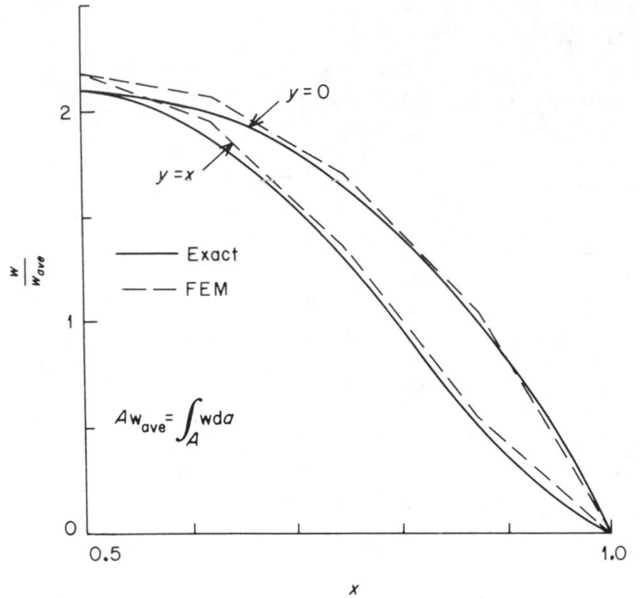

Fig. 11.22 Element flow rates

```
1AREA INTEGRAL OF DEPENDENT VARIABLE

 ELEMENT         INTEGRAL
       1         0.88335898E-02
       2         0.78338342E-02
       3         0.56968624E-02
       4         0.21592110E-02
       5         0.78338340E-02
       6         0.69732384E-02
       7         0.51094872E-02
       8         0.19498458E-02
       9         0.56968622E-02
      10         0.51094872E-02
      11         0.38125746E-02
      12         0.14814687E-02
      13         0.21592110E-02
      14         0.19498458E-02
      15         0.14814687E-02
      16         0.58637981E-03
0TOTAL AREA INTEGRAL =    0.68667202E-01
```

11.5 Potential flow

11.5.1 Introduction

Another common class of problem which can be formulated in terms of the Poisson equation is that of potential flow of ideal fluids. The diffusion coefficients, K_x and K_y, become unity and the source term, G, is usually zero so that the problem reduces to a solution of Laplace's equation. Potential flow can be formulated in terms of the velocity potential, ϕ, or the stream function, ψ. The latter sometimes yields simpler boundary conditions but ϕ will be utilized here since it can be extended to three dimensions. Thus, the governing equation is

$$\frac{\partial^2 \phi}{\partial x^2} + \frac{\partial^2 \phi}{\partial y^2} = 0. \tag{11.13}$$

In this section it is desirable to consider potential flow applications, which involve curved boundaries. This provides the opportunity of illustrating how the previous program developed for the straight sided isoparametric elements can be easily extended to curved elements. Select the eight sided quadratic quadrilateral element shown in Fig. 6.8. It has been shown that its interpolation functions are

$$\begin{aligned}
H_i(r, s) &= \tfrac{1}{4}(1 + rr_i)(1 + ss_i)(rr_i + ss_i - 1), && 1 \leqslant i \leqslant 4, \\
H_i(r, s) &= \tfrac{1}{2}(1 - r^2)(1 + ss_i), && i = 5 \text{ and } 7, \\
H_i(r, s) &= \tfrac{1}{2}(1 - s^2)(1 + rr_i) && i = 6 \text{ and } 8.
\end{aligned} \tag{11.14}$$

These shape functions are implemented in subroutine SHP8Q and the local derivatives of these relations are presented in subroutine DER8Q. These two subroutines are illustrated in Fig. 6.8. To modify the previous set of isoparametric subroutines to utilize this element one simply replaces the original references to SHP4Q and DER4Q with calls to SHP8Q and DER8Q, respectively.

Recall that the potential flow problem is formulated in terms of either the stream function, ψ, or the velocity potential, ϕ. These quantities are usually of secondary interest and the analyst generally requires information on the velocity components. They are defined by the global derivatives of ψ and ϕ. For example, having formulated the problem in terms of ϕ one has

$$u \equiv \frac{\partial \phi}{\partial x}, \quad v \equiv \frac{\partial \phi}{\partial y} \tag{11.15}$$

where u and v denote the x- and y-components of the velocity vector.

Thus, although the program will yield the nodal values of ϕ one must also calculate the above global derivatives. This can be done economically since these derivative quantities must be generated at each quadrature point during construction of the element square matrix, **S**. Hence, one can simply store this derivative information, i.e. matrix **GLOBAL**, and retrieve it later for calculating the global derivatives of ϕ or ψ.

The new element square matrix subroutine and the new post-solution routine POSTEL are shown in Figs. 11.23 and 11.24. There are some new features in each of these routines. They include the option to store, within ELSQ, the global derivatives at each integration point as well as the global coordinates of the point. Thus ELPOST is a dummy routine in this case. The square matrix calculations also utilize ISOPAR. Subroutine ELSQ also recovers installation-dependent plotting data through the miscellaneous properties.

Subroutine POSTEL also executes some CALCOMP plots of the velocity vectors at the integration points. This is done only if the user has made unit NTAPE2 available for storing the data to be plotted. The plots are begun only after the gradients have been evaluated in all elements. Thus it was necessary to utilize the value of the total number of elements, NE, to flag the call on which the plots are generated. Of course, the velocity vectors are also printed by POSTEL.

Fig. 11.23 **(a)** Subroutine ELSQ and **(b)** subroutine LAPL8Q

(a)

```
      SUBROUTINE  ELSQ  (N,NELFRE,NSPACE,IP5,JP2,KP1,KP2,
     1                KP3,COORD,D,ELPROP,LPROP,PRTLPT,FLTMIS,
     2                S,NTAPE1,NTAPE2,NTAPE3,NTAPE4)
C      * * * * * * * * * * * * * * * * * * * * * * * * * *
C                GENERATE ELEMENT SQUARE MATRIX
C      * * * * * * * * * * * * * * * * * * * * * * * * * *
CDP   IMPLICIT REAL*8(A-H,O-Z)
      DIMENSION COORD(N,NSPACE), D(NELFRE), ELFRCP(KP2),
     1          LPROP(JP2), S(NELFRE,NELFRE),
     2          PRTLPT(IP5,KP1), FLTMIS(KP3)
C     N = NUMBER OF NODES PER ELEMENT
C     NELFRE = NUMBER OF DEGREES OF FREEDOM PER ELEMENT
C     NSPACE = DIMENSION OF SPACE
C     IP5 = NUMBER OF ROWS IN ARRAY PRTLPT
C     JP2 = NUMBER OF COLUMNS IN ARRAYS LPFIX AND LPROP
C     KP1 = NUMBER OF COLUMNS IN FLTNP & PRTLPT & PTPROP
C     KP2 = NUMBER OF COLUMNS IN ARRAYS FLTEL AND ELPROP
C     KP3 = NUMBER OF COLUMNS IN ARRAY FLTMIS
C     COORD = SPATIAL COORDINATES OF ELEMENT'S NODES
C     D = NODAL PARAMETERS ASSOCIATED WITH AN ELEMENT
C     ELPRCP = ELEMENT ARRAY OF FLOATING PT PROPERTIES
C     LPROP = ARRAY OF FIXED POINT ELEMENT PROPERTIES
C     PRTLPT = FLOATING POINT PROP FOR ELEMENT'S NODES
C     FLTMIS = SYSTEM STORAGE OF FLOATING PT MISC PROP
C     S = ELEMENT SQUARE MATRIX
C     NTAPE1 = UNIT FOR POST SOLUTION MATRICES STORAGE
C     NTAPE2,3,4 = OPTIONAL UNITS FOR USER (USED WHEN > 0)
C
C     ..........................................................
C      ***  ELSQ PROBLEM DEPENDENT STATEMENTS FOLLOW ***
C     ..........................................................
C     SOLUTION OF LAPLACE EQUATION IN TWO DIMENSIONS
C     USING 8 NODE ISOPARAMETRIC QUADRILATERIAL ELEMENT
C     NSPACE = 2, N = 8, NG = 1, NELFRE = 8
      COMMON /ELARG/ GPT(4),GWT(4),PT(2,16),WT(16),NIP
      COMMON /ELARG2/ XY(2),H(8),GRAD(2),AJ(2,2),AJINV(2,2),
     1                DELTA(2,8),GLOBAL(2,8),COL(8)
```

Fig. 11.23—continued

```
      COMMCN /ELARG3/ XK, YK
      EXTERNAL  LAPL8Q
C     LAPL8Q = SUBR TC FORM PROB DEPENENT INTEGRAN
C     NIP = TOTAL NUMBER OF INTEGRATION POINTS
C     PT, WT = 2-D INTEGRATION POINTS ANC WEIGHTS
C     XY = GLOBAL COORDINATES OF QUADRATURE PCINT
C     H(I) = INTERPOLATION FUNCTION FOR NODE I    1.LE.I.LE.N
C     GRAD = GLOBAL COMPONENTS OF GRADIENT VECTOR
C     AJ = JACOBIAN MATRIX, INVERSE - AJINV
C     DELTA(J,I) = J TH LOCAL DERIVATIVE OF H(I)
C     GLOBAL(J,I) = J TH GLOBAL DERIVATIVE OF H(I)
C     ELPRCP(1) = CCNDUCTIVITY IN X-DIRECTICN  = XK
C     ELPROP(2) = CCNDUCTIVITY IN Y-DIRECTION  = YK
C     LPRCP(1) = GAUSSIAN INTEGRATION ORDER IN EACH DIM
      CATA KALL, NCPOLD /1, 0/
      IF ( KALL.EQ.0 )  GO TO 10
C-->  CN FIRST CALL EXTRACT CALCOMP PLOT DATA (IF ANY)
      KALL = 0
      IF ( NTAPE2.GT.0 )  CALL PLTSET (FLTMIS)
C-->  DEFINE PROPERTIES
   10 XK = ELPROP(1)
      YK = ELPROP(2)
      NQP = LPROP(1)
C     DEFAULT NQP = 2 SINCE LPROP(1) = 0 IF NLPFIX = 0

      IF ( NQP.LT.2 )  NQP = 2
      IF ( NQP.GT.4 )  NQP = 4
      IF ( NQPOLD.EQ.NQP )  GO TO 20
C-->  CALCULATE QUADRATURE DATA ( IF RECUIRED )
      NQPOLD = NQP
      CALL  GAUS2D (NQP,GPT,GWT,NIP,PT,WT)
C     STORE NUMBER OF POINTS FOR GRADIENT CALCULATICNS
   20 IF ( NTAPE1.GT.0 )  WRITE (NTAPE1)  NIP
C-->  NUMERICAL INTEGRATION LOOP
      CALL ISOPAR (N,NSPACE,NELFRE,NIP,S,CCL,PT,WT,H,
     1     DELTA,GLOBAL,COORD,XY,AJ,AJINV,NTAPE1,LAPL8Q)
      RETURN
      END
```

(**b**)

```
      SUBRCUTINE  LAPL8Q (WT,DET,H,DGH,XPT,N,NSPACE,
     1                    NELFRE,COL,SQ,NTAPE1)
C     * * * * * * * * * * * * * * * * * * * * * * * * *
C     PROBLEM DEPENDENT INTEGRAND EVALUATICN IN
C     AN ISOPARAMETRIC ELEMENT
C     * * * * * * * * * * * * * * * * * * * * * * * * *
CDP   IMPLICIT REAL*8 (A-H,O-Z)
      DIMENSION COL(NELFRE), SQ(NELFRE,NELFRE),
     1          H(N), DGH(NSPACE,N), XPT(NSPACE)
C     N = NUMBER OF NODES PER ELEMENT
C     NSPACE = NUMBER OF SPATIAL DIMENSIONS
C     NELFRE = NO OF ELEMENT DEGREES OF FREEDOM
C     H = ELEMENT INTERPOLATION FUNCTICNS
C     DGH = GLOBAL DERIVATIVES OF H
C     XPT = GLOBAL COORDS OF THE POINT
C     WT = QUADRATURE WEIGHT AT POINT
C     DET = JACOBIAN DETERMINATE AT POINT
C     COL = PROB DEP COLUMN MATRIX INTEGRAND
C     SQ = PROB DEP SQUARE MATRIX INTEGRANC
C     NTAPE1 = STORAGE UNIT FOR POST SOLUTICN DATA
C
C     ...................................................
C     *** LAPL8Q PROBLEM DEPENDENT STATEMENTS FCLLCW ***
C     ...................................................
      COMMCN /ELARG3/ XK, YK
C     STORE COORDINATES AND DERIVATIVE MATRIX AT THE PCINT
      IF ( NTAPE1 .GT.0 )  WRITE (NTAPE1)  XPT, DGH
C     EVALUATE PRODUCTS
      DETWT = DET*WT
      DO 40  J = 1,N
      DO 30  I = 1,J
      SQ(I,J) = SQ(I,J) + DETWT*( XK*DGH(1,I)*DGH(1,J)
     1                          + YK*DGH(2,I)*DGH(2,J) )
   30 SQ(J,I) = SQ(I,J)
   40 CONTINUE
      RETURN
      END
```

Fig. 11.24 Subroutine POSTEL

```
      SUBROUTINE POSTEL (NTAPE1,NELFRE,D,IE,NTAPE2,
     1                   NTAPE3,NTAPE4,IT,NITER,NE,M)
C     * * * * * * * * * * * * * * * * * * * * * * * * * * * *
C          ELEMENT LEVEL POST SOLUTION CALCULATICNS
C     * * * * * * * * * * * * * * * * * * * * * * * * * * * *
C     NTAPE1 = UNIT FOR POST SOLUTION MATRICES STCRAGE
C     NELFRE = NUMBER OF DEGREES OF FREEDOM PER ELEMENT
C     D = NODAL PARAMETERS ASSOCIATED WITH THE ELEMENT
C     IE = ELEMENT NUMBER
C     NTAPE2,3,4 = OPTIONAL UNITS FOR USER (USED WHEN > 0)
C     IT = CURRENT ITERATION NUMBER
C     NITER = MAXIMUM NUMBER OF ITERATICNS
C     NE = TOTAL NUMBER OF ELEMENTS
C     M = TOTAL NUMBER OF NODES
CDP   IMPLICIT REAL*8 (A-H,O-Z)
      DIMENSION  D(NELFRE)
C     ..........................................................
C      *** POSTEL PROBLEM DEPENDENT STATEMENTS FCLLCW ***
C     ..........................................................
C-->       GRADIENT CALCULATION AND PLCTS FCR
C     SOLUTION OF LAPLACE EQUATION IN TWO CIMENSIONS
C     USING 8 NODE ISOPARAMETRIC QUADRILATERIAL ELEMENT
      COMMON /ELARG2/ XY(2),H(8),GRAD(2),AJ(2,2),AJINV(2,2),
     1                DELTA(2,8),GLOBAL(2,8),COL(8)
C     CONNECT TO PLOTTER COMMONS
      COMMCN /PLTDIM/ XLIMIT, IBUF(64)
      COMMCN /PLTKAL/ XLEN, YLEN, FIRSTX, FIRSTY, DELTAX,
     1                DELTAY, XLAST, YLAST
C     NE = TOTAL NUMBER OF ELEMENTS
C     XY = GLOBAL COORDINATES OF QUADRATURE PCINT
C     GRAD = GLOBAL COMPONENTS OF GRADIENT VECTOR
C     H(I) = INTERPOLATION FUNCTION FOR NODE I    1<=I<=N
C     GLOBAL(J,I) = J TH GLOBAL DERIVATIVE OF H(I)
      DATA KALL, SIZE, DERMAX /1, 0.07, 0.0/
CDP   ABS(Z) = DABS(Z)
      IF ( KALL.EQ.0 )  GO TO 10
C-->  PRINT TITLES CN THE FIRST CALL
      KALL = 0
      WRITE (6,5000)
 5000 FORMAT ('1 *** GLOBAL DERIVATIVES AT INTEGRATICN'
     1 ' POINTS ***',/,' POINT        X                Y',
     2 '          DP/DX             DP/DY',/)
   10 CONTINUE
C-->  BEGIN ELEMENT POST SOLUTION ANALYSIS
      WRITE (6,5010)   IE
 5010 FORMAT (1H0,20X,'ELEMENT NUMBER ',I3)
C     READ NUMBER OF POINTS TO BE CALCULATED
      READ (NTAPE1)   NIP
C     STORE FOR PLOT USE
      IF ( NTAPE2.GT.0 )   WRITE (NTAPE2)   NIP
      DO 20   J = 1,NIP
C-->  READ COORDS. AND DERIVATIVE MATRIX
      READ (NTAPE1)   XY, GLOBAL
C     CALCULATE DERIVATIVES,   GRAD=GLOBAL*D
      CALL  MMULT (GLOBAL,D,GRAD,2,NELFRE,1)
C-->  PRINT COORDINATES AND GRADIENT AT THE POINT
      WRITE (6,5020)   J, XY, GRAD
 5020 FORMAT ( I7,2E12.4,2E15.5)
C-->  STORE RESULTS TO BE PLOTED LATER
      IF ( NTAPE2.EQ.0 )   GO TO 20
      WRITE (NTAPE2)   XY, GRAD
      IF ( ABS(GRAD(1)).GT.DERMAX ) DERMAX = ABS(GRAD(1))
      IF ( ABS(GRAD(2)).GT.DERMAX ) DERMAX = ABS(GRAD(2))
   20 CONTINUE
C-->  ARE GRADIENT CALCULATIONS COMPLETE FCR ALL ELEMENTS
      IF ( IE.LT.NE )   RETURN
C-->  PRODUCE GRADIENT VECTOR PLOTS   (IF DESIRED)
      IF ( NTAPE2.LT.1 )   RETURN
      WRITE (6,5030)
 5030 FORMAT ('   BEGIN PLOT')
C     START NEW PLOT PAGE
      CALL PAGNEW
C     READ AND DRAW GRADIENT VECTORS
      CALL VECT2D (NTAPE2,XLEN,YLEN,FIRSTX,FIRSTY,DELTAX,DELTAY,
     1             XLAST,YLAST,DERMAX,SIZE,NE)
      WRITE (6,5040)
 5040 FORMAT ('   END PLOT')
      RETURN
      END
```

11.5.2 Patch test

Originally based on engineering judgement, the *patch test* has been proven to be a mathematically valid convergence test [47]. Consider a patch (or sub-assembly) of finite elements containing at least one internal node. An internal node is one completely surrounded by elements. Let the problem be formulated by an integral statement containing derivatives of order 'n'. Assume an arbitrary function, $P(x)$, whose n^{th} order derivatives are constant. Use this function to prescribe the dependent variable on all external nodes of the patch (i.e. $\theta_e = P(x_e)$). Solve for the internal nodal values of the dependent variable, θ_I, and its nth order derivatives in each element. To be a convergent formulation:

1. The internal nodal values must agree with the assumed function evaluated at the internal points (i.e. $\phi_I = P(x_I)$?); and
2. The calculated n^{th} order derivatives *must* agree with the assumed constant values.

It has been found that some non-conforming elements will yield convergent solutions for only one particular mesh pattern. Thus the patch mesh should be completely arbitrary for a valid numerical test. The patch test is very important from the engineering point of view since it can be executed numerically. Thus one obtains a numerical check of the entire program used to formulate the patch test. The patch of elements shown in Fig. 11.25 was utilized. It was assumed that

$$\phi(x, y) \equiv 1 + 3x - 4y \tag{11.16}$$

such that the derivatives

$$\phi_{,x} = 3, \quad \phi_{,y} = -4$$

are constant everywhere. All sixteen points on the exterior boundary were assigned values by substituting their coordinates into Eqn. (11.16). That is, the boundary conditions that $\phi_1 \equiv \phi(x_1, y_1)$, etc., were applied on the exterior boundary. Then the problem was solved numerically to determine the value of ϕ at the interior points (7, 10, 11, 12, 15) and the values of its global derivatives at each integration point. The input data and output results of this patch test are shown in Figs. 11.26 and 11.27. The output shows clearly that the global derivatives at all integration points have the assumed values. It is also easily verified that all the interior nodal values of ϕ are consistent with the assumed form. Thus, the patch test is satisfied and the subroutines pass a necessary numerical test.

Fig. 11.25 Patch test mesh

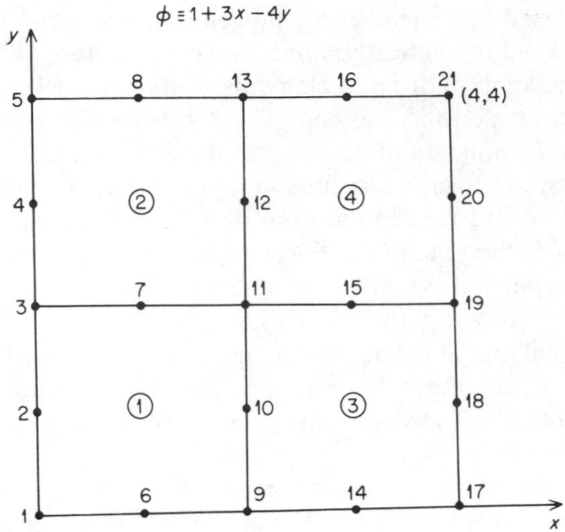

Fig. 11.26 Quadratic quadrilateral patch test (Laplace equation); test problem data

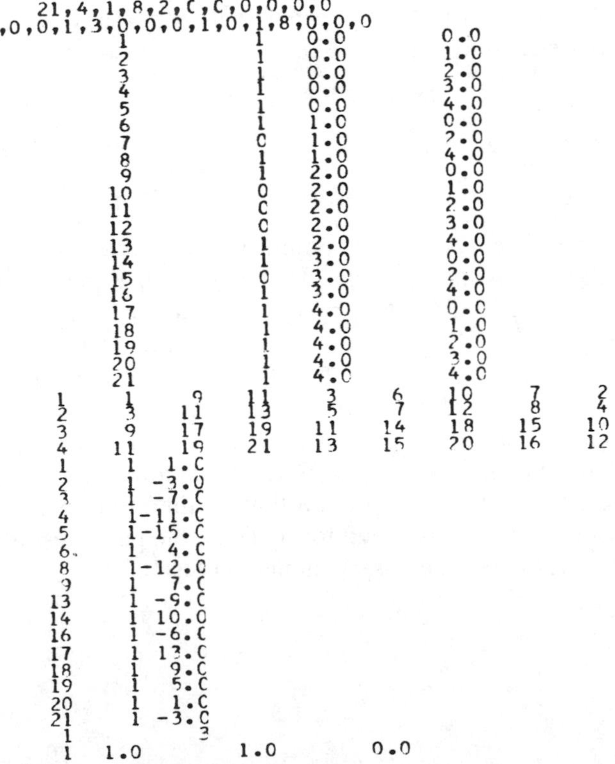

```
  21,4,1,8,2,0,0,0,0,0,0
,0,0,1,3,0,0,0,1,0,1,8,0,0,0
         1               1    0.0          0.0
         2               1    0.0          1.0
         3               1    0.0          2.0
         4               1    0.0          3.0
         5               1    0.0          4.0
         6               1    1.0          0.0
         7               0    1.0          2.0
         8               1    1.0          4.0
         9               1    2.0          0.0
        10               0    2.0          1.0
        11               0    2.0          2.0
        12               0    2.0          3.0
        13               1    2.0          4.0
        14               1    3.0          0.0
        15               0    3.0          2.0
        16               1    3.0          4.0
        17               1    4.0          0.0
        18               1    4.0          1.0
        19               1    4.0          2.0
        20               1    4.0          3.0
        21               1    4.0          4.0
  1      1     9   11     3      6     10     7     2
  2      3    11   13     5      7     12     8     4
  3      9    17   19    11     14     18    15    10
  4     11    19   21    13     15     20    16    12
  1      1   1.0
  2      1  -3.0
  3      1  -7.0
  4      1-11.0
  5      1-15.0
  6      1   4.0
  8      1-12.0
  9      1   7.0
 13      1  -9.0
 14      1  10.0
 16      1  -6.0
 17      1  13.0
 18      1   9.0
 19      1   5.0
 20      1   1.0
 21      1  -3.0
  1
  1    1.0        1.0         0.0
```

Fig. 11.27 Quadratic quadrilateral patch test (Laplace equation); test problem output

```
*****  PROBLEM PARAMETERS  *****
NUMBER CF NODAL POINTS IN SYSTEM =...........21
NUMBER CF ELEMENTS IN SYSTEM =............... 4
NUMBER CF NODES PER ELEMENT =................ 8
NUMBER CF PARAMETERS PER NCDE =.............. 1
DIMENSICN OF SPACE =......................... 2
NUMBER CF BOUNDARIES WITH GIVEN FLUX=........ 0
NUMBER OF NODES CN BOUNDARY SEGMENT=......... 0
NUMBER CF FIXED PT PROP PER NODE =........... 0
NUMBER CF FLOATING PT PRCP PER NCDE =........ 0
NUMBER CF FIXED PT PROP PER ELEMENT =........ 1
NUMBER OF FLOATING PT PROP PER ELEMENT =..... 3
NUMBER CF FIXED PT MISC PRCP =............... 0
NUMBER CF FLOATING PT MISC PROP =............ 0

ELEMENT PROPERTIES ARE HCMCGENECUS.
OPTICNAL UNIT NUMBERS (UTILIZED IF > 0)
NTAPE1=8 NTAPE2=0 NTAPE3=0 NTAPE4=0
NODAL PARAMETERS TO BE LISTED BY NODES

*** NCDAL POINT DATA ***
NODE, CONSTRAINT INDICATOR, 2 COCRDINATES
       1          1     0.0000     0.0000
       2          1     0.00C0     1.0000
       3          1     0.0000     2.0000
       4          1     0.0000     3.0000
       5          1     C.0CCC     4.0000
       6          1     1.000C     0.0000
       7          0     1.00C0     2.0000
       8          1     1.0CCC     4.0000
       9          1     2.00C0     0.0000
      10          0     2.0CCC     1.0000
      11          0     2.0000     2.0000
      12          0     2.0000     3.0000
      13          1     2.00CC     4.0C00
      14          1     3.0000     0.0000
      15          0     3.0000     2.0000
      16          1     3.0000     4.0000
      17          1     4.0000     0.0000
      18          1     4.0C00     1.0000
      19          1     4.000C     2.0000
      20          1     4.0000     3.0000
      21          1     4.0000     4.0000

*** ELEMENT CONNECTIVITY DATA ***
ELEMENT NO.,  8 NODAL INCIDENCES.
   1    1    9   11    3    6   10    7    2
   2    3   11   13    5    7   12    8    4
   3    9   17   19   11   14   18   15   10
   4   11   19   21   13   15   20   16   12

*** CONSTRAINT EQUATION DATA ***
CONSTRAINT TYPE ONE
EQ. NC.   NODE1   PAR1         A1
     1       1       1    .1CCC000E+01
     2       2       1   -.3C0C000E+01
     3       3       1   -.7CCC000E+01
     4       4       1   -.11C0C00E+02
     5       5       1   -.150CCC0E+02
     6       6       1    .4CCCC00E+01
     7       8       1   -.120CCC0E+02
     8       9       1    .7CCCC00E+01
     9      13       1   -.9CCCCC0E+01
    10      14       1    .1C0C000E+02
    11      16       1   -.6CCC000E+01
    12      17       1    .130CC00E+02
    13      18       1    .9C00C00E+01
    14      19       1    .5CCCCC0E+01
    15      20       1    .1000C00E+01
    16      21       1   -.3CCC000E+01

***  ELEMENT   PROPERTIES     ***
ELEMENT NO.  PRCPERTY NC.     VALUE
     1            1             3
END OF FIXED POINT PROPERTIES CF ELEMENTS
     1            1        0.1CCC0000E+01
     1            2        C.1C000000E+01
     1            3        C.C0000000E+00
END OF FLOATING PT PROPERTIES CF ELEMENTS
```

Fig. 11.27—continued

```
***  OUTPUT OF RESULTS  ***
NODE,  2 COORDINATES,   1 PARAMETERS.
   1      0.0000E+00      0.000CE+00     0.1000E+01
   2      0.0000E+00      0.100CE+01    -C.300CE+01
   3      C.0000E+00      0.2000E+01    -0.700CE+01
   4      C.0000E+00      0.300CE+01    -0.110CE+02
   5      0.0000E+00      0.4000E+01    -C.150CE+02
   6      0.1000E+01      0.000CE+00     0.4000E+01
   7      0.1000E+01      0.200CE+01    -C.3999E+01
   8      0.1000E+01      0.4000E+01    -C.120CE+02
   9      0.2000E+01      0.000CE+00     C.7000E+01
  10      0.2000E+01      0.100CE+01     C.300CE+01
  11      0.2000E+01      0.2000E+01    -C.100CE+01
  12      0.2000E+01      C.300CE+C1    -C.500CE+01
  13      0.2000E+01      0.400CE+01    -C.900CE+01
  14      C.3000E+01      0.000CE+00     C.1000E+02
  15      0.3000E+01      0.2000E+01     C.200CE+01
  16      C.3000E+01      0.400CE+01    -0.6000E+01
  17      0.4000E+01      0.000CE+00     C.130CE+02
  18      0.4000E+01      0.100CE+01     C.900CE+01
  19      C.4000E+01      C.200CE+01     0.5000E+01
  20      0.4000E+01      0.300CE+01     0.100CE+01
  21      C.4000E+01      0.40CCE+01    -0.3000E+01
** GLOBAL DERIVATIVES AT INTEGRATION POINTS **
POINT       X            Y           DP/DX         DP/DY
                   ELEMENT NUMBER    1
 1   0.178E+01   0.178E+01   0.300E+01   -0.400E+01
 2   0.100E+01   C.178E+01   C.300E+01   -0.4C0E+01
 3   0.225E+00   C.178E+01   C.300E+01   -0.4C0E+01
 4   0.178E+01   0.100E+01   0.300E+01   -0.400E+01
 5   0.100E+01   0.100E+01   C.300E+01   -0.4C0E+01
 6   0.225E+00   0.100E+01   0.300E+01   -0.4C0E+01
 7   0.178E+01   0.225E+0C   C.300E+01   -0.400E+01
 8   0.10CE+01   0.225E+00   C.300E+01   -C.4C0E+01
 9   0.225E+00   0.225E+00   C.300E+01   -0.400E+01
                   ELEMENT NUMBER    2
 1   0.178E+01   0.378E+01   C.300E+01   -C.4C0E+01
 2   0.100E+01   C.378E+C1   C.300E+01   -0.400E+01
 3   0.225E+00   C.378E+01   C.300E+01   -0.4C0E+01
 4   0.178E+01   0.300E+01   C.300E+01   -0.400E+01
 5   0.1C0E+01   C.300E+C1   C.300E+01   -0.400E+01
 6   0.225E+00   C.300E+01   0.300E+01   -0.400E+01
 7   0.178E+01   0.223E+01   C.300E+01   -0.400E+01
 8   0.100E+01   0.223E+C1   C.300E+01   -0.4C0E+01
 9   0.225E+00   0.223E+01   C.300E+01   -0.400E+01
                   ELEMENT NUMBER    3
 1   0.378E+01   0.178E+01   0.300E+01   -0.400E+01
 2   0.3C0E+01   0.178E+01   0.300E+01   -0.4C0E+01
 3   0.223E+01   C.178E+C1   C.300E+01   -C.4C0E+01
 4   0.378E+01   0.100E+01   0.300E+01   -0.4C0E+01
 5   0.3C0E+01   C.100E+01   C.30CE+01   -0.400E+01
 6   0.223E+01   0.100E+C1   C.300E+01   -C.400E+01
 7   0.378E+01   0.225E+00   C.300E+01   -0.400E+01
 8   0.3C0E+01   C.225E+0C   C.30CE+01   -0.4C0E+01
 9   0.223E+01   0.225E+00   0.300E+01   -0.4C0E+01
                   ELEMENT NUMBER    4
 1   0.378E+01   0.378E+C1   C.300E+01   -0.400E+01
 2   0.300E+01   0.378E+01   0.300E+01   -0.400E+01
 3   0.223E+01   C.378E+C1   C.300E+01   -0.400E+01
 4   0.378E+01   0.300E+01   0.300E+01   -0.4C0E+01
 5   0.300E+01   0.300E+01   0.300E+01   -0.400E+01
 6   0.223E+01   C.300E+C1   C.300E+01   -C.400E+01
 7   0.378E+01   0.223E+01   0.300E+01   -0.400E+01
 8   0.3C0E+01   0.223E+01   C.300E+01   -0.400E+01
 9   0.223E+01   0.223E+01   C.300E+01   -0.400E+01
NORMAL ENDING OF MODEL PROGRAM.
```

11.5.3 Example—Flow Around a Cylinder

Martin and Carey [51] were among the first to publish a numerical example of a finite element potential flow analysis. This same example is also discussed by others [23, 29]. The problem considers the flow around a cylinder in a finite rectangular channel with a uniform inlet flow. The geometry is shown in Fig. 11.28. The present mesh is compared with that of Martin and Carey in Fig. 11.29 (p. 269). By using centreline symmetry and midstream antisymmetry it is possible to employ only one fourth of the flow field. The stream function boundary conditions are discussed by Martin and Carey [51] and Chung [23]. For the velocity potential one has four sets of Neumann (boundary flux) conditions and one set of Dirichlet (nodal parameter) conditions. The first involve zero normal flow, $q_n = \phi_{,n} \equiv 0$, along the centreline *ab* and the solid surfaces *bc* and *de*, and a uniform unit inflow, $q_n \equiv -1$, along *ad*. At the mid-section, *ce*, antisymmetry requires that $v = 0$. Thus $\phi_{,y} = 0$ so that $\phi = \phi(x)$, but in this special case x is constant along that line so we can set ϕ to any desired constant, say zero, along *ce*.

The input data and output results are illustrated in Figs. 11.30 and 11.31. The velocity vector plots are shown in Fig. 11.32. By changing only the specified inlet flux conditions other problems can be considered. If one keeps the same total flow but introduces a parabolic boundary flow one obtains the vector plots shown in Fig. 11.33.

As before, the presence of a boundary flux, q_n, makes it necessary to evaluate the flux column matrix

$$\mathbf{C}^{\mathrm{b}} = \int_{l^{\mathrm{b}}} \mathbf{H}^{\mathrm{b}T} q_n \, \mathrm{d}s. \tag{11.17}$$

The variation of q_n along the boundary segment is assumed to be defined by the nodal (input) values and the segment interpolation equations, i.e.

$$q_n(s) \equiv \mathbf{H}^{\mathrm{b}}(s)\, \mathbf{q}_n^{\mathrm{b}}.$$

Therefore the segment column matrix (for a straight segment) becomes

$$\mathbf{C}^{\mathrm{b}} = \left(\int_{l^{\mathrm{b}}} \mathbf{H}^{\mathrm{b}T}\mathbf{H}^{\mathrm{b}} \, \mathrm{d}s \right) \mathbf{q}_n^{\mathrm{b}} = \frac{l^{\mathrm{b}}}{30} \begin{bmatrix} 4 & 2 & -1 \\ 2 & 16 & 2 \\ -1 & 2 & 4 \end{bmatrix} \mathbf{q}_n^{\mathrm{b}}. \tag{11.18}$$

These calculations are executed in subroutine BFLUX of Fig. 11.34.

Since the second set of post-solution calculations involved vector plots using standard CALCOMP, the routines for this purpose are shown for completeness in Figs. 11.35 and 11.36. Labelled element COMMON beginning with PLT are used to transmit CALCOMP plot data. The parameter definitions used in the previous examples are given in Table 11.3. The values of the centre velocity estimates for the flow around the cylinder are shown in Fig. 11.37. With two elements across the section there was about a 10% error in the maximum velocity. Since ϕ in each element is quadratic the corresponding element velocity estimates are linear, as illustrated by the dashed lines.

Fig. 11.28 Flow around cylinder between parallel walls

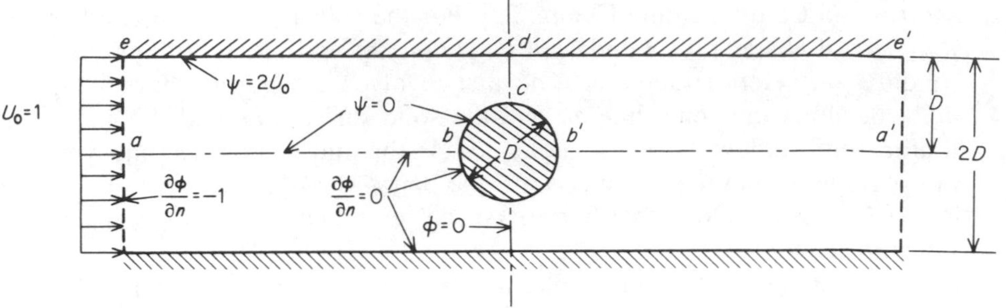

Fig. 11.30 Potential flow around cylinder in a rectangular channel; test problem data

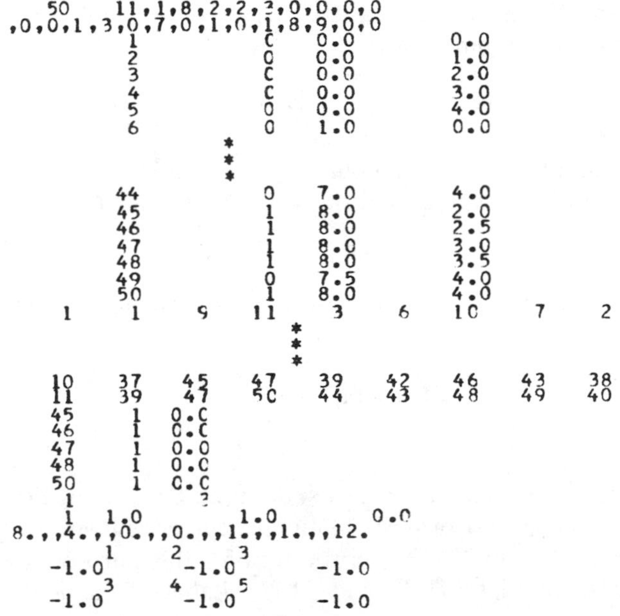

```
       50      11,1,8,2,2,3,0,0,0,0
  ,0,0,1,3,0,7,0,1,0,1,8,9,0,0
              1               0    0.0          0.0
              2               0    0.0          1.0
              3               0    0.0          2.0
              4               0    0.0          3.0
              5               0    0.0          4.0
              6               0    1.0          0.0
                        *
                        *
                        *
             44               0    7.0          4.0
             45               1    8.0          2.0
             46               1    8.0          2.5
             47               1    8.0          3.0
             48               1    8.0          3.5
             49               0    7.5          4.0
             50               1    8.0          4.0
       1      1       9      11      3      6      10      7      2
                                 *
                                 *
                                 *
      10     37      45      47     39     42      46     43     38
      11     39      47      50     44     43      48     49     40
      45      1     0.0
      46      1     0.0
      47      1     0.0
      48      1     0.0
      50      1     0.0
       1            3
       1    1.0             1.0            0.0
  8.,,4.,,0.,,0.,,1.,,1.,,12.
              1        2       3
      -1.0          -1.0          -1.0
              3        4       5
      -1.0          -1.0          -1.0
```

Fig. 11.29 Finite element idealization of cylinder flow problem

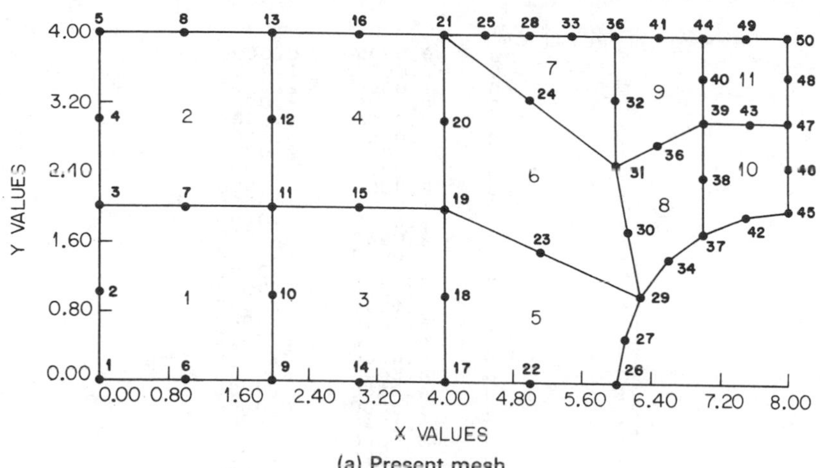

(a) Present mesh

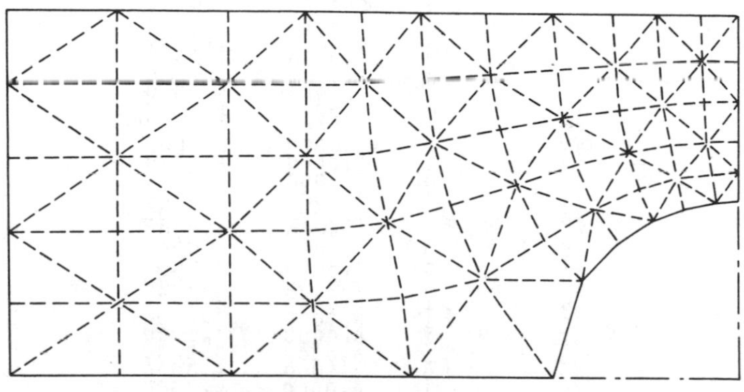

(b) Mesh of Martin and Carey

Fig. 11.31 Potential flow around cylinder in a rectangular channel; test problem output

```
*****  PROBLEM PARAMETERS  *****
NUMBER OF NODAL POINTS IN SYSTEM =..........50
NUMBER OF ELEMENTS IN SYSTEM =.............11
NUMBER OF NODES PER ELEMENT =.............. 8
NUMBER OF PARAMETERS PER NODE =............ 1
DIMENSICN OF SPACE =....................... 2
NUMBER CF BOUNDARIES WITH GIVEN FLUX=...... 2
NUMBER CF NODES CN BOUNCARY SEGMENT=....... 3
NUMBER CF FIXED PT PROP PER NODE =......... 0
NUMBER CF FLOATING PT PROP PER NCDE =...... 0
NUMBER CF FIXED PT PROP PER ELEMENT =...... 1
NUMBER OF FLOATING PT PROP PER ELEMENT =... 3
NUMBER OF FIXED PT MISC PROP =............. 0
NUMBER CF FLOATING PT MISC PROP =.......... 7

ELEMENT PROPERTIES ARE HCMCGENECUS.
OPTIONAL UNIT NUMBERS (UTILIZED IF > 0)
NTAPE1=8 NTAPE2=9 NTAPE3=0 NTAPE4=0
NODAL PARAMETERS TO BE LISTED BY NODES
NODE, CCNSTRAINT INDICATOR, 2 CCORDINATES
```

Fig. 11.31—continued

```
*** NODAL POINT DATA ***
       1         0     0.0000     0.0000
       2         0     0.0000     1.0000
       3         0     0.0000     2.0000
       4         0     0.0000     3.0000
       5         0     0.0000     4.0000
       6         0     1.0000     0.0000
       7         0     1.0000     2.0000
       8         0     1.0000     4.0000
       9         0     2.0000     0.0000
      10         0     2.0000     1.0000
      11         0     2.0000     2.0000
      12         0     2.0000     3.0000
      13         0     2.0000     4.0000
      14         0     3.0000     0.0000
      15         0     3.0000     2.0000
      16         0     3.0000     4.0000
      17         0     4.0000     0.0000
      18         0     4.0000     1.0000
      19         0     4.0000     2.0000
      20         0     4.0000     3.0000
      21         0     4.0000     4.0000
      22         0     5.0000     0.0000
      23         0     5.1340     1.5000
      24         0     5.0000     3.2500
      25         0     4.5000     4.0000
      26         0     6.0000     0.0000
      27         0     6.1340     0.5000
      28         0     5.0000     4.0000
      29         0     6.2680     1.0000
      30         0     6.1340     1.7500
      31         0     6.0000     2.5000
      32         0     6.0000     3.2500
      33         0     5.5000     4.0000
      34         0     6.5858     1.4142
      35         0     6.5000     2.7500
      36         0     6.0000     4.0000
      37         0     7.0000     1.7320
      38         0     7.0000     2.3660
      39         0     7.0000     3.0000
      40         0     7.0000     3.5000
      41         0     6.5000     4.0000
      42         0     7.4842     1.9319
      43         0     7.5000     3.0000
      44         0     7.0000     4.0000
      45         1     8.0000     2.0000
      46         1     8.0000     2.5000
      47         1     8.0000     3.0000
      48         1     8.0000     3.5000
      49         0     7.5000     4.0000
      50         1     8.0000     4.0000

*** ELEMENT CONNECTIVITY DATA ***
 ELEMENT NO.,   8 NODAL INCIDENCES.
   1     1     9    11     3     6    10     7     2
   2     3    11    13     5     7    12     8     4
   3     9    17    19    11    14    18    15    10
   4    11    19    21    13    15    20    16    12
   5    17    26    29    19    22    27    23    18
   6    19    29    31    21    23    30    24    20
   7    21    31    36    28    24    32    33    25
   8    29    37    39    31    34    38    35    30
   9    31    39    44    36    35    40    41    32
  10    37    45    47    39    42    46    43    38
  11    39    47    50    44    43    48    49    40

 *** CONSTRAINT EQUATION DATA ***
 CONSTRAINT TYPE ONE
 EQ. NO.    NODE1    PAR1          A1
     1        45       1    .00000000E+00
     2        46       1    .00000000E+00
     3        47       1    .00000000E+00
     4        48       1    .00000000E+00
     5        50       1    .00000000E+00

 ***   ELEMENT    PROPERTIES       ***
 ELEMENT NO.   PROPERTY NO.      VALUE
       1             1             3
 END OF FIXED POINT PROPERTIES OF ELEMENTS
       1             1        0.10000000E+01
       1             2        0.10000000E+01
       1            -3        0.00000000E+00
 END OF FLOATING PT PROPERTIES OF ELEMENTS
```

Fig. 11.31—continued

```
***   MISCELLANEOUS SYSTEM PROPERTIES   ***
PROPERTY NO.     VALUE
            1  0.80000000E+01
            2  0.40000000E+01
            3  0.00000000E+00
            4  0.00000000E+00
            5  0.10000000E+01
            6  0.10000000E+01
            7  0.12000000E+02
END OF FLOATING PT PROPERTIES OF SYSTEM

*** ELEMENT BOUNDARY FLUXES ***
SEGMENT          3 NODES,/,
                 1 FLUX COMPONENTS PER NODE
      1         1         2         3
        -1.0000
        -1.0000
        -1.0000
      2         3         4         5
        -1.0000
        -1.0000
        -1.0000

***  OUTPUT OF RESULTS   ***
NODE,   2 COORDINATES,   1 PARAMETERS.
  1   0.00000E+00  0.00000E+00  -0.99793769E+01
  2   0.00000E+00  0.10000E+01  -0.99743004E+01
  3   0.00000E+00  0.20000E+01  -0.99642311E+01
  4   0.00000E+00  0.30000E+01  -0.99553375E+01
  5   0.00000E+00  0.40000E+01  -0.99492269E+01
  6   0.10000E+01  0.00000E+00  -0.89830562E+01
  7   0.10000E+01  0.20000E+01  -0.89648184E+01
  8   0.10000E+01  0.40000E+01  -0.89468280E+01
  9   0.20000E+01  0.00000E+00  -0.80016869E+01
 10   0.20000E+01  0.10000E+01  -0.79889559E+01
 11   0.20000E+01  0.20000E+01  -0.79651409E+01
 12   0.20000E+01  0.30000E+01  -0.79394547E+01
 13   0.20000E+01  0.40000E+01  -0.79300071E+01
 14   0.30000E+01  0.00000E+00  -0.70398853E+01
 15   0.30000E+01  0.20000E+01  -0.69620699E+01
 16   0.30000E+01  0.40000E+01  -0.68889508E+01
 17   0.40000E+01  0.00000E+00  -0.61443051E+01
 18   0.40000E+01  0.10000E+01  -0.60760562E+01
 19   0.40000E+01  0.20000E+01  -0.59539673E+01
 20   0.40000E+01  0.30000E+01  -0.58568555E+01
 21   0.40000E+01  0.40000E+01  -0.57917362E+01
 22   0.50000E+01  0.00000E+00  -0.53645467E+01
 23   0.51340E+01  0.15000E+01  -0.49577944E+01
 24   0.50000E+01  0.32500E+01  -0.46820984E+01
 25   0.45000E+01  0.40000E+01  -0.52340845E+01
 26   0.60000E+01  0.00000E+00  -0.49598156E+01
 27   0.61340E+01  0.50000E+00  -0.47837601E+01
 28   0.50000E+01  0.40000E+01  -0.46229193E+01
 29   0.62680E+01  0.10000E+01  -0.43496357E+01
 30   0.61340E+01  0.17500E+01  -0.38255329E+01
 31   0.60000E+01  0.25000E+01  -0.35636267E+01
 32   0.60000E+01  0.32500E+01  -0.33679454E+01
 33   0.55000E+01  0.40000E+01  -0.39923956E+01
 34   0.65858E+01  0.14142E+01  -0.35853996E+01
 35   0.65000E+01  0.27500E+01  -0.27438956E+01
 36   0.60000E+01  0.40000E+01  -0.32887312E+01
 37   0.70000E+01  0.17320E+01  -0.25711218E+01
 38   0.70000E+01  0.23660E+01  -0.20747071E+01
 39   0.70000E+01  0.30000E+01  -0.18557452E+01
 40   0.70000E+01  0.35000E+01  -0.17658045E+01
 41   0.65000E+01  0.40000E+01  -0.25443422E+01
 42   0.74842E+01  0.19319E+01  -0.13367652E+01
 43   0.75000E+01  0.30000E+01  -0.95113970E+00
 44   0.70000E+01  0.40000E+01  -0.17446758E+01
 45   0.80000E+01  0.20000E+01  -0.33411568E-12
 46   0.80000E+01  0.25000E+01  -0.62366505E-12
 47   0.80000E+01  0.30000E+01  -0.27132199E-12
 48   0.80000E+01  0.35000E+01  -0.52458473E-12
 49   0.75000E+01  0.40000E+01  -0.88477812E+00
 50   0.80000E+01  0.40000E+01  -0.25556061E-12
```

Fig. 11.31—continued

```
* GLOBAL DERIVATIVES AT INTEGRATION POINTS *
POINT      X         Y        DP/DX          DP/DY
                  ELEMENT NUMBER   1
  1     1.775     1.775    0.99661E+00    0.23599E-01
  2     1.000     1.775    0.99773E+00    0.15345E-01
  3     0.225     1.775    0.99885E+00    0.11656E-01
  4     1.775     1.000    0.98698E+00    0.15545E-01
  5     1.000     1.000    0.99267E+00    0.91189E-02
  6     0.225     1.000    0.99836E+00    0.72572E-02
  7     1.775     0.225    0.97919E+00    0.74919E-02
  8     1.000     0.225    0.98944E+00    0.28926E-02
  9     0.225     0.225    0.99970E+00    0.28585E-02
                  ELEMENT NUMBER   2
  1     1.775     3.775    0.10198E+01    0.36130E-02
  2     1.000     3.775    0.10098E+01    0.16280E-02
  3     0.225     3.775    0.99989E+00    0.38903E-02
  4     1.775     3.000    0.10136E+01    0.15017E-01
  5     1.000     3.000    0.10079E+01    0.89952E-02
  6     0.225     3.000    0.10023E+01    0.72207E-02
  7     1.775     2.225    0.10035E+01    0.26421E-01
  8     1.000     2.225    0.10020E+01    0.16362E-01
  9     0.225     2.225    0.10006E+01    0.10551E-01
                  ELEMENT NUMBER   3
  1     3.775     1.775    0.99032E+00    0.11735E+00
  2     3.000     1.775    0.99264E+00    0.64052E-01
  3     2.225     1.775    0.99497E+00    0.32132E-01
  4     3.775     1.000    0.93275E+00    0.79377E-01
  5     3.000     1.000    0.95645E+00    0.38908E-01
  6     2.225     1.000    0.98015E+00    0.19814E-01
  7     3.775     0.225    0.88801E+00    0.41406E-01
  8     3.000     0.225    0.93308E+00    0.13763E-01
  9     2.225     0.225    0.97816E+00    0.74962E-02
          *         *          *              *
          *         *          *              *
          *         *          *              *
                  ELEMENT NUMBER   9
  1     6.887     3.881    0.16400E+01    0.16317E-01
  2     6.500     3.859    0.15505E+01    0.44829E-01
  3     6.113     3.837    0.14617E+01    0.60230E-01
  4     6.887     3.472    0.16667E+01    0.12511E+00
  5     6.500     3.375    0.15622E+01    0.15964E+00
  6     6.113     3.278    0.14615E+01    0.17945E+00
  7     6.887     3.063    0.16798E+01    0.23390E+00
  8     6.500     2.891    0.15581E+01    0.27446E+00
  9     6.113     2.719    0.14437E+01    0.29867E+00
                  ELEMENT NUMBER  10
  1     7.887     2.887    0.19517E+01    0.30902E-01
  2     7.498     2.880    0.18769E+01    0.13206E+00
  3     7.112     2.863    0.17978E+01    0.20899E+00
  4     7.884     2.498    0.21353E+01    0.78004E-01
  5     7.492     2.466    0.20304E+01    0.33100E+00
  6     7.110     2.394    0.18818E+01    0.52337E+00
  7     7.882     2.109    0.24696E+01    0.12416E+00
  8     7.486     2.052    0.23094E+01    0.52809E+00
  9     7.107     1.925    0.20231E+01    0.83746E+00
                  ELEMENT NUMBER  11
  1     7.887     3.887    0.17858E+01    0.48338E-02
  2     7.500     3.887    0.17434E+01    0.13060E-01
  3     7.113     3.887    0.17011E+01    0.82928E-02
  4     7.887     3.500    0.18211E+01    0.16848E-01
  5     7.500     3.500    0.17658E+01    0.66362E-01
  6     7.113     3.500    0.17105E+01    0.10288E+00
  7     7.887     3.113    0.18978E+01    0.28863E-01
  8     7.500     3.113    0.18295E+01    0.11966E+00
  9     7.113     3.113    0.17612E+01    0.19747E+00
BEGIN PLOT
END PLOT
NORMAL ENDING OF MODEL PROGRAM.
```

Fig. 11.32 Velocities for uniform inflow

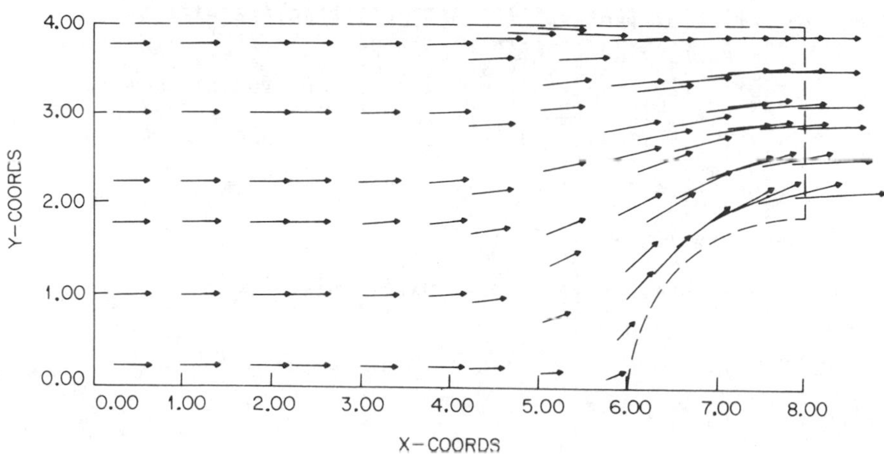

Fig. 11.33 Velocities for parabolic inflow

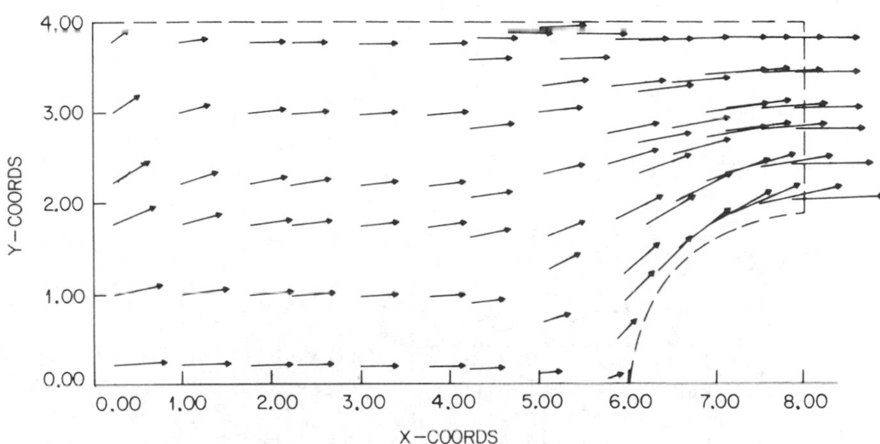

Fig. 11.34 Subroutine BFLUX

```
      SUBROUTINE  BFLUX (FLUX,COORD,LBN,NSPACE,NFLUX,
     1                   NG,C,S,IOPT)
C     * * * * * * * * * * * * * * * * * * * * * * * * * *
C          PROBLEM DEPENDENT BOUNDARY FLUX CONTRIBUTIONS
C     * * * * * * * * * * * * * * * * * * * * * * * * * *
CDP   IMPLICIT REAL*8 (A-H,O-Z)
      DIMENSION COORD(LBN,NSPACE), FLUX(LBN,NG), C(NFLUX),
     1          S(NFLUX,NFLUX)
C     FLUX = SPECIFIED BOUNDARY FLUX COMPONENTS
C     COORD = SPATIAL COORDINATES OF SEGMENT NODES
C     LBN = NO. OF NODES ON AN ELEMENT BOUNDARY SEGMENT
C     NSPACE = DIMENSION OF SOLUTION SPACE
C     NFLUX = LBN*NG = MAXIMUM NUMBER OF FLUX CONTRIBUTIONS
C     C = BOUNDARY FLUX COLUMN MATRIX CONTRIBUTIONS
C     S = BOUNDARY FLUX SQUARE MATRIX
C     NG = NUMBER OF PARAMETERS PER NODE POINT
C     IOPT = PROBLEM MATRIX REQUIREMENTS
C          = 1, CALCULATE C ONLY
C          = 2, CALCULATE S ONLY
C          = 3, CALCULATE BOTH C AND S
```

Fig. 11.34—continued

```
C       ...................................................
C         ** BFLUX PROBLEM DEPENDENT STATEMENTS FOLLOW **
C       ...................................................
C       QUADRATIC NORMAL FLUX ON QUADRATIC BOUNDARY SEGMENT
C        LBN = 3, NSPACE = 2, NG = 1, NFLUX = 2
C       FLUX(K,1) = SPECIFIED OUTWARD NORMAL FLUX AT NODE K
C       TANGENTIAL FLUX NOT USED         1....2....3 ---->  S
C       BL = BOUNDARY SEGMENT LENGTH     I
C       ( MUST BE STRAIGHT )             N        SEGMENT SKETCH
CDP     SQRT(Z) = DSQRT(Z)
        IOPT = 1
C       FIND DISTANCE BETWEEN FIRST AND THIRD POINTS
        DX = COORD(3,1) - COORD(1,1)
        DY = COORD(3,2) - COORD(1,2)
        BL = SQRT(DX*DX+DY*DY)
C       CALCULATE SEGMENT COLUMN MATRIX
        C(1) = BL*(FLUX(1,1)*4.+FLUX(2,1)*2.-FLUX(3,1))/30.
        C(2) = BL*(FLUX(1,1)*2.+FLUX(2,1)*16.+FLUX(3,1)*2.)/30.
        C(3) = BL*(-FLUX(1,1)+FLUX(2,1)*2.+FLUX(3,1)*4.)/30.
        RETURN
        END
```

Fig. 11.35 Subroutine PLTSET

```
        SUBROUTINE PLTSET (FLTMIS)
C       ****************************************************
C       EXTRACT CALCOMP PARAMETERS FROM MISC. DATA STORAGE
C       ****************************************************
C       REFER TO STANDARD CALCOMP MANUALS
CDP     IMPLICIT REAL*8 (A-H,O-Z)
        DIMENSION FLTMIS(7)
        COMMON /PLTKAL/ XLEN, YLEN, FIRSTX, FIRSTY, DELTAX,
     1                  DELTAY, XLAST, YLAST
        COMMON /PLTDIM/ XLIMIT, IBUFF(644)
        DATA KALPLT /1/
C       XLEN, YLEN = PLOT LENGTH IN INCHES
C       FIRSTX, FIRSTY = GLOBAL COORDS. OF PLOT ORGIN
C       DELTAX, DELTAY = GLOBAL COORD. PER INCH OF PLOT
C       XLIMIT = MAXIMUM LENGTH OF PAPER
C       IBUF = SCRATCH ARRAY ( INSTALLATION DEPENDENT )
        XLEN = FLTMIS(1)
        YLEN = FLTMIS(2)
        FIRSTX = FLTMIS(3)
        FIRSTY = FLTMIS(4)
        DELTAX = FLTMIS(5)
        DELTAY = FLTMIS(6)
        XLIMIT = FLTMIS(7)
        IF ( XLIMIT.LE.0.0 )  XLIMIT = XLEN + 4.
        XLAST = FIRSTX + XLEN*DELTAX
        YLAST = FIRSTY + YLEN*DELTAY
        WRITE (6,5000) XLEN, YLEN, FIRSTX, FIRSTY, DELTAX,
     1                 DELTAY
 5000   FORMAT ('0 SUPPLIED PLOT PARAMETERS:',/,
     1'X-LENGTH...',E15.6,' Y-LENGTH...',E15.6,/,
     2'FIRST-X....',E15.6,' FIRST-Y....',E15.6,/,
     3'DELTA-X....',E15.6,' DELTA-Y....',E15.6,/)
        IF (KALPLT.EQ.0 ) RETURN
C-->    ON THE FIRST CALL OPEN THE PLOT FILE
        KALPLT = 0
C       INSTALLATION DEPENDENT STATEMENTS FOLLOW
        CALL PLOTS (IBUF,664,XLIMIT,'AKIN#','REMOTE#')
        RETURN
        END
```

Fig. 11.36 Subroutine VECT2D

```
      SUBROUTINE VECT2D (NTAPE2,XLEN,YLEN,FIRSTX,FIRSTY,DELTAX,
     1                  DELTAY,XLAST,YLAST,SCALIT,SIZE,NE)
C     ****************************************************************
C     CONSTRUCT 2-D CALCOMP VECTORS TABULATED ON NTAPE2
C     ****************************************************************
CDP   IMPLICIT REAL*8 (A-H,O-Z)
CDP   ATAN2(Z1,Z2) = DATAN2(Z1,Z2)
CDP   SQRT(Z) = DSQRT(Z)
C     XLEN, YLEN = PLOT LENGTH IN INCHES
C     FIRSTX, FIRSTY = GLOBAL COORD. OF PLOT ORGIN
C     DELTAX, DELTAY = GLOBAL COORD. PER INCH OF PLOT
C     NE =NO. ELEMENTS, NIP = NO. PLOT POINTS IN ELEMENT
C     SIZE = SIZE OF SYMBOLS, IN INCHES
      REWIND NTAPE2
      CALL AXIS (0.,0.,8HX-COORDS,-8,XLEN,0.0,FIRSTX,DELTAX)
      CALL AXIS (0.,0.,8HY-COORDS, 8,YLEN,90.,FIRSTY,DELTAY)
C     LOOP OVER ELEMENTS
      DO 20 J = 1,NE
      READ (NTAPE2) NIP
C     LOOP OVER VECTOR POINTS
      DO 10 K = 1,NIP
      READ (NTAPE2) X,Y,DX,DY
C     IS POINT IN REGION OF INTREST
      IF ( X.LT.FIRSTX .OR. X.GT.XLAST ) GO TO 10
      IF ( Y.LT.FIRSTY .OR. Y.GT.YLAST ) GO TO 10
C     MOVE PEN TO POINT (CONVERT GLOBAL COORD. TO INCHES)
      X = (X-FIRSTX)/DELTAX
      Y = (Y-FIRSTY)/DELTAY
      CALL PLOT (X,Y,3)
C     SCALE MAX. COMPONENT TO 1 INCH & FIND ANGLE
      DX = DX/SCALIT
      DY = DY/SCALIT
      ANG = ATAN2(DY,DX)*57.3 - 90.
      X = X + DX
      Y = Y + DY
      VECTOR = SQRT(DX*DX+DY*DY)
C     DRAW LINE AND ARROW
      CALL SYMBOL (X,Y,SIZE,6,ANG,-2)
   10 CONTINUE
   20 CONTINUE
      RETURN
      END
```

Fig. 11.37 Velocity estimates

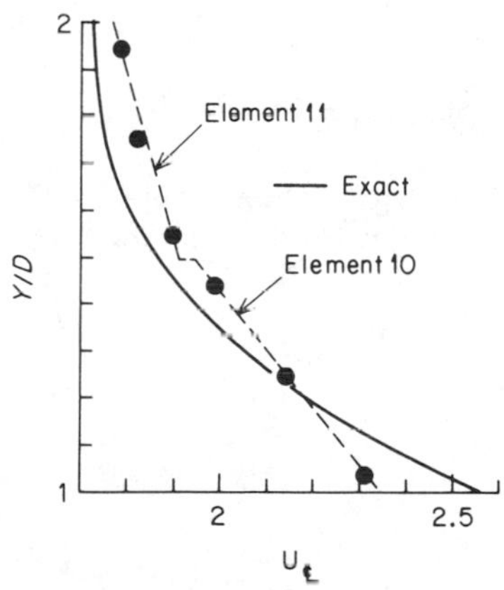

Table 11.3 Parameter Definitions for Quadratic Quadrilateral Potential Flow

CONTROL:

NSPACE = 2	NLPFIX = 1	NSEG = 0*
N = 8	NLPFLO = 2	LBN = 0*, or 3**
NG = 1	MISCFX = 0*, or 7**	NTAPE2 = 0*, or 9**
NTAPE1 = 0*, or 8**		

DEPENDENT VARIABLES:
- 1 = velocity potential, ϕ

PROPERTIES:
- Element level:
  - Fixed point:
    - 1 = Gaussian quadrature order in each dimension, (2*)
  - Floating point:
    - 1 = conductivity in x-direction, K_x
    - 2 = conductivity in y-direction, K_y
- Miscellaneous (NTAPE2 $\neq$ 0, Plot Data):**
  - Floating point:
    - 1 = length of plot in x-direction
    - 2 = length of plot in y-direction
    - 3 = x-coordinate of plot origin
    - 4 = y-coordinate of plot origin
    - 5 = change in x-coordinate per inch of plot
    - 6 = change in y-coordinate per inch of plot
    - 7 = maximum length of plot

BOUNDARY FLUX DATA (when NSEG $\neq$ 0):
- Fixed point:
  - Three nodal points (domain on left)
- Floating point:
  - Three nodal outward flux values

CONSTRAINTS:
- Type 1: Specified potential values

POST-SOLUTION CALCULATIONS (when NTAPE1 $\neq$ 0):
1. Global coordinates and velocities at each quadrature point
2. (NTAPE2 $\neq$ 0) Velocity vector plots**

* Default
** Optional

11.6 Electromagnetic Waveguides

The calculation of electrical and magnetic fields for practical geometries often leads to problems which contain known boundary singularities of the type discussed in Section 6.8. For example, waveguides often contain re-entrant corners that introduce point singularities that affect the calculation of the cut-off frequency and mode shapes. Similarly, the transverse electromagnetic line (T.E.M.) analysis of microstrips or coaxial lines often have re-entrant corners. For a uniform mesh it has been shown analytically and experimentally that the error due to the singularity predominates the standard mesh errors.

The analysis of T.E.M. line again involves the use of Laplace's equation for the potential, ϕ. Thus, again one could employ the formulation presented in Section 11.5. However, it is necessary to change the program to correct for the known point singularity. If the solution domain contains a re-entrant corner with an internal angle $\pi < \alpha \leq 2\pi$ then near the corner ϕ is of order $0(\rho^{1-a})$, where $a = \pi/\alpha$ and ρ is the distance from the corner. Its derivatives are $0(\rho^{-a})$ and it is this loss of regularity of the derivatives that causes the singularity error to predominate.

The minor changes required in the programming are illustrated in Fig. 11.38. The modifications of the shape functions and their local derivatives are made by the routine SINGLR given in Fig. 6.23. It is also necessary to flag the presence of the singularity. One way of doing this is to let the order, a, be an element property. If it is zero then the element is not adjacent to a singular point and no changes are required. Of course it would also be necessary to input the incidences of the singularity elements so that the singular node is listed first.

As an example application consider a microstrip consisting of a centre strip within a rectangular case. Let the boundary conditions be $\phi = 0$ on the outer case and $\phi = 1$ on the inner strip. At the edge of the inner strip one has $\phi = 0(\rho^{1/2})$ and $\partial\phi/\partial\rho = 0(\rho^{-1/2})$. Thus, there is a strong singularity at such a point. This problem has been analyzed by several authors including Daly [27] and Akin [4]. It was modelled using quarter symmetry, assuming an isotropic, homogeneous dielectric as shown in Fig. 11.39.
In addition to ϕ, the capacitance per unit length, c, where

$$c = \varepsilon_0 \kappa \iint_A |(\phi_{,x})^2 + (\phi_{,y})^2| \, da, \tag{11.19}$$

was obtained from the post-solution calculations. Here ε_0 and κ denote the free-space and dielectric permittivity, respectively. The exact value of c is known and was compared with values calculated on a uniform grid. The accuracy of standard higher order elements and singularity elements are compared in Fig. 11.39.

To illustrate a milder $r^{-1/3}$ singularity, and compare local grid refinement with singularity modifications, consider an L-shaped region as did Whiteman and Akin [80] and Akin [4], as in Fig. 11.40. Both used bilinear quadrilaterals in a finite element solution of the Laplace equation. The standard finite element results were compared against singularity modifications, using Eqn. (5.28), and local grid refinement near the corner. The latter employed five node bilinear transition quadrilaterals developed by Whiteman. Full details are given in the references. Here, only the solutions near the singularity will be considered. They are compared in Fig. 11.40 with an *analytic* solution obtained by numerical conformed mapping.

As expected, both singularity modifications and local grid refinement give significant improvements in the accuracy of the finite element solutions. In this case, the modification procedure is more cost-effective. In general it would probably be best to have limited local grid refinement, say two or three levels, combined with a singularity modification.

Fig. 11.38 Typical modifications for point singularities

```
C       SET SINGULARITY ORDER (NONE IF=0)
        ORDER = ELPROP(4)
        . . .
C-->    NUMERICAL INTEGRATION
        DO 60  IP = 1,NIP
C       FORM LOCAL DERIV, JACOBIAN, INVERSE
        CALL DER4Q (U(IP),V(IP),DELTA)
        CALL JACOB (N,NSPACE,DELTA,COORD,AJ)
        CALL I2BY2 (AJ,AJINV,DET)
C       TEST DET, SET WEIGHT
        IF ( DET.LE.0.0 ) WRITE (6,1000)
 1000   FORMAT (' ERROR, NEGATIVE JACOBIAN.')
        DETWT = DET*WT(IP)
C       *** MODIFY FOR POINT SINGULARITY ***
        IF ( ORDER.EQ.0.0 )  GO TO 35
        CALL SHP4Q (U(IP),V(IP),H)
        CALL SINGLR (ORDER,H,DELTA,N,NSPACE)
C       ***   MODIFICATION COMPLETE   ***
   35   CONTINUE
C       FORM GLOBAL DERIV, ELEM MATRICES
        CALL GDERIV (NSPACE,N,AJINV,DELTA,GLOBAL)
        . . .
```

Fig. 11.39 A comparison of singularity effects

Percent error in capacitance

Corner nodes	Number of singularity elements	Types of element			
		LT	QT	CT	BQ
3 × 5	0	—	4.02	—	—
3 × 5	4	—	0.76	—	—
5 × 9	0	7.00	4.01	—	3.73
5 × 9	2	—	—	—	3.03
7 × 13	0	4.57	2.66	2.01	—
8 × 15	0	—	—	—	2.15
8 × 15	2	—	—	—	1.75
9 × 17	0	3.39	1.29	—	—

Uniform grid
$B = 2H = 2W$
LT = Linear Triangle
QT = Quadratic Triangle
CT = Cubic Triangle
BQ = Bilinear Quadrilateral

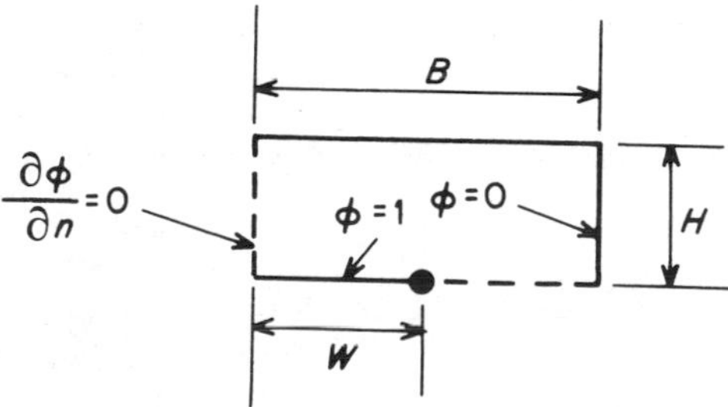

Fig. 11.40 Comparison of standard elements, local refinement, and singularity modifications near a singularity of $O(r^{-1/3})$

P	Q	SQ	RQ	NCM
26	8725	8820	8824	8843
37	8409	8565	8553	8586
48	8496	8467	8448	8487
27	8440	8528	8535	8553
38	7953	8095	8112	8154
49	7928	7948	7923	7961
28	8135	8200	8199	8210
39	7456	7557	7553	7565
50	6635	6667	6667	6667
56	5118	4887	4894	4870
62	3674	3601	3603	3580
29	7864	7901	7895	7898
40	7043	7070	7077	7066
51	6048	6026	6039	6020
57	4891	4835	4811	4881
63	3637	3571	3569	3550
30	7658	7673	7677	7672
41	6768	6772	6787	6772
52	5780	5766	5776	5757
58	4693	4663	4659	4642
64	3541	3507	3499	3486

P = Point Number
Q = Bilinear Quadrilaterals, 79 dof
RQ = Q with five local refinements, 144 dof
SQ = Q with three modified Q, 79 dof
NCM = Numerical Conformal Mapping

P	
Q	SQ
RQ	NCM

$h = 1/10$

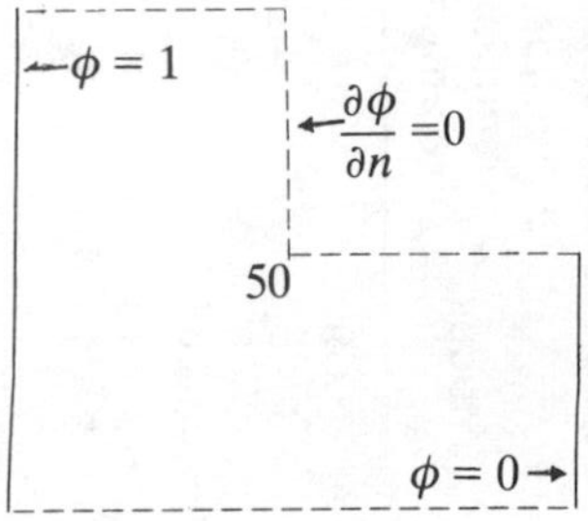

11.7 Axisymmetric plasma equilibria

Nuclear fusion is being developed as a future source of energy. The heart of the fusion reactors will be a device for confining the reacting plasma and heating it to thermonuclear temperatures. This confinement problem can be solved through the use of magnetic fields of the proper geometry which generate a so-called 'magnetic bottle'. The tokamak containment concept employs three magnetic field components to confine the plasma. An externally applied toroidal magnetic field, B_T, is obtained from coils through which the torus passes. A second field component is the polodial magnetic field, B_P, which is produced by a large current flowing in the plasma itself. This current is induced in the plasma by transformer action and assists in heating the plasma. Finally a vertical (axial) field, B_V, is also applied. These typical fields are illustrated in Fig. 11.41. For many purposes a very good picture of the plasma behaviour can be obtained by treating it as an ideal magnetohydrodynamic (MHD) media. The equations governing the steady state flow of an ideal MHD plasma are:

$$
\begin{aligned}
\text{grad}\,\mathbf{B} &= 0 \\
\nabla \mathrm{P} &= \mathbf{J} \times \mathbf{B} \qquad (11.20) \\
\text{curl}\,\mathbf{B} &= \mu \mathbf{J}
\end{aligned}
$$

where P is the pressure, $\mathbf{B}$ the magnetic flux density vector, $\mathbf{J}$ the current density vector, and μ a constant that depends on the system of units being employed. Consider an axisymmetric equilibria defined in cylindrical coordinates (ρ, z, θ) so that $\partial/\partial\theta = 0$. Equation (11.20) implies the existence of a vector potential, $\mathbf{A}$, such that curl $\mathbf{A} = \mathbf{B}$. Assuming that $\mathbf{A} = \mathbf{A}(\rho, z)$ and $A_\theta = \psi/\rho$, where ψ is a stream function, we obtain

$$
\begin{aligned}
B_\rho &= -\psi_{,z}/\rho \\
B_z &= \psi_{,\rho}/\rho \qquad (11.21) \\
B_\theta &= A_{\rho,z} - A_{z,\rho} = B_T
\end{aligned}
$$

Therefore Eqn. (11.18) simplifies to

$$
\frac{\partial^2 \psi}{\partial \rho^2} - \frac{1}{\rho}\frac{\partial \psi}{\partial \rho} + \frac{\partial^2 \psi}{\partial z^2} = -\mu \rho^2 P' - xx' = J_\theta \rho \qquad (11.22)
$$

The above is the governing equation for the steady equilibrium flow of a plasma. For certain simple choices of P and B_θ, Eqn. (11.22) will be linear but in general it is nonlinear. The essential boundary condition on the limiting surface, Γ_1, is

$$\psi = K + \tfrac{1}{2}r^2 B_V \quad \text{on } \Gamma_1$$

where K is a constant and B_V is a superimposed direct current vertical (z) field. On planes of symmetry one also has vanishing normal gradients of ψ, i.e.

$$\frac{\partial \psi}{\partial n} = 0 \text{ on } \Gamma_2.$$

The right-hand side of Eqn. (11.22) can often be written as

$$H_{\theta\rho} = g\psi + h \tag{11.23}$$

where, for the above special cases, $g = g(\rho, z)$ and $h = h(\rho, z)$ but where in general h is a nonlinear function of ψ, i.e. $h = h(\rho, z, \psi)$. Equations (11.22) and (11.23) are those for which we wish to establish the finite element model.

A finite element formulation of this problem has been presented by Akin and Wooten [8]. They recast Eqn. (11.22) in self-adjoint form, applied the Galerkin criterion, and integrated by parts. This defines the governing variational statement

$$I = \iint_\Omega \left[\frac{1}{2}\{(\psi_{,\rho})^2 + (\psi_{,z})^2 + g\psi^2\} + h\psi\right]\frac{1}{\rho}\,\mathrm{d}\rho\,\mathrm{d}z \tag{11.24}$$

which, for the linear problem, yields Eqn. (11.22) as the Euler equation when I is stationary, i.e. $\delta I = 0$. When $g = h = 0$, Eqn. (11.24) also represents the case of axisymmetric inviscid fluid flow. Flow problems of this type were considered by Chung [23] using a similar procedure.

For a typical element with N nodes the element contributions for Eqn. (11.23) are

$$\begin{aligned} \mathbf{S}^e &= \int_{\Omega^e} [\mathbf{H}_{,\rho}^T\mathbf{H}_{,\rho} + \mathbf{H}_{,z}^T\mathbf{H}_{,z} + g(\rho, z)\mathbf{H}^T\mathbf{H}]\frac{1}{\rho}\,\mathrm{d}\rho\,\mathrm{d}z, \\ \mathbf{C}^e &= \int_\Omega h(\rho, z)\mathbf{H}^T\frac{1}{\rho}\,\mathrm{d}\rho\,\mathrm{d}z, \end{aligned} \tag{11.25}$$

where the N element interpolation functions, $\mathbf{H}$, define the value of ψ within an element by interpolating from its nodal values, ψ^e. Several applications of this model to plasma equilibria are given in [8]. The major advantage of the finite element formulation over other methods such as finite differences is that it allows the plasma physicist to study arbitrary geometries. Some feel that the fabrication of the toroidal field coils may require the use of a circular plasma, while others recommend the use of dee-shaped plasmas. The current model has been applied to both of these geometries.

Figure 11.42 illustrates typical results for a dee-shape toruns cross-section where the bilinear isoparametric quadrilateral was employed. Biquadratic or bicubic elements would be required for some formulations which require post-solution calculations using the first and second derivatives of ψ. Figure 11.43 shows the integral calculations (called by ISOPAR).

Fig. 11.41 Schematic of tokamak fields

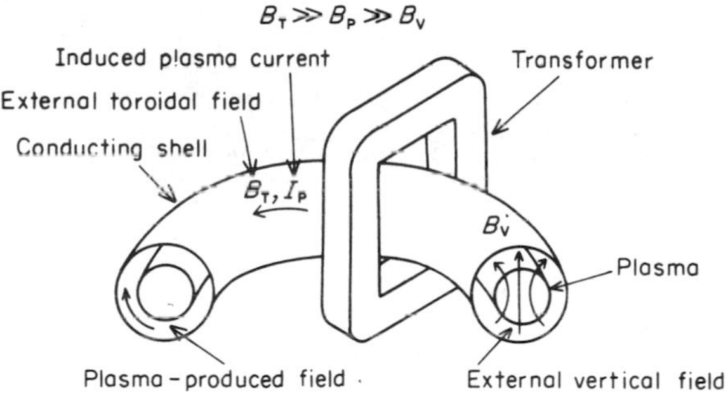

Fig. 11.42 Results for a dee-shaped toruns

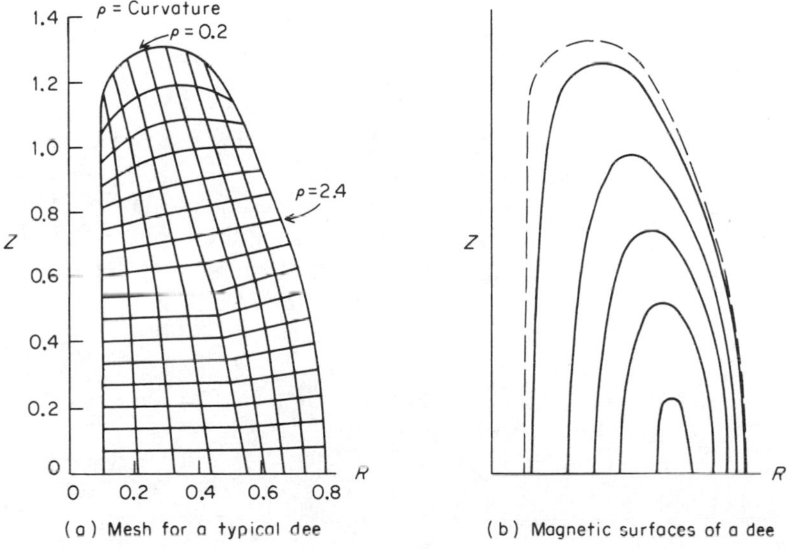

Fig. 11.43 Plasma equilibria calculation

```
      SUBROUTINE  PLASMA (WT,DET,H,DGH,XPT,N,NSPACE,
     1                    NELFRE,COL,SQ,NTAPE1)
C     * * * * * * * * * * * * * * * * * * * * * * * * * * * *
C     PROBLEM DEPENDENT INTEGRAND EVALUATION IN
C     AN ISOPARAMETRIC ELEMENT      (NGRAND)
C     * * * * * * * * * * * * * * * * * * * * * * * * * * * *
CDP   IMPLICIT REAL*8 (A-H,O-Z)
      DIMENSION COL(NELFRE), SQ(NELFRE,NELFRE),
     1          H(N), DGH(NSPACE,N), XPT(NSPACE)
C     N = NUMBER OF NODES PER ELEMENT
C     NSPACE = NUMBER OF SPATIAL DIMENSIONS
C     NELFRE = NO OF ELEMENT DEGREES OF FREEDOM
C     H = ELEMENT INTERPOLATION FUNCTIONS
C     DGH = GLOBAL DERIVATIVES OF H
C     XPT = GLOBAL COORDS OF THE POINT
C     WT = QUADRATURE WEIGHT AT POINT
C     DET = JACOBIAN DETERMINATE AT POINT
C     COL = PROB DEP COLUMN MATRIX INTEGRAND
C     SQ = PROB DEP SQUARE MATRIX INTEGRAND
C     NTAPE1 = STORAGE UNIT FOR POST SOLUTION DATA
C     ..........................................................
C     *** PLASMA PROBLEM DEPENDENT STATEMENTS FOLLOW ***
C     ..........................................................
      COMMON /ELARG3/  GCONST,HCONST
C     STORE COORDS AND GLOBAL DERIVATIVES AT PT
      IF ( NTAPE1.GT.0 )  WRITE(NTAPE1) XPT, DGH
C     EVALUATE CONSTANT
      R=XPT(1)
      IF ( R.GT.0.0 )  GOTO 10
      R = 0.00001
      WRITE (6,20)
   20 FORMAT (34H WARNING QUADRATURE POINT NEAR R=0   )
   10 WJDR = WT*DET/R
C     UPDATE SQUARE AND COLUMN MATRICES
      DO 40  J=1,N
      DO 30  I=1,J
      SQ(I,J) = SQ(I,J) + WJDR*( DGH(1,I)*DGH(1,J)
     1          + DGH(2,I)*DGH(2,J) + GCONST*H(I)*H(J) )
   30 SQ(J,I) = SQ(I,J)
   40 COL(J) = COL(J) + WJDR*H(J)*HCONST
      RETURN
      END
```

11.8 Exercises

1. Show that Eqn. (11.6) could be written as

$$\mathbf{S}^e = \iint \mathbf{B}^T\mathbf{D}\mathbf{B}\,|J|\,du\,dv$$

   where **D** is a diagonal matrix and **B** contains the global derivatives of the interpolation functions (i.e., **B** = **d**).
2. Rewrite the element calculations in Fig. 11.14 using subroutine ISOPAR. Use the argument list for NGRAND to define the problem dependent integrands in a routine called HEAT2D. Remember to set HEAT2D as an 'external' variable and include it in the call to ISOPAR.
3. Rewrite the square matrix integrand calculations in Fig. 11.14 by using a call to BTDIAB.
4. Write the element routines to evaluate Eqns. (11.8) to (11.11) for the linear triangle viscous flow problem.
5. Use numerical integration to evaluate $\mathbf{C}^b$ of Eqn. (11.18) when the side is not straight.
6. Use the exact integrals of Section 6.2 to verify Eqn. (11.18).
7. Prepare a table of parameters for the viscous duct flow four node isoparametric quadrilateral.

12

Three-dimensional applications

12.1 Introduction

Another advantage of the finite element procedure is the relative ease of extension to three-dimensional applications. Several three-dimensional solutions have been presented in the literature. The earliest applications involved the four node linear tetrahedral element. The simplicity of the interpolation functions for this element make it possible to derive the element matrices in closed form. For example, Zienkiewicz [82] presented the expressions for three-dimensional stress analysis using that element.

The disadvantage of the tetrahedral elements is the great burden of data preparation that it places on the user. One can generally obtain a more accurate solution with less data preparation by utilizing a hexahedral element. The simplest element in this family is the eight node *linear* hexahedron. This element will be illustrated in the following section on heat conduction so as to reduce the number of new concepts to be considered.

12.2 Heat conduction

The isoparametric formulation of the two-dimensional Poisson equation was presented in section 11.3. The governing differential equation in three dimensions becomes

$$(K_x T_{,x})_{,x} + (K_y T_{,y})_{,y} + (K_z T_{,z})_{,z} + g = 0 \tag{12.1}$$

and the element matrices are

$$\mathbf{S}^e = \int_{V^e} [K_x \mathbf{H}^T_{,x} \mathbf{H}_{,x} + K_y \mathbf{H}^T_{,y} \mathbf{H}_{,y} + K_z \mathbf{H}^T_{,z} \mathbf{H}_{,z}] \, dv, \tag{12.2}$$

and

$$\mathbf{C}^e = \int_{V^e} g\mathbf{H}^T \, dv. \tag{12.3}$$

Of course, if there are convective or normal heat flux boundary conditions on the surface of the domain then the element matrices will contain additional terms. Comparing these equations with the corresponding two-dimensional equations one notes some important differences. First, the integrals are three-dimensional, which makes the relative numerical integration costs go from n^2 to n^3 where n is the number of quadrature points in each dimension. For this element one can utilize a value of $n = 2$ unless the element has a greatly distorted shape. The element square matrix contains an additional set of matrix products. Also, the size of that matrix has gone from 4×4 to 8×8 which also represents a significant increase in the number of terms to be calculated and assembled into the system equations.

The problem-dependent subroutines for his element are shown in Fig. 12.1 and 12.2. Although the extensions of the isoparametric elements to three-dimensional applications was outlined at the end of Section 11.3, the above routines do contain some new concepts. For example, the floating point properties for this element are identified by the use of a material number code. The integer material code number is contained in the fixed point data array LPROP. Once the material number is identified subroutine MATPRT is called to extract the four floating point properties from the array FLTMIS. The array LPROP also contains the desired integration order for the element. This element is summarized in Table 12.1.

As a simple test problem consider the temperature distribution in an infinite cylinder. Figure 12.3 shows the mesh which is a one degree wedge segment divided into four elements. The inner radius has a specified temperature of 100°F and the outer surface has a temperature of 0°F. Due to symmetry, the normal heat flux on the other four surfaces is zero. This is a natural boundary condition that is automatically satisfied. For a homogeneous material the exact solution for this problem is shown by Myers [56] to be $t(r) = t_i - (t_i - t_o)(\ln(r/r_i))/\ln(r_o/r_i)$ where the subscripts i and o denote inner and outer surfaces, respectively. Figure 12.4 shows the comparison with this solution. The maximum error is 0.6%. The effect of non-homogeneous properties is also shown in that figure. The input data for the latter problem, where the inner and outer elements have conductivities that are ten times as large as the other two, are shown in Fig. 12.5 and a typical output is shown in Fig. 12.6.

Before leaving this section, it may be useful to comment on three-dimensional problems that involve surface integrals. Recall that if one has a specified normal heat flux on a surface Γ then the element column matrix will contain a term

$$\mathbf{C}_q = \int_\Gamma q_n \mathbf{H}^T \, \mathrm{d}a. \tag{12.4}$$

Evaluation of da on the three-dimensional surface in terms of the two local coordinates, say ξ and η, requires the use of differential geometry. It can be shown that $\mathrm{d}a = \sqrt{A}\,\mathrm{d}\xi\,\mathrm{d}\eta$ where A is the determinate of the first fundamental magnitudes of the surface. The numerical integration of the above equation gives

$$\mathbf{C}_q = \sum_{i=1}^{\mathrm{NQP}} W_i q_i \mathbf{H}(\xi_i, \eta_i)^T \sqrt{A_i}. \tag{12.5}$$

To evaluate the A_i term at an integration point one could utilize subroutine FMONE. In other applications one would also need the unit normal vector at the point to determine the global components of a given traction vector. The latter operation is performed by subroutine SNORMV. These two routines are shown in Figs. 12.7 and 12.8, respectively.

Fig. 12.1 Subroutine ELSQ

```
      SUBROUTINE  ELSQ   (N,NELFRE,NSPACE,IP5,JP2,KP1,KP2,
     1                    KP3,COORD,D,ELPROP,LPROP,PRTLPT,FLTMIS,
     2                    S,NTAPE1,NTAPE2,NTAPE3,NTAPE4)
C     * * * * * * * * * * * * * * * * * * * * * * * * * * * * *
C                   GENERATE ELEMENT SQUARE MATRIX
C     * * * * * * * * * * * * * * * * * * * * * * * * * * * * *
CDP   IMPLICIT REAL*8(A-H,O-Z)
      DIMENSION COORD(N,NSPACE), D(NELFRE), ELPROP(KP2),
     1          LPROP(JP2), S(NELFRE,NELFRE),
     2          PRTLPT(IP5,KP1), FLTMIS(KP3)
C     N = NUMBER OF NODES PER ELEMENT
C     NELFRE = NUMBER OF DEGREES OF FREEDOM PER ELEMENT
C     NSPACE = DIMENSION OF SPACE
C     IP5  = NUMBER OF ROWS IN ARRAY PRTLPT
C     JP2  = NUMBER OF COLUMNS IN ARRAYS LPFIX AND LPROP
C     KP1  = NUMBER OF COLUMNS IN FLTNP & PRTLPT & PTPROP
C     KP2  = NUMBER OF COLUMNS IN ARRAYS FLTEL AND ELPROP
C     KP3  = NUMBER OF COLUMNS IN ARRAY FLTMIS
C     COORD = SPATIAL COORDINATES OF ELEMENT'S NODES
C     D = NODAL PARAMETERS ASSOCIATED WITH AN ELEMENT
C     ELPROP = ELEMENT ARRAY OF FLOATING PT PROPERTIES
C     LPROP = ARRAY OF FIXED POINT ELEMENT PROPERTIES
C     PRTLPT = FLOATING POINT PROP FOR ELEMENT'S NODES
C     FLTMIS = SYSTEM STORAGE OF FLOATING PT MISC PROP
C     S = ELEMENT SQUARE MATRIX
C     NTAPE1 = UNIT FOR POST SOLUTION MATRICES STORAGE
C     NTAPE2,3,4 = OPTIONAL UNITS FOR USER (USED WHEN > 0)
C     ..........................................................
C       ***  ELSQ PROBLEM DEPENDENT STATEMENTS FOLLOW ***
C     ..........................................................
C     SOLUTION OF POISSON EQ IN THREE DIMENSIONS
C     USING AN 8 NODE ISOPARAMETRIC HEXAHEDRON ELEMENT
C     NSPACE = 3, N = 8, NG = 1, NELFRE = 8
C     PRTMAT(1) = CONDUCTIVITY IN X-DIRECTION
C     PRTMAT(2) = CONDUCTIVITY IN Y-DIRECTION
C     PRTMAT(3) = CONDUCTIVITY IN Z-DIRECTION
C     PRTMAT(4) = SOURCE PER UNIT VOLUME
C     LPROP(1) = GAUSSIAN INTEGRATION ORDER IN EACH DIM
C     LPROP(2) = ELEMENT MATERIAL NUMBER
      COMMON /ELARG/ GPT(4),GWT(4),PT(3,64),WT(64),NIP
```

Fig. 12.1—continued

```
      COMMON /ELARG2/ XYZ(3),AJ(3,3),AJINV(3,3),DELTA(3,8),
     1                GLOBAL(3,8),H(8),COL(8)
      COMMON /ELARG3/ PRTMAT(4),MATNO
      DATA NQPOLD /0/
      EXTERNAL  LAPL8H
C     LAPL8H = EXTERNAL SUBR FOR INTEGRAN DEFINITICN
C-->  DEFINE INTEGRATION ORDER
      NQP = LPROP(1)
C     DEFAULT NQP = 2 SINCE LPROP(1) = 0 INITIALLY
      IF ( NQP.LT.2 )  NQP = 2
      IF ( NQP.GT.4 )  NQP = 4
C-->  CALCULATE QUADRATURE DATA ( IF REQUIRED )
      IF ( NQPOLD.EQ.NQP )  GO TO 40
      NQPOLD = NQP
      CALL GAUS3D (NQP,GPT,GWT,NIP,PT,WT)
C-->  DEFINE PROPERTIES
   40 MATNC = LPROP(2)
C     DEFAULT MATNO = 1 SINCE LPROP(2) = 0 INITIALLY
      IF ( MATNO.LT.1 )  MATNO = 1
      CALL  MATPRT (MATNO,4,KP3,FLTMIS,PRTMAT)
C-->  NUMERICAL INTEGRATION
      CALL ISOPAR (N,NSPACE,NELFRE,NIP,S,CCL,PT,WT,H,
     1     DELTA,GLOBAL,COORD,XYZ,AJ,AJINV,NTAPE1,LAPL8H)
      RETLRN
      END
```

Fig. 12.2 **(a)** Subroutine LAPL8H, and **(b)** subroutine ELCOL

```
      SUBROUTINE  LAPL8H (WT,DET,H,DGH,XPT,N,NSPACE,
     1                    NELFRE,COL,SQ,NTAPE1)
C     * * * * * * * * * * * * * * * * * * * * * * * * *
C     PROBLEM DEPENDENT INTEGRAND EVALUATICN IN
C     AN ISOPARAMETRIC ELEMENT
C     * * * * * * * * * * * * * * * * * * * * * * * * *
CDP   IMPLICIT REAL*8 (A-H,O-Z)
      DIMENSION COL(NELFRE), SQ(NELFRE,NELFRE),
     1          H(N), DGH(NSPACE,N), XPT(NSPACE)
C     N = NUMBER OF NODES PER ELEMENT
C     NSPACE = NUMBER OF SPATIAL DIMENSIONS
C     NELFRE = NO OF ELEMENT DEGREES OF FREEDCM
C     H = ELEMENT INTERPOLATION FUNCTIONS
C     DGH = GLOBAL DERIVATIVES OF H
C     XPT = GLOBAL COORDS OF THE POINT
C     WT = QUADRATURE WEIGHT AT POINT
C     DET = JACOBIAN DETERMINATE AT POINT
C     COL = PROB DEP COLUMN MATRIX INTEGRAND
C     SQ = PROB DEP SQUARE MATRIX INTEGRANC
C     NTAPE1 = STORAGE UNIT FOR POST SOLUTION DATA
C     .....................................................
C     *** LAPL8H PROBLEM DEPENDENT STATEMENTS FOLLCW ***
C     .....................................................
      CCMMCN /ELARG3/ PRTMAT(4),MATNO
C     DEFINE PROPERTIES
      XK = PRTMAT(1)
      YK = PRTMAT(2)
      ZK = PRTMAT(3)
      G = PRTMAT(4)
C     EVALUATE PRODUCTS
      DETWT = DET*WT
      DO 20  I = 1,N
      DO 1C  J = 1,I
      SQ(I,J) = SQ(I,J) + DETWT*( XK*DGH(1,I)*DGH(1,J)
     1                          + YK*DGH(2,I)*DGH(2,J)
     2                          + ZK*DGH(3,I)*DGH(3,J) )
   10 SQ(J,I) = SQ(I,J)
      IF ( G.EQ.0.0 )  GO TO 20
      COL(I) = COL(I) + DETWT*G*H(I)
   20 CONTINUE
      RETURN
      END
```

(a)

Fig. 12.2—continued

```
      SUBROUTINE  ELCOL (N,NELFRE,NSPACE,IP5,JP2,KP1,KP2,
     1            KP3,COORD,D,ELPROP,LPROP,PRTLPT,FLTMIS,
     2            C,NTAPE1,NTAPE2,NTAPE3,NTAPE4)
C     * * * * * * * * * * * * * * * * * * * * * * * * * *
C               GENERATE ELEMENT COLUMN MATRIX
C     * * * * * * * * * * * * * * * * * * * * * * * * * *
CDP   IMPLICIT REAL*8(A-H,O-Z)
      DIMENSION COORD(N,NSPACE), D(NELFRE), ELPROP(KP2),
     1          LPROP(JP2), C(NELFRE), PRTLPT(IP5,KP1),
     2          FLTMIS(KP3)
C     N = NUMBER OF NODES PER ELEMENT
C     NELFRE = NUMBER OF DEGREES OF FREEDOM PER ELEMENT
C     NSPACE = DIMENSION OF SPACE
C     IP5 = NUMBER OF ROWS IN ARRAY PRTLPT
C     JP2 = NUMBER OF COLUMNS IN ARRAYS LPFIX AND LPROP
C     KP1 = NUMBER OF COLUMNS IN  FLTNP & PRTLPT & PTPROP
C     KP2 = NUMBER OF COLUMNS IN ARRAYS FLTEL AND ELPROP
C     KP3 = NUMBER OF COLUMNS IN ARRAY FLTMIS
C     COORD = SPATIAL COORDINATES OF ELEMENT'S NODES
C     D = NODAL PARAMETERS ASSOCIATED WITH A ELEMENT
C     ELPROP = ELEMENT ARRAY OF FLOATING POINT PROPERTIES
C     LPRCP = ARRAY OF FIXED POINT ELEMENT PROPERTIES
C     PRTLPT = FLOATING POINT PROP OF ELEMENT'S NODES
C     FLTMIS = SYSTEM STORAGE OF FLOATING POINT MISC. PROP
C     C = ELEMENT COLUMN MATRIX
C     NTAPE1 = UNIT FOR POST SOLUTION MATRICES STORAGE
C     NTAPE2,3,4 = OPTIONAL UNITS FOR USER  (USED WHEN > 0)
C     ..........................................................
C      ***  ELCOL PROBLEM DEPENDENT STATEMENTS FOLLOW ***
C     ..........................................................
C     RECOVER COLUMN MATRIX FROM ELSQ
      COMMON /ELARG2/ XYZ(3),AJ(3,3),AJINV(3,3),DELTA(3,8),
     1                GLOBAL(3,8),H(8),COL(8)
      DO 10  I = 1,NELFRE
   10 C(I) = COL(I)
      RETURN
      END
```

(b)

Fig. 12.3 Mesh of one degree wedge segment divided into four elements

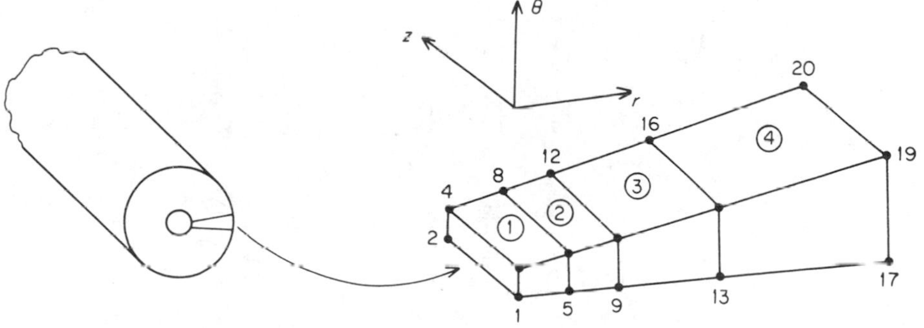

Fig. 12.4 Comparison of exact and finite element solutions

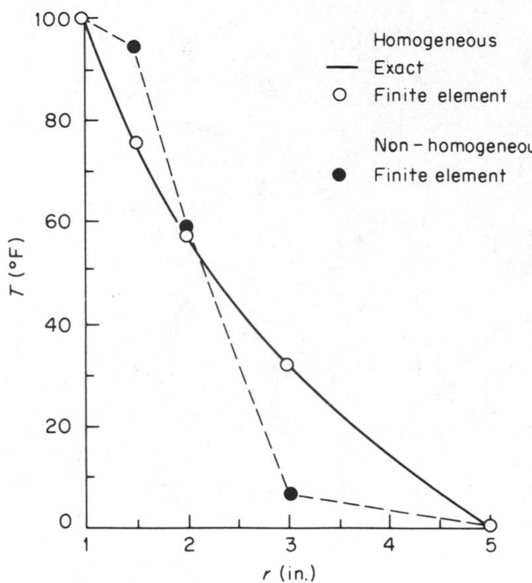

Fig. 12.5 Conduction through a 1-deg segment of an infinite cylinder; test problem data

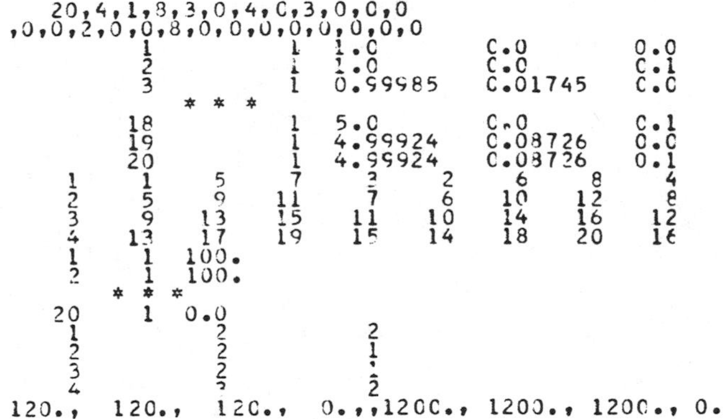

```
    20,4,1,8,3,0,4,C,3,0,0,0
,0,0,2,0,0,8,0,0,0,0,0,0,0,0
          1         1  1.0       C.0       0.0
          2         1  1.0       C.0       C.1
          3         1  0.99985   C.01745   C.C
              *  *  *
         18         1  5.0       C.0       C.1
         19         1  4.99924   C.08726   0.C
         20         1  4.99924   C.08726   0.1
    1     1    5    7    3    2    6    8    4
    2     5    9   11    7    6   10   12    8
    3     9   13   15   11   10   14   16   12
    4    13   17   19   15   14   18   20   16
    1     1  100.
    2     1  100.
        *  *  *
   20     1  0.0
    1          2         2
    2          2         1
    3          2         .
    4          2         2
120.,  120.,  120.,  0.,,1200., 1200., 1200., 0.
```

Fig. 12.6 Conduction through a 1-deg segment of an infinite cylinder; test problem output

```
*****  PROBLEM PARAMETERS  *****
NUMBER CF NODAL POINTS IN SYSTEM =..........20
NUMBER CF ELEMENTS IN SYSTEM =............. 4
NUMBER CF NODES PER ELEMENT =.............. 8
NUMBER CF PARAMETERS PER NCDE =............ 1
DIMENSICN OF SPACE =....................... 3

NUMBER CF BOUNDARIES WITH GIVEN FLUX=...... 0
NUMBER CF NODES CN BOUNCARY SEGMENT=....... 4
NUMBER OF CONTOUR CURVES BETWEEN
  5TH & 95TH PERCENTILE CF EACH PARAMETER =  3
NUMBER OF FIXED PT PROP PER NODE =......... 0
NUMBER CF FLOATING PT PRCP PER NCDE =...... 0
NUMBER CF FIXED PT PROP PER ELEMENT =...... 2
NUMBER CF FLOATING PT PRCP PER ELEMENT =... 0
NUMBER OF FIXED PT MISC PRCP =............. 0
NUMBER CF FLOATING PT MISC PROP =.......... 8
NODAL PARAMETERS TO BE LISTEC BY NODES
NODAL PARAMETERS TO BE LISTEC BY ELEMENTS
ALL ELEMENT COLUMN MATRICES ARE ZERO.

*** NCDAL POINT DATA ***
NODE, CCNSTRAINT INDICATCR, 3 CCCRDINATES
     1        1     1.0000     0.0000      0.0000
     2        1     1.0000     C.CCCC      0.1000
     3        1     0.9998     0.0174      0.00C0
     4        1     0.9998     C.0174      0.1000
     5        0     1.5000     C.CCCC      0.0000
     6        0     1.5000     C.0000      0.1000
     7        0     1.4998     C.0262      0.0000
     8        0     1.4998     C.0262      0.1000
     9        0     2.0000     C.0000      0.0000
    10        0     2.0000     C.0CC0      0.1000
    11        0     1.9997     0.0349      0.0000
    12        0     1.9997     0.0349      0.1000
    13        0     3.0000     C.CCCC      0.0000
    14        0     3.0000     C.0000      0.1000
    15        0     2.9995     0.0524      0.0000
    16        0     2.9995     0.0524      0.1000
    17        1     5.0000     C.00CC      0.0000
    18        1     5.0000     C.0000      0.1000
    19        1     4.9992     C.0873      0.0000
    20        1     4.9992     0.0873      0.1000

*** ELEMENT CONNECTIVITY CATA ***
 ELEMENT NO.,   8 NCDAL INCIDENCES.
   1     1     5     7     3     2     6     8     4
   2     5     9    11     7     6    10    12     8
   3     9    13    15    11    10    14    16    12
   4    13    17    19    15    14    18    20    16

 *** CCNSTRAINT EQUATICN CATA ***
 CONSTRAINT TYPE ONE
 EQ. NC.     NODE1     PAR1            A1
     1          1         1    .1CCC0000E+03
     2          2         1    .1CCCCC00E+03
     3          3         1    .1C00C000E+03
     4          4         1    .1CCCCC00E+03
     5         17         1    .CCCCCC00E+00
     6         18         1    .CCCCC000E+00
     7         19         1    .CCCCCC00E+00
     8         20         1    .CCCCC000E+00

 ***   ELEMENT     PROPERTIES      ***
 ELEMENT NO.   PRCPERTY NC.        VALUE
     1               1            2
     2               1            2
     3               1            2
     4               1            3
     1               2            2
     2               2            1
     3               2            1
     4               2            2
 END OF FIXED POINT PROPERTIES OF ELEMENTS
```

Fig. 12.6—continued

```
***  MISCELLANEOUS SYSTEM PROPERTIES   ***
PROPERTY NO.      VALUE
            1  0.12000000E+03
            2  0.12000000E+03
            3  0.12000000E+03
            4  0.00000000E+00
            5  0.12000000E+04
            6  0.12000000E+04
            7  0.12000000E+04
            8  0.00000000E+00

***   OUTPUT OF RESULTS    ***
NODE,   3 COORDINATES,   1 PARAMETERS.
 1   1.0000E+0   0.0000E+0   0.0000E+0   0.1000E+03
 2   1.0000E+0   0.0000E+0   1.0000E-1   0.1000E+03
 3   9.9985E-1   1.7450E-2   0.0000E+0   0.1000E+03
 4   9.9985E-1   1.7450E-2   1.0000E-1   0.1000E+03
 5   1.5000E+0   0.0000E+0   0.0000E+0   0.9484E+02
 6   1.5000E+0   0.0000E+0   1.0000E-1   0.9484E+02
 7   1.4998E+0   2.6180E-2   0.0000E+0   0.9484E+02
 8   1.4998E+0   2.6180E-2   1.0000E-1   0.9484E+02
 9   2.0000E+0   0.0000E+0   0.0000E+0   0.5801E+02
10   2.0000E+0   0.0000E+0   1.0000E-1   0.5801E+02
11   1.9997E+0   3.4910E-2   0.0000E+0   0.5801E+02
12   1.9997E+0   3.4910E-2   1.0000E-1   0.5801E+02
13   3.0000E+0   0.0000E+0   0.0000E+0   0.6446E+01
14   3.0000E+0   0.0000E+0   1.0000E-1   0.6446E+01
15   2.9995E+0   5.2360E-2   0.0000E+0   0.6446E+01
16   2.9995E+0   5.2360E-2   1.0000E-1   0.6446E+01
17   5.0000E+0   0.0000E+0   0.0000E+0   0.0000E+00
18   5.0000E+0   0.0000E+0   1.0000E-1   0.0000E+00
19   4.9992E+0   8.7260E-2   0.0000E+0   0.0000E+00
20   4.9992E+0   8.7260E-2   1.0000E-1   0.0000E+00

*** EXTREME VALUES OF THE NODAL PARAMETERS ***
PARAMETER NO.   MAX.VALUE,NODE    MIN.VALUE,NODE
 1    0.10000000E+03 , 4   0.00000000E+00 , 20

*** CALCULATION OF CONTOUR CURVES ***
PARAMETER NUMBER =  1
 5TH PERCENTILE VALUE  =  0.5000E+01
95TH PERCENTILE VALUE  =  0.9500E+02

CONTOUR NUMBER =       1
VALUE OF CONTOUR =   0.5000E+01
POINT NO.,   3 COORDINATES
   1  0.3449E+01  0.0000E+00  0.0000E+00
   2  0.3448E+01  0.6019E-01  0.0000E+00
   3  0.3449E+01  0.0000E+01  0.1000E+00
   4  0.3448E+01  0.6019E-01  0.1000E+00
CONTOUR NUMBER =       2
VALUE OF CONTOUR =   0.5000E+02
POINT NO.,   3 COORDINATES
   1  0.2155E+01  0.0000E+00  0.0000E+00
   2  0.2155E+01  0.3762E-01  0.0000E+00
   3  0.2155E+01  0.0000E+00  0.1000E+00
   4  0.2155E+01  0.3762E-01  0.1000E+00
CONTOUR NUMBER =       3
VALUE OF CONTOUR =   0.9500E+02
POINT NO.,   3 COORDINATES
   1  0.1485E+01  0.0000E+00  0.0000E+00
   2  0.1485E+01  0.2591E-01  0.0000E+00
   3  0.1485E+01  0.0000E+00  0.1000E+00
   4  0.1485E+01  0.2591E-01  0.1000E+00
NORMAL ENDING OF MODEL PROGRAM.
```

Fig. 12.7 Subroutine FMONE

```
      SUBROUTINE  FMONE (N,DELTA,COORD,RJ,A,FFM)
C     * * * * * * * * * * * * * * * * * * * * * * * * * *
C     FIND 1ST FUNDAMENTAL MAGNITUDES ON A 3-D SURFACE
C     * * * * * * * * * * * * * * * * * * * * * * * * * *
CDP   IMPLICIT REAL*8 (A-H,O-Z)
      DIMENSION  A(2,2), RJ(2,3), DELTA(2,N), COORD(N,3)
C     N = NUMBER OF NODES DEFINING THE SURFACE
C     DELTA = LOCAL DERIVATIVES OF THE SHAPE FUNCTIONS
C     COORD = THREE SPATIAL COORDINATES OF THE NODES
C     RJ = REDUCED JACOBIAN MATRIX
C     A = 1ST FUND. MAG. (METRIC TENSOR),  FFM = DET(A)
C     INDEX NOTATION:  I=1,2,3,   R,S=1,2,  /=SUBSCRIPTS
C                A/RS = X/I,R * X/I,S  RJ/RI = X/I,R
C     FORM REDUCED JACOBIAN
      ZERO = 0.D0
      DO 30  I = 1,2
      DO 20  J = 1,3
      SUM = ZERO
      DO 10  K = 1,N
   10 SUM = SUM + DELTA(I,K)*COORD(K,J)
   20 RJ(I,J) = SUM
   30 CONTINUE
C     FORM METRIC TENSOR
      A(1,1) = RJ(1,1)*RJ(1,1) + RJ(1,2)*RJ(1,2)
     1        + RJ(1,3)*RJ(1,3)
      A(1,2) = RJ(1,1)*RJ(2,1) + RJ(1,2)*RJ(2,2)
     1        + RJ(1,3)*RJ(2,3)
      A(2,2) = RJ(2,1)*RJ(2,1) + RJ(2,2)*RJ(2,2)
     1        + RJ(2,3)*RJ(2,3)
      FFM = A(1,1)*A(2,2) - A(1,2)*A(1,2)
C     SURFACE AREA DA = SQRT(FFM)*DR*DS
C     ANGLE R-S    COS (ANG) = A12/SQRT( A11*A22 )
C     LENGTHS  DL1 = SQRT(A11)*DR  DL2 = SQRT(A22)*DS
      IF ( FFM.GT. ZERO )  RETURN
C     DETERMINATE OF A IS NOT POSITIVE DEFINITE
      WRITE (6,5000)
 5000 FORMAT (' FATAL DATA ERROR IN SUBR FMONE')
      RETURN
      END
```

Fig. 12.8 Subroutine SNORMV

```
      SUBROUTINE  SNORMV (RJ,FFM,VECTOR)
C     * * * * * * * * * * * * * * * * * * * * * * * * * *
C     FIND UNIT NORMAL VECTOR TO A 3-D SURFACE AT A POINT
C     * * * * * * * * * * * * * * * * * * * * * * * * * *
CDP   IMPLICIT REAL*8 (A-H,O-Z)
      DIMENSION  RJ(2,3), VECTOR(3)
C     VECTOR = GLOBAL COMPONENTS OF UNIT NORMAL VECTOR
C     THE FOLLOWING QUANTITIES ARE OBTAINED IN SUBR FMONE
C     RJ = REDUCED JACOBIAN MATRIX
C     A = 1ST FUND MAGNITUDES(METRIC TENSOR), FFM = DET(A)
CDP   SQRT(Z) = DSQRT(Z)
      ROOT = SQRT(FFM)
      VECTOR(1) = (RJ(1,2)*RJ(2,3) - RJ(1,3)*RJ(2,2))/ROOT
      VECTOR(2) = (RJ(1,3)*RJ(2,1) - RJ(1,1)*RJ(2,3))/ROOT
      VECTOR(3) = (RJ(1,1)*RJ(2,2) - RJ(1,2)*RJ(2,1))/ROOT
      RETURN
      END
```

Table 12.1 Parameter Definitions for Three-dimensional Heat Conduction

---

CONTROL

NSPACE = 3	NLPFIX = 2
N = 8	MISFIX = 8
NG = 1	LBN = 0*, or 4**
NCURVE = 3	

DEPENDENT VARIABLES:
1 = temperature, T

BOUNDARY FLUX COMPONENTS AT NODES**:
1 = specified normal outward heat flux per unit area, q_n

PROPERTIES:
Element level:
Fixed point:
1 = Gaussian quadrature order in each dimension, (2*)
2 = Element material number
Miscellaneous Level:
K + 1 = thermal conductivity in x-direction, k_x
K + 2 = thermal conductivity in y-direction, k_y
K + 3 = thermal conductivity in z-direction, k_z
K + 4 = heat generation per unit volume
where for material number I, K = 4*(I − 1)

CONSTRAINTS:
Type 1: specified nodal temperatures

---

* Default value
** Optional

13

Automatic mesh generation

13.1 Introduction

As we have seen the practical application of the finite element method requires extensive amounts of input data. Analysts quickly learned that special mesh generation programs could help reduce this burden. Today the most powerful commercial codes, like PAFEC [39], offer extensive mesh generation and data supplementation options. Most such codes use a mapping, or transformation, method to generate the mesh data.

Mesh generation by mapping is a technique where the element connectivity (topology) is simplified to a square or triangular grid system which is then mapped into the actual shape of the domain of interest. Only the coordinates of the perimeter points must be known in advance since they are used to execute the mapping process. The spatial coordinates of all internal nodal points are automatically computed. The node numbers and element incidence data are also automatically generated. The two most common mapping methods are *isoparametric mapping* and the *I–J mapping*. Both of these methods impose certain restrictions on the layout of the nodal points but the reduction in the effort required to effect the solution of a given problem more than compensates for these restrictions. Since isoparametric elements were discussed earlier, the latter method will be considered here.

A typical two-dimensional mapping program locates and numbers the nodal points, numbers the elements and determines the element incidences. In addition, it usually allows for assignment of codes to each node and each element. Thus, it also has limited capability to generate boundary condition codes and element material codes.

To layout a nodal point system for the body to be analyzed, the region of the (x, y)-plane intersecting the body is covered, insofar as any curved boundaries will permit, with an array of quadrilaterals (see Fig. 13.1).

Each vertex of a quadrilateral is taken to be a nodal point. In the mapping program each nodal point is identified by a pair of positive integers, denoted by (I, J). Nodes with common values of J are said to lie in the same row. This implies that they will be on a common curve in the (x, y)-plane.

The scheme for mesh generation may be thought of as representing a mapping of points from the (I, J)-plane into the (x, y)-plane. As illustrated in Fig. 13.2, each quadrilateral in the (x, y)-plane is a square in the (I, J)-plane and may be identified by the (I, J) coordinates of its bottom left-hand corner in the (I, J)-plane. An integer nodal point number is assigned to each (I, J) point, and an integer element number is assigned to each (I, J) point that corresponds to the bottom left-hand corner of the squares in the (I, J)-plane. Another typical mapping is illustrated in Fig. 13.3. In this figure, the body to be analyzed is shown in part (a), the points in the (I, J)-plane are in (b), and their mapped locations in the (x, y)-plane are shown in (c).

The mesh generation is accomplished in the following manner. Data containing (among other things) the values of I, J and the x- and y-coordinates are input for each node whose coordinates are to be specified. Such nodes must include at least all nodes on the boundary of the region of interest, as well as on any interfaces between regions with different element properties. As many other interior nodal points as the user may desire may have their coordinates specified, but no others are required. As the data cards are read, a list is compiled of the minimum and maximum values of I for each J, and each node for which coordinates have been input is identified and the coordinates are stored. After all the desired nodal points cards have been input, the coordinates for all unspecified nodes which have I in the interval $I\text{MIN} \leqslant I \leqslant I\text{MAX}$ for the proper J, are calculated for all $J (1 \leqslant J \leqslant J\text{MAX})$. The calculation, or mapping, of the unspecified coordinates can be accomplished several ways. Four options are outlined here.

Fig. 13.1 Mapping from (I,J) to (x, y) space (six node triangle)

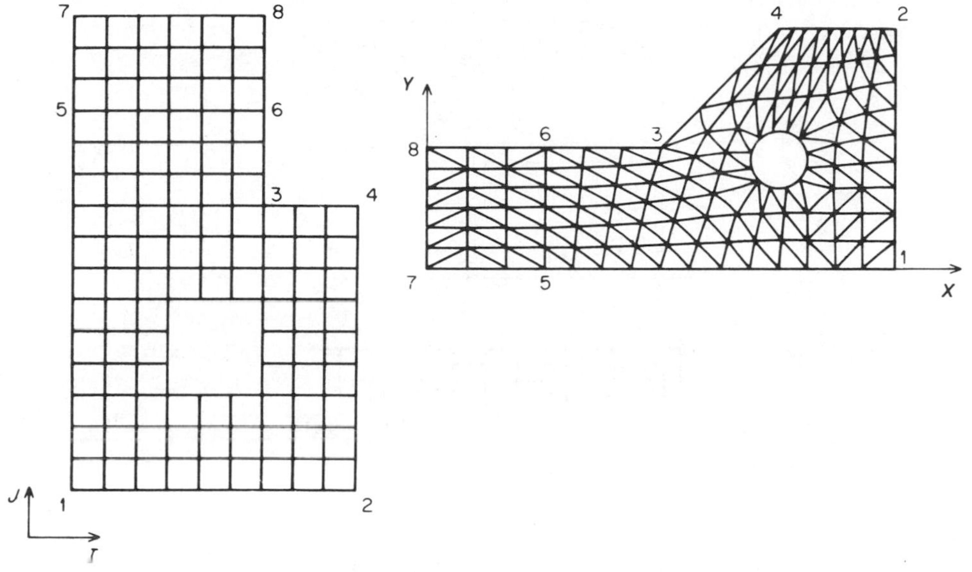

Fig. 13.2 Mapping of a typical element

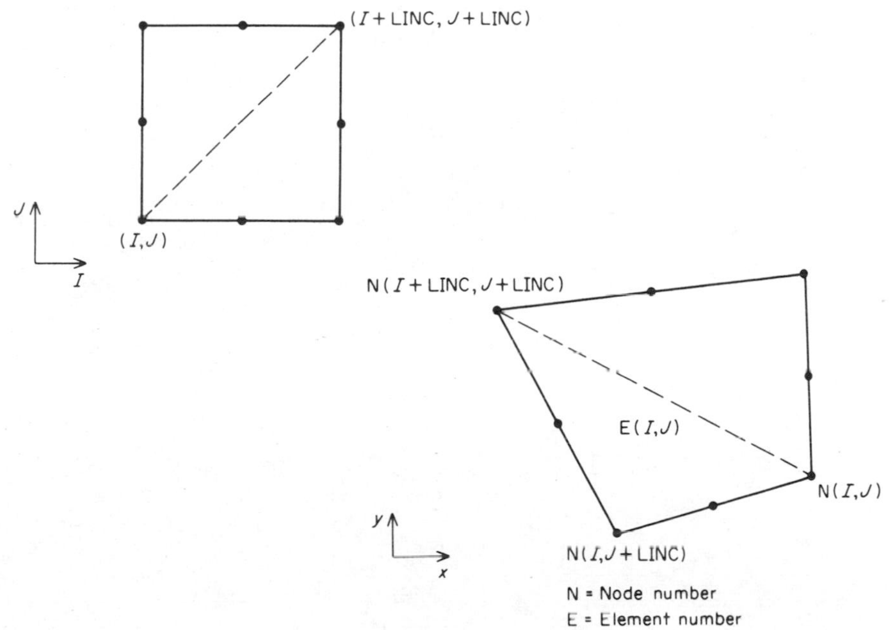

Fig. 13.3 Typical mapping

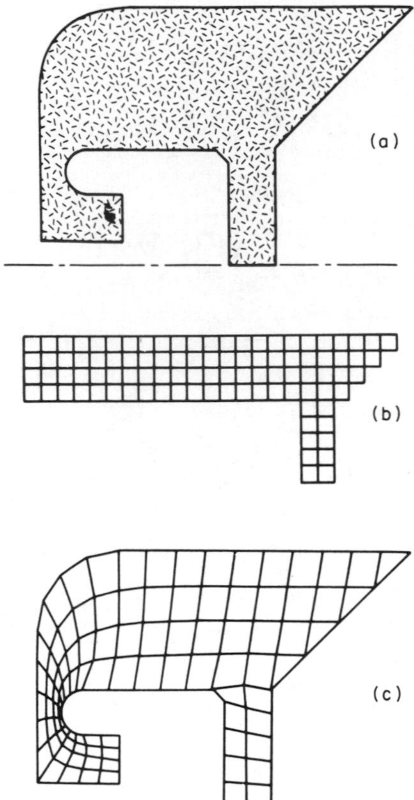

13.2 Mapping functions

The first method is an iterative one that generates a mesh by approximately solving Laplace's equation using the given perimeter (and interior) points as boundary conditions. For example, the x-coordinates are obtained from an approximate solution of

$$\frac{\partial^2 x}{\partial I^2} + \frac{\partial^2 x}{\partial J^2} = 0,$$

where x is specified on the perimeter of the (I, J) domain. First the x-coordinates of the boundary points are used as boundary values of the unknown harmonic function, and the values obtained on the interior points, and $x(I, J)$ are taken as the x-coordinates of the corresponding points in the (x, y)-plane. A similar procedure yields the y-coordinates of the unspecified interior nodes. It should be noted that, in general, this method tends to yield nodal points with uniform curvilinear spacing. If this is not deemed desirable, some nodal points interior to the region may have their coordinates specified to control the distribution of the remaining points. This last point is illustrated by comparing Fig. 13.4 with Fig. 13.3. In Fig. 13.4 some interior nodes have been specified to make desired changes in the mapped mesh. The interior nodes which have been specified are shown with the blocked-in symbols.

The Laplace mapping method takes the largest amount of computer time but it is still quite fast and usually the preferred method. The other three methods to be discussed (see Fig. 13.5) are direct interpolation techniques, and therefore require little computer time. The two point linear interpolation method is the fastest but it can easily lead to overlapping mesh lines. It utilizes an interpolation between two (given) bounding points on the J line. A four point interpolation is also used at times. This option determines the two closest (given) bounding points in the I direction and the two in the J direction. The coordinates of point (I, J) are then calculated by a quadratic interpolation between the coordinates of these four points (see Fig. 13.6).

An eight point interpolation is also commonly used, but this routine has some minor restrictions on the shape of the mesh in the (I, J)-plane. If the mesh is not a rectangle then it *must* consist of sub-areas, defined by interior points, that are all rectangular (see Fig. 13.7). The two closest bounding points in both the I and J directions are determined. These four points are used to define the four *corner* points of the rectangular sub-area. Then the coordinates of point (I, J) are calculated by nonlinear interpolation between the given coordinates of the above eight points. The eight node interpolation uses a set of blending functions that also leads to smooth variations in nodal locations. In addition, it tends to make the calculated lines orthogonal to the four boundary curves used in the interpolation.

Usually the program calculates the location of each point and assigns a nodal number to each point (except those that are interior to a hole). After this, the program forms quadrilateral or pairs of triangular elements (except those interior to a hole) by assigning the proper nodal point numbers to the corners of the elements. After the elements (and any element codes) have been generated, they are checked for possible errors involving negative areas. The output of the mesh generation program is designed so that it may be fed, as input data, directly to the finite element analysis program MODEL, or similar codes.

The user should note that the bandwidth of the equations in the finite element codes will be directly proportional to the maximum difference in *I* coordinates in the mesh. The maximum bandwidth generated by the mapping is usually calculated and output for the user's information.

All of these mapping methods work well for convex domains. However, for concave domains the nodes may become too crowded and it is usually best to divide concave domains into nearly convex by specifying various interior lines.

It is possible to extend all of the above techniques to surface generation problems that might be required in the analysis of general shell structures. They may also be generalized to three-dimensional solids but the isoparametric method appears to be the simplest for such problems.

Fig. 13.4 Effect of two interior lines

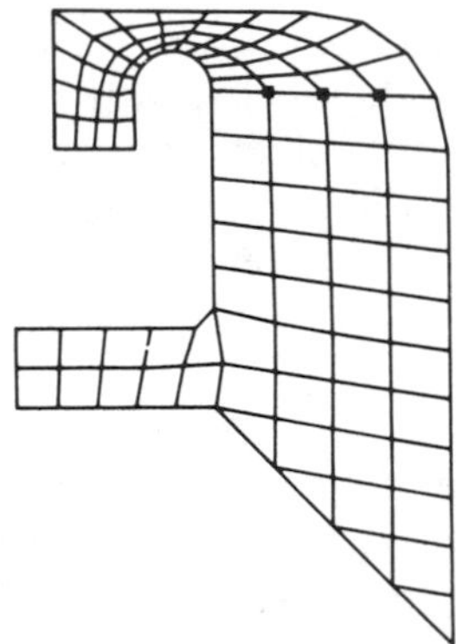

Fig. 13.5 Common mapping options

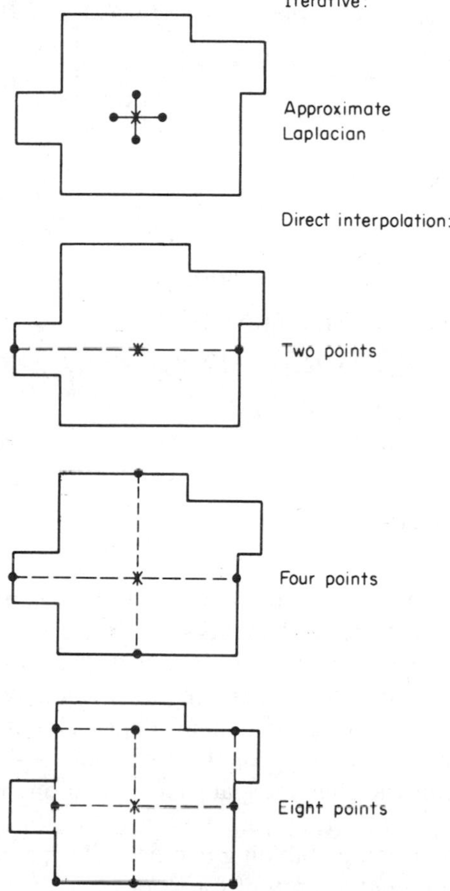

Fig. 13.6 A four point interpolation

$(I, J2)$

$(I1, J)$ — × — $(I2, J)$

$(I, J1)$

$a = I2 - I$

$b = J2 - J$

$c = I - I1$

$d = J - J1$

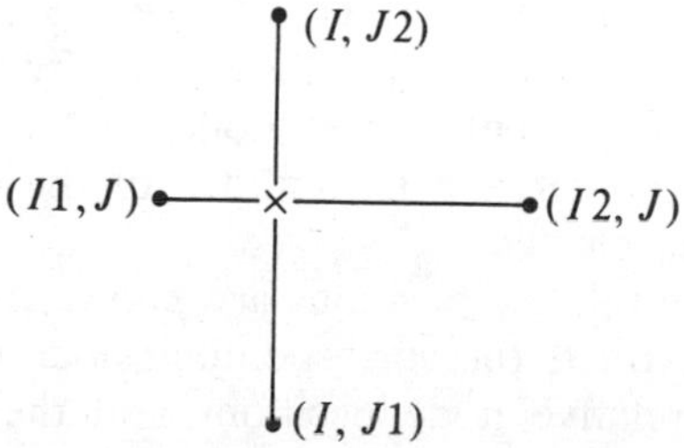

$$X(I,J) = \frac{abcd}{ac + bd}\left\{\frac{X(I2,J)}{a(a+c)} + \frac{X(I,J2)}{b(b+d)} + \frac{X(I1,J)}{c(a+c)} + \frac{X(I,J1)}{d(b+d)}\right\}$$

Fig. 13.7 An eight point interpolation

$$f_0(z) \equiv 10z^3 - 15z^4 + 6z^5$$

$$f_1(z) \equiv 1 - f_0(z)$$

$$u = \frac{I - I1}{I2 - I1}$$

$$v = \frac{J - J1}{J2 - J1}$$

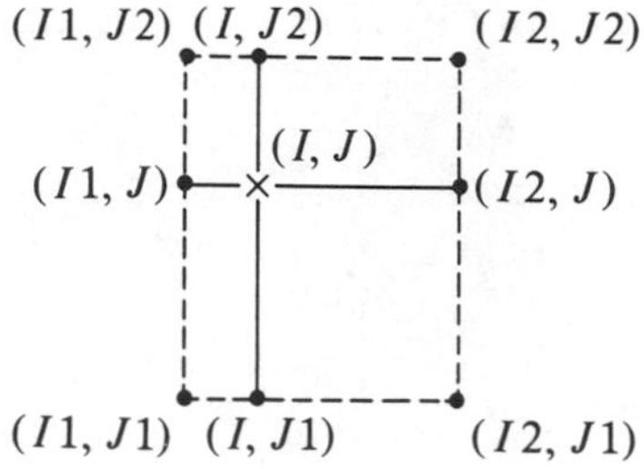

$$X(I,J) = f_1(u)X(I1,J) + f_0(u)X(I2,J) + f_1(v)X(I,J1) + f_0(v)X(I,J2) - f_1(u)f_1(v)X(I1,J1) - f_1(u)f_1(v)X(I2,J1) - f_1(u)f_0(v)X(I1,J2) - f_0(u)f_0(v)X(I2,J2)$$

13.3 Higher order elements

For the higher order elements there are additional considerations. One is that for the 8, 9, and 12 node elements some internal nodes are omitted. The simple procedure is to map all points in the usual manner but omit them in the output state while they are being numbered. The second problem is that of identifying the element incidences. As usual, the elements are identified by the (I, J) values of their bottom left-hand corner. For higher order elements the element spans more than one row and column. It is necessary to store the (I, J) values of all nodes of a typical element type. This is accomplished by building two lists which contain numbers to be added to (I, J) to identify each of the nodes.

For example, for the simple four node quadrilateral one could define $IQ = [0, 1, 1, 0]$ and $JQ = [0, 0, 1, 1]$ to identify the four relative positions $(I + IQ, J + JQ)$, i.e. (I, J), $(I + 1, J)(I + 1, J + 1)$, and $(I, J + 1)$. Similar arrays exist for each type but they range from zero to LINC, the number of increments between corner nodes. Of course, for the triangles it is necessary to have two sets to allow for the different diagonal directions which can be selected.

Once these relative positions are known one simply loops over each I, J pair and extracts the corresponding node numbers. Figure 13.8 shows the typical relative node positions and the omitted internal nodes. It illustrates only half of the triangular options.

Each element has an assigned code (material number) for use in the output. Consider an element associated with point (I, J). The element's nodes not lying on $I +$ LINC or $J +$ LINC are assigned the same code as (I, J). It is important in non-homogeneous problems to verify that one has correctly input the various element codes. As an aid the example program prints a sketch of the input element codes. The codes are initialized to unity, and negative values imply elements in a hole to be omitted from both the point and element output.

It is also possible to use a mesh generator to also generate constraint or flux data. However, this tends to become rather specialized. It is relatively easy to do since after the mesh generation is complete one has the coordinate arrays, boundary codes, and node numbers for all (I, J) points in the mapped space. Thus it is just a matter of combining these data in a particular form. For example, to generate flux data one could input two (I, J) points and their corresponding flux components, and then all nodes along the line could be recovered and the flux obtained by linear interpolation.

The previously outlined procedures have been implemented in a program, MESH2D, which is discussed here. The program allows the user the option of selecting elements that are linear, quadratic, or cubic. Either quadrilaterals or triangles can be generated. The triangles are obtained by dividing the quadrilaterals across their shortest diagonal. Points being used to define the physical shape can be input in either cartesian or polar coordinates.

The minimum data consist of a title, control card, the nodal definition cards, and any material code changes (default value is unity). These data are explained in the table of input formats. Some of the parameters have their meanings illustrated in Figs. 13.9 to 13.12. The nodal point definition cards can generate points in either cartesian (default) or polar coordinates. The number of points generated by such a card depends on the parameter KDELTA. Lines or arcs are generated by linear interpolation between the two limiting values defined on the card.

The location of points in the integer space are defined by the (I, J) coordinates of the first point, the number of additional points, KDELTA, and direction of motion flag, INTEST. Data can be for a single point (INTEST $= 0$) but are generally for motion parallel to the I-axis, J-axis, $I = -J$ diagonal, or $I = J$ diagonal. If the end point (R, Z) coordinates are being input the respective flags are 1, 2, 3, and 4. However, if the previously defined (R, Z) coordinates are to be used to generate a straight line then the flags are 5, 6, 7, and 8 and one must only supply the integer definitions $(I, J,$ KDELTA$)$. The generation must be in the direction of positive I (or positive J when I is constant).

Holes in the mesh and any changes in element codes are defined by the element code generation cards. These assign element codes for rectangular regions as illustrated in Figs. 13.11 and 13.12. Negative codes are interpreted to denote holes which are omitted from the point and element output phases. In practical problems one must usually carefully define any interior lines that divide material regions.

The output of the code usually consists of print, punch, and plot files. For punched data the MODEL code formats are default but a provision has been made for a user supplied set of formats. The plotting logic is quite simple. First any existing J-lines are plotted, then the I-lines are plotted and if the elements are triangles then one must insert the required diagonals on each J-row of elements. The print file echoes the given data and lists data generated by the point and element definition cards. As a debugging aid it also prints a list of given integer points and the element codes. The controlling program is given in Fig. 13.13 while the input routines for the points and material codes are given in Figs. 13.14 and 13.15, respectively.

A typical four point mapping routine for the nodes is shown in Fig. 13.16. After the coordinates have been found subroutine OMITP, Fig. 13.17, identifies any points that will be omitted from the higher order elements. Then the nodal points are numbered and output by PCOUNT which is given in Fig. 13.18.

The element incidences, material number, etc., are established by subroutine LCOUNT. As shown in Fig. 13.19 it calls IJQUAD or IJTRI to obtain the list of relative (I, J) coordinates of the element nodes. These two routines supply the data illustrated in Fig. 13.8. Subroutine LCOUNT also outputs the element data.

Now that the points and elements are completely defined in the (I, J) database they can be plotted using PLOTIJ which is given in Fig. 13.20. Hardware-dependent calls have been hidden in the calls to dummy routines. These include WINDO, MOVETO, LINETO, STARTP and FINSHP.

Figure 13.21 shows a set of data chosen to illustrate all of the input options. The generated (I, J) and (R, Z) plots are shown in Fig. 13.22. The data formats are given in the Appendix.

Fig. 13.8 Location of nodes relative to defining point (I,J)

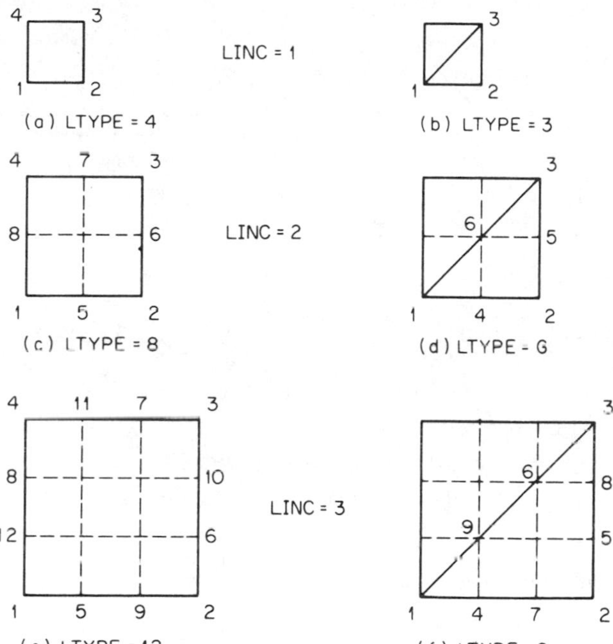

Fig. 13.9 Interpolation of physical coordinates

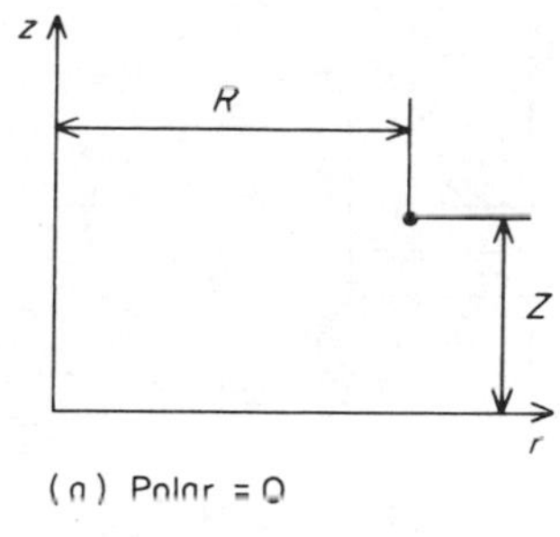

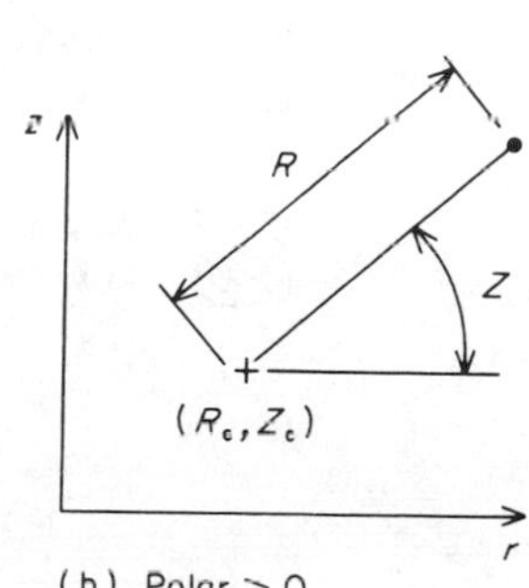

Fig. 13.10 Meaning of point input generators

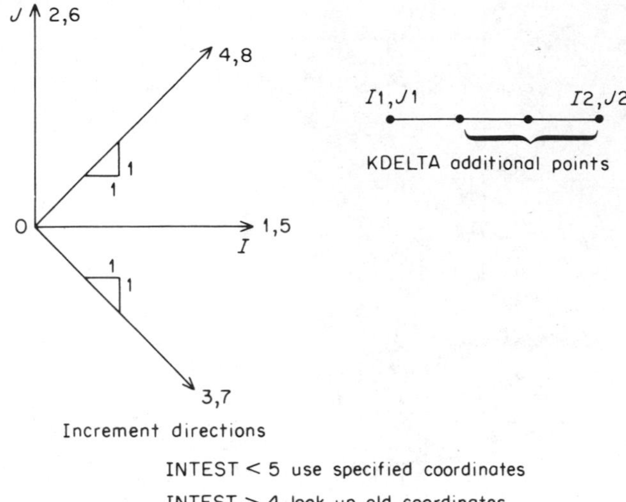

Fig. 13.11 Identifying holes

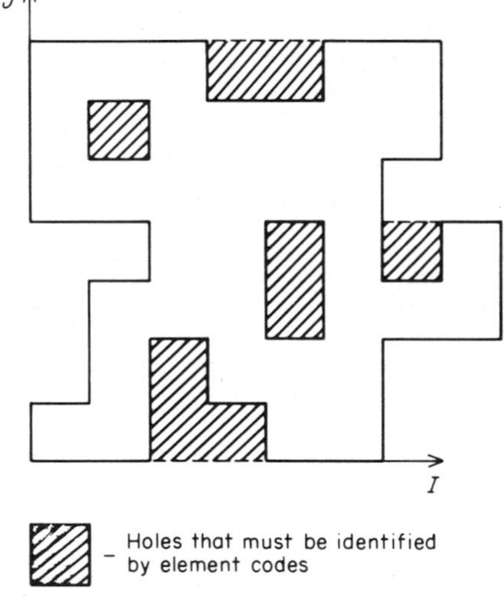

Fig. 13.12 A block of elements with material code MM

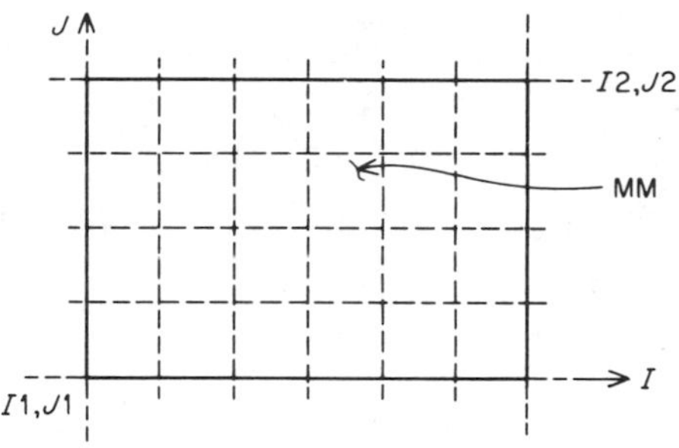

Fig. 13.13 Controlling program

```
      * * * * * * * * * * * * * * * * * * * * * * * * * * * *
      DIMENSION  AD(1800), ID(2000)
C     AD(KFLT), ID(KFIX) = FLOATING, FIXED PT ARRAY STORAGE
C     MINNPT, MINEL = STARTING NODE AND ELEMENT NUMBERS
      DATA KFLT, KFIX /1800,2000/
      DATA MINNPT, MINEL /1,1/
      CALL MESH2D (KFLT,KFIX,AD,ID,MINNPT,MINEL)
      CALL EXIT
      STOP
      END
      SUBROUTINE  MESH2D (KFLT,KFIX,AD,ID,MINNPT,MINEL)
C     * * * * * * * * * * * * * * * * * * * * * * * * * * * *
C     FEM MESH GENERATION FOR TRIANGLES AND QUADRILATERALS
C     * * * * * * * * * * * * * * * * * * * * * * * * * * * *
C
C     TITLE IS THE NAME OF THE PROBLEM,  ETC.
C     NPTCD = NUMBER OF NODAL POINT INPUT CARDS
C     MAXR = MAX. VALUE OF J IN INPUT DATA
C     MAXC = MAX. VALUE OF I IN INPUT DATA
C     FORMATS FOR PUNCHED OUTPUT (IF ANY)
C         IOPT = 0, AKIN'S MODEL CODE
C              ¬ 0, USER SUPPLIED FORMATS
C     OUTPT AND OUTEL GOVERN THE METHOD OF OUTPUT FOR
C         THE POINTS AND ELEMENTS. THE CPTICNS ARE:
C         = 0, PRINT
C         = 1, PRINT AND PUNCH
C         = 2, PUNCH CNLY
C         = 3, NO OUTPUT
C     LTYPE = ELEMENT TYPE SELECTION
C         = 3, 3 NODE TRIANGLE
C         = 4, 4 NODE QUAD
C         = 6, 6 NODE TRIANGLE
C         = 8, 8 NODE QUAD
C         = 9, 9 NODE TRIANGLE
C         = 12, 12 NODE QUAD
C     THE UNSPECIFIED POINTS ARE DETERMINED BY:
C         MAP = 0 , MAPPING USING LAPLACE'S EQUATION (ITERATIVE)
C             = 1 , TWO POINT LINEAR INTERPOLATION BETWEEN
C                   BOUNDING POINTS ON THE J LINE
C             = 2 , FOUR POINT INTERPOLATION BETWEEN GIVEN
C                   POINTS ON THE I AND J LINES
C             = 3 , EIGHT POINT INTERPOLATION BETWEEN GIVEN
C                   POINTS ON THE I AND J LINES PLUS FOUR
C                   'CORNER' POINTS
C     NMTCD = NUMBER OF ELEMENT CODE INPUT CARDS
C     IF ECHO = 0 THEN INPUT DATA ARE LISTED
C     LPLOT = PLOTTING OPTIONS
C         = 0, NO PLOTS
C         = 1, PLOT R,Z MESH
C         = 2, PLOT I,J MESH
C         = 3, PLOT BOTH R,Z AND I,J MESHES
C     NEDCD = NO. CARDS FOR EDGE FLUX CALCULATIONS = 0
C-----------------------------------------------------------------
C     LOGICAL UNITS USED BY MESH2D:
C         5 = CARD READER,  6 = LINE PRINTER
C         7 = CARD PUNCH, WHEN OUTPT OR OUTEL =  1 OR 2
C-----------------------------------------------------------------
```

Fig. 13.13—continued

```
      INTEGER  OUTPT, OUTEL, ECHO
      DIMENSION  TITLE(19), AD(KFLT), ID(KFIX)
      CATA IJFLAG /987654/
C             *** PRINT AUTHOR CREDIT ***
      CALL  TOOT
C-->      *** READ,PRINT, AND CHECK INPUT DATA ***
      READ (5,5000) TITLE
      READ (5,5010) MAXR, MAXC, NPTCD, NMTCD, LTYPE, LPLOT,
     1              MAP, OUTPT, OUTEL, IOPT, ECHO, NEDCD
      IF ( LTYPE.LT.3 )  LTYPE = 3
      WRITE (6,5020)  TITLE
      WRITE (6,5030) NPTCD, NMTCD, NEDCD, MAXC, MAXR, OUTPT,
     1               MAP, OUTEL, LPLOT, IOPT, ECHO, LTYPE
C     DETERMINE I,J INCREMENTS FOR LTYPE
      LINC = 1
      IF ( LTYPE.EQ.6 .OR. LTYPE.EQ.8 )  LINC = 2
      IF ( LTYPE.EQ.9 .OR. LTYPE.EQ.12 )  LINC = 3
C
C     ARRAYS ARE DEFINED AT FIRST USE
C     FIXED PT ARRAY STORAGE LI: 1-INNER 2-MAT 3-IMAX
C              4-IMIN 5-LIMIN 6-LIMAX 7-IGIVEN 8=1-NPTNUM
C     FLOATING PT ARRAY STORAGE MI: 1-R 2-Z 3-CODE 4-ALINE
C
C-->           *** CALCULATE ARRAY POINTERS ***
      CALL PTER (L1,L2,L3,L4,L5,L6,L7,L8,LTOTAL,
     1           M1,M2,M3,M4,MTOTAL,MAXR,MAXC)
C              *** POINTERS COMPLETED ***
      WRITE (6,5040)  KFLT,MTOTAL,KFIX,LTOTAL
      IF (KFLT.GE.MTOTAL .AND. KFIX.GE.LTOTAL)  GO TO 10
      WRITE (6,5050)
      CALL EXIT
   10 CONTINUE
C                    *** INITIALIZE ARRAYS ***
      CALL  INITAL (KFLT,KFIX,AD,ID,MAXR,MAXC,L2,L3,L4,L6)
C-->            *** READ NODAL POINT DEFINITIONS ***
      CALL  INPTS (MAXC,MAXR,AD(M1),AD(M2),AD(M3),ID(L1),
     1            ID(L4),ID(L3),ID(L5),ID(L6),ECHC,NPTCD,MAP)
C               *** PRINT MAP OF DEFINED POINTS ***
      IF (ECHO.EQ.0)  CALL IJMESH (MAXC,MAXR,AD(M4),
     1                  ID(L1),ID(L3),ID(L4))
C-->        *** READ CHANGES IN ELEMENT CODES ***
      IF (NMTCD.GT.0)  CALL  INEL (MAXC,MAXR,ID(L2),ID(L5),
     1                              ID(L6),ECHO,NMTCD)
      IF (ECHO.EQ.0) CALL IJMAT (MAXC,MAXR,AD(M4),ID(L5),
     1                           ID(L6),ID(L2),ID(L1))
C-->                      *** MAP POINTS ***
      IF (LPLOT.EQ.2 .AND. OUTPT.EQ.3)  GO TO 15
      IF (MAP.EQ.0)  CALL LAPLAC (MAXC,MAXR,AD(M1),AD(M2),
     1                      ID(L1),ID(L4),ID(L3),ID(L7),N1)
      IF (MAP.EQ.1)  CALL TWOPT (MAXC,MAXR,AD(M1),AD(M2),
     1                      ID(L1),ID(L4),ID(L3),N1)
      IF (MAP.EQ.2)  CALL FOURPT (MAXC,MAXR,AD(M1),AD(M2),
     1                      ID(L1),ID(L4),ID(L3),ID(L7),N1)
      IF (MAP.EQ.3)  CALL EIGHTP (MAXC,MAXR,AD(M1),AD(M2),ID(L1),
     1                      ID(L4),ID(L3),ID(L7),N1)
      IF (OUTPT.LT.2)  WRITE (6,5060)  N1
C-->                   *** NUMBER POINTS ***
   15 IF (LTYPE.GT.6)  CALL OMITP (MAXR,MAXC,ID(L2),LTYPE,
     1                     LINC,IJFLAG,AD(M1),AD(M2))
      CALL  PCOUNT (MAXC,MAXR,AD(M1),AD(M2),AD(M3),ID(L2),
     1      ID(L8),ID(L4),ID(L3),IJFLAG,IOPT,OUTPT,MINNPT)
C-->             *** GENERATE ELEMENTS ***
      IF (OUTEL.LT.4)  CALL  LCOUNT (MAXR,MAXC,AD(M1),AD(M2),
     1      ID(L6),ID(L5),ID(L2),ID(L8),LTYPE,LINC,MINEL,OUTEL,IC
C                *** PLOT RESULTS ***
      IF (LPLOT.EQ.0)  GO TO 20
C     INITIALIZE PLOT FILE AND SET FIRST ORIGIN
      CALL  STARTP
C     PLOT R-Z MESH
      IF (LPLOT.EQ.1 .OR. LPLOT.EQ.3)  CALL  PLOTIJ (MAXC,MAXR,
     1          AD(M1),AD(M2),ID(L8),ID(L2),ID(L3),ID(L4),ID(L5),
     2          ID(L6),LINC,LTYPE,LPLOT)
      IF (LPLOT.LT.2)  GO TO 20
C     REPLACE R,Z BY I,J VALUES AND PLOT
      CALL  IJTORZ (MAXC,MAXR,AD(M1),AD(M2))
      CALL  PLOTIJ (MAXC,MAXR,AD(M1),AD(M2),ID(L8),ID(L2),ID(L3),
     1             ID(L4),ID(L5),ID(L6),LINC,LTYPE,LPLOT)
```

Fig. 13.13—continued

```
   20 CONTINUE
C     CLOSE PLOT FILE
      IF (LPLOT.GT.0)  CALL  FINSHP
      WRITE (6,5070)
C     .........................   FORMATS  ....................
 5000 FORMAT (20A4)
 5010 FORMAT (16I5)
 5020 FORMAT (1H1,12(5H*****),/,1X,19A4,/,1X,12(5H*****))
 5030 FORMAT (' ****   M E S H 2 D   ****',//,
     1' NUMBER OF INPUT CARDS FOR:',/,
     2'                NODAL POINTS                ',I5,/,
     3'               ELEMENT CODES                ',I5,/,
     4'               EDGE CONDITIONS              ',I5,///,
     5' MAXIMUM INPUT VALUES|   I =',I3,', J =',I3,///,
     6' CONTROL PARAMETERS:',/,
     7'       OUTPT     ',I3,10X,'MAP        ',I3,/,
     8'       OUTEL     ',I3,10X,'LPLOT      ',I3,/,
     9'       IOPT      ',I3,10X,'ECHO       ',I3,/,
     1'       ELTYPE    ',I3,//)
 5040 FORMAT ('1  ARRAY STORAGE :',/,'                      AVAILABLE
     1ED',/,' FLOATING POINT',I10,1X,I10,/,' FIXED POINT    ',I10,
 5050 FORMAT (' DIMENSION STATEMENT EXCEEDED. ABNORMAL ENDING.')
 5060 FORMAT ('OCOORDINATES CALCULATED AFTER ',I3,' ITERATIONS')
 5070 FORMAT (' NORMAL ENDING OF MESH2D')
      RETURN
      END
```

Fig. 13.14 Subroutine INPTS

```
      SUBROUTINE  INPTS (K1,K2,R,Z,CODE,INNER,IMIN,IMAX,
     1                   LIMIN,LIMAX,ECHO,NPTCD,MAP)
C     . . . . . . . . . . . . . . . . . . . . . . . . . .
C     INPUT COORDINATES AND NODAL POINT CODES
C     . . . . . . . . . . . . . . . . . . . . . . . . . .
      INTEGER ECHO, POLAR
      DIMENSION   R(K1,K2),Z(K1,K2),CODE(K1,K2),INNER(K1,K2),
     1            IMIN(K2),LIMIN(K2),LIMAX(K2),IMAX(K2)
C     R(I,J),Z(I,J) = COORDS. OF PT. CORRESPONDING TO (I,J)
C     CODE(I,J) = PT. BOUNDARY CODE ASSIGNED TO (I,J)
C     INNER(I,J) > 0 IMPLIES PT (I,J) WAS INPUT
C     IMIN(J),IMAX(J) = MIN   MAX I FOR ANY PT IN ROW J
C     LIMIN(J),LIMAX(J) = MIN, MAX I OF PT IN ELEM IN ROW J
   10 IF ( ECHO.EQ.0 )  WRITE (6,5000)
 5000 FORMAT (1H1,23X,'NODAL POINT INPUT DATA'/3X,1HI,3X,
     1 1HJ,6X,6HR(I,J),6X,6HZ(I,J),3X,10HANGLE(I,J),6X,
     2 5HRCENT,6X,5HZCENT,2X,9HCODE(I,J)   )
C     POLAR = 0 INDICATES INPUT IS IN CARTESIAN COORDS.
C     *** THEN  R= RADIUS,  AND Z=AXIAL POSITION
C     POLAR NOT = 0 INDICATES INPUT IS IN POLAR COORDS.
C       RELATIVE TO A CIRCLE WHOSE CENTER IS (RC,ZC)
C     THEN R=RADIUS OF CIRCLE, Z=ANGLE (DEG) FROM R-AXIS
C      2,6 J        / 4,8        MEANING OF INTEST IS TO LEFT.
C          J     /               DIRECTION IS AWAY FROM POINT.
C          J  /                  IF > 4 FIND COORDS OF END PTS AND
C        0 J/IIIIII 1,5          USE OLD CODES. I-J INCREMENTS:
C          .                    1,5 = POS I,  4,8 = POS I    J
C            .                  2,6 = POS J,  3,7 = POS I    NEG J
C             . 3,7             0 = SINGLE POINT MODE
C     KDELTA = NO. ADDITIONAL POINTS BEYOND I,J
C     R1,Z1 = COORD OF (I,J), R2,Z2 - COORD OF FINAL PT
      DO 160  N = 1,NPTCD
      READ (5,5010) INTEST,I,J,KDELTA,POLAR,CC,R1,Z1,
     1              R2,Z2,RC,ZC
 5010 FORMAT ( 5I3, F5.0, 6F10.0 )
C     SET I-J LIMITS AND INCREMENTS
      IP = 1
      IF (INTEST.EQ.0.OR.INTEST.EQ.2.OR.INTEST.EQ.6) IP = 0
      JP = 1
      IF ( INTEST.LT.2 .OR. INTEST.EQ.5 )  JP = 0
      IF ( INTEST.EQ.3 .OR. INTEST.EQ.7 )  JP = -1
      KDELTA = IABS(KDELTA)
      LOOP = KDELTA + 1
      IF ( J.LT.1 .OR. J.GT.K2 )  WRITE (6,5040) I,J
      IF ( I.LT.1 .OR. I.GT.K1 )  WRITE (6,5040) I,J
```

Fig. 13.14—continued

```
 5040 FORMAT ('  INVALID DATA FOR I =',I3,' J =',I3)
      I1 = I
      J1 = J
      I2 = I + KDELTA*IP
      J2 = J + KDELTA*JP
      IF ( KDELTA.LT.1)  KDELTA = 1
      IF ( INTEST.LT.5 )  GO TO 5
C     USE PREVIOUSLY DEFINED COORDINATES
      R1 = R(I1,J1)
      Z1 = Z(I1,J1)
      R2 = R(I2,J2)
      Z2 = Z(I2,J2)
    5 DR = (R2 - R1)/KDELTA
      DZ = (Z2 - Z1)/KDELTA
      IF ( POLAR.NE.0 .AND. DR.NE.0.0 .AND. DZ.NE.0.0 )
     1    WRITE (6,5070)
 5070 FORMAT (' WARNING,BOTH RADIUS ANGLE ARE VARYING')
      I = I1 - IP
      J = J1 - JP
C     LOOP OVER GENERATED POINTS
      DO 110  L = 1,LOOP
      I = I + IP
      J = J + JP
      IF ( I.LT.IMIN(J) )  IMIN(J) = I
      IF ( I.GT.IMAX(J) )  IMAX(J) = I
      INNER(I,J) = 1
      IF ( INTEST.GT.4 )  CC = CODE(I,J)
      CODE(I,J) = CC
      IF ( POLAR.EQ.0 )  GO TO 20
      IF ( ECHO.EQ.0 )  WRITE (6,5080) I,J,R1,Z1,RC,ZC,CC
 5080 FORMAT (2I4, 1PE12.4, 12X, 3E12.4, 0PF10.0 )
      R(I,J) = R1*COS(Z1/57.29578) + RC
      Z(I,J) = R1*SIN(Z1/57.29578) + ZC
      GO TO 25
   20 IF ( ECHO.EQ.0 ) WRITE (6,5090) I, J, R1, Z1, CC
 5090 FORMAT (2I4, 1PE12.4, E12.4, 36X, F10.0)
      R(I,J) = R1
      Z(I,J) = Z1
   25 R1 = R1 + DR
      Z1 = Z1 + DZ
  110 CONTINUE
      IF ( ECHO.EQ.0 )  WRITE (6,5100)
 5100 FORMAT (1H0,15(5H----+),2H--)
  160 CONTINUE
C     FIND BEGINNING & END OF PTS & ELEMS FOR EACH J
      IF ( K2.LE.2 )  GO TO 175
      J2 = K2 - 1
      DO 170  J = 1,J2
      LIMIN(J) = MAX0 ( IMIN(J), IMIN(J+1) )
  170 LIMAX(J) = MIN0 ( IMAX(J), IMAX(J+1) )
  175 LIMIN(K2) = IMIN(K2)
      LIMAX(K2) = IMAX(K2)
C     DETERMINE MAXIMUM NODAL CONNECTIVITY
      MAX = 0
      DO 180  J = 1,J2
      INTEST = IMAX(J) - IMIN(J)
      IF ( INTEST.GT.MAX )  MAX = INTEST
  180 CONTINUE
      MAX = MAX + 2
      IF ( ECHO.EQ.0 )  WRITE (6,5120)  MAX
 5120 FORMAT ( '0NODAL POINT CONNECTIVITY: ',/,
     1 '        MAXIMUM DIFFERENCE IN NODE NUMBERS, ND =',I3,/,
     2 '        HALF BANDWIDTH = (ND+1)*(DOF PER NODE)',//)
      RETURN
      END
```

Fig. 13.15 Subroutine INEL

```
      SUBROUTINE  INEL (K1,K2,MAT,LIMIN,LIMAX,ECHO,NMTCD)
C     . . . . . . . . . . . . . . . . . . . . . . . . . . . . .
C     INPUT CHANGES IN ELEMENT CODES
C     . . . . . . . . . . . . . . . . . . . . . . . . . . . . .
      INTEGER ECHO
      DIMENSION  MAT(K1,K2), LIMIN(K2), LIMAX(K2)
C     MAT(I,J) = CODE OF QUAD ELEM WITH BLC AT I,J
C     IF MAT    0, ELEMENT IS PART OF A HOLE
      IF ( ECHO.EQ.0 )  WRITE (6,5000)
 5000 FORMAT (1H1, ' CHANGES IN BLOCKS OF ELEMENT CODES',/,
     1 ' BLOCK   MAT.NO.   LIMITS:    I1     J1     I2     J2',/)
C     .........< I2,J2     MM IS ASSIGNED TO ELEMENT
C     .        .           CODES FOR ELEMENTS WITH THEIR
C     . /I1,I2 .           BOTTOM LHC WITHIN THE BLOCK
C     .........            MAT(I,J)=MM
      DO 80  N = 1,NMTCD
      READ (5,5010)  I1, J1, I2, J2, MM
 5010 FORMAT ( 5I3 )
      IF ( ECHO.EQ.0 )  WRITE (6,5020) N, MM, I1, J1, I2, J2
 5020 FORMAT (I5, I8, 11X, 4I5)
      IF ( J1.LT.1 .OR. J2.GT.K2 )  WRITE (6,5040)
 5040 FORMAT ('0WARNING, DATA INCLUDES NONEXISTANT ELEMENT')
      IF ( I2.GT.I1 )  I2 = I2 - 1
      IF ( J2.GT.J1 )  J2 = J2 - 1
      DO 70  J = J1,J2
      IF (I1.LT.LIMIN(J) .OR. I2.GE.LIMAX(J)) WRITE (6,5040)
      DO 60  I = I1,I2
      MAT(I,J) = MM
   60 CONTINUE
   70 CONTINUE
   80 CONTINUE
      RETURN
      END
```

Fig. 13.16 Subroutine FOURPT

```
      SUBROUTINE FOURPT (K1,K2,R,Z,INNER,IMIN,IMAX,
     1                   IGIVEN,N1)
C     . . . . . . . . . . . . . . . . . . . . . . . . . . . . .
C        FOUR POINT NONLINEAR INTERPOLATION
C     . . . . . . . . . . . . . . . . . . . . . . . . . . . . .
      DIMENSION  R(K1,K2), Z(K1,K2), INNER(K1,K2), IMIN(K2),
     1           IMAX(K2), IGIVEN(K1)
      JLESS = K2 - 1
      N1 = 1
      IF ( JLESS.LE.1 )  RETURN
      DO 80  J=2,JLESS
C     COUNT GIVEN POINTS IN ROW
      NG = 0
      I1 = IMIN(J)
      I2 = IMAX(J)
      DO 10 I = I1,I2
      IF ( INNER(I,J).NE.1 )  GO TO 10
      NG = NG + 1
      IGIVEN(NG) = I
   10 CONTINUE
C     SKIP TO NEXT ROW IF ALL POINTS ARE GIVEN
      IF ( NG.EQ.(I2-I1+1) )  GO TO 80
      NGLESS = NG - 1
      DO 70 IG = 1,NGLESS
      IGPLUS = IG + 1
C     FIND LIMITING I POINTS
      IF ( (IGIVEN(IGPLUS)-IGIVEN(IG)).EQ.1 )  GO TO 70
      IB = IGIVEN(IG)
      IF = IGIVEN(IGPLUS)
      I1 = IB + 1
      I2 = IF - 1
      DO 60  I = I1,I2
C     FIND LIMITING J POINTS
      J1 = J + 1
      DO 20  JJ = J1,K2
      IF ( INNER(I,JJ).EQ.1 )  GO TO 30
   20 CONTINUE
   30 JT = JJ
      J2 = J - 1
      DO 40  JJ = 1, J2
      JJJ = J - JJ
```

Fig. 13.16—continued

```
C     CALCULATE COORDINATES
      IF ( INNER(I,JJJ).EQ.1 )  GO TO 50
   40 CONTINUE
   50 JB = JJJ
      A = IF - I
      B = JT - J
      C = I - IB
      D = J - JB
      E = A*B*C*D
C     CALCULATE R COORDINATES OF POINT I,J
      R(I,J) = ( R(IF,J)/(A*(A+C)) + R(I,JT)/(B*(B+D))
     1 + R(IB,J)/(C*(C+A)) + R(I,JB)/(D*(D+B)) )*E/(A*C+B*D)
C     CALCULATE Z COORDINATE OF POINT I,J
      Z(I,J) = ( Z(IF,J)/(A*(A+C)) + Z(I,JT)/(B*(B+D))
     1 + Z(IB,J)/(C*(C+A)) + Z(I,JB)/(D*(D+B)) )*E/(A*C+B*D)
   60 CONTINUE
   70 CONTINUE
   80 CONTINUE
      RETURN
      END
```

Fig. 13.17 Subroutine OMITP

```
      SUBROUTINE OMITP (MAXR,MAXC,MAT,LTYPE,LINC,IJFLAG,
     1                  R,Z)
C     . . . . . . . . . . . . . . . . . . . . . . . . . . . .
C     FLAG NODES TO BE OMITTED IN HIGH ORDER ELEMENTS
C     . . . . . . . . . . . . . . . . . . . . . . . . . . . .
      DIMENSION R(MAXC,MAXR), Z(MAXC,MAXR), MAT(MAXC,MAXR)
      IF ( LTYPE.LT.8 )  RETURN
      I2 = MAXC - LINC
      J2 = MAXR - LINC
      DO 10  I = 1,I2,LINC
      IL = I + LINC
      DO 20  J = 1,J2,LINC
      IF ( MAT(I,J).LT.0 )  GO TO 20
      JL = J + LINC
      IF (LTYPE.EQ.8.OR.LTYPE.EQ.12)  MAT(I+1,J+1) = IJFLAG
      IF ( LTYPE.LT.12 )  GO TO 25
      MAT(I+2,J+2) = IJFLAG
      MAT(I+1,J+2) = IJFLAG
      MAT(I+2,J+1) = IJFLAG
   25 IF ( LTYPE.NE. 9)  GO TO 20
      D13SQ = (R(I,J)-R(IL,JL))**2+(Z(I,J)-Z(IL,JL))**2
      D24SQ = (R(I,JL)-R(IL,J))**2+(Z(I,JL)-Z(IL,J))**2
      D24SQ = D24SQ*1.2
      IF ( D24SQ.LT.D13SQ )  GO  ] 5
      MAT(I+2,J+1) = IJFLAG
      MAT(I+1,J+2) = IJFLAG
      GO TO 20
    5 MAT(I+1,J+1) = IJFLAG
      MAT(I+2,J+2) = IJFLAG
   20 CONTINUE
   10 CONTINUE
      RETURN
      END
```

Fig. 13.18 Subroutine PCOUNT

```
      SUBROUTINE  PCOUNT (K1,K2,R,Z,CODE,MAT,NPTNO,IMIN,
     1                    IMAX,IJFLAG,ICPT,OUTPT,MINNPT)
C     . . . . . . . . . . . . . . . . . . . . . . . . . . . .
C     NUMBER AND OUTPUT NODAL COORDINATES AND POINT CODES
C     . . . . . . . . . . . . . . . . . . . . . . . . . . . .
      INTEGER  OUTPT
      DIMENSION  R(K1,K2),Z(K1,K2),CODE(K1,K2),MAT(K1,K2),
     1           NPTNO(K1,K2),IMIN(K2),IMAX(K2)
      NODE = MINNPT - 1
      K = OUTPT + 1
      DO 70  J = 1,K2
      I1 = IMIN(J)
      I2 = IMAX(J)
      DO 60 I = I1,I2
C     CHECK FOR OMITTED POINTS IN HIGH ORDER ELEMENTS
      IF ( MAT(I,J).EQ.IJFLAG )  GO TO 60
```

Fig. 13.18—continued

```
C       DO NOT NUMBER THE POINT IF INTERIOR TO A HOLE.
        IF ( MAT(I,J).GE.0 .OR. I.EQ.I1 .OR. I.EQ.I2 )  GO TO 10
        IF ( MAT(I-1,J).GE.0 )  GO TO 10
        IF (J.GT.1.AND.MAT(I-1,J-1).LT.0.AND.MAT(I,J-1).LT.0)
     1   GO TO 60
   10 NODE = NODE + 1
      NPTNO(I,J) = NODE
C     OUTPUT POINTS
      CALL POUT(NODE,R(I,J),Z(I,J),CODE(I,J),I,J,OUTPT,IOPT)
   60 CONTINUE
   70 CONTINUE
      RETURN
      END
```

Fig. 13.19 Subroutine LCOUNT

```
      SUBROUTINE  LCOUNT (MAXR,MAXC,R,Z,LIMAX,LIMIN,MAT,
     1                   NPTNO,LTYPE,LINC,MINEL,OUTEL, IOPT)
C     . . . . . . . . . . . . . . . . . . . . . . . . . . . .
C     COUNT AND OUTPUT GENERATED ELEMENTS
C     . . . . . . . . . . . . . . . . . . . . . . . . . . . .
      INTEGER OUTEL
      DIMENSION  R(MAXC,MAXR), Z(MAXC,MAXR), LIMAX(MAXR),
     1          LIMIN(MAXR), MAT(MAXC,MAXR), NPTNO(MAXC,MAXR)
     2         ,IQ(12), JQ(12), I13(18), J13(18),
     3          I24(18), J24(18), NODES(18)
      EQUIVALENCE (I13(1),IQ(1)), (J13(1),JQ(1))
      L = MINEL - 1
      M = LTYPE + 1
C     EXTRACT I,J VALUES OF NODES
      IF ( LTYPE.EQ.3 .OR. LTYPE.EQ.6 .OR. LTYPE.EQ.9 )
     1    CALL   IJTRI  (LTYPE,I13,J13,I24,J24)
      IF ( LTYPE.EQ.4 .OR. LTYPE.EQ.8 .OR. LTYPE.EQ.12 )
     1    CALL   IJQUAD  (LTYPE,IQ,JQ)
      JLESS = MAXR - LINC
      DO 50  J = 1,JLESS,LINC
      I1 = LIMIN(J)
      I2 = LIMAX(J) - LINC
      DO 40  I = I1,I2,LINC
C     DO NOT NUMBER ELEMENTS IN A HOLE
      IF (  MAT(I,J).LT.0 )  GO TO 40
      MIJ = MAT(I,J)
      L = L + 1
C     CHECK AREAS
      IL = I + LINC
      JL = J + LINC
      A123 = R(I,J)*(Z(IL,J)-Z(IL,JL)) + R(IL,J)*(Z(IL,JL)
     1  -Z(I,J)) + R(IL,JL)*(Z(I,J)-Z(IL,J))
      A134 = R(I,J)*(Z(IL,JL)-Z(I,JL)) + R(IL,JL)*(Z(I,JL)
     1  -Z(I,J)) + R(I,JL)*(Z(I,J)-Z(IL,JL))
      IF ( A123.LE.0.0 .OR. A134.LE.0.0 )  WRITE (6,1000)  L
 1000 FORMAT ('0POSSIBLE NEGATIVE AREA NEAR ELEMENT ', I5)
      IF (LTYPE.EQ.3.OR.LTYPE.EQ.6.OR.LTYPE.EQ.9) GO TO 60
C     QUADRILATERALS
      CALL LNODES (I,J,NPTNO,MAXR,MAXC,LTYPE,IQ,JQ,NODES)
      CALL  LOUT (L,I,J,LTYPE,OUTEL,IOPT,NODES,MIJ)
      GO TO 40
C     TRIANGLES
   60 D13SQ = (R(I,J)-R(IL,JL))**2+(Z(I,J)-Z(IL,JL))**2
      D24SQ = (R(I,JL)-R(IL,J))**2+(Z(I,JL)-Z(IL,J))**2
      D24SQ = D24SQ*1.2
      IF ( D24SQ.LT.D13SQ ) GO TO 70
      CALL LNODES (I,J,NPTNO,MAXR,MAXC,LTYPE,I13,J13,NODES)
      CALL  LOUT (L,I,J,LTYPE,OUTEL,IOPT,NODES,MIJ)
      L = L + 1
      CALL LNODES (I,J,NPTNO,MAXR,MAXC,LTYPE,I13(M),J13(M),
     1            NODES)
      CALL  LOUT (L,I,J,LTYPE,OUTEL,IOPT,NODES,MIJ)
      GO TO 40
   70 CALL LNODES (I,J,NPTNO,MAXR,MAXC,LTYPE,I24,J24,NODES)
      CALL  LOUT (L,I,J,LTYPE,OUTEL,IOPT,NODES,MIJ)
      L = L + 1
      CALL LNODES (I,J,NPTNO,MAXR,MAXC,LTYPE,I24(M),J24(M),
     1            NODES)
      CALL  LOUT (L,I,J,LTYPE,OUTEL,IOPT,NODES,MIJ)
   40 CONTINUE
   50 CONTINUE
      RETURN
      END
```

Fig. 13.20 Subroutine PLOTIJ

```
      SUBROUTINE  PLOTIJ (K1,K2,R,Z,NPTNO,MAT,IMAX,IMIN,
     1                    LIMIN,LIMAX,LINC,LTYPE,IPLOT)
C     . . . . . . . . . . . . . . . . . . . . . . . . . . . .
C     PLOT A GRID OF POINTS DEFINED BY  I,J COORDINATES
C     . . . . . . . . . . . . . . . . . . . . . . . . . . .
      DIMENSION  NPTNO(K1,K2),MAT(K1,K2),IMIN(K2),IMAX(K2),
     1           LIMIN(K2),LIMAX(K2)
      DIMENSION  R(K1,K2), Z(K1,K2)
      DATA KOUNT /0/
C     FIND LIMITS OF COORDINATES ALONE EDGES OF MESH
      CALL  RZSIZE (K1,K2,R,Z,IMIN,IMAX,NPTNO,RMIN,RMAX,
     1               ZMIN,ZMAX)
C     ADD A BORDER TO REGION
      RDIFF = RMAX - RMIN
      ZDIFF = ZMAX - ZMIN
      BORDER = AMAX1(RDIFF,ZDIFF)/100.
      BORDER = ABS(BORDER)
      RMIN = RMIN - BORDER
      ZMIN = ZMIN - BORDER
      RMAX = RMAX + BORDER
      ZMAX = ZMAX + BORDER
C     ADVANCE FRAME AND SET WINDOW LIMITS
      CALL WINDO (RMIN,RMAX,ZMIN,ZMAX)
C     PLOT THE ROW LINES
      DO 10  J = 1,K2,LINC
      I1 = IMIN(J)
      I2 = IMAX(J) - 1
      JLL = J - LINC
      DO 20  I = I1,I2
      IF (NPTNO(I,J).EQ.0.OR.NPTNO(I+1,J).EQ.0)  GO TO 20
      ILL = I - LINC
C     POSSIBLE LINE, MOVE TO POINT, PEN UP
      CALL MOVETO (R(I,J),Z(I,J))
C     IS AN ELEMENT ON EITHER SIDE
      IF ( MAT(I,J).GE.0 )  GO TO 5
      IF ( J.GT.1 .AND. MAT(I,JLL).LT.0 )  GO TO 20
C     LINE EXISTS, MOVE TO NEXT POINT, PEN DOWN
    5 CALL LINETO (R(I+1,J),Z(I+1,J))
   20 CONTINUE
   10 CONTINUE
C     PLOT THE COLUMN LINES
      JLESS = K2 - 1
      DO 40  I = 1,K1,LINC
      ILL = I - LINC
      DO 50  J = 1,JLESS
      IF ( NPTNO(I,J).EQ.0.OR.NPTNO(I,J+1).EQ.0 )  GO TO 50
C     POSSIBLE LINE, MOVE TO POINT, PEN UP
      CALL MOVETO (R(I,J),Z(I,J))
C     IS AN ELEMENT ON EITHER SIDE
      IF ( MAT(I,J).GE.0 )  GO TO 35
      IF ( I.GT.1 .AND. MAT(ILL,J).LT.0 )  GO TO 50
C     LINE EXISTS, MOVE TO NEXT POINT, PEN DOWN
   35 CALL LINETO (R(I,J+1),Z(I,J+1))
   50 CONTINUE
   40 CONTINUE
      IF ( IPLOT.EQ.2 )  RETURN
      KOUNT = KOUNT + 1
      IF ( IPLOT.EQ.3 .AND. KOUNT.EQ.2 )  RETURN
C     CHECK FOR TRIANGLES
      IF ( LTYPE.EQ.4.OR.LTYPE.EQ.8.OR.LTYPE.EQ.12 ) RETURN
C     LOOP OVER QUADRILATERALS AND ADD DIAGONAL LINES
      JLESS = K2 - LINC
      DO 60  J = 1,JLESS,LINC
      I1 = LIMIN(J)
      I2 = LIMAX(J) - LINC
      JL = J + LINC
      DO 80  I = I1,I2,LINC
      IF ( MAT(I,J).LT.0 )  GO TO 80
      IL = I + LINC
C     ELEMENT EXISTS, FIND SHORTEST DIAGONAL
      D13SQ = (R(I,J)-R(IL,JL))**2+(Z(I,J)-Z(IL,JL))**2
      D24SQ = (R(I,JL)-R(IL,J))**2+(Z(I,JL)-Z(IL,J))**2
      IF ( D13SQ.GT.D24SQ )  GO TO 65
C     MOVE TO CORNER 1, PEN UP
      CALL MOVETO (R(I,J),Z(I,J))
```

Fig. 13.20—continued

```
C       MOVE UP DIAGONAL, PEN DOWN
        CALL LINETO (R(I+1,J+1),Z(I+1,J+1))
        IF ( LINC.GT.1 ) CALL LINETO (R(I+2,J+2),Z(I+2,J+2))
        IF ( LINC.EQ.3 ) CALL LINETO (R(I+3,J+3),Z(I+3,J+3))
        GO TO 80
C       MOVE TO CORNER 2, PEN UP
     65 CALL MOVETO (R(IL,J),Z(IL,J))
C       MOVE UP DIAGONAL, PEN DOWN
        CALL LINETO (R(IL-1,J+1),Z(IL-1,J+1))
        IF ( LINC.GT.1 ) CALL LINETO (R(IL-2,J+2),Z(IL-2,J+2))
        IF ( LINC.EQ.3 ) CALL LINETO (R(I,J+3),Z(I,J+3))
     80 CONTINUE
     60 CONTINUE
        RETURN
        END
```

Fig. 13.21 Mesh grading example data

```
000090001000013000010000400000000000000000000000000000000000
   1  1  1  9  1   11 2.236       120.        2.236      60.        1.  -2.
   2  1  1  2  0   00 0.          0.          0.         0.8
   2 10  1  2  0   00 2.          0.          2.         0.8
   2  4  6  3  0   00 0.          2.          0.         4.
   2  7  6  3  0   00 2.          2.          2.         3.
   3  7  5  2  0   00 2.          1.6         2.         1.
   4  2  3  2  0   00 0.          1.          0.         1.6
   5  4  9  3
   5  4  6  3
   6  6  6  3
   7  7  6  3
   8  1  3  3
   0  4  6  0  0   10 0.          2.
   4  6  7  9  2


   J   IMIN  IMAX      GIVEN I-J POINTS

   9     4     7          ****
   8     4     7          *.**
   7     4     7          *.**
   6     4     7          ****
   5     3     8         **..**
   4     2     9        **....**
   3     1    10       **......**
   2     1    10       *........*
   1     1    10       **********
 CHANGES IN BLOCKS OF ELEMENT CODES
BLOCK   MAT.NO.  LIMITS:    I1     J1    I2    J2

  1       2                  4      6     7     9
   J   IMIN  IMAX                GIVEN MATERIAL CODE

   9     4     7          ----
   8     4     7          222-
   7     4     7          222-
   6     4     7          222-
   5     4     7          111-
   4     3     8         11111-
   3     2     9        1111111-
   2     1    10       111111111-
   1     1    10       111111111-
```

Fig. 13.22 A typical mesh grading

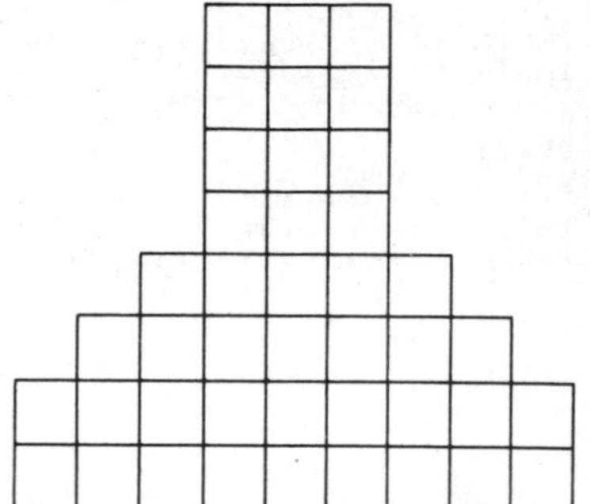

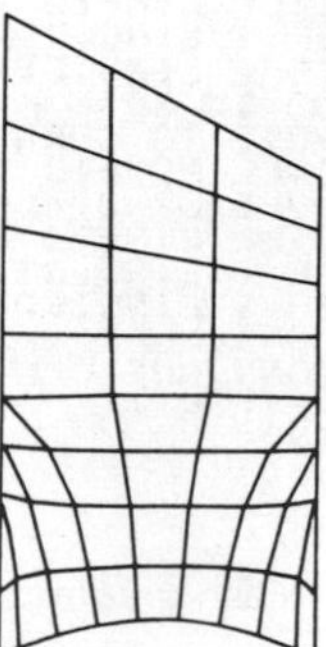

14

Initial value problems

14.1 Introduction

Many problems require the solution of time-dependent equations. In this context, there are numerous theoretical topics that an analyst should investigate before selecting a computational algorithm. These include the stability limits, amplitude error, phase error, etc. A large number of implicit and explicit procedures have been proposed. Several of these have been described in the texts by Bathe and Wilson [14], Chung [23], and Zienkiewicz [87]. It is even possible to combine these procedures as suggested by Hughes and Liu [46]. A recent review of the stability considerations was given by Park [62].

The purpose here is to cite typical additional computational procedures that arise in the time integration problems. Both simple explicit and implicit algorithms will be illustrated. Applications involving second order spatial derivatives and first order time derivatives will be referred to as parabolic or transient, while those with second order time derivatives will be referred to as hyperbolic or dynamic. For these classes of problems it is common to select various temporal operators to approximate the time derivatives. Common difference operators for this purpose include the forward difference,

$$\dot{\mathrm{R}} \approx (\mathrm{R}(t + \Delta t) - \mathrm{R}(t))/\Delta t,$$

the backward difference,

$$\dot{\mathrm{R}} \approx (\mathrm{R}(t) - \mathrm{R}(t - \Delta t))/\Delta t,$$

and the central difference

$$\dot{\mathrm{R}} \approx (\mathrm{R}(t + \Delta t) - \mathrm{R}(t - \Delta t))/(2\Delta t),$$

where Δt denotes a small time difference. Similar expressions can be developed to estimate the second derivative, $\ddot{R}$. The actual time integration algorithm is determined by the choices of the difference operators and the ways that they are combined.

14.2 Parabolic equations

The governing system equations for a transient application are generally ordinary differential equations of the form

$$\mathbf{A}\dot{\mathbf{R}}(t) + \mathbf{B}\mathbf{R}(t) = \mathbf{P}(t) \tag{14.1}$$

where $(\dot{}) = \mathrm{d}(\ )/\mathrm{d}t$ denotes the derivative with respect to time. Generally, one or more of the coefficients in $\mathbf{R}$, say R_i, will be defined in a boundary condition as a function of time, i.e. $R_i = g(t)$. Also, the initial values, $\mathbf{R}(0)$, must be known to start the transient solution.

Note that the governing equations now involve two square matrices, $\mathbf{A}$ and $\mathbf{B}$, at the system level. Thus in general it will be necessary to apply the previously discussed square matrix assembly procedure twice. This is,

$$\begin{aligned}\mathbf{A} &= \sum_{e=1}^{NE} \mathbf{A}^e \\ \mathbf{B} &= \sum_{e=1}^{NE} \mathbf{B}^e\end{aligned} \tag{14.2}$$

where $\mathbf{A}^e$ and $\mathbf{B}^e$ are generated from the Boolean assembly matrices and the corresponding element contributions, say $\mathbf{a}^e$ and $\mathbf{b}^e$. In a nonlinear problem the system matrices usually depend on the values of $\mathbf{R}(t)$ and an iterative solution is required. For example, a heat transfer problem may involve material conductivities which are temperature-dependent. For the sake of simplicity such nonlinear applications will not be considered at this point.

Only the direct step-by-step time integration of Eqn. (14.1) will be considered. There are many such procedures published in the literature. When selecting a computational procedure from the many available algorithms one must consider the relative importance of storage requirements, stable step size, input–output operations, etc. The text by Myers [56] examines in detail many of the aspects of simple time integration procedures for linear transient applications. He utilizes both finite difference and finite element spatial approximations and illustrates how their transient solutions differ.

14.2.1 Simple approximations

The accuracy, stability, and relative computational cost of a transient integration scheme depend on how one approximates the 'velocity', $\dot{\mathbf{R}}$, during the time step. For example, one could assume that the velocity during the time step is (a) constant, (b) equal to the average value at the beginning and end of the step, (c) varies linearly during the step, etc.

To illustrate cases (a) and (c) consider a time interval of $k = \Delta t$ and assume a Taylor series for $\mathbf{R}(t)$ in terms of the value at the previous time step, $\mathbf{R}(t-k)$:

$$\mathbf{R}(t) = \mathbf{R}(t-k) + k\dot{\mathbf{R}}(t-k) + k^2/2\, \ddot{\mathbf{R}}(t-k) + \ldots \tag{14.3}$$

Then, as illustrated in Fig. 14.1, the first assumption gives $\ddot{\mathbf{R}} = 0$ and the above equation yields $\mathbf{R}(t)$. The standard Euler integration procedure is obtained by multiplying by $\mathbf{A}$:

$$\mathbf{AR}(t) = \mathbf{AR}(t-k) + k\mathbf{A}\dot{\mathbf{R}}(t-k) \tag{14.4}$$

and substituting Eqn. (14.1) at $(t-k)$

$$\mathbf{A}\dot{\mathbf{R}}(t-k) = \mathbf{P}(t-k) - \mathbf{BR}(t-k)$$

to obtain the final result that

$$\mathbf{AR}(t) = k\mathbf{P}(t-k) + \{\mathbf{A} - k\mathbf{B}\}\mathbf{R}(t-k). \tag{14.5}$$

One can make the general observation that the governing ordinary differential equations have been reduced to a new set of algebraic equations of the form

$$\mathbf{SR}(t) = \mathbf{F}(t) \tag{14.6}$$

which must be solved at each time step. In the present case of the Euler method one has system matrices

$$\mathbf{S} = \mathbf{A}$$

and

$$\mathbf{F}(t) = k\mathbf{P}(t-k) + (\mathbf{A} - k\mathbf{B})\mathbf{R}(t-k). \tag{14.7}$$

As shown in Table 14.1, all integrations can be reduced to the form of Eqn. (14.6). When the problem is linear and the time step, k, is held constant, the system square matrix does not change with time. Thus it need be assembled and 'inverted' only once. Then at each time step it is only necessary to evaluate $\mathbf{F}(t)$ and solve for $\mathbf{R}(t)$.

Before considering the practical significance of the alternate forms of

Eqn. (14.7), let us return to the assumption that $\dot{\mathbf{R}}$ is linear during the time step. From Fig. 14.1 one notes that

$$\dot{\mathbf{R}}(t) = \dot{\mathbf{R}}(t-k) + k\ddot{\mathbf{R}}(t-k). \tag{14.8}$$

Solving Eqn. (14.3) for $\ddot{\mathbf{R}}$ and substituting into the above equation leads to

$$\dot{\mathbf{R}}(t) = 2[\mathbf{R}(t) - \mathbf{R}(t-k)]/k - \dot{\mathbf{R}}(t-k). \tag{14.9}$$

Substituting into Eqn. (14.1) at time t yields the system equations

$$\mathbf{SR}(t) = \mathbf{F}(t),$$

where now

$$\begin{aligned} \mathbf{S} &= \mathbf{B} + 2\mathbf{A}/k \\ \mathbf{F}(t) &= \mathbf{P}(t) + \mathbf{A}(2\mathbf{R}(t-k)/k + \dot{\mathbf{R}}(t-k)). \end{aligned} \tag{14.10}$$

This is referred to as the linear velocity algorithm. A comparison of Eqns. (14.7) and (14.10) is useful. The Euler method is known as an explicit method since $\mathbf{R}(t)$ is obtained explicitly from $\mathbf{R}(t-k)$. The linear velocity formulation involves an implicit dependence on $\dot{\mathbf{R}}(t-k)$ and is one of many implicit algorithms. Note that the Euler form requires no additional storage while the linear velocity algorithm must store the array $\dot{\mathbf{R}}(t-k)$, and perform the calculations necessary to update its value at each time step. The necessary recurrence relation which utilizes the above calculated values for $\mathbf{R}(t)$ is obtained from Eqn. (14.9). Also note that the implicit procedure requires one to have initial starting values for $\dot{\mathbf{R}}(0)$. These can be obtained from Eqn. (14.1) as

$$\dot{\mathbf{R}}(0) = \mathbf{A}^{-1}(\mathbf{P}(0) - \mathbf{BR}(0)). \tag{14.11}$$

However, this is a practical approach only so long as $\mathbf{A}$ is a diagonal matrix. The system matrices [84] resulting from Crank–Nicolson, Galerkin, and least-squares approximations in the time interval are summarized in Table 14.1. A typical subroutine, EULER, for executing the simple Euler integration algorithm is given in Fig. 14.2. Typical numerical results will be considered later.

Fig. 14.1 Typical integration assumptions

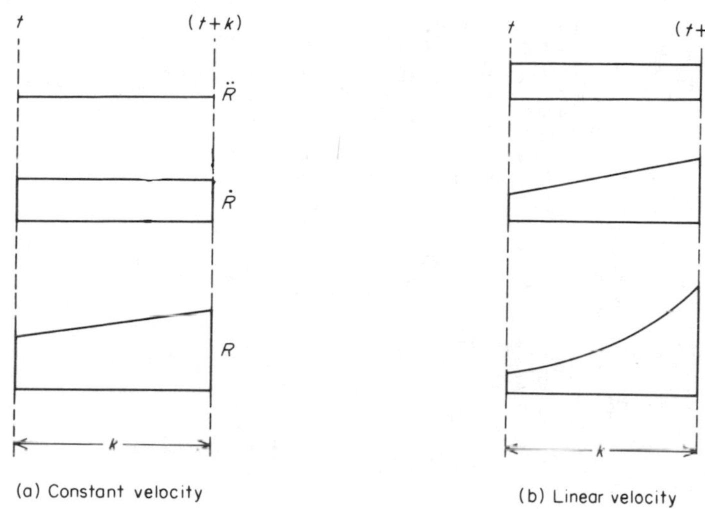

Fig. 14.2 An explicit Euler integration

```
      SUBROUTINE  EULER (NDFREE,IBW,A,B,P,R,DT,NSTEPS,
     1                   IPRINT,NBC,IBC)
C     * * * * * * * * * * * * * * * * * * * * * * * * * *
C         STEP BY STEP INTEGRATION OF MATRIX EQUATIONS
C                   A*DR(T)/DT  +  B*R(T)  =  P(T)
C                        BY THE EULER METHOD
C     * * * * * * * * * * * * * * * * * * * * * * * * * *
CDP   IMPLICIT REAL*8 (A-H,O-Z)
      DIMENSION A(NDFREE,IBW), B(NDFREE,IBW), P(NDFREE),
     1          R(NDFREE), IBC(1)
C     INITIAL VALUES OF R ARE PASSED THRU ARGUMENT LIST
C     NBC= NO. D.O.F. WITH SPECIFIED VALUES OF ZERO
C     IBC  = ARRAY CONTAINING NBC DOF NUMBERS WITH ZERO BC
C     DIM  IBC(NBC)
C     NDFREE = TOTAL NUMBER OF SYSTEM DOF
C     IBW  = MAX HALF BANDWIDTH, INCLUDING DIAGONAL
C     NSTEPS = NO. OF INTEGRATION STEPS
C     IPRINT = NO. OF INTEGRATION STEPS BETWEEN PRINTING
C                  *** INITIAL CALCULATIONS ***
      ZERO  = 0.D0
      WRITE (6,5000)
 5000 FORMAT ('1STEP BY STEP INTEGRATION BY THE EULER '
     1  'METHOD',/)
      IF ( IPRINT.LT.1 )  IPRINT = 1
      NSTEPS = (NSTEPS/IPRINT)*IPRINT
      IF ( NSTEPS.EQ.0 )  NSTEPS = IPRINT
      IF ( NBC.GT.0 )   GO TO 10
      WRITE (6,5010)
 5010 FORMAT (' WARNING, NO CONSTRAINTS IN SUBR. EULER')
   10 DO 30  J = 1,IBW
      DO 20  I = 1,NDFREE
   20 B(I,J)  = -DT*B(I,J) + A(I,J)
   30 CONTINUE
      ICOUNT = IPRINT - 1
C     PRINT INITIAL VALUES OF R
      ISTEP = 0
      T  =  ZERO
      WRITE (6,5020)   ISTEP,T
 5020 FORMAT (/,' STEP NUMBER = ',I5,5X,' TIME = ',E14.8,/,
     1 '            I              R(I)')
      WRITE (6,5030) (K,R(K),K=1,NDFREE)
```

Fig. 14.2—continued

```
 5030 FORMAT (I10,2X,E14.8)
C        *** APPLY BOUNDARY CONDITIONS TO A ***
      DO 40  I = 1,NBC
      CALL  MODFY1 (NDFREE,IBW,IBC(I),ZERO,B,P)
   40 CALL  MODFY1 (NDFREE,IBW,IBC(I),ZERO,A,P)
C        *** TRIANGULARIZE A ***
      CALL  FACTOR (NDFREE,IBW,A)
      IBWL1 = IBW - 1
C        *** END INITIAL CALCULATIONS ***
C        *** CALCULATE SOLUTION AT TIME T ***
      DO 80  ISTEP = 1,NSTEPS
      ICOUNT = ICOUNT + 1
      T = T*DT
      TLESS = T - DT
      IF ( ICOUNT.EQ.IPRINT )  WRITE (6,5020) ISTEP,T
C     FORCER DEFINES THE FORCING FUNCTION P(T)
C     FORM MODIFIED FORCING FUNCTION
      DO 60  I = 1,NDFREE
      SUM = ZERO
C     CONSIDER ONLY MATRIX BAND
      J1 = I - IBWL1
      J2 = I + IBWL1
      J1 = MAX0 (1,J1)
      J2 = MIN0 (J2,NDFREE)
      DO 50  J = J1,J2
C     CONVERT SUBSCRIPTS FOR COMPRESSED STORAGE
      CALL  BANSUB (I,J,IB,JB)
      BIJ = B(IB,JB)
      IF ( BIJ.EQ.ZERO )  GO TO 50
      SUM = SUM + BIJ*R(J)
   50 CONTINUE
      P(I) = P(I)*DT + SUM
   60 CONTINUE
C        *** APPLY BOUNDARY CONDITIONS TO P ***
      DO 70  I = 1,NBC
      IN = IBC(I)
   70 P(IN) = ZERO
C     SOLVE FOR R AT TIME T
      CALL SOLVE (NDFREE,IBW,A,P,R)
C     OUTPUT RESULTS FOR TIME T
      IF (ICOUNT.NE.IPRINT) GO TO 80
      WRITE (6,5030) (K,R(K),K=1,NDFREE)
      ICOUNT=0
   80 CONTINUE
      RETURN
      END
```

Table 14.1 System Matrices for Linear Transients

1. Euler (forward difference), $k = \Delta t$
   $\mathbf{S} = \mathbf{A}/k$
   $\mathbf{F} = \mathbf{P}(t - k) + (\mathbf{A}/k - \mathbf{B})\mathbf{R}(t - k)$
2. Crank–Nicolson (mid-difference), $h = \Delta t/2$,
   $\mathbf{S} = \mathbf{A}/k + \mathbf{B}/2$
   $\mathbf{F} = \mathbf{P}(t - h) + (\mathbf{A}/k - \mathbf{B}/2)\mathbf{R}(t - k)$
3. Linear velocity
   $\mathbf{S} = 2\mathbf{A}/k + \mathbf{B}$
   $\mathbf{F} = \mathbf{P}(t) + \mathbf{A}(2\mathbf{R}(t - k)/k - \dot{\mathbf{R}}(t - k))$ and
   $\dot{\mathbf{R}}(t) = 2(\mathbf{R}(t) - \mathbf{R}(t - k))/k + \dot{\mathbf{R}}(t - k)$
4. Galerkin
   $\mathbf{S} = \mathbf{A}/k + 2\mathbf{B}/3$
   $\mathbf{F} = (\mathbf{A}/k - \mathbf{B}/3)\mathbf{R}(t - k) + 2\int_0^k P(\tau)\tau d\tau/k^2$
5. Least-squares
   $\mathbf{S} = \mathbf{B}^T\mathbf{B}k/3 + (\mathbf{B}^T\mathbf{A} + \mathbf{A}^T\mathbf{B})/2 + \mathbf{A}^T\mathbf{A}/k$
   $\mathbf{F} = (\mathbf{A}^T\mathbf{A}/k + (\mathbf{B}^T\mathbf{A} - \mathbf{A}^T\mathbf{B})/2 - \mathbf{B}^T\mathbf{B}k/6)\mathbf{R}(t - k)$
   $+ \mathbf{B}^T\int_0^k \mathrm{P}(\tau)\, \tau d\tau + \mathbf{A}^T\int_0^k \mathrm{P}(\tau)\, d\tau)/k$

14.2.2. Diagonal Matrices for Transient Solutions

When one uses a finite difference spatial formulation, the system matrix $\mathbf{A}$ is a diagonal matrix. However, if one utilizes a consistent finite element formulation it is not a diagonal matrix. Thus, the consistent finite element form introduces an implied coupling of some of the coefficients in the time derivative matrix, $\dot{\mathbf{R}}$. As shown by Myers [56], this tends to result in a less stable time integration algorithm. The mathematical implications of this coupling are unclear, but experience shows it to be undesirable.

Clearly, converting $\mathbf{A}$ to a diagonal matrix would also save storage and make the evaluation of equations such as (14.11) much more economical. Some engineering approaches for modifying $\mathbf{A}$ have been shown to be successful. To illustrate these, consider the form of a typical element contribution, say $\mathbf{a}^e$. The consistent definition for constant properties is

$$\mathbf{a}^e \equiv q \int_{V^e} \mathbf{H}^{eT}\mathbf{H}^e \, dv, \tag{14.12}$$

where q is some constant property per unit volume and **H** denotes the element interpolation functions. This generally can be written as

$$\mathbf{a}^e = Q\mathbf{M}, \tag{14.13}$$

where $Q = qV^e$, and **M** is a symmetric full matrix. In some cases it is possible to obtain a diagonal form by using a nodal quadrature rule [33]. Since that procedure often leads to negative terms other procedures will be considered here. Let the sum of the coefficients of the matrix **M** be T, that is

$$T \equiv \sum_i \sum_j M_{ij}. \tag{14.14}$$

In most cases T will be unity but this is not true for axisymmetric integrals. Another quantity of interest is the sum of the diagonal terms of **M**, i.e.

$$d \equiv \sum_i M_{ii}. \tag{14.15}$$

The most common engineering solution to defining a diagonal matrix is to *lump*, or sum, all the terms in a given row onto the diagonal of the row and then set the off-diagonal terms to zero. That is, the *lumped matrix* **ML** is defined such that

$$\begin{aligned} ML_{ij} &= 0 \quad \text{if} \quad i \neq j \\ ML_{ii} &= \sum_j M_{ij}. \end{aligned} \tag{14.16}$$

Note that doing this does not alter the value of T. Another diagonal matrix, **MD**, with the same value of T can be obtained by simply extracting the diagonal of **M** and scaling it by a factor of T/d. That is,

$$\begin{aligned} MD_{ij} &= 0 \quad \text{if} \quad i \neq j \\ MD_{ii} &= M_{ii}T/d. \end{aligned} \tag{14.17}$$

The matrix **MD** will be called the diagonalized matrix and **ML** the lumped matrix. The corresponding matrices **AD** and **AL** are referred to as 'condensed' matrices. For linear simplex elements in two and three dimensions both procedures yield identical diagonal matrices. However, for axisymmetric problems and higher order elements they yield different results and the diagonalized matrix appears to be best in general. This is because for higher order elements the lumped form can introduce zeros or negative numbers on the diagonal. The matrices **M**, **ML**, and **MD** are illustrated for a quadratic triangle in Fig. 14.3 and for an axisymmetric line element in Fig. 14.4. The effects of the condensed matrices on transient calculations will be considered in the next section.

Fig. 14.3 Consistent and diagonal matrices for a quadratic triangle

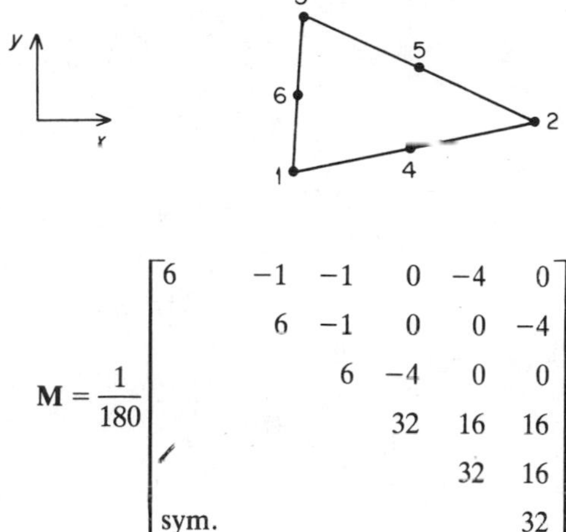

$$\mathbf{M} = \frac{1}{180}\begin{bmatrix} 6 & -1 & -1 & 0 & -4 & 0 \\ & 6 & -1 & 0 & 0 & -4 \\ & & 6 & -4 & 0 & 0 \\ & & & 32 & 16 & 16 \\ & & & & 32 & 16 \\ \text{sym.} & & & & & 32 \end{bmatrix}$$

(a) Consistent

$$\mathbf{ML} = \frac{1}{180}\begin{bmatrix} 0 & & & & & \text{zero} \\ & 0 & & & & \\ & & 0 & & & \\ & & & 60 & & \\ & & & & 60 & \\ \text{zero} & & & & & 60 \end{bmatrix}$$

(b) Lumped

$$\mathbf{MD} = \frac{1}{114}\begin{bmatrix} 6 & & & & & \text{zero} \\ & 6 & & & & \\ & & 6 & & & \\ & & & 32 & & \\ & & & & 32 & \\ \text{zero} & & & & & 32 \end{bmatrix}$$

(c) Diagonalized

Fig. 14.4 Consistent and diagonal matrices for an axisymmetric line element

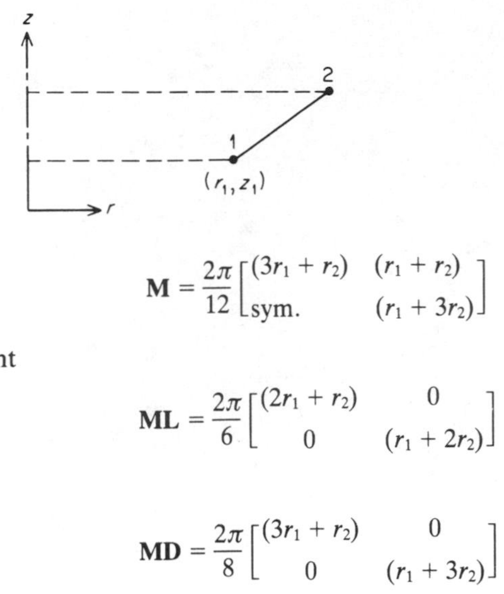

$$\mathbf{M} = \frac{2\pi}{12}\begin{bmatrix}(3r_1 + r_2) & (r_1 + r_2) \\ \text{sym.} & (r_1 + 3r_2)\end{bmatrix}$$

(a) Consistent

$$\mathbf{ML} = \frac{2\pi}{6}\begin{bmatrix}(2r_1 + r_2) & 0 \\ 0 & (r_1 + 2r_2)\end{bmatrix}$$

(b) Lumped

$$\mathbf{MD} = \frac{2\pi}{8}\begin{bmatrix}(3r_1 + r_2) & 0 \\ 0 & (r_1 + 3r_2)\end{bmatrix}$$

(c) Diagonalized

14.2.3 Examples

The one-dimensional problem presented by Myers [56] will serve as a useful example application of subroutine EULER and the effects of condensed matrices. Myers presents finite difference, finite element, and exact analytic solutions for the following transient problem. Consider a rod having a uniform initial temperature of unity. Suddenly both ends of the rod have their temperature reduced to zero. The object is to determine the time history of the temperature at interior points.

By utilizing half symmetry one can apply a simpler finite element model and use the natural boundary of zero thermal gradient at the centre. The analytic solution is illustrated in Fig. 14.5. Myers considered finite element solutions involving 1, 2, 3, and 4 elements. The 1, 2, and 3 element models yielded consistent matrices which could be integrated analytically and these results are also in reference [56]. The latter two solutions gave small time solutions that had the centre-line temperature higher than its initial value. This is a clear violation of the basic laws of thermodynamics that was not present in the finite difference solutions. This response is due to the off-diagonal terms in the **A** matrix. Myers also presented the **A** and **B** matrices for a four element solution. The matrices are

$$\mathbf{A} = \begin{bmatrix} 2 & 1 & 0 & 0 & 0 \\ & 4 & 1 & 0 & 0 \\ & & 4 & 1 & 0 \\ & & & 4 & 1 \\ \text{sym.} & & & & 2 \end{bmatrix}$$

and (14.18)

$$\mathbf{B} = \frac{6k}{\rho c l^2} \begin{bmatrix} 1 & -1 & 0 & 0 & 0 \\ & 2 & -1 & 0 & 0 \\ & & 2 & -1 & 0 \\ & & & 2 & -1 \\ \text{sym.} & & & & 1 \end{bmatrix}$$

where in this problem $6k/(\rho c l^2) = 96$, and the forcing function is zero, i.e. $\mathbf{P}(t) = \mathbf{0}$. These equations were stored in upper half-bandwidth form and integrated by subroutine EULER.

The results at the centre point from $t = 0$ to $t = 1$ for the consistent and condensed **A** matrices for a step size of $k = 1/100$ are shown in Fig. 14.6. To the scale shown they are in agreement with each other and the exact solution. However, Myers shows that other integration algorithms can give good results with $k = 1/16$. Thus it is useful to observe what happens to the Euler procedure when k is increased. The consistent formulation was found to be unstable for $k = 1/75$. The condensed form was stable at $k = 1/75$ and $k = 1/50$ but was unstable at $k = 1/30$. A comparison of these results at selected small and large times is presented in Table 14.2. Since **P** was zero the problem-dependent forcing function, subroutine FORCER, simply set **P** to zero on each call. The version of EULER shown in Fig. 14.2 is written for zero boundary values for one or more parameters. Only minor changes are required for time-dependent boundary values.

In the above example the lumped and diagonalized matrices were identical. As an example of a problem where the results are different the following equations were obtained for a two degree of freedom axisymmetric conduction problem.

$$\mathbf{A} = \frac{1}{12}\begin{bmatrix} 9 & 5 \\ 5 & 24 \end{bmatrix}$$

$$\mathbf{B} = \begin{bmatrix} 9 & -5 \\ -5 & 12 \end{bmatrix} \qquad (14.19)$$

Both condensed forms yielded the exact steady state results but as shown in Fig. 14.7 the small time results were different. The consistent EULER was unstable for the same step size.

Fig. 14.5 Typical linear transient problem

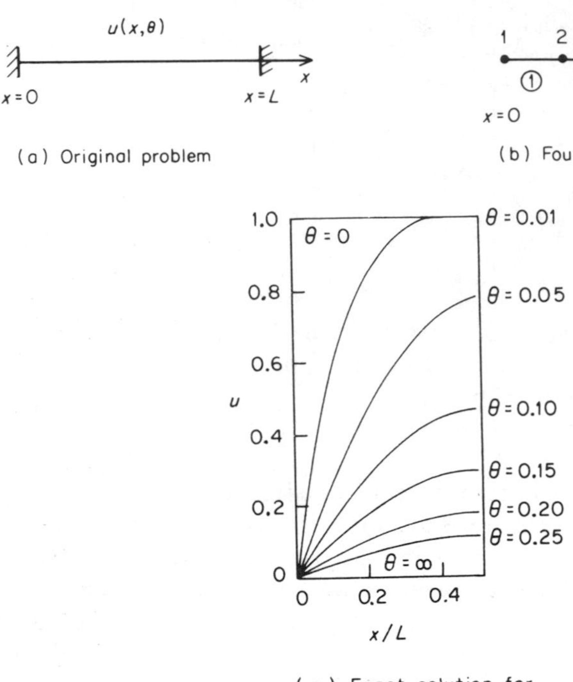

Fig. 14.6 Transient response of centre point

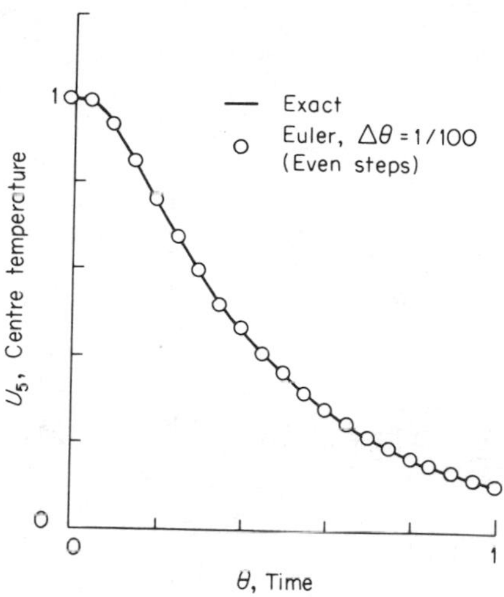

Table 14.2 Stable Solution Comparisons at Various Steps

	$\theta = 0.04$				$\theta = 0.96$			
	100 C	100 D	75 D	50 D	100 C	100 D	75 D	50 D
u_2	0.550	0.594	0.585	0.565	0.042	0.045	0.045	0.044
u_3	0.878	0.903	0.902	0.898	0.078	0.083	0.082	0.081
u_4	1.016	0.988	0.990	1.000	0.102	0.109	0.108	0.106
u_5	1.004	1.000	1.000	1.000	0.111	0.118	0.117	0.114

C = Consistent
D = Diagonalized
θ = Time

Fig. 14.7 Euler solution of single element example

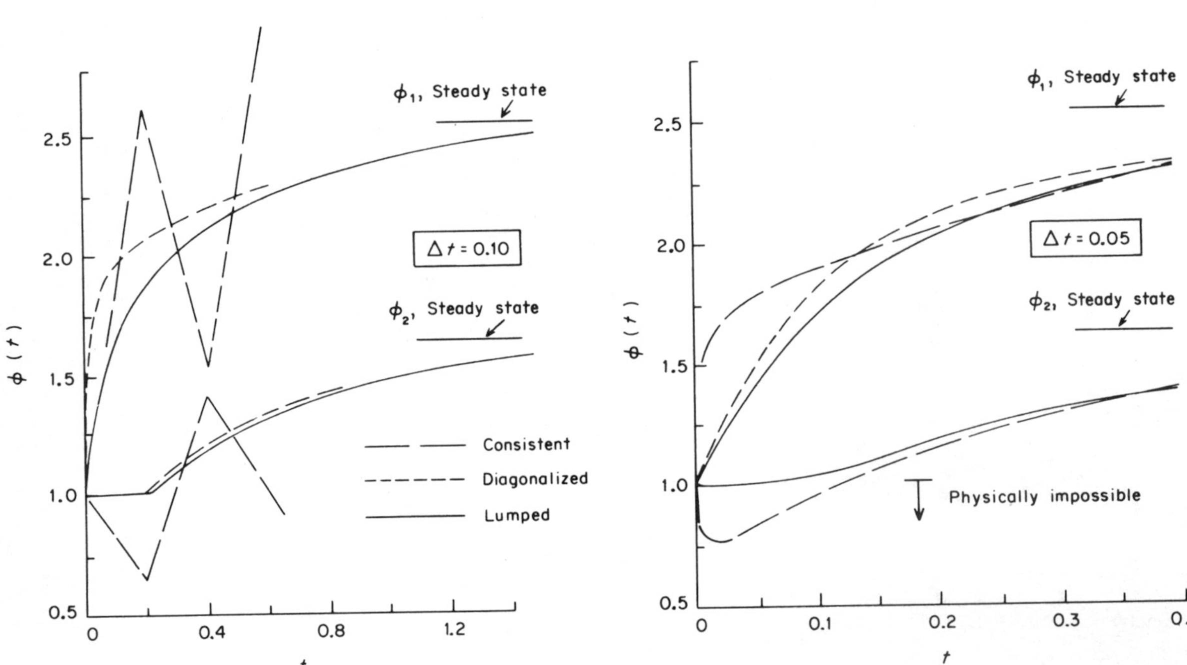

14.3 Hyperbolic equations

Many solid mechanics problems and wave problems involve the solution of dynamic equations. The text by Bathe and Wilson [14] presents detailed discussions of both linear and nonlinear dynamic problems. They present both direct integration and modal (eigen) algorithms. In this section a typical direct integration procedure will be illustrated.

The governing system equations for dynamic applications are generally ordinary differential equations of the form

$$\mathbf{A}\ddot{\mathbf{R}} + \mathbf{B}\dot{\mathbf{R}} + \mathbf{C}\mathbf{R} = \mathbf{P}(t) \tag{14.20}$$

where $\mathbf{A}$, $\mathbf{B}$, and $\mathbf{C}$ are square banded matrices, $\mathbf{P}(t)$ is a time-dependent forcing function, and $\mathbf{R}$ is the vector of unknown nodal parameters. The system square matrices are again assembled from corresponding element matrices. Generally the initial values of $\mathbf{R}$, and $\dot{\mathbf{R}}$ are known at time $t =$ 0 and it is necessary to solve for the initial values of $\ddot{\mathbf{R}}$ from Eqn. (14.20). That is,

$$\ddot{\mathbf{R}}(0) = \mathbf{A}^{-1}(\mathbf{P}(0) - \mathbf{B}\dot{\mathbf{R}}(0) - \mathbf{C}\mathbf{R}(0)). \tag{14.21}$$

If $\mathbf{A}$ is a diagonal matrix this is easily evaluated. Otherwise, it may not be computationally efficient to exactly calculate the values of $\ddot{\mathbf{R}}(0)$. In the algorithm to be presented the initial values of $\ddot{\mathbf{R}}(0)$ will be approximated by a simple technique.

Assume that $\mathbf{R}(t)$, $\dot{\mathbf{R}}(t)$, and $\ddot{\mathbf{R}}(t)$ are the known values of the above quantities at some time t. We desire to obtain the solution of Eqn. (14.20) at time $(t + h)$ where h represents the size of the time increments to be used. There are numerous procedures for accomplishing this goal. The procedure to be presented herein has been successfully applied to many problems. It can be programmed so as to reduce the storage requirements and solution time.

14.3.1 Simple approximations

The basic assumption of the algorithm is that within a time increment, $0 \leq t \leq k = \alpha h$, the second time derivative (acceleration) varies with t^n. That is,

$$\ddot{\mathbf{R}}(t) = \mathbf{f}_1(t) = \ddot{\mathbf{R}}(0) + [\ddot{\mathbf{R}}(k) - \ddot{\mathbf{R}}(0)]t^n/k^n. \tag{14.22}$$

The first time derivative is obtained by integrating with respect to time:

$$\dot{\mathbf{R}}(t) \equiv \mathbf{f}_2(t) = \dot{\mathbf{R}}(0) + \int \mathbf{f}_1(t)\,\mathrm{d}t$$

or (14.23)

$$\dot{\mathbf{R}}(t) = \dot{\mathbf{R}}(0) + \ddot{\mathbf{R}}(0)t + [\ddot{\mathbf{R}}(k) - \ddot{\mathbf{R}}(0)]t^{n-1}/[(n+1)k^n].$$

In a similar manner the value of $\mathbf{R}(t)$ is obtained from:

$$\mathbf{R}(t) \equiv \mathbf{f}_3(t) = \mathbf{R}(0) + \int \mathbf{f}_2(t)\,\mathrm{d}t$$

$$= \mathbf{R}(0) + \dot{\mathbf{R}}(0)t + \mathbf{R}(0)t^2/2 + [\mathbf{R}(k) - \mathbf{R}(0)]t^{n+2}/[(n+1)(n+2)k^n]. \quad (18.24)$$

It is useful to note that if one evaluates Eqns. (14.23) and (14.24) at time $t = k$ one obtains:

$$\dot{\mathbf{R}}(k) = \dot{\mathbf{R}}(0) + (1-\gamma)k\ddot{\mathbf{R}}(0) + \gamma k\ddot{\mathbf{R}}(k) \quad (14.25)$$

and

$$\mathbf{R}(k) = \mathbf{R}(0) + k\dot{\mathbf{R}}(0) + (\tfrac{1}{2} - \beta)k^2\ddot{\mathbf{R}}(0) + \beta k^2\ddot{\mathbf{R}}(k) \quad (14.26)$$

where the constants γ and β have been defined as

$$\gamma = \frac{1}{(n+1)}$$

$$\beta = \frac{1}{(n+1)(n+2)} = \frac{\gamma}{(n+2)}$$

These equations represent the basic equations of the algorithm and they will be briefly discussed before proceeding with the details of the algorithm. Of course Eqns. (14.25) and (14.26) could be taken as a starting point with β and γ being arbitrary constants [57].

Note that Eqns. (14.22) and (14.23) imply that the values of $\mathbf{R}$, $\dot{\mathbf{R}}$, and $\ddot{\mathbf{R}}$ are known at time $t = 0$ and that $\ddot{\mathbf{R}}$ is known at time $t = k$. As a point of fact, the latter quantity is not known and must be estimated. Of course, iteration could be used to improve the accuracy of the estimate if necessary. The choice of n in Eqn. (14.22) determines the relative weight that is assigned to the two quantities $\ddot{\mathbf{R}}(0)$ and $\ddot{\mathbf{R}}(k)$. If $n = 1$ they are assigned equal weights; if $n < 1$ then $\ddot{\mathbf{R}}(k)$ is more important; and if $n > 1$ then $\ddot{\mathbf{R}}(0)$ is more important. This is clearly illustrated by the limiting cases ($n = 0$ and $n = \infty$). The most commonly used values are $n = 1$ and $n = 0$.

Newmark [57] has considered the application of Eqns. (14.25) and (14.26) to the solution of Eqn. (14.20) in some detail. Starting with these equations he assumed that the parameters γ and β were independent. That assumption does not apply in the present analysis; nevertheless, portions of his analysis can be utilized here. By considering the known difference solution for the simple harmonic motion of a single degree of freedom system, Newmark concluded that to avoid the introduction of erroneous damping in the solution one must set $\gamma = 1/2$. For the present assumptions this result requires that $n = 1$ which in turn implies that the second derivative is linear during the time interval, k. Of course, if $n = 1$ then $\beta = 1/6$. For $\beta = 1/6$ and $k = h$ ($\alpha = 1$), Newmark shows that for the above problem the solution will theoretically be stable and converge if $h/T \leqslant 0.389$ where T represents the smallest period of the system. For multi degree of freedom systems subjected to forced motion a much smaller ratio must be used in practice. The actual algorithm will be developed using $n = 1$. For $n = 1$ Eqns. (14.25) and (14.26) can be generalized to

$$\dot{\mathbf{R}}(t + k) = \dot{\mathbf{R}}(t) + [\ddot{\mathbf{R}}(t + k) + \ddot{\mathbf{R}}(t)]k/2 \tag{14.27}$$

$$\mathbf{R}(t + k) = \mathbf{R}(t) + k\dot{\mathbf{R}}(t) + k^2[\ddot{\mathbf{R}}(t + k) + 2\ddot{\mathbf{R}}(t)]/6. \tag{14.28}$$

For $t = h$ ($\alpha = 1$) the above equations correspond to the standard "linear acceleration" method. This technique, which requires iterations within each time step to establish $\ddot{\mathbf{R}}(t + h)$, has been outlined in detail (including a flow chart) by Fenves [31]. A similar study, with examples, has been given by Biggs [19].

To avoid these iterations, and at the same time maintain numerical stability, modified integration schemes are usually preferred. A typical modified *linear acceleration* integration scheme will now be considered [14]. This extrapolation algorithm utilizes Eqns. (14.27) and (14.28). These equations can be arranged to make any one of the quantities $\ddot{\mathbf{R}}(t + k)$, $\dot{\mathbf{R}}(t + k)$, or $\mathbf{R}(t + k)$ the independent unknown. The best choice of the three is not yet known; however, the "displacement", $\mathbf{R}(t + k)$, formulation is the most common.

Since the eventual objective is to solve for $\mathbf{R}(t + k)$ in terms of $\mathbf{R}(t)$, $\dot{\mathbf{R}}(t)$, and $\ddot{\mathbf{R}}(t)$, it is desirable to solve Eqns. (14.27) and (14.28) simultaneously for $\dot{\mathbf{R}}(t + k)$ and $\ddot{\mathbf{R}}(t + k)$. That is, (for $n = 1$),

$$\ddot{\mathbf{R}}(t + k) = 6[\mathbf{R}(t + k) - \mathbf{R}(t)]/k^2 - 6\dot{\mathbf{R}}(t)k - 2\ddot{\mathbf{R}}(t) \tag{14.29}$$

and

$$\dot{\mathbf{R}}(t + k) = 3[\mathbf{R}(t + k) - \mathbf{R}(t)]/k - k\ddot{\mathbf{R}}(t)/2. \tag{14.30}$$

To establish the relations for $\mathbf{R}(t + k)$ we return to Eqn. (14.20). Evaluating that equation at time $t + k$ yields

$$\mathbf{A}\ddot{\mathbf{R}}(t + k) + \mathbf{B}\dot{\mathbf{R}}(t + k) + \mathbf{C}\mathbf{R}(t + k) = \mathbf{P}(t + k) \qquad (14.31)$$

Substituting Eqns. (14.29) and (14.30) into Eqn. (14.31) and collecting like terms gives

$$[6\mathbf{A}/k^2 + 3\mathbf{B}/k + \mathbf{C}]\mathbf{R}(t + k) = \mathbf{P}(t + k) + [6\mathbf{A}/k^2 + 3\mathbf{B}/k]\mathbf{R}(t)$$

$$+ [6\mathbf{A}/k + 2\mathbf{B}]\dot{\mathbf{R}}(t) + [2\mathbf{A} + k\mathbf{B}/2]\mathbf{R}(t) \qquad (14.32)$$

where the only unknown is $\mathbf{R}(t + k)$. Experience indicates that if k is to be held constant throughout the integration then the computational efficiency can be increased by defining a square matrix

$$\mathbf{D} = 2\mathbf{A}/k + \mathbf{B}$$

and rewriting Eqn. (14.32) as

$$[3\mathbf{D}/k + \mathbf{C}]\mathbf{R}(t + k) = \big(\mathbf{P}(t + k) + \mathbf{D}(3\mathbf{R}(t)/k$$

$$+ 3\dot{\mathbf{R}}(t) + k\mathbf{R}(t)) - \mathbf{B}(\dot{\mathbf{R}}(t) + k\mathbf{R}(t)/2)\big) \qquad (14.33)$$

One notes that the right hand side of Eqn. (14.33) is a column matrix so that the final result is analogous to Eqn. (14.6), i.e.

$$\mathbf{S}(t + k)\mathbf{R} = \mathbf{F}(t + k) \qquad (14.34)$$

where $\mathbf{S}$ is the resultant square matrix and $\mathbf{F}$ is the resultant forcing function. In the following programs, $\mathbf{D}$ is stored in the original location of $\mathbf{A}$ while $\mathbf{S}$ is stored in the original location of $\mathbf{C}$. If $\mathbf{C}$ is zero, and/or $\mathbf{A}$ is diagonal, then one would use alternative storage schemes. The above equations would yield the values of $\mathbf{R}$ at time $t + k$ but we desire the values at a smaller time $t + h$ where $k = \theta h$, $\theta \geqslant 1$. Once the above equation has been solved the required parameters are obtained by interpolation from the old t and new $t + k$ values. This is accomplished by utilizing Eqns. (14.22), (14.23), and (14.24) evaluated at the intermediate time. That is,

$$\ddot{\mathbf{R}}(t + h) = \ddot{\mathbf{R}}(t) + [\ddot{\mathbf{R}}(t + k) - \ddot{\mathbf{R}}(t)]/\theta$$

$$\dot{\mathbf{R}}(t + h) = \dot{\mathbf{R}}(t) + h\ddot{\mathbf{R}}(t) + h[\ddot{\mathbf{R}}(t + k) - \ddot{\mathbf{R}}(t)]/(2\theta) \qquad (14.35)$$

$$\mathbf{R}(t + h) = \mathbf{R}(t) + h\dot{\mathbf{R}}(t) = h^2\ddot{\mathbf{R}}(t)/2 + h^2[\ddot{\mathbf{R}}(t + k) - \ddot{\mathbf{R}}(t)]/(6\theta),$$

where $\ddot{\mathbf{R}}(t + k)$ and $\dot{\mathbf{R}}(t + k)$ are given by Eqns. (14.29) and (14.30), respectively.

These concepts are illustrated in Fig. 14.8. The linear acceleration algorithm is implemented in subroutine DIRECT and is shown in Fig. 14.9.

Since the forcing function $\mathbf{P}(t)$ is problem-dependent in general, it must be supplied to subroutine DIRECT by the function program FORCER. Subroutine DIRECT assumes that $\mathbf{A}$ is not a diagonal matrix; thus Eqn. (14.21) is not utilized to initialize $\ddot{\mathbf{R}}(0)$. Instead, this routine starts the solution at $t = -h$ using $\mathbf{R}(-h) = \mathbf{R}(0)$, $\dot{\mathbf{R}}(-h) = \dot{\mathbf{R}}(\mathbf{0})$, and $\ddot{\mathbf{R}}(-h) = \mathbf{0}$. It solves the standard equations and interpolates for the value of $\ddot{\mathbf{R}}(0)$. At this point the entire integration solution is begun. The above approximation gives the exact value of $\ddot{\mathbf{R}}(0)$ for many forcing functions, $\mathbf{P}(t)$.

Fig. 14.8 The linear acceleration procedure

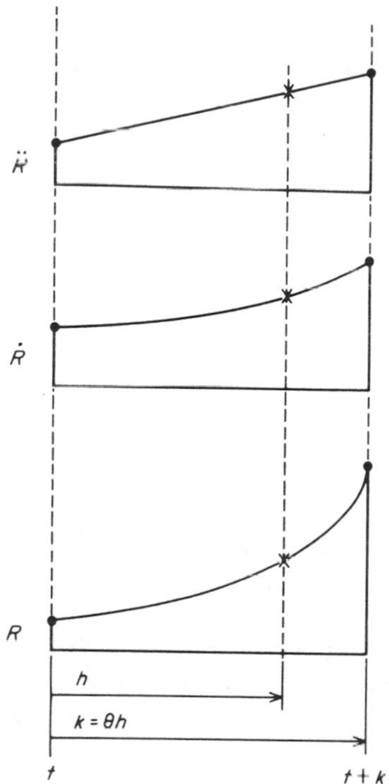

Fig. 14.9 Subroutine DIRECT

```
      SUBROUTINE  DIRECT (NDFREE,MBW,A,B,C,DRP,R,DR,D2R,P,
     1                   OMEGA,DT,NSTEPS,IPRINT,NBC,IBC)
C     * * * * * * * * * * * * * * * * * * * * * * * * * *
C     STEP BY STEP INTEGRATION OF MATRIX EQUATIONS:
C         A*D2R(T)/DT2 + B*DR(T)/DT + C*R(T) = P(T)
C     * * * * * * * * * * * * * * * * * * * * * * * * * *
CDP   IMPLICIT  REAL*8(A-H,O-Z)
      DIMENSION  A(NDFREE,MBW), B(NDFREE,MBW), IBC(1),
     1        R(NDFREE), DR(NDFREE), D2R(NDFREE), P(NDFREE),
     2        DRP(NDFREE), C(NDFREE,MBW)
```

Fig. 14.9—continued

```
C       INITIAL VALUES OF R AND DR ARE PASSEC THRU ARGUMENTS
C       NBC= NO. D.O.F. WITH SPECIFIED VALUES CF ZERO
C       IBC = ARRAY CONTAINING THE NBC DCF NCS WITH ZERO BC
C       R,DR,D2R = 0,1,2 ORDER DERIV. OF R W.R.T. T AT TIME=T
C       DRP = VALUE OF DR AT TIME = T + DELT
C       NSTEPS = NO. OF INTEGRATION STEPS
C       IPRINT = NO. OF INTEGRATION STEPS BETWEEN PRINTING
C       SUGGEST OMEGA = 1.25
C                ** INITIAL  CALCULATIONS **
        ZERC = 0.D0
        WRITE (6,5000)
 5000   FORMAT ( '1STEP BY STEP INTEGRATICN',//)
        IF ( IPRINT.LT.1 )  IPRINT = 1
        NSTEPS = (NSTEPS/IPRINT)*IPRINT
        IF ( NSTEPS.EQ.0 )  NSTEPS = IPRINT
        DELT = (OMEGA-1.)*DT
        TAU = OMEGA*DT
        MBWL1 = MBW - 1
        IF ( NBC.GT.0 )   GO TO 10
        WRITE (6,5010)
 5010   FORMAT (' WARNING, NO CONSTRAINTS IN DIRECT <<<')
   10   CONTINUE
        DO 30  I = 1,NDFREE
        P(I) = ZERO
        DRP(I) = ZERO
        D2R(I) = ZERO
        DO 20  J = 1,MBW
   20   B(I,J) = B(I,J) + 2.*A(I,J)/TAU + TAU*C(I,J)/3.
   30   CONTINUE
        ICOUNT = IPRINT - 1
C               ** APPLY BOUNDARY CONDITIONS ( TO B ) **
        DO 40  I = 1,NBC
   40   CALL MODFY1 (NDFREE,MBW,IBC(I),ZERO,B,P)
C                              ** TRIANGULARIZE B **
        CALL  FACTOR (NDFREE,MBW,B)
C       APPROXIMATE THE INITIAL VALUE OF D2R
        TPLUS = ZERO
        ISTEP = 0
C          ( THE FOLLOWING IS AN EXTRA LEGAL STATEMENT )
        GO TO 60
C                      ** END OF INITIAL CALCULATICNS **
C                     ** CALCULATE SOLUTION AT TIME T **
   50   DO 120  ISTEP = 1,NSTEPS
        ICOUNT = ICCUNT + 1
        T = DT*FLOAT(ISTEP-1)
        TPLUS = T + DELT
        IF ( ISTEP.EQ.1 )  TPLUS = ZERO
        IF ( ICOUNT.EQ.IPRINT )  WRITE(6,5020)  ISTEP,T
 5020   FORMAT ( //,' STEP NUMBER = ',I5,5X,'TIME = ',E14.8,/,
     1 '      I            R(I)                DR/DT          ',
     2 'D2R/DT2',/)
C       FORCER IS A SUBR TO DEFINE THE FCRCING FUNCTION P(T)
   60   CALL  FORCER ( TPLUS,P,NDFREE)
C       FORM MODIFIED FORCING FUNCTION AT T + DELT
        DO 90  I = 1,NDFREE
        SUM = 0.
        J1 = I - MBWL1
        J2 = I + MBWL1
        J1 = MAX0 (1,J1)
        J2 = MIN0 (J2,NDFREE)
        DO 80  J = J1,J2
        CALL  BANSUB (I,J,IB,JB)
        AIJ = A(IB,JB)
        IF ( AIJ.EQ.ZERO )  GO TO 70
        SUM = SUM + AIJ*( 2.*DR(J)/TAU+D2R(J) )
   70   CIJ = C(IB,JB)
        IF ( CIJ.EQ.ZERO )  GO TO 80
        SUM = SUM - CIJ*( R(J) + 2.*TAU*DR(J)/3.
     1           + TAU*TAU*D2R(J)/6. )
   80   CONTINUE
        P(I) = P(I) + SUM
   90   CCNTINUE
C            ** APPLY BOUNDARY CONDITIONS ( TO P ) **
        DO 100  I = 1,NBC
        IN = IBC(I)
  100   P(IN) = ZERO
```

Fig. 14.9—continued

```
C      SOLVE FOR DRP AT TIME T+DELT
       CALL  SOLVE  (NDFREE,MBW,B,P,DRP)
C      USING DATA AT T-DT AND T+DELT CALCULATE VALUES AT T
       DO 110   I = 1,NDFREE
       DRDT = (1.-1./OMEGA)*DR(I) + DRP(I)/OMEGA
       D2RDT2 = D2R(I)*(1.-2./OMEGA)+2.*(DRDT-DR(I))/OMEGA/DT
C      APPROXIMATE THE INITIAL VALUE OF D2R
       IF ( ISTEP.LT.2 )  GO TO 110
       R(I) = R(I) + DT*(2.*DR(I)+DRDT)/3. + DT*DT*D2R(I)/6.
       DR(I) = DRDT
  110  D2R(I) = D2RDT2
       IF ( ISTEP.EQ.0 )  GO TO 50
C      OUTPUT RESULTS FOR TIME T
       IF ( ICOUNT.NE.IPRINT )  GO TO 120
       WRITE (6,5030)  (K,R(K),DR(K),D2R(K),K=1,NDFREE)
 5030  FORMAT ( I10,2X,E14.8,2X,E14.8,2X,E14.8 )
       ICOUNT = 0
  120  CONTINUE
       RETURN
       END
```

14.3.2 Lumped mass forced vibration example

To illustrate the step by step integration procedure consider the forced vibration of the lumped mass system given in Fig. 14.10. The exact solution of this problem is given by Biggs [19]. The loads are linearly decreasing ramp functions. The function FORCER for these loads is shown in Fig. 14.11. The calculated displacements, velocities, and accelerations of node three are compared with the exact values in Figs. 14.12 to 14.14, respectively. Figure 14.15 shows the effect of increased time step size.

Fig. 14.10 Test problem for direct integration

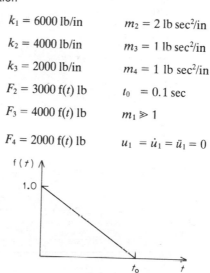

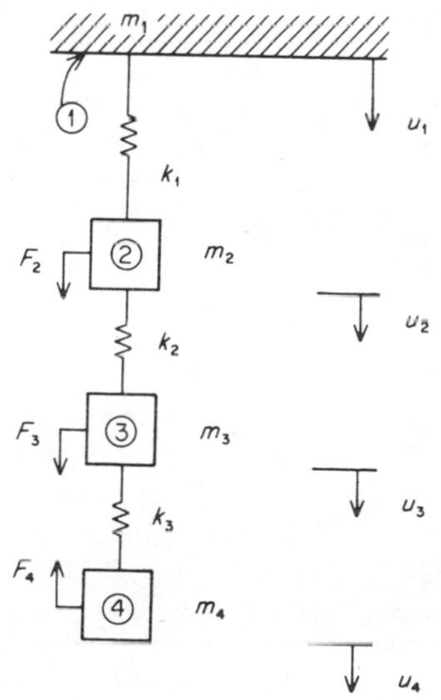

$k_1 = 6000$ lb/in	$m_2 = 2$ lb sec^2/in
$k_2 = 4000$ lb/in	$m_3 = 1$ lb sec^2/in
$k_3 = 2000$ lb/in	$m_4 = 1$ lb sec^2/in
$F_2 = 3000\ \mathrm{f}(t)$ lb	$t_0 = 0.1$ sec
$F_3 = 4000\ \mathrm{f}(t)$ lb	$m_1 \gg 1$
$F_4 = 2000\ \mathrm{f}(t)$ lb	$u_1 = \dot{u}_1 = \ddot{u}_1 = 0$

f(t)
1.0
t_0
t

Fig. 14.11 Subroutine FORCER

```
      SUBROUTINE  BIGGS
C     TEST OF DIRECT INTEGRATION PROGRAMS
      IMPLICIT REAL*8 (A-H,O-Z)
      DIMENSION AK(4,2),AM(4,2),U(4),F(4), VEL(4),ACC(4),IBC(1)
      DIMENSION  AC(4,2),UP(4)
      DATA  AC/8*0./
C     TRANSIENT TEST PROBLEM  BIGG'S P. 123
      MBW=2
      NDFREE=4
      NBC=1
      IBC(1)=1
      KPRINT = 1
      DT = 0.002500
      NSTEPS = 120
      DO 20 J =1,13,4
C      TEST OF EFFECT OF OMEGA
      OMEGA = J
      OMEGA = (OMEGA+20.)*0.05
      WRITE (6,5000) OMEGA
 5000 FORMAT ('0  OMEGA = ',E15.8)
      AK(1,1)=6000.
      AK(1,2)=-6000.
      AK(2,1)=10000.
      AK(2,2)=-4000.
      AK(3,1)=6000.
      AK(3,2)=-2000.
      AK(4,1)=2000.
      AK(4,2)=0.
      AM(1,1)=100.
      AM(1,2)=0.
      AM(2,1)=2.
      AM(2,2)=0.
      AM(3,1)=1.
      AM(3,2)=0.
      AM(4,1)=1.
      AM(4,2)=0.
      DO 10  I = 1,NDFREE
      U(I) = 0.
   10 VEL(I) = 0.
      CALL DIRECT (NDFREE,MBW,AM,AC,AK,UP,U,VEL,ACC,F,OMEGA,
     1               DT,NSTEPS,KPRINT,NBC,IBC)
   20 CONTINUE
      RETURN
      END
```

Fig. 14.14 Acceleration response of node three

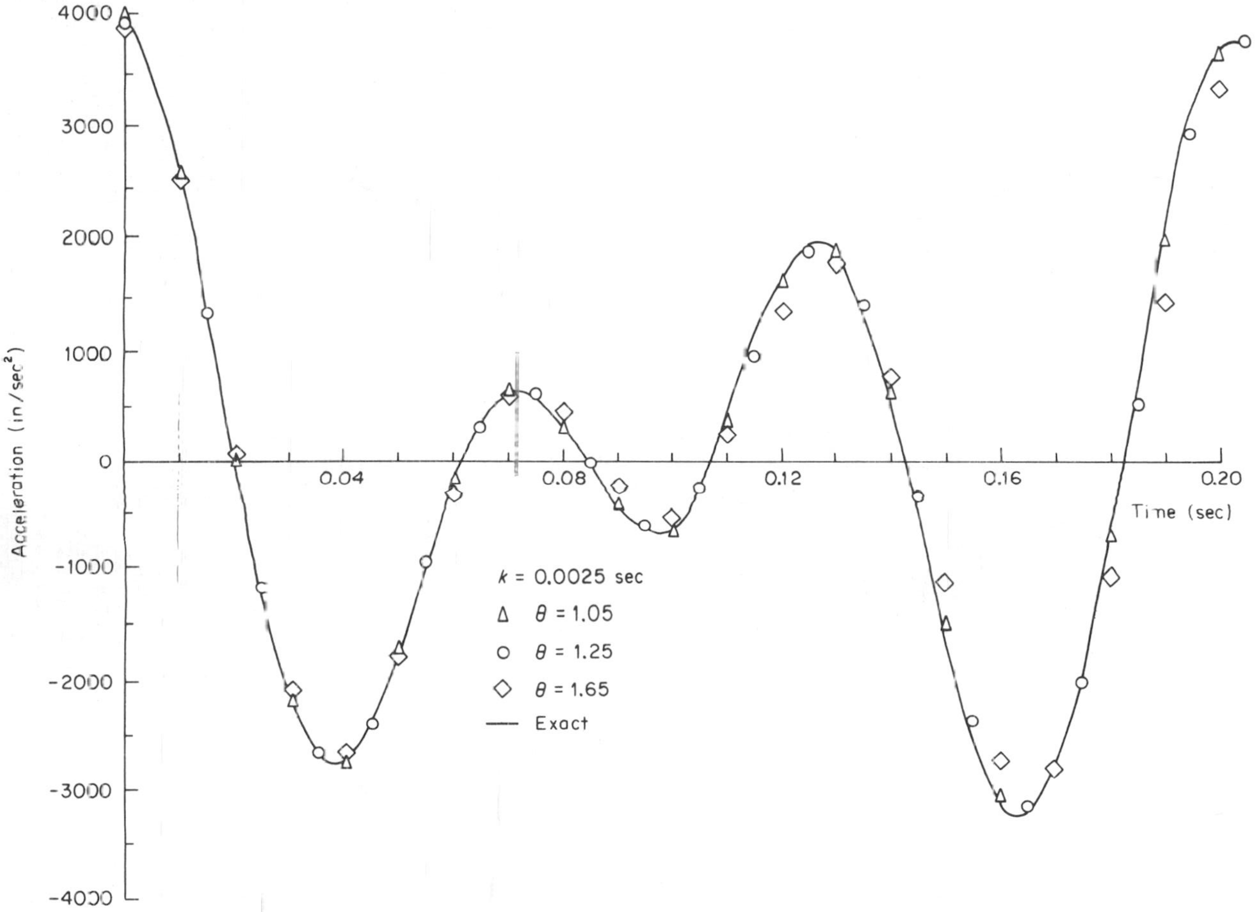

Acceleration (in/sec^2)

4000
3000
2000
1000
0
-1000
-2000
-3000
-4000

0.04
0.08
0.12
0.16
0.20

Time (sec)

k = 0.0025 sec

Δ θ = 1.05

○ θ = 1.25

◇ θ = 1.65

— Exact

Fig. 14.13 Velocity response of node three

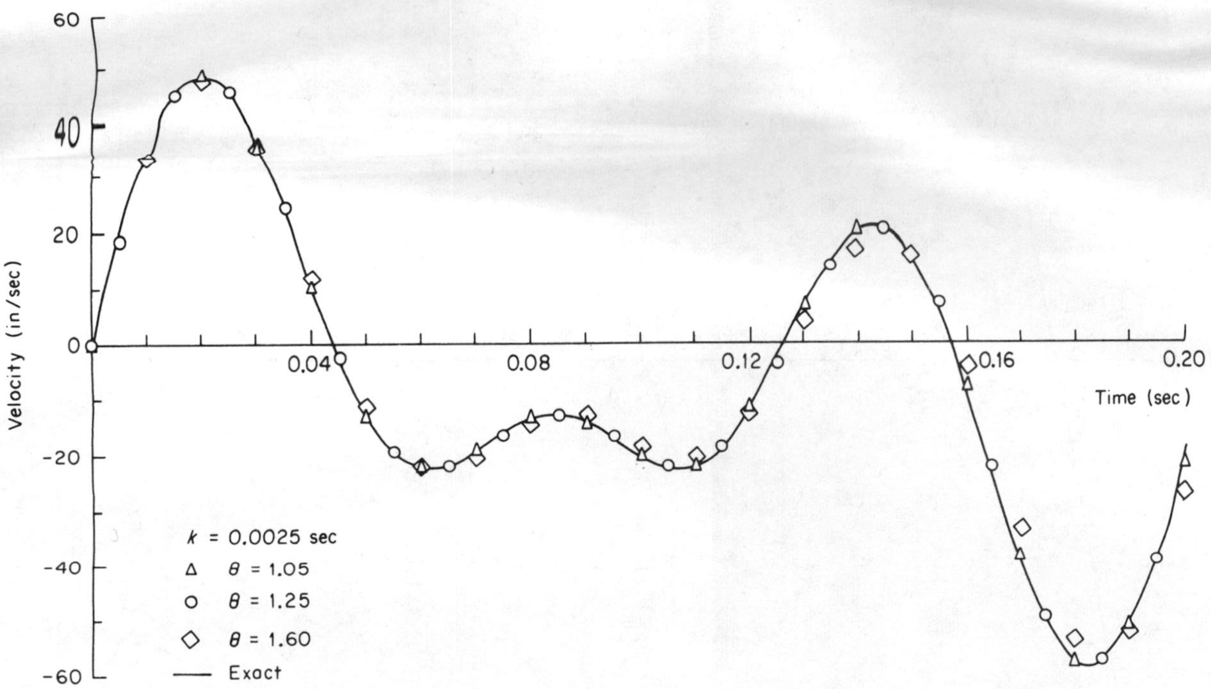

Fig. 14.12 Displacement response of node three

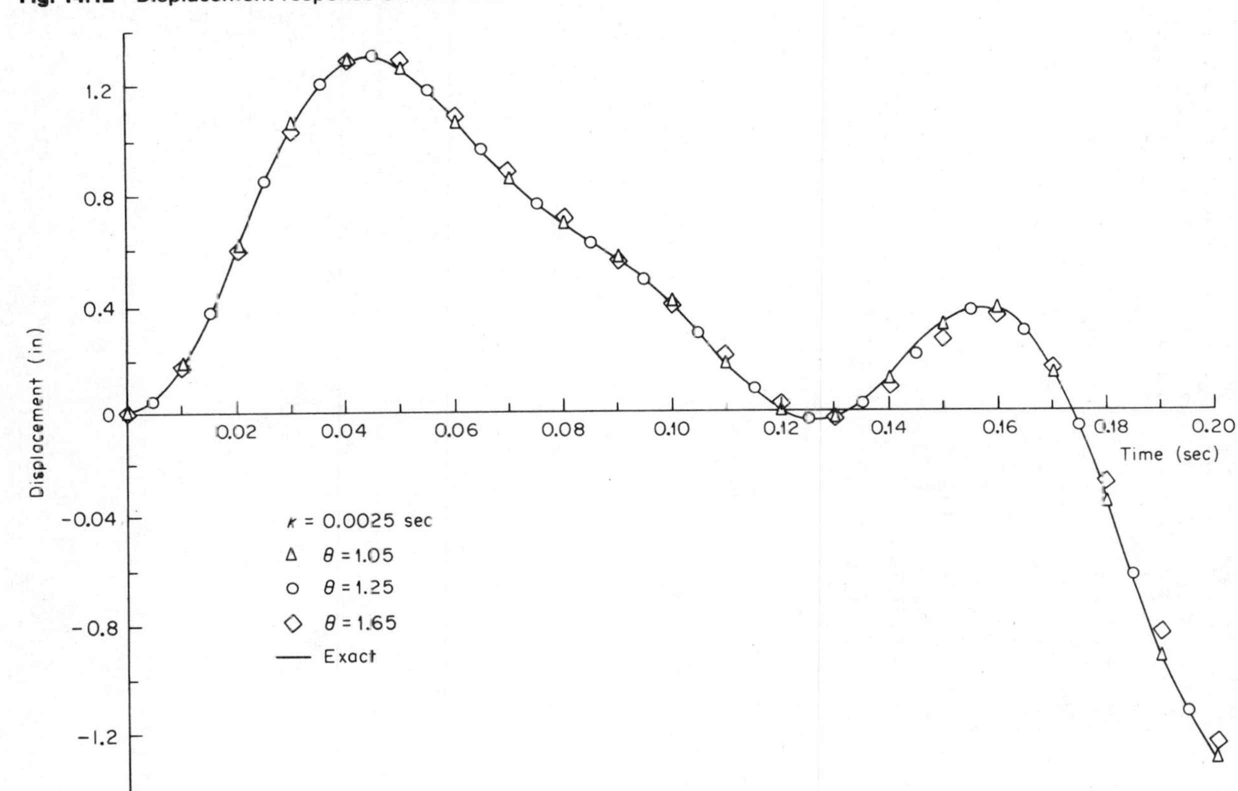

Fig. 14.11—continued

```
      SUBROUTINE  FORCER (T,P,NDFREE)
C     * * * * * * * * * * * * * * * * * * * * * * * * * * * * * *
C     DEFINE FORCING FUNCTION FOR STEP BY STEP INTEGRATION
C     * * * * * * * * * * * * * * * * * * * * * * * * * * * * * *
CDP   IMPLICIT REAL*8(A-H,O-Z)
      DIMENSION  P(NDFREE)
C     TRANSIENT TEST PROBLEM  BIGG'S P. 123
      TONE = 0.1
      FOFT = 0.0
      F2= 3000.
      F3 = 4000.
      F4 = -2000.
C     LINEAR RAMP (DECREASING)
      IF ( T-TONE )  10,20,20
   10 FOFT = 1.-T/TONE
   20 P(1) = 0.
      P(2) = F2*FOFT
      P(3) = F3*FOFT
      P(4) = F4*FOFT
      RETURN
      END
```

Fig. 14.15 Effect of increased step size

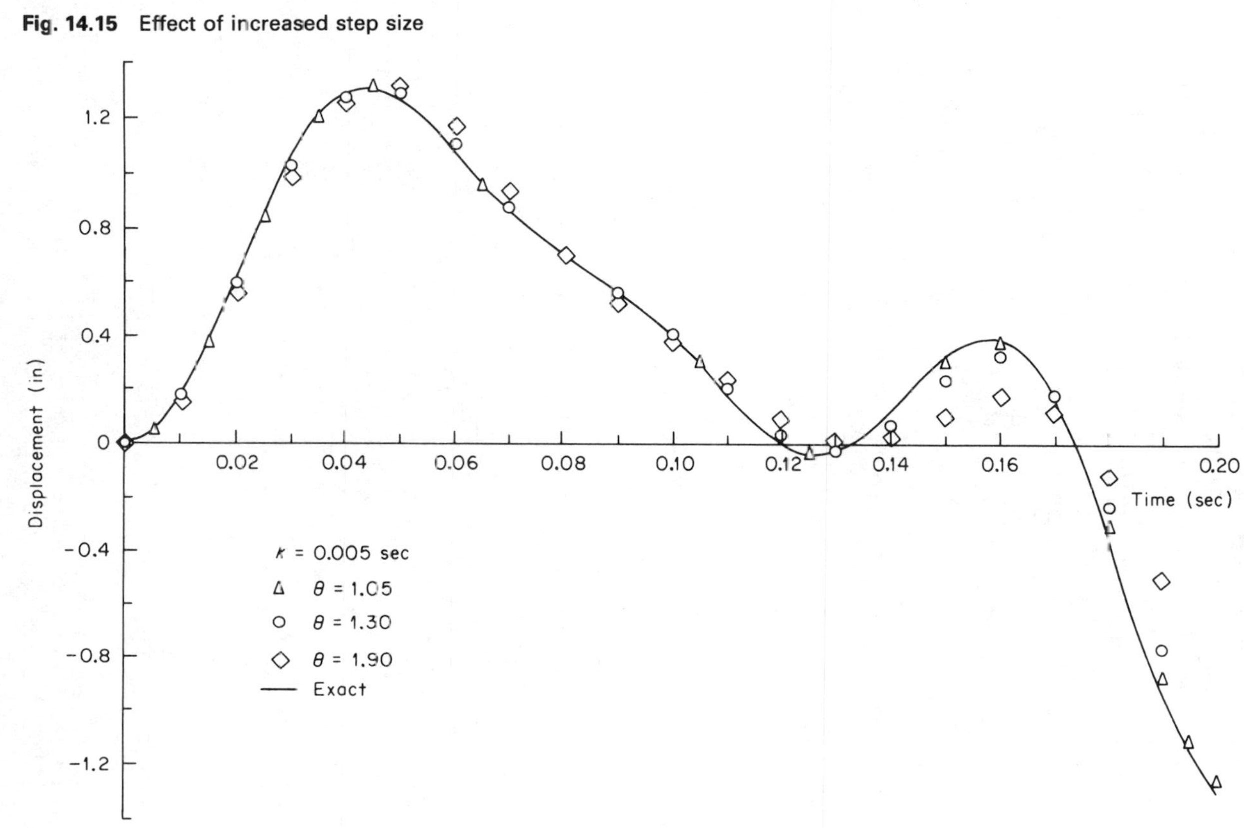

14.4 Exercises

1. Modify subroutine EULER to estimate the initial velocity by using the scaled diagonal of the system equations.
2. Modify subroutine DIRECT to establish the initial acceleration by using the scaled diagonal of the system equations.
3. Modify subroutine EULER for the case where **A** is a diagonal matrix.
4. Discuss the changes necessary to include time-dependent boundary conditions on the nodal parameters.
5. Modify subroutine DIRECT where **A** is diagonal and **B** is zero.
6. Modify subroutine DIRECT where **A** is diagonal and $\mathbf{B} = \alpha\mathbf{A} + \beta\mathbf{C}$ where α and β are known constants.

References and bibliography

1. Akin, J. E., A Least Squares Finite Element Solution of Nonlinear Equations, in *The Mathematics of Finite Elements and Applications*, J. R. Whiteman (Ed.), Academic Press, London, 153–62, 1973.
2. Akin, J. E., The Generation of Elements with Singularities, *Int. J. Num. Meth. Eng*, **10**, 1249–59, 1976.
3. Akin, J. E. and Gray, W. H., Finite Element Stress Analysis of High Field Solenoids, *Sixth Symp. Eng. Prob. Fusion Research*, IEEE, 240–2, 1976.
4. Akin, J. E., Finite Element Analysis of Fields with Boundary Singularities, *Int. Conf. Num. Meth. in Electric and Magnetic Field Problems*, May 1976.
5. Akin, J. E. and Pardue, R. M., Element Resequencing for Frontal Solutions, in *The Mathematics of Finite Elements and Applications*, Vol. II, J. R. Whiteman (Ed.), Academic Press, 535–41, 1976.
6. Akin, J. E. and Gray, W. H., Contouring on Isoparametric Surface, *Int. J. Num. Meth. Eng.*, Dec. 1977.
7. Akin, J. E., Elements for the Analysis of Line Singularities, in *The Mathematics of Finite Elements and Applications, Vol. III*, J. R. Whiteman (Ed.), Academic Press, 1978.
8. Akin, J. E. and Wooten, J. W., Tokamak Plasma Equilibria by Finite Elements, R. H. Gallagher (Ed.), *Finite Elements in Fluids III*, John Wiley, New York, Chapter 21, 1978.
9. Allan, T., The Application of Finite Element Analysis to Hydrodynamic and Externally Pressurized Pocket Bearings, *Wear*, **19**, 169–206, 1972.
10. Aziz, A. K., *The Mathematical Foundations of the Finite Element Method with Applications to Partial Differential Equations*, Academic Press, New York, 1972.
11. Barnhill, R. E. and Whiteman, J. R., Error Analysis of Finite Element Methods with Triangles for Elliptic Boundary Value Problems, in *The Mathematics of Finite Elements and Applications*, J. R. Whiteman (Ed.), Academic Press, London, 1973.
12. Barlow, J., Optimal Stress Locations in Finite Element Models, *Int. J. Num. Meth. Eng.*, **10**, 243–51, 1976.
13. Barsoum, R. S., On the Use of Isoparametric Finite Elements in Linear Fracture Mechanics, *Int. J. Num. Meth. Eng.*, **10**, 25–37, 1976.
14. Bathe, K. H. and Wilson, E. L., *Numerical Methods for Finite Element Analysis*, Prentice-Hall, 1976.
15. Belytischko, T., Chiapetta, R. L. and Bartel, H. D., Efficient Large Scale Non-Linear Transient Analysis by Finite Elements, *Int, J. Num. Meth. Eng.*, **10**, 579–96, 1976.
16. Benzley, S. E., Representation of Singularities with Isoparametric Finite Elements, *Int. J. Num. Meth. Eng.*, **8**, 537–45, 1974.

17. Berztiss, A. T., *Data Structures, Theory and Practice*, 2nd ed, Academic Press, New York, 1975.
18. Bettess, P., Infinite Elements, *Int. J. Num. Meth. Eng.*, **11**, 53–64, 1977.
19. Biggs, J. M., *Introduction to Structural Dynamics*, McGraw-Hill, 1964.
20. Blackburn, W. S., Calculation of Stress Intensity Factors at Crack Tips Using Special Finite Elements, in *The Mathematics of Finite Elements and Applications*, J. R. Whiteman (Ed.), Academic Press, London, 327–36, 1973.
21. Campbell, J. S., A Penalty Function Approach to the Minimization of Quadratic Functionals in Finite Element Analysis, *Finite Element Methods in Engineering*, Univ. of NSW, Australia, 33–54, 1977.
22. Christie, I., Griffiths, D. F., Mitchell, A. R. and Zienkiewicz, O. C., Finite Element Methods for Second Order Equations with Significant First Derivatives, *Int. J. Num. Meth. Eng.*, **10** 1389–96, 1976.
23. Chung, T. J., *Finite Element Analysis in Fluid Dynamics*, McGraw-Hill, New York, 1978.
24. Connor, J. C. and Brebbia C. A., *Finite Element Techniques for Fluid Flow*, Butterworth, 1976.
25. Cook, R. D., *Concepts and Applications of Finite Element Analysis*, John Wiley, New York, 1974.
26. Curiskis, J. I. and Valliappan, S., Solution Algorithm for Linear Constraint Equations in Finite Element Analysis, *Computers and Structures*, **8**, 117–24, 1978.
27. Daly, P., Singularities in Transmission Lines, in *The Mathematics of Finite Elements and Applications*, J. R. Whiteman (Ed.), Academic Press, London, 337–50, 1973.
28. DeBoor, C. (Ed.), *Mathematical Aspects of Finite Elements in Partial Differential Equations*, Academic Press, 1974.
29. Desai, C. S. and Abel, J. F., *Introduction to the Finite Element Method*, Van Nostrand–Reinhold Co., New York, 1972.
30. Fenner, R. T., *Finite Element Methods for Engineers*, Macmillan Ltd., London, 1975.
31. Fenves, S. J., *Computer Methods in Civil Engineering*, Prentice-Hall, 1967.
32. Fried, I., Finite Element Analysis of Incompressible Materials by Residual Energy Balancing, *Int. J. Solids Struct.*, **10**, 993–1002, 1974.
33. Fried, I. and Melkus, D. S., Finite Element Mass Matrix Lumping by Numerical Integration with the Convergence Rate Loss, *Int. J. Solids Struct.*, **11**, 461–65, 1975.
34. Gallagher, R. H., *Finite Element Analysis Fundamentals*, Prentice-Hall, Englewood Cliffs, 1975.
35. Gallagher R. H., *et al.* (Eds.), *Finite Elements in Fluids*, John Wiley, New York, Vol. 1, 1973, Vol. 2, 1975; Vol. 3, 1978.
36. Gartling, D. and Becker, E. B., Computationally Efficient Finite Element Analysis of Viscous Flow Problems, *Computational Methods in Nonlinear Mechanics*, J. T. Oden *et al.* (Eds.), T.I.C.O.M., Austin, Texas, 1974.
37. Gupta, A. K., A Finite Element for Transition from a Fine to a Course Grid, *Int. J. Num. Meth. Eng.*, **12**, 35–46, 1978.
38. Henshell, R. D. and Shaw, K. G., Crack Tip Elements are Unnecessary, *Int. Num. Meth. Eng.*, **9**, 495–509, 1975.
39. Henshell, R. D., *PAFEC 75 Theory, Results*, PAFEC Ltd., Nottingham. 1975.
40. Hinton, E. and Campbell, J., Local and Global Smoothing of Discontinuous

Finite Element Functions Using a Least Square Method, *Int. J. Num. Meth. Eng.*, **8**, 461–80, 1974.

41. Hinton, E., Scott, F. C. and Ricketts, R. E., Local Least Squares Stress Smoothing for Parabolic Isoparametric Elements, *Int. J. Num. Meth. Eng.*, **9**, 235–8, 1975.
42. Hinton, E. and Owen, D. R. J., *Finite Element Programming*, Academic Press, London, 1977.
43. Huebner, K. H., *Finite Element Method for Engineers*, John Wiley, New York, 1974.
44. Hughes, T. J. R., Cohen, M. and Haroun, M., Reduced and Selective Integration Techniques in the Finite Element Analysis of Plates, *Nuclear Eng. Design*, **46**, 1, 203–22, 1978.
45. Hughes, T. J. R., A Simple Scheme for Developing 'Upwind' Finite Elements, *Int. J. Num. Meth. Eng.*, **12**, 1359–67, 1978.
46. Hughes, T. J. R. and Liu, W. K., Implicit–Explicit Finite Elements in Transient Analysis: Implementations and Numerical Examples, *J. Appl. Mech.*, **45**, 2, 375–8, 1978.
47. Irons, B. M. and Razzaque, A., Experience with the Patch Test, in *Mathematical Foundations of the Finite Element Method*, A. R. Aziz (Ed.), Academic Press, New York, 1972.
48. Lynn, P. P. and Arya, S. K., Use of Least Square Criterion in Finite Element Formulation, *Int. J. Num. Meth. Eng.*, **6**, 75–88, 1973.
49. Lynn, P. P. and Arya, S. K., Finite Elements Formulation by the Weighted Discrete Least Squares Method, *Int. J. Num. Meth. Eng.*, **8**, 71–90, 1974.
50. Malkus, D. S. and Hughes, T. J. R., Mixed Finite Element Methods—Reduced and Selective Integration Techniques, *Comp. Meth. Appl. Mech. Eng.*, **15**, 1, 63–81, 1978.
51. Martin, H. C. and Carey, G. F., *Introduction to Finite Element Analysis*, McGraw-Hill, New York, 1974.
52. Meek, J. L., *Matrix Structural Analysis*, McGraw-Hill, New York, 1972.
53. Meek, J. L., Field Problems Solutions by Finite Element Methods, *Civil Eng. Trans.*, *Inst. Eng. Aust.*, 173–80, *Oct.* 1968.
54. Mitchell, A. R. and Wait, R., *The Finite Element Method in Partial Differential Equations*, John Wiley, London, 1977.
55. Moan, T., Orthogonal Polynomials and Best Numercial Integration Formulas on a Triangle, *Zamm*, **54**, 501–8, 1974.
56. Myers, G. E., *Analytical Methods in Conduction Heat Transfer*, McGraw-Hill, New York, 1971.
57. Newmark, N. M., A Method of Computation for Structural Dynamics, *J. Eng. Mech. Div.*, *ASCE*, **85**, EM3, 67–94, July 1959.
58. Norrie, D. H. and DeVries, G., *Finite Element Method: Fundamentals and Applications*, Academic Press, New York, 1973.
59. Norrie, D. H. and DeVries, G., *Finite Element Bibliography*, Plenum Press, New York, 1976.
60. Oden, J. T., *Finite Elements of Nonlinear Continua*, McGraw-Hill, New York, 1972.
61. Oden, J. T. and Reddy, J. N., *An Introduction to the Mathematical Theory of Finite Elements*, John Wiley, 1976.
62. Park, K. C., Practical Aspects of Numerical Time Integration, *Computers and Structures*, **7**, 2, 343–54, 1977.

63. Reddi, M. M., Finite Element Solution of the Incompressible Lubrication Problem, *J. Lubrication Technology*, 534–43, July 1969.
64. Rockey, K. C., *et al.*, *Finite Element Method—A Basic Introduction*, Halsted Press, 1975.
65. Schechter, R. S., *The Variational Method in Engineering*, McGraw-Hill, New York, 1967.
66. Silvester, P. P., Lowther, D. A., Carpenter, C. J. and Wyatt, E. A., Exterior Finite Elements for 2-Dimensional Field Problems with Open Boundaries, *Proc. IEE*, 124, 12, 1267–70, 1977.
67. Stern, M., Families of Consistent Conforming Elements with Singular Derivative Fields, *Int. J. Num. Meth. Eng.*, **14**, 409–21, 1979.
68. Strang, W. G. and Fix, G. J., *An Analysis of the Finite Element Method*, Prentice-Hall, 1973.
69. Taylor, R. L., Computer Procedures for Finite Element Analysis, in *The Finite Element Method*, O. C. Zienkiewicz (Ed.), McGraw-Hill, London, 1977.
70. Tracey, D. M. and Cook, T. S., Analysis of Power Type Singularities Using Finite Elements, *Int. J. Num. Meth. Eng.*, **11**, 1225–34, 1977.
71. Ural, O., *Matrix Operations and Use of Computers in Structural Engineering*, International Textbook Co., 1971.
72. Ural, O., *Finite Element Method: Basic Concepts and Application*, Intext, 1973.
73. Wada, S. and Hayashi, H., Application of Finite Element Method to Hydrodynamic Lubrication Problems, *Bulletin of Japanese Soc. Mech. Eng.*, **14**, 77, 1222–44, 1971.
74. Weaver, W. F., *Computer Programs for Structural Analysis*, Van Nostrand Co., Princeton, 1967.
75. Whiteman, J. R. (Ed.), *The Mathematics of Finite Elements and Applications*, Academic Press, London, 1973.
76. Whiteman, J. R., *A Bibliography of Finite Elements*, Academic Press, London, 1975.
77. Whiteman, J. R., Numerical Solution of Steady State Diffusion Problems Containing Singularities, *Finite Elements in Fluids II*, R. H. Gallagher (Ed.), 101–20, John Wiley, 1975.
78. Whiteman, J. R., Some Aspects of the Mathematics of Finite Elements, in *The Mathematics of Finite Elements and Applications*, *Vol. II*, J. R. Whiteman (Ed.), 25–42, Academic Press, London, 1976.
79. Whiteman, J. R., (Ed.), *The Mathematics of Finite Elements and Applications*, *Vol. III*, Academic Press, London, 1978.
80. Whiteman, J. R. and Akin, J. E., Finite Elements, Singularities and Fracture, in *The Mathematics of Finite Elements and Applications*, *Vol. III*, J. R. Whiteman (Ed.), Academic Press, London, 1978.
81. Yamada, Y. and Gallagher, R. H., *Theory and Practice in Finite Element Structural Analysis*, University of Tokyo Press, 1973.
82. Zienkiewicz, O. C., *The Finite Element Method in Structural and Continuum Mechanics*, McGraw-Hill, New York, 1967.
83. Zienkiewicz, O. C., *The Finite Element Method in Engineering Science*, McGraw-Hill, New York, 1971.
84. Zienkiewicz, O. C. and Lewis, R. W., An Analysis of Various Time-Stepping Schemes for Initial Value Problems, *Earthquake Engineering and Structural Dynamics*, **1**, 407–8, 1973.

85. Zienkiewicz, O. C. and Hinton, E., Reduced Integration smoothing and Non-Conformity, *J. Franklin Inst.*, 302, 6, 443–61, 1976.
86. Zienkiewicz, O. C., Heinrich, J. C., Huyakorn, P. S. and Mitchell, A. R., An Upwind Finite Element Scheme for Two-Dimensional Convective Transport, *Int. J. Num. Meth. Eng.*, **11**, 131–44, 1977.
87. Zienkiewicz, O. C., *The Finite Element Method*, McGraw-Hill, London, 1977.

Appendix

A Summary of Input Formats and Supporting Programs

1

Input Data Requirements for the Model Program

```
A typical problem will require: 1. a dummy  main code
to set the maximum dimensions, 2. the problem dependent
element subroutines, 3. the problem data.

This file contains information sections on:
1. Input data formats
2. Dummy main program

++++ ++++ ++++  SECTION 1  ++++ ++++ ++++

      **** INPUT DATA ****

-->> CARD TYPE 1    (Subroutine MODEL)

TITLE, A general description of the problem.(20A4)

-->> CARD TYPE 2    (Subroutine MODEL)
Control parameters. (16I5)

M      = Number of nodal points in the system.
NE     = Number of elements in the system.
NG     = Number of parameters per node.
N      = Number of nodes per element.
NSPACE = Dimension of solution space.
NSEG   = No. of elem boundary segments with given flux.
LBN    = Number of nodes on an elem. boundary segment.
NITER  = Number of iterations to be run. (If any.)
INRHS  = Initial system column matrix flag.
         = 0, Not to be input. (Default)
         > 0, Will be input by CARD TYPE 16.
ISAY   = Number of user remark cards to be read and
         printed. See CARD TYPE 4.

-->> CARD TYPE 3   (Subroutine MODEL)
Control parameters.  (16I5)

NNPFIX = Number of fixed pt. properties per node.
NNPFLO = Number of floating pt. properties per node.
NLPFIX = Number of fixed pt. proterties per elem.
NLPFLO = Number of floating pt. properties per elem.
MISCFX = Number of miscellaneous fixed pt. properties.
MISCFL = Number of miscellaneous floating pt. properties
NHOMO  = Nodal properties flag.
         > 0, Properties are homogeneous. Only
              values for the first node will be input.
         = 0, Non-homogeneous properties. (Default)
LHOMO  = Element properties flag.
         > 0, Properties are homogeneous. Only
              values for the first element will be input.
         = 0, Non-homogeneous properties. (Default)
NPTWRT = Nodal parameter print flag.
         = 0, List calculated values by nodes. (Default)
         Otherwise omit nodal list.
LEMWRT = Nodal parameter print flag.
         = 0, List node and calculated nodal values for
         each element. Otherwise omit list.
```

```
NTAPE1 = I/O Unit for post solution calculation data.
        = 0, No post solution calculations. (Default)
        > 0, Unit to be utilized.
NTAPE2 = Extra scratch unit when > 0.
NTAPE3 = Extra scratch unit when > 0.
NTAPE4 = Extra scratch unit when > 0. (Note: the above
        units are for the analyst's problem dependent
        data.)
NULCOL = Element column matrix flag.
        = 0, Problem requires the matrix. (Default)
        > 0, The matrix is always a null vector.

-->> CARD TYPE 4   (Subroutine IREMRK)
User comments. Utilized only when ISAY > 0 on
CARD TYPE 2. (20A4)

REMARK = Additional information to be printed with
         problem output. Used to describe meaning
         of nodal prameters, properties, etc.

*** NOTE *** There will be ISAY TYPE 4 cards.

-->> CARD TYPE 5    (Subroutine INPUT)
Nodal point data. (2I10,(6F10.0))

J      = System node number.
IBC(J) = Parameters constraint indicators at
        node J. (Right justified) Digit K equals
        constraint type of parameter K.
X(J,K) = Spatial coordinates (K=1,NSPACE) of node J

*** NOTE *** There will be one TYPE 5 card for each
nodal point in the system.
         Example, IBC = 12213 and IBC = 10330
         implies 3 Type 1, 1 Type 2, 1 Type 3.
   Mesh generation programs are available to
supply TYPE 5 and TYPE 6 cards.

-->> CARD TYPE 6    (Subroutine INPUT)
Element Connectivity Data  (I5, (15I5))

J           = Element number.
NODES(J,K)  = Nodal incidences (K=1,N) of elem. J

*** NOTE *** There is one TYPE 6 card for each elem
   Mesh generation programs are available to
supply TYPE 5 and TYPE 6 cards.

-->> CARD TYPE 7    (Subroutine INCEQ)
Input data for a Type 1 nodal constraint, if any.
(2I5,F10.0)
NODE1 = system nodal point number.
IPAR1 = Constrainted parameter number, 1<=IPAR1<=NG
A1    = Value assigned to the parameter.

*** NOTES *** These constraints are of the form
D(J) = A1, where D(J) is the system degree of
freedom corresponding to parameter IPAR1 at node
NODE1. That is, J = NG*(NODE1-1)+IPAR1.
    There will be one TYPE 7 card for each Type 1
constraint flagged in IBC(NODE1). See  CARD TYPE 5.

-->> CARD TYPE 8    (Subroutine INCEQ)
Input data for a Type 2 nodal constraint, if any.
(4I5,3F10.0)

NODE1 = First system node number.
IPAR1 = Constrained parameter number at NODE1.
NODE2 = System node number of second node.
IPAR2 = Constrained parameter number at NODE2.
A1    = Coefficient of first parameter.
A2    = Coefficient of second parameter.
A3    = Coefficient of right hand side.
```

```
*** NOTES *** These constraints are of the form
A1*D(J1)+A2*D(J2)=A3.
    There will be one TYPE 8 card for each Type 2
constraint equation.

-->> CARD TYPE 9    (Subroutine INCEQ)
Input data for a Type 3 nodal constraint, if any.
(6I5,4F10.0)

NODE1 = First node point number.
IPAR1 = First constrained parameter number.
NODE2 = Second nodal point.
IPAR2 = Second constrained parameter number.
NODE3 = Third nodal number.
IPAR3 = Third constrained parameter.
A1    = Coefficient of first parameter.
A2    = Coefficient of second parameter.
A3    = Coeffieient of third parameter.
A4    = Constant on right hand side.

*** NOTES *** These constraints are of the form
A1*D(J1)+A2*D(J2)+A3*D(J3)=A4.
    There will be one TYPE 9 card for each Type 3
nodal parameter constraint equation.
    The program is easily extended up to Type 9
constraints.

-->> CARD TYPE 10    (Subroutine INPROP)
Fixed point nodal properties. Utilized only when
NNPFIX > 0 on CARD TYPE 3.  (I5,(7I10))

J            = Nodal point number.
NPFIX(J,K)   = Fixed point properties. K=1,NNPFIX

*** NOTE *** Reading terminated when J=M, unless
NHOMO > 0 . Then only J=1 is read.
   Omitted values default to zero when NHOMO=0.

-->> CARD TYPE 11    (Subroutine INPROP)
Floating point nodal properties. Utilized only when
NNPFLO > 0 on CARD TYPE 3.  (I5,(7F10.0))

J              = Nodal point number.
FLTNP(J,K)   = Floating point properties. K=1,NNPFLO

*** NOTE *** Reading terminated when J=M, unless
NHOMO > 0. Then only J=1 is read.
   Omitted values default to zero when NHOMO=0.

-->> CARD TYPE 12    (Subroutine INPROP)
Fixed point elem properties.  Utilizied only when
NLPFIX > 0 on CARD TYPE 3.  (I5,(7I10))

J            = Element number.
LPFIX(J,K)   = Fixed point properties. K=1,NLPFIX

*** NOTE *** Reading terminated when J=NE, unless
LHOMO > 0. Then only J=1 is read.
   Omitted values default to zero when LHOMO=0.

-->> CARD TYPE 13    (Subroutine INPROP)
Floating point element properties. Utilized only
when NLPFLO > 0 on CARD TYPE 1.  (I5,7(F10.0))

J            = Element number.
FLTEL(J,K)   = Floating point properties. K=1,NLPFLO

*** NOTE *** Reading terminated when J=NE, unless
LHOMO > 0. Then only J=1 is read.
   Omitted values default to zero when LHOMO=0.

-->> CARD TYPE 14    (Subroutine INPROP)
Fixed point miscellaneous system properties. Utilized
only if MISCFX > 0 on CARD TYPE 3.  (8I10.0)
```

MISFIX(K) = Misc. fixed point properties.K=1,MISCFX

-->> CARD TYPE 15 (Subroutine INPROP)
Floating point miscellaneous system properties. Utilized only when MISCFL > 0 on CARD TYPE 3. (8F10.0)

FLTMIS(K) = Misc. floating point properties.K=1,MISCFL

-->> CARD TYPE 16 (Subroutine INVECT)
Specified initial values of the system column matrix. Utilized only if INRHS > 0 on CARD TYPE 2. (2I5,F10.0)

NODE = System node number of the point.
IPARM = Parameter number associated with the value.
VALUE = Specified value to be added to column matrix.

*** NOTES *** Addition is of the form of
C(J) = C(J) + VALUE where J=NG*(NODE-1)+IPARM.
Reading is terminated when NODE=M and IPARM=NG.
Omitted values default to zero.

-->> CARD SET 17 (Subroutine INFLUX)
Specified boundary flux or surface tractions.
Utilized only when NSEG > 0 and LBN > 0 on CARD TYPE 2.

TYPE 17A (17I5)
LNODE(K) = List of nodes on the segment. K=1,LBN.

TYPE 17B (8F10.4)
(FLUX(K,L),L=1,NG) = The NG components of flux at each of the above nodes. K=1,LBN.

*** NOTES *** These data are related to problem dependent calculations. Thus the user assigns meanings to the nodal order and component order.
There will be NSEG type 17 CARD SETs.

***** END OF INPUT INSTRUCTIONS *****

++++ ++++ ++++ SECTION 2 ++++ ++++ ++++

**** MAIN PROGRAM ****

MAIN , a dummy driver to set array sizes.

The main program should have the following form. It sets various control parameters and calls MODEL. The user sets the numbers, #1 and #2, as desired for the computer in use.

```
      DIMENSION AD(#1),ID(#2)
      DATA KFLOAT, KFIXED / #1, #2/
      DATA CUTOFF,MAXTIM /1.D-7, 0/
C     AD STORES ALL FLOATING POINT ARRAYS
C     ID STORES ALL FIXED POINT ARRAYS
C     KFLOAT = DIMENSION OF AD
C     KFIXED = DIMENSION OF ID
C     MAXTIM = MAX. NO. CALLS TO CPU CLOCK
C     CUTOFF = MAX. RESIDUAL FOR ITERATIVE SOLUTIONS
      CALL MODEL (KFLOAT,KFIXED,MAXTIM,AD,ID,CUTOFF)
      STOP
      END
```

The MAIN should be followed at this point with any required problem dependent subroutines or functions. That is, ELSQ, ELCOL, ELPOST, POSTEL, ETC.

------ END OF INPUT DATA SUMMARY -------

2
Supporting Programs Listings

```
SUBROUTINE  MODEL (KFLOAT,KFIXED,MAXTIM,AD,ID,CUTOFF)
C       ******************************************************
C       *                                                    *
C       *                  -M-O-D-E-L-                       *
C       *     MODULAR PROGRAMS FOR FINITE ELEMENT ANALYSES   *
C       *                                                    *
C       ******************************************************
CDP     IMPLICIT REAL*8 (A-H,O-Z)
        DIMENSION TITLE(20) , TIME(7), AD(KFLOAT), ID(KFIXED)
C            PRINT AUTHOR CREDITS
        CALL  TOOT
C                 ZERO STORAGE ARRAYS
        CALL  ZEROA (KFLOAT,AD)
        CALL  ZEROI (KFIXED,ID)
        IF ( MAXTIM.GT.0 )  CALL  ZEROA (MAXTIM,TIME)
C                         INITIALIZE TIME CALLS
        IF ( MAXTIM.GT.0 )  CALL  CPUTIM (TZERO)
C-->       ** READ AND PRINT TITLE AND CONTROL DATA **
        READ (5,5000)   TITLE
        WRITE (6,5010)  TITLE
        READ (5,5020) M, NE, NG, N, NSPACE, NSEG, LBN, NITER,
       1              NCURVE, INRHS, ISAY
        WRITE (6,5030) M, NE, N, NG, NSPACE
        WRITE (6,5040) NSEG, LBN
        IF ( LBN.GT.N )    WRITE (6,5050)
        IF ( NITER.GT.1 )  WRITE (6,5060)   NITER
        IF ( NITER.LT.1 )  NITER = 1
        IF ( NCURVE.GT.0 ) WRITE (6,5150)   NCURVE
        IF ( INRHS.NE.0 )  WRITE (6,5160)
        READ (5,5070)  NNPFIX, NNPFLO, NLPFIX, NLPFLO, MISCFX,
       1               MISCFL, NHOMO, LHOMO, NPTWRT, LEMWRT,
       2               NTAPE1, NTAPE2, NTAPE3, NTAPE4, NULCOL
        WRITE (6,5080) NNPFIX, NNPFLO, NLPFIX, NLPFLO,
       1               MISCFX, MISCFL
        IF ( NHOMO.EQ.1 )  WRITE (6,5090)
        IF ( LHOMO.EQ.1 )  WRITE (6,5100)
        NSUM = NTAPE1 + NTAPE2 + NTAPE3 + NTAPE4
        IF ( NSUM.GT.0 )  WRITE (6,5180)   NTAPE1, NTAPE2,
       1                                   NTAPE3, NTAPE4
        IF ( NPTWRT.EQ.0 )   WRITE (6,5190)
        IF ( LEMWRT.EQ.0 )   WRITE (6,5200)
        IF ( NTAPE1.GT.0 )   REWIND NTAPE1
        IF ( NTAPE2.GT.0 )   REWIND NTAPE2
        IF ( NTAPE3.GT.0 )   REWIND NTAPE3
        IF ( NTAPE4.GT.0 )   REWIND NTAPE4
        IF ( NULCOL.NE.0 )   WRITE (6,5210)
        IF ( ISAY.GT.0 )  CALL  IREMRK (ISAY)
C          SET INITIAL ARRAY DIMENSIONS AND CONSTANTS
        CALL   SET (NNPFIX,NNPFLO,NLPFIX,NLPFLO,MISCFX,MISCFL,
       1            IPTEST,LPTEST,IP1,IP2,IP3,IP4,IP5,JP1,
       2            JP2,JP3,KP1,KP2,KP3,NHOMO,LHOMO,M,N,NE,
       3            RATIO,MAXTYP,NELFRE,NDFREE,NFLUX,NG,LBN)
C-->          *** CALCULATE ARRAY POINTERS ***
C
C       FIXED POINT ARRAY POINTERS (NI):
C       N1-IBC 2-NODES 3-LNODE 4-KODES 5-NFEQ 6-INDEX
C       8-NDXC 17-NPFIX 18-LPFIX 19-MISFIX 20-LPROP
C       21-NPROP 22-NRANGE
C
C       FLOATING POINT ARRAY POINTERS (MI):
C       M1-X 1A-CP 1B-SP 2-CEQ 11-FLTNP 12-FLTEL 13-FLTMIS
C       14-ELPROP 15-PTPROP 16-COORD 17-S 18-C 19-SS 20-CC
C       21-DD 22-D 23-DDOLD 24-RANGE 25-XPT 26-PRTLPT 27-FLUX
C
C                       BEGIN POINTERS
        CALL   PT1 (M,N,NE,NG,NELFRE,NDFREE,N1,N2,N3,N4,N5,
       1            N6,N8,M1,M1A,M1B,M2,NSPACE,MAXTYP)
        MTOTAL = M2
```

```
      NTOTAL = N8
      IF (MTOTAL.GT.KFLOAT .OR. NTOTAL.GT.KFIXED) GO TO 10
C-->                *** READ MESH DATA ***
      CALL  INPUT (M,N,NE,NG,NSPACE,AD(M1),ID(N1),ID(N2))
C-->   *** COUNT NUMBER OF NODAL PARAMETER CONSTRAINTS ***
      CALL  CCOUNT (M,NG,ID(N5),ID(N1),ID(N4),
     1              MAXACT,NUMCE,MAXTYP)
C                    CONTINUE POINTERS
      CALL  PT2 (N,NG,NSPACE,NELFRE,MAXACT,NUMCE,
     2           IP1,IP2,IP3,IP4,JP1,JP2,JP3,KP1,KP2,
     4           KP3,N17,N18,N19,N20,N21,N22,N23,
     5           M11,M12,M13,M14,M15,
     6           M16,M17,M18,M19,N8,M2)
      MTOTAL = M19
      NTOTAL = N23
      IF (MTOTAL.GT.KFLOAT .OR. NTOTAL.GT.KFIXED)  GO TO 10
C-->    *** READ NODAL PARAMETER CONSTRAINT EQUATIONS ***
      CALL INCEQ (NG,MAXACT,NUMCE,ID(N5),AD(M2),ID(N8),M)
C-->                *** READ PROPERTIES ***
      IF ( IPTEST.GT.0 )  CALL  INPROP (M,NE,NNPFIX,
     1                    NNPFLO,NLPFIX,NLPFLO,MISCFX,
     2                    MISCFL,IP1,IP2,IP3,IP4,JP1,JP2,
     3                    JP3,KP1,KP2,KP3,AD(M11),AD(M12),
     4                    AD(M13),ID(N17),ID(N18),ID(N19))
C            DETERMINE SYSTEM HALF-BANDWIDTH
      CALL SYSBAN (NE,N,NG,IBW,ID(N2),ID(N3),LID)
      JBW = 1
      IF ( MAXACT.GT.1 )  CALL  CEQBAN (JBW,ID(N5),MAXACT,
     1                          NUMCE,ID(N8),NDFREE)
      MAXBAN = MAX0 ( JBW,IBW )
      WRITE (6,5110)  IBW, LID, JBW, MAXBAN, NDFREE, NELFRE
      NOCOEF = NDFREE*MAXBAN
C                    COMPLETE POINTERS
      CALL  PT3 (NOCOEF,NDFREE,NELFRE,NITER,NG,NSPACE,IP5,
     1      KP1,M20,M21,M22,M23,M24,M25,M26,M27,M28,LBN,M19)
      NNEXT = N23
      MNEXT = M28
C               *** POINTERS COMPLETED ***
      MTOTAL = MNEXT
      NTOTAL = NNEXT
   10 WRITE (6,5120) MTOTAL, KFLOAT, NTOTAL, KFIXED
C          CHECK DATA AGAINST DIMENSION STATEMENTS
      IF (MTOTAL.LE.KFLOAT .AND. NTOTAL.LE.KFIXED) GO TO 20
      WRITE (6,5130)
   20 CONTINUE
C-->          ** INPUT INITIAL FORCING VECTOR **
      IF ( INRHS.GT.0 ) CALL INVECT (NDFREE,NG,AD(M20),M)
C-->   ** READ FLUX BOUNDARY COND. AND ADD TO SYSTEM EQS **
      IF ( NSEG.GT.0 ) CALL INFLUX (NSEG,LBN,ID(N3),AD(M27),
     1                 NG,AD(M16),NSPACE,AD(M1),M,ID(N6),
     2                 AD(M18),AD(M20),NDFREE,AD(M17),AD(M19),
     3                 NFLUX)
      IF ( MAXTIM.GT.0 )  CALL  CPUTIM (TIME(1))
C        INITIALIZE SYSTEM DOF FOR ITERATIVE SOLUTION
      IF ( NITER.GT.1 ) CALL  DSTART (1,M,NG,NSPACE,NDFREE,
     1                  ID(N6),AD(M1),AD(M16),AD(M23))
C                 *** BEGIN ITERATION LOOP ***
      DO 30  IT = 1,NITER
C-->   *** CALCULATE AND ASSEMBLE ELEMENT MATRICES ***
C-->   *** GENERATE POST SOLUTION MATRICES & STORE ***
      CALL  ASYMBL (M,N,NE,NG,NSPACE,NELFRE,MAXBAN,NDFREE,
     1             NITER,LPTEST,LHOMO,NNPFLO,IP1,IP3,
     2             IP4,IP5,JP2,KP1,KP2,NTAPE1,ID(N2),
     3             ID(N3),ID(N6),ID(N18),ID(N20),AD(M1),
     4             AD(M11),AD(M12),AD(M13),AD(M14),AD(M16),
     5             AD(M17),AD(M18),AD(M19),AD(M20),AD(M22),
     6             AD(M23),AD(M26),KP3,NULCOL,NTAPE2,
     7             NTAPE3,NTAPE4,NHOMO)
      IF ( MAXTIM.GT.0 )  CALL  CPUTIM (TIME(2))
C                *** ASSEMBLY COMPLETED ***
C-->  ** APPLY BOUNDARY CONSTRAINTS TO NODAL PARAMETERS **
      CALL  PENMOD (MAXACT,NUMCE,ID(N5),ID(N8),AD(M2),AD(M1A),
     1             AD(M1B),AD(M20),AD(M19),NDFREE,MAXBAN)
```

```
      IF ( MAXTIM.GT.0 )  CALL  CPUTIM (TIME(3))
C-->     *** SOLVE FOR UNKNOWN NODAL PARAMETERS ***
      CALL  FACTOR (NDFREE,MAXBAN,AD(M19))
      CALL  SOLVE  (NDFREE,MAXBAN,AD(M19),AD(M20),AD(M21))
      IF ( MAXTIM.GT.0 )  CALL  CPUTIM (TIME(4))
C-->           *** SOLUTION COMPLETE ***
C              *** PRINT RESULTS ***
      IF ( NPTWRT.EQ.0 )  CALL  WRTPT ( M,NG,NDFREE,
     1                     NSPACE,AD(M1),AD(M21),ID(N6))
      IF ( LEMWRT.EQ.0 ) CALL WRTELM (NE,N,NG,NDFREE,NELFRE,
     1                   AD(M21),ID(N6),ID(N2),ID(N3))
      IF ( MAXTIM.GT.0 )  CALL  CPUTIM (TIME(5))
C-->        *** POST SOLUTION CALCULATIONS ***
      IF ( NTAPE1.GT.0 )  CALL  POST (NTAPE1,NE,N,NG,NELFRE,
     1                  NDFREE,ID(N2),ID(N3),ID(N6),AD(M21),
     2                  AD(M22),ID(N20),NTAPE2,NTAPE3,NTAPE4,
     3                  IT,NITER,M)
      IF ( MAXTIM.GT.0 )  CALL  CPUTIM (TIME(6))
      IF ( NITER.LT.2 )  GO TO 30
C    *** UPDATE VALUES FOR NEXT ITERATION (IF ANY) ***
      WRITE (6,5140)  IT
      CALL  CHANGE  (NDFREE,AD(M21),AD(M23),TOTAL,
     1              RATIO,1)
      IF ( IT.EQ.1 )  RTEST = RATIO
      IF ( (RATIO/RTEST).LT.CUTOFF )  GO TO 35
      CALL  CORECT  (NDFREE,AD(M21),AD(M23))
      CALL  ZEROA  (NOCOEF,AD(M19))
      CALL  ZEROA  (NDFREE,AD(M20))
C     PRINT SEGMENT CPU TIMES
      IF ( MAXTIM.GT.0 )  CALL  TYMLOG (TIME,MAXTIM)
 30   CONTINUE
 35   CONTINUE
      IF ( NCURVE.EQ.0 )  GO TO 40
C      ** CALCULATE CONTOUR CURVES FOR NODAL PARAMETERS **
      CALL  MAXMIN  (M,NG,NDFREE,1,AD(M24),AD(M21),ID(N6),
     1              ID(N22))
      CALL  CONTUR  (M,NE,N,NG,NSPACE,NDFREE,NELFRE,NCURVE,
     1              AD(M1),AD(M16),AD(M25),AD(M21),AD(M22),
     2              AD(M24),ID(N2),ID(N3),ID(N6))
      IF ( MAXTIM.GT.0 )  CALL  CPUTIM (TIME(7))
 40   CONTINUE
C                    *** PROBLEM COMPLETED ***
      IF ( MAXTIM.GT.0 .AND. NCURVE.EQ.0 )
     1    CALL TYMLOG (TIME,MAXTIM)
      WRITE (6,5220)
C     ******************************************************
C                          F O R M A T S
C     ******************************************************
 5000 FORMAT ( 20A4 )
 5010 FORMAT (1H1,15(4H****),/,1X,20A4,/,1X,15(4H****),//)
 5020 FORMAT ( 16I5 )
 5030 FORMAT('0*****  PROBLEM PARAMETERS  *****',/,
     1 ' NUMBER OF NODAL POINTS IN SYSTEM =...........',I5,/,
     2 ' NUMBER OF ELEMENTS IN SYSTEM =...............',I5,/,
     3 ' NUMBER OF NODES PER ELEMENT =................',I5,/,
     4 ' NUMBER OF PARAMETERS PER NODE =..............',I5,/,
     5 ' DIMENSION OF SPACE =.........................',I5,/)
 5040 FORMAT (' NUMBER OF BOUNDARIES WITH GIVEN',
     1 ' FLUX=.......',I5,/,' NUMBER OF NODES ON',
     2 ' BOUNDARY SEGMENT=........',I5)
 5050 FORMAT (' INCONSISTANT VALUES OF LBN AND N.')
 5060 FORMAT (' NUMBER OF ITERATIONS TO BE RUN',
     1 '=.............',I5,/)
 5070 FORMAT ( 16I5 )
 5080 FORMAT ('0NUMBER OF FIXED PT ',
     A 'PROP PER NODE =..........',I5,/,
     1 ' NUMBER OF FLOATING PT PROP PER NODE =.......',I5,/,
     2 ' NUMBER OF FIXED PT PROP PER ELEMENT =.......',I5,/,
     3 ' NUMBER OF FLOATING PT PROP PER ELEMENT =....',I5,/,
     4 ' NUMBER OF FIXED PT MISC PROP =..............',I5,/,
     5 ' NUMBER OF FLOATING PT MISC PROP =...........',I5 /)
 5090 FORMAT ('0NODAL POINT PROPERTIES ARE HOMOGENEOUS.')
 5100 FORMAT (' ELEMENT PROPERTIES ARE HOMOGENEOUS.')
```

```
5110 FORMAT ('1EQUATION HALF BANDWIDTH =.........',I5,/,
    1 ' AND OCCURS IN ELEMENT NUMBER',I5,'.',/,
    2 ' CONSTRAINT HALF BANDWIDTH =..............',I5,/,
    3 ' MAXIMUM HALF BANDWIDTH OF SYSTEM =.......',I5,/,
    4 ' TOTAL NUMBER OF SYSTEM EQUATIONS =.......',I5,/,
    5 ' NUMBER OF ELEMENT DEGREES OF FREEDOM =...',I5,//)
5120 FORMAT ('0*** ARRAY STORAGE ***',/,
    1 '        TYPE            REQUIRED         AVAILABLE',/,
    2 ' FLOATING POINT',6X,I5,9X,I5,/,
    3 ' FIXED POINT',9X,I5,9X,I5,//)
5130 FORMAT ('0 STORAGE EXCEEDED, ABNORMAL PROGRAM END')
5140 FORMAT ('0ITERATION NUMBER = ',I5)
5150 FORMAT (' NUMBER OF CONTOUR CURVES BETWEEN',/,
    1 '    5TH & 95TH PERCENTILE OF EACH PARAMETER = ',I5)
5160 FORMAT (' INITIAL FORCING VECTOR TO BE INPUT' )
5180 FORMAT (' OPTIONAL UNIT NUMBERS (UTILIZED IF > 0)',/,
    1 ' NTAPE1 = ',I2,' NTAPE2 = ',I2,' NTAPE3 = ',I2,
    2 ' NTAPE4 = ',I2)
5190 FORMAT (' NODAL PARAMETERS TO BE LISTED BY NODES')
5200 FORMAT (' NODAL PARAMETERS TO BE LISTED BY ELEMENTS')
5210 FORMAT (' ALL ELEMENT COLUMN MATRICES ARE ZERO.')
5220 FORMAT (' NORMAL ENDING OF MODEL PROGRAM.')
     RETURN
     END
```

```
      SUBROUTINE  SET (NNPFIX,NNPFLO,NLPFIX,NLPFLO,MISCFX,
     1                 MISCFL,IPTEST,LPTEST,IP1,IP2,IP3,IP4,
     2                 IP5,JP1,JP2,JP3,KP1,KP2,KP3,NHOMO,LHOMO,
     3                 M,N,NE,RATIO,MAXTYP,NELFRE,NDFREE,
     4                 NFLUX,NG,LBN)
C     * * * * * * * * * * * * * * * * * * * * * * * * * * *
C               SET DIMENSIONS OF PROPERTIES ARRAYS
C     * * * * * * * * * * * * * * * * * * * * * * * * * * *
C     IF LPTEST.GT.0 ELEMENT PROPERTIES ARE DEFINED
C     IF IPTEST.GT.0 SOME PROPERTIES ARE DEFINED
C     IP1 = NUMBER OF ROWS IN ARRAY FLTNP
C     IP2 = NUMBER OF ROWS IN ARRAY NPFIX
C     IP3 = NUMBER OF ROWS IN ARRAY FLTEL
C     IP4 = NUMBER OF ROWS IN ARRAY LPFIX
C     IP5 = NUMBER OF ROWS IN ARRAY PRTLPT
C     JP1 = NUMBER OF COLUMNS IN ARRAYS NPFIX AND NPROP
C     JP2 = NUMBER OF COLUMNS IN ARRAYS LPFIX AND LPROP
C     JP3 = NUMBER OF COLUMNS IN ARRAY MISFIX
C     KP1 = NUMBER OF COLUMNS IN FLTNP & PRTLPT & PTPROP
C     KP2 = NUMBER OF COLUMNS IN ARRAYS FLTEL AND ELPROP
C     KP3 = NUMBER OF COLUMNS IN ARRAY FLTMIS
C     NLPFIX = NUMBER OF FIXED PT ELEMENT PROP
C     NLPFLO = NUMBER OF FLOATING PT ELEMENT PROP
C     NNPFIX = NUMBER OF FIXED PT NUMBER PROP
C     NNPFLO = NUMBER OF FLOATING PT NUMBER PROP
C     MISCFL = NUMBER OF MISC FLOATING PT SYSTEM PROP
C     MISCFX = NUMBER OF MISC FIXED  PT SYSTEM PROP
C     IF LHOMO=1 ELEMENT PROPERTIES ARE HOMOGENEOUS
C     IF NHOMO=1 NODAL   PROPERTIES ARE HOMOGENEOUS
C     MAXTYP = MAX ALLOWED CONSTRAINT TYPE
C     RATIO = CONSTANT FOR ITER CONTROL,SEE MODEL
      RATIO = 1.D0
      MAXTYP = 5
      NELFRE = N*NG
      NDFREE = M*NG
      NFLUX =  LBN*NG
      IF ( NFLUX.LT.1 ) NFLUX = 1
      IPTEST = NNPFIX + NNPFLO + NLPFIX + NLPFLO
     1        + MISCFX + MISCFL
      LPTEST = NLPFIX + NLPFLO
      IP1 = 1
      IF ( NNPFLO.GT.0 .AND. NHOMO.EQ.0 ) IP1 = M
      IP2 = 1
      IF ( NNPFIX.GT.0 .AND. NHOMO.EQ.0 ) IP2 = M
      IP3 = 1
      IF ( NLPFLO.GT.0 .AND. LHOMO.EQ.0 ) IP3 = NE
```

```
      IP4 = 1
      IF ( NLPFIX.GT.0 .AND. LHOMO.EQ.0 ) IP4 = NE
      IP5 = 1
      IF ( NNPFLO.GT.0 ) IP5 = N
      JP1 = MAX0 (NNPFIX,1)
      JP2 = MAX0 (NLPFIX,1)
      JP3 = MAX0 (MISCFX,1)
      KP1 = MAX0 (NNPFLO,1)
      KP2 = MAX0 (NLPFLO,1)
      KP3 = MAX0 (MISCFL,1)
      RETURN
      END


      SUBROUTINE  PT1 (M,N,NE,NG,NELFRE,NDFREE,N1,N2,N3,
     1                 N4,N5,N6,N8,M1,M1A,M1B,M2,NSPACE,MAXTYP)
C     * * * * * * * * * * * * * * * * * * * * * * * * * * * * *
C     INITIAL CALCULATIONS OF POINTERS FOR ARRAY STORAGE
C     * * * * * * * * * * * * * * * * * * * * * * * * * * * * *
C     M = NO. OF SYSTEM NODES
C     M1 TO MNEXT = POINTERS FOR FLOATING POINT ARRAYS
C     MAXTYP = MAX NO OF CONSTRAINT TYPES
C     N = NUMBER OF NODES PER ELEMENT
C     NDFREE = TOTAL NUMBER OF SYSTEM DEGREES OF FREEDOM
C     NE = NUMBER OF ELEMENTS IN SYSTEM
C     NELFRE = NUMBER OF DEGREES OF FREEDOM PER ELEMENT
C     NG = NUMBER OF NODAL PARAMETERS PER NODE POINT
C     NSPACE = DIMENSION OF SPACE
C     N1 TO NNEXT = POINTERS FOR FIXED POINT ARRAYS
      N1 = 1
C     IBC(M)
      N2 = N1 + M
C     NODES(NE,N)
      N3 = N2 + NE*N
C     LNODE(N)
      N4 = N3 + N
C     KODES(NG)
      N5 = N4 + NG
C     NREQ(MAXTYP)
      N6 = N5 + MAXTYP
C     INDEX(NELFRE)
      N8 = N6 + NELFRE
C
      M1 = 1
C     X(M,NSPACE)
      M1A = M1 + (M+2)*NSPACE
C     CP(MAXTYP)
      M1B = M1A + MAXTYP
C     SP(MAXTYP,MAXTYP)
      M2 = M1B + MAXTYP*MAXTYP
      RETURN
      END


      SUBROUTINE  PT2 (N,NG,NSPACE,NELFRE,MAXACT,NUMCE,
     1                 IP1,IP2,IP3,IP4,JP1,JP2,JP3,KP1,KP2,
     2                 KP3,N17,N18,N19,N20,N21,N22,N23,M11,
     3                 M12,M13,M14,M15,M16,M17,M18,M19,N8,M2)
C     * * * * * * * * * * * * * * * * * * * * * * * * * * * * *
C     CONTINUE CALCULATIONS OF POINTERS FOR ARRAY STORAGE
C     * * * * * * * * * * * * * * * * * * * * * * * * * * * * *
C     IP1 = NUMBER OF ROWS IN ARRAY FLTNP
C     IP2 = NUMBER OF ROWS IN ARRAY NPFIX
C     IP3 = NUMBER OF ROWS IN ARRAY FLTFL
C     IP4 = NUMBER OF ROWS IN ARRAY LPFIX
C     JP1 = NUMBER OF COLUMNS IN ARRAYS NPFIX AND NPROP
C     JP2 = NUMBER OF COLUMNS IN ARRAYS LPFIX AND LPROP
C     JP3 = NUMBER OF COLUMNS IN ARRAY MISFIX
C     KP1 = NO OF COLS IN ARRAYS FLTNP & PRTLPT & PTPROP
C     KP2 = NUMBER OF COLUMNS IN ARRAYS FLTEL AND ELPROP
```

```
C       KP3 = NUMBER OF COLUMNS IN ARRAY FLTMIS
C       N = NUMBER OF NODES PER ELEMENT
C       NELFRE = NUMBER OF DEGREES OF FREEDCM PER ELEMENT
C       NG = NUMBER OF NODAL PARAMETERS PER NODE POINT
C       NSPACE = DIMENSION OF SPACE
C       N1 TO NNEXT = POINTERS FOR FLIXED POINT ARRAYS
C       M1 TO MNEXT = POINTERS FOR FLOATING POINT ARRAYS
C       MAXACT = NO ACTIVE CONSTRAINT TYPES
C       NUMCE = NUMBER OF CONSTRAINT EQUATICNS
C       N8-NDXC(MAXACT,NUMCE)
        N17 = N8 + MAXACT*NUMCE + 1
C       NPFIX(IP2,JP1)
        N18 = N17 + IP2*JP1
C       LPFIX(IP4,JP2)
        N19 = N18 + IP4*JP2
C       MISCFX(JP3)
        N20 = N19 + JP3
C       LPROP(JP2)
        N21 = N20 + JP2
C       NPROP(JP1)
        N22 = N21 + JP1
C       NRANGF(NG,2)
        N23 = N22 + NG*2
C
C       M2-CEQ(MAXACT,NUMCE)
        M11 = M2 + MAXACT*NUMCE + 1
C       FLTNP(IP1,KP1)
        M12 = M11 + IP1*KP1
C       FLTEL(IP3,KP2)
        M13 = M12 + IP3*KP2
C       FLTMIS(KP3)
        M14 = M13 + KP3
C       ELPROP(KP2)
        M15 = M14 + KP2
C       PTPROP(KP1)
        M16 = M15 + KP1
C       COORD(N,NSPACE)
        M17 = M16 + (N+2)*NSPACE
C       S(NELFRE,NELFRE)
        M18 = M17 + NELFRE*NELFRE
C       C(NELFRE)
        M19 = M18 + NELFRE
        RETURN
        END


        SUBROUTINE  PT3 (NBANCO,NDFREE,NELFRE,NITER,NG,
       1                 NSPACE,IP5,KP1,M20,M21,M22,M23,
       2                 M24,M25,M26,M27,M28,LBN,M19)
C       * * * * * * * * * * * * * * * * * * * * * * * * *
C       COMPLETE CALCULATION OF POINTERS FOR ARRAY STOPAGE
C       * * * * * * * * * * * * * * * * * * * * * * * * *
C       NBANCO = NDFREE*MAXBAN
C       NDFREE = TOTAL NUMBER OF SYSTEM DEGREES OF FREEDOM
C       NELFRE = NUMBER OF DEGREES OF FREEDCM PER ELEMENT
C       NITER = NO. OF ITERATIONS TO BE RUN
C       NG = NUMBER OF NODAL PARAMETERS PER NODE POINT
C       IP5 = NUMBER OF ROWS IN ARRAY PRTLPT
C       KP1 = NO OF COLS IN ARRAYS FLTNP & PRTLPT & PTPROP
C       LBN = NO OF NODES ON AN ELEMENT BOUNDARY SEGMENT
C       M1 TO MNEXT = POINTERS FOR FLOATING POINT ARRAYS
C       N1 TO NNEXT = POINTERS FOR FIXED POINT ARRAYS
C
C       SS(NDFREE,MAXBAN)
        M20 = M19 + NBANCO
C       CC(NDFREE)
        M21 = M20 + NDFREE
C       DD(NDFREE)
        M22 = M21 + NDFREE + 2
C       D(NELFRE)
        M23 = M22 + NELFRE + 2
```

```
C      DDOLD(NDFREE)
       M24 = M23 + 1
       IF ( NITER.GT.1 )  M24 = M23 + NDFREE + 2
C      RANGE(NG,2)
       M25 = M24 + (NG+2)*2
C      XPT(NSPACE)
       M26 = M25 + NSPACE
C      PRTLPT(IP5,KP1)
       M27 = M26 + IP5*KP1
C      FLUX(LBN,NSPACE)
       M28 = M27 + 1
       IF ( LBN.GT.0 )  M28 = M27 + LBN*NSPACE
       RETURN
       END

       SUBROUTINE  INPUT (M,N,NE,NG,NSPACE,X,IBC,NODES)
C      * * * * * * * * * * * * * * * * * * * * * * * * *
C                READ BASIC PROBLEM DATA
C      * * * * * * * * * * * * * * * * * * * * * * * * *
CDP    IMPLICIT REAL*8 (A-H,O-Z)
       DIMENSION  X(M,NSPACE), IBC(M), NODES(NE,N)
C      M = NUMBER OF NODES IN SYSTEM
C      N = NUMBER OF NODES PER ELEMENT
C      NE = NUMBER OF ELEMENTS IN SYSTEM
C      NG = NUMBER OF PARAMETERS (DOF) PER NODE
C      NSPACE = DIMENSION OF SOLUTION SPACE
C      X = SYSTEM COORDINATES OF ALL NODES
C      IBC = PACKED NODAL CONSTRAINT INDICATOR
C      NODES = SYSTEM ARRAY OF ELEMENT INCIDENCES
C-->   READ NODAL POINT DATA
       WRITE (6,5000)  NSPACE
 5000  FORMAT ('1*** NODAL POINT DATA ***',/,
     1 ' NODE, CONSTRAINT INDICATOR,',I2,' COORDINATES',/)
       DO 10  I = 1,M
       READ (5,5010)  J,IBC(J), (X(J,K),K=1,NSPACE)
 5010  FORMAT (2I10,(6F10.4))
       WRITE(6,5010) J,IBC(J),(X(J,K),K=1,NSPACE)
       IF ( J.GT.M )  WRITE (6,5020)
 5020  FORMAT (' INVALID NODE NUMBER IN SUBROUTINE INPUT.')
 10    CONTINUE
C-->   READ ELEMENT DATA
       WRITE (6,5030)   N
 5030  FORMAT ('*** ELEMENT CONNECTIVITY DATA ***',/,
     1' ELEMENT NO., ', I2 ,' NODAL INCIDENCES.',/)
       DO 20 I = 1,NE
       READ (5,5050)  J,  (NODES(J,K),K=1,N)
       WRITE(6,5050)  J,  (NODES(J,K),K=1,N)
       IF ( J.GT.NE )  WRITE (6,5040)
 5040  FORMAT (' INVALID ELEMENT NUMBER IN SUBR INPUT.')
 20    CONTINUE
 5050  FORMAT ( I5, (15I5) )
       RETURN
       END

       SUBROUTINE PTCODE (JPT,NG,KODE,KODES)
C      * * * * * * * * * * * * * * * * * * * * * * * * *
C      EXTRACT B.C. INDICATORS AT NODE NUMBER JPT
C      * * * * * * * * * * * * * * * * * * * * * * * * *
       DIMENSION  KODES(NG)
C      JPT=NODE NO.,  NG=NO. PARAMETERS PER NODE
C      KODE=(NG) DIGIT INTEGER CONTAINING BC INDICATORS
C      KODES=VECTOR CONTAINING NG INTEGER CODES (0 OR I)
C         0  IMPLIES NO B. C.
C         I  IMPLIES A B. C.  OF TYPE I
       NGPLUS = NG + 1
       IOLD = KODE
       ISUM = 0
       DO 10  I = 1,NG
       II = NGPLUS - I
       INEW = IOLD/10
       IK = IOLD - INEW*10
```

```
      ISUM = ISUM + IK*10**(I-1)
      IOLD = INEW
 10   KODES(II) = IK
C     WAS DATA RIGHT JUSTIFIED?
      IF ( KODE.GT.ISUM )  WRITE (6,5000) JPT
 5000 FORMAT ('OBC NOT RIGHT JUSTIFIED AT NODE',I5)
      RETURN
      END


      SUBROUTINE  CCOUNT (M,NG,NRES,IBC,KODES,
     1                    MAXACT,NUMCE,MAXTYP)
C     * * * * * * * * * * * * * * * * * * * * * * * * * * * * *
C     CALCULATE NO OF CONSTRAINT FLAGS OF EACH TYPE
C     * * * * * * * * * * * * * * * * * * * * * * * * * * * * *
      DIMENSION  IBC(M), NRES(MAXTYP), KODES(NG)
C     M = TOTAL NUMBER OF SYSTEM NODES
C     NG = NO. OF PARAMETERS (DOF) PER NODE
C     IBC = NODAL POINT BOUNDARY RESTRAINT INDICATOR
C     KODES = LIST OF RESTRAINT INDICATORS AT A NODE
C     NRES = LIST OF NUMBER OF FLAGS OF EACH TYPE
C          = NO OF CONSTR EQS ON EXIT
C     MAXTYP = MAX NO OF DIFFERENT CONSTR TYPES
C     MAXACT = ACTIVE NO OF TYPES
C     NUMCE = TOTAL NO OF CONSTR EQUATIONS
C     INITIALIZATION
      DO 10  I = 1,MAXTYP
 10   NRES(I) = 0
      DO 30  I = 1,M
C     DOES NODE I HAVE A NODAL PARAMETER CONSTRAINT
      ITEST = IABS( IBC(I) )
      IF ( ITEST.EQ.0 )  GO TO 30
C     EXTRACT PARAMETER CODES
      CALL  PTCODE (I,NG,ITEST,KODES)
      DO 20  J = 1,NG
      K = KODES(J)
      IF ( K.EQ.0 )  GO TO 20
C     UPDATE CONSTRAINT COUNTERS
      NRES(K) = NRES(K) + 1
 20   CONTINUE
 30   CONTINUE
C     CONVERT TO EQUATION COUNTERS
      NUMCE = 0
      MAXACT = 1
      WRITE (6,5000)
 5000 FORMAT ('1***  NODAL PARAMETER CONSTRAINT LIST',
     1 '  ***',/,' CONSTRAINT        NUMBER OF',/,
     2 '     TYPE           EQUATIONS',/ )
      DO 40  I = 1,MAXTYP
      K = NRES(I)
      IF ( K.GT.0 )  MAXACT = I
      IF ( ((K/I)*I).LT.K )  WRITE (6,5010)  I
 5010 FORMAT (' INVALID DATA FOR TYPE',I3)
      NRES(I) = NRES(I)/I
      WRITE (6,5020)  I, NRES(I)
 5020 FORMAT ( I8, I16 )
 40   NUMCE = NUMCE + NRES(I)
      RETURN
      END


      SUBROUTINE INCEQ (NG,MAXACT,NUMCE,NREQ,CEQ,
     1                  NDXC,M)
C     * * * * * * * * * * * * * * * * * * * * * * * * * * * * *
C        READ NODAL PARAMETER CONSTRAINT EQUATION DATA
C     * * * * * * * * * * * * * * * * * * * * * * * * * * * * *
CDP   IMPLICIT REAL*8 (A-H,O-Z)
      DIMENSION CEQ(MAXACT,NUMCE), NDXC(MAXACT,NUMCE),
     1          NREQ(MAXACT)
C     NG = NO. PARAMETERS PER NODE
```

```
C      NREQ(I) = NUMBER OF CONSTR OF TYPE I
C      CEQ(I,J) = CONSTR COEFF I OF EQ J
C      NDXC(I,J) = CONSTR DOF NO I OF EQ J
       WRITE (6,5000)
 5000  FORMAT ('1*** CONSTRAINT EQUATION DATA ***',/)
       IEQ = 0
       DO 130  IN = 1,MAXACT
       NTEST = NREQ(IN)
       IF ( NTEST.EQ.0 )  GO TO 130
       GO TO (10,30,50,70), IN
C-->       TYPE 1       D(L1) = A1
 10    WRITE (6,5010)
 5010  FORMAT ('0CONSTRAINT TYPE ONE',/,
      1' EQ. NO.    NODE1    PAR1                A1')
       DO 20  NEQ = 1,NTEST
       IEQ = IEQ + 1
       READ (5,5020)  NODE1, IPAR1, A1
 5020  FORMAT ( 2I5, F10.0 )
       WRITE (6,5030)  IEQ, NODE1, IPAR1, A1
 5030  FORMAT ( 3I8, 2X, E14.8 )
       IF ( NODE1.GT.M .OR. IPAR1.GT.NG )  WRITE (6,5040)
 5040  FORMAT (' DATA ERROR IN SUBROUTINE INCEQ.')
       NDXC(1,IEQ) = NG*(NODE1 - 1) + IPAR1
 20    CEQ(1,IEQ) = A1
       GO TO 130
C-->       TYPE 2    A1*D(L1)+A2*D(L2)=A3
 30    WRITE (6,5050)
 5050  FORMAT ('0CONSTRAINT TYPE TWO',/,
      1' EQ. NO.    NODE1    PAR1    NODE2    PAR2',
      2'            A1              A2              A3')
       DO 40  NEQ = 1,NTEST
       IEQ = IEQ + 1
 5060  FORMAT ( 4I5, 3F10.0 )
       READ (5,5060)  NODE1, IPAR1, NODE2, IPAR2, A1, A2, A3
       WRITE (6,5070)   IEQ,NODE1,IPAR1,NODE2,IPAR2,A1,A2,A3
 5070  FORMAT (5I7,3(2X,E14.8))
       IF ( NODE1.GT.M .OR. NODE2.GT.M .OR. IPAR1.GT.NG .OR.
      1     IPAR2.GT.NG )  WRITE (6,5040)
       NDXC(1,IEQ) = NG*(NODE1 - 1) + IPAR1
       NDXC(2,IEQ) = NG*(NODE2 - 1) + IPAR2
       CEQ(1,IEQ) = A2/A1
 40    CEQ(2,IEQ) = A3/A1
       GO TO 130
C          TYPE 3    A1*D(L1)+A2*D(L2)+A3*D(L3)=A4
 50    WRITE (6,5080)
 5080  FORMAT ('0CONSTRAINT TYPE THREE',/,
      1' EQ. NO.    NODE1    PAR1    NODE2    PAR2    NODE3',
      2'    PAR3            A1              A2              A3',
      3'            A4')
       DO 60  NEQ = 1,NTEST
       IEQ = IEQ + 1
       READ (5,5090)  NODE1,IPAR1,NODE2,IPAR2,NODE3,IPAR3,
      1               A1,A2,A3,A4
 5090  FORMAT ( 6I5, 4F10.0 )
       WRITE (6,5100)  IEQ,NODE1,IPAR1,NODE2,IPAR2,NODE3,
      1                IPAR3,A1,A2,A3,A4
 5100  FORMAT ( 7I7, 4(2X,E14.8))
       IF ( NODE1.GT.M .OR. NODE2.GT.M .OR. NODE3.GT.M .OR.
      1     IPAR1.GT.NG .OR. IPAR2.GT.NG .OR. IPAR3.GT.NG )
      2     WRITE (6,5040)
       NDXC(1,IEQ) = NG*(NODE1 - 1) + IPAR1
       NDXC(2,IEQ) = NG*(NODE2 - 1) + IPAR2
       NDXC(3,IEQ) = NG*(NODE3 - 1) + IPAR3
       CEQ(1,IEQ) = A2/A1
       CEQ(2,IEQ) = A3/A1
 60    CEQ(3,IEQ) = A4/A1
       GO TO 130
C      OTHER TYPES NOT TREATED
 70    CONTINUE
 130   CONTINUE
       RETURN
       END
```

```
      SUBROUTINE  INPROP (M,NE,NNPFIX,NNPFLO,NLPFIX,NLPFLO,
     1                MISCFX,MISCFL,IP1,IP2,IP3,IP4,JP1,JP2,JP3,
     2                KP1,KP2,KP3,FLTNP,FLTEL,FLTMIS,NPFIX,
     3                LPFIX,MISFIX)
C     * * * * * * * * * * * * * * * * * * * * * * * * * * * * *
C       INPUT NODAL POINT, ELEMENT, AND MISCELLANEOUS
C            SYSTEM PROPERTIES
C     * * * * * * * * * * * * * * * * * * * * * * * * * * * * *
CDP   IMPLICIT REAL*8 (A-H,O-Z)
      DIMENSION  FLTNP(IP1,KP1),FLTEL(IP3,KP2),FLTMIS (KP3),
     1           NPFIX(IP2,JP1),LPFIX(IP4,JP2),MISFIX(JP3)
C     M = NO. SYSTEM NODES, NE = NO. ELEMENTS IN SYSTEM
C     FLTNP, FLTEL, FLTMIS = FLOATING POINT PROP. OF SYSTEM
C        NODES, ELEMENTS, AND MISC.
C     NPFIX, LPFIX, MISFIX = FIXED POINT PROP. OF SYSTEM
C        NODES, ELEMENTS, AND MISC.
C     NNPFLO = NO.  FLOATING POINT NODAL PROPERTIES
C     NNPFIX = NO.  FIXED POINT NODAL PROPERTIES
C     NLPFIX = NO.    FIXED  POINT ELEMENT PROPERTIES
C     MISCFL = NO. MISC. FLOATING POINT SYSTEM PROPERTIES
C     MISCFX = NO. MISC.   FIXED  POINT SYSTEM PROPERTIES
C     IP1 = NUMBER OF ROWS IN ARRAY FLTNP
C     IP2 = NUMBER OF ROWS IN ARRAY NPFIX
C     IP3 = NUMBER OF ROWS IN ARRAY FLTEL
C     IP4 = NUMBER OF ROWS IN ARRAY LPFIX
C     IP5 = NUMBER OF ROWS IN ARRAY PRTLPT
C     JP1 = NUMBER OF COLUMNS IN ARRAYS NPFIX AND NPROP
C     JP2 = NUMBER OF COLUMNS IN ARRAYS LPFIX AND LPROP
C     JP3 = NUMBER OF COLUMNS IN ARRAY MISFIX
C     KP1 = NUMBER OF COLUMNS IN FLTNP & PRTLPT & PTPROP
C     KP2 = NUMBER OF COLUMNS IN ARRAYS FLTEL AND ELPROP
C     KP3 = NUMBER OF COLUMNS IN ARRAY FLTMIS
      IF ( NNPFIX.LT.1 .AND. NNPFLO.LT.1 )  GO TO 60
C-->   READ NODAL POINT PROPERTIES
      WRITE (6,5000)
 5000 FORMAT ('1***   NODAL  POINT  PROPERTIES   ***',/,
     1 ' NODE NO.  PROPERTY NO.    VALUE',/)
      IF ( NNPFIX.LT.1 )  GO TO 30
      DO 10  I = 1,IP2
      READ (5,5010) J, (NPFIX(J,K),K=1,NNPFIX)
 5010 FORMAT ( I5, (12I10) )
      IF ( J.EQ.M )  GO TO 20
 10   CONTINUE
 20   WRITE (6,5030)  ((J,K,NPFIX(J,K),J=1,IP2),K=1,NNPFIX)
 5030 FORMAT (I8, I13, I10)
      WRITE (6,5040)
 5040 FORMAT (' END OF FIXED POINT PROPERTIES OF NODES')
 30   IF ( NNPFLO.LT.1 )  GO TO 60
      DO 40  I = 1,IP1
      READ (5,5050) J, (FLTNP(J,K),K=1,NNPFLO)
 5050 FORMAT ( I5, (12F10.4) )
      IF ( J.EQ.M )  GO TO 50
 40   CONTINUE
 50   WRITE (6,5060)  ((J,K,FLTNP(J,K),J=1,IP1),K=1,NNPFLO)
 5060 FORMAT (I8,I13,E20.8)
      WRITE (6,5070)
 5070 FORMAT (' END OF FLOATING POINT PROPERTIES OF NODES')
 60   IF ( NLPFIX.LT.1 .AND. NLPFLO.LT.1 )  GO TO 120
C-->   READ ELEMENT PROPERTIES
      WRITE (6,5080)
 5080 FORMAT ('1 ***  ELEMENT  PROPERTIES   ***',/,
     1 ' ELEMENT NO.   PROPERTY NO.      VALUE',/ )
      IF ( NLPFIX.LT.1 )  GO TO 90
      DO 70  I = 1,IP4
      READ  (5,5010)  J, (LPFIX(J,K),K=1,NLPFIX)
      IF ( J.EQ.NE )  GO TO 80
 70   CONTINUE
 80   WRITE (6,5030)  (  (J,K,LPFIX(J,K),J=1,IP4),K=1,NLPFIX )
      WRITE (6,5100)
 5100 FORMAT (' END OF FIXED POINT PROPERTIES OF ELEMENTS')
 90   IF ( NLPFLO.LT.1 )  GO TO 120
      DO 100  I = 1,IP3
      READ (5,5050)   J, (FLTEL(J,K),K=1,NLPFLO)
```

```
      IF ( J.EQ.NE )  GO TO 110
 100  CONTINUE
 110  WRITE (6,5060)  ( (J,K,FLTEL(J,K),J=1,IP3),K=1,NLPFLO )
      WRITE (6,5110)
 5110 FORMAT (' END OF FLOATING PT PROPERTIES OF ELEMENTS')
 120  IF ( MISCFX.LT.1 .AND. MISCFL.LT.1 )  GO TO 140
C-->  READ MISC. SYSTEM PROPERTIES
      WRITE (6,5120)
 5120 FORMAT ('1***  MISCELLANEOUS SYSTEM PROPERTIES',
     1 '  ***',/, ' PROPERTY NO.    VALUE',/)
      IF ( MISCFX.LT.1 )  GO TO 130
      READ (5,5130)   (MISFIX(K),K=1,MISCFX)
 5130 FORMAT ( (8I10) )
      WRITE (6,5140)  ((K,MISFIX(K)),K=1,MISCFX)
 5140 FORMAT ( I12,I10 )
 5150 FORMAT (' END OF FIXED POINT PROPERTIES OF SYSTEM')
      WRITE (6,5150)
 130  IF ( MISCFL.LT.1 )  GO TO 140
      READ (5,5160)   (FLTMIS(K),K=1,MISCFL)
 5160 FORMAT ( (8F10.4) )
      WRITE (6,5170)  ((K,FLTMIS(K)),K=1,MISCFL)
 5170 FORMAT ( I12, E16.8 )
      WRITE (6,5180)
 5180 FORMAT (' END OF FLOATING PT PROPERTIES OF SYSTEM')
 140  CONTINUE
      RETURN
      END


      SUBROUTINE  INVECT (NDFREE,NG,CC,M)
C     * * * * * * * * * * * * * * * * * * * * * * * * *
C     INPUT SPECIFIED VALUES IN FORCING VECTOR, CC
C     * * * * * * * * * * * * * * * * * * * * * * * * *
CDP   IMPLICIT REAL*8 (A-H,O-Z)
      DIMENSION  CC(NDFREE)
C     NDFREE = TOTAL NUMBER OF SYSTEM DEGREES OF FREEDOM
C     NG = NUMBER OF PARAMETERS PER NODE
C     CC = SYSTEM EQUATIONS COLUMN MATRIX
      WRITE (6,5000)
 5000 FORMAT ('1  *** INITIAL FORCING VECTOR DATA ***',/,
     1'   NODE   PARAMETER        VALUE                DOF',//)
      DO  10  I = 1,NDFREE
      READ (5,5010)  NODE, IPARM, VALUE
 5010 FORMAT  ( 2I5,F10.4 )
C     FIND CORRESPONDING DEGREE OF FREEDOM NUMBER
      CALL  DEGPAR (NODE,IPARM,NG,J)
      CC(J) = VALUE
C     LIST INPUT DATA
      WRITE (6,5020) NODE, IPARM, VALUE, J
 5020 FORMAT ( 2I8,E18.8,I8)
      IF ( NODE.GT.M .OR. IPARM.GT.NG )  WRITE (6,5030)
 5030 FORMAT (' DATA ERROR IN SUBROUTINE INVECT.')
      IF ( J.EQ.NDFREE )  GO TO 20
 10   CONTINUE
 20   CONTINUE
      RETURN
      END


      SUBROUTINE  DEGPAR (IPT,JPARM,NG,INDEX)
C     * * * * * * * * * * * * * * * * * * * * * * * * *
C      DETERMINE THE DEGREE OF FREEDOM NUMBER
C      OF NODAL PARAMETER JPARM AT NODE POINT IPT
C     * * * * * * * * * * * * * * * * * * * * * * * * *
C     NG = NUMBER OF PARAMETERS PER NODE
      INDEX = NG*(IPT-1) + JPARM
      RETURN
      END
```

```
      SUBROUTINE  INFLUX (NSEG,LBN,LNODE,FLUX,NG,COORD,
     1                    NSPACE,X,M,INDEX,C,CC,NDFREE,
     2                    S,SS,MAXBAN,NFLUX)
C     * * * * * * * * * * * * * * * * * * * * * * * * * * *
C     READ FLUX BOUNDARY COND. AND APPLY TO SYSTEM EQS
C     * * * * * * * * * * * * * * * * * * * * * * * * * * *
CDP   IMPLICIT REAL*8 (A-H,O-Z)
      DIMENSION X(M,NSPACE), COORD(LBN,NSPACE),
     1          FLUX(LBN,NG), CC(NDFREE), C(NFLUX),
     2          S(NFLUX,NFLUX), SS(NDFREE,MAXBAN),
     3          LNODE(LBN), INDEX(NFLUX)
C     C = BOUNDARY SEGMENT COLUMN MATRIX
C     CC = COLUMN MATRIX OF SYSTEM EQUATIONS
C     COORD = SPATIAL COORDINATES OF SEGMENT NODES
C     INDEX = SYSTEM DEGREE OF FREEDOM NUMBERS ARRAY
C     M = NO. OF SYSTEM NODES, NG = NO. OF DOF PER NODE
C     NDFREE = TOTAL NUMBER OF SYSTEM DEGREES OF FREEDOM
C     NSPACE = DIMENSION OF SPACE
C     X = COORDINATES OF SYSTEM NODES
C     NSEG = NO. OF ELEMENT BOUNDARY SEGMENTS IN SYSTEM
C     LBN = NO. OF NODES ON AN ELEMENT BOUNDARY SEGMENT
C     LNODE = INCIDENCES OF SEGEMENT ISEG
C     FLUX = SPECIFIED COMPONENTS OF FLUX AT NODES
C     NFLUX = LBN*NG = NO OF SEGMENT DOF
C     S = BOUNDARY SEGMENT SQ MATRIX
C     SS = SYSTEM SQUARE MATRIX UPPER BAND
C     MAXBAN = SYSTEM BANDWIDTH
C     IOPT = PROBLEM MATRIX REQUIREMENT FLAG (RETURNED)
C          = 1, BFLUX CALCULATES C ONLY
C          = 2, BFLUX CALCULATES S ONLY
C          = 3, BFLUX GIVES BOTH C AND S
      IOPT = 0
      WRITE (6,5000) LBN, NG
 5000 FORMAT ('1 *** ELEMENT BOUNDARY FLUXES ***',/,
     2' SEGMENT          ',I3,' NODES,',/,' ',/,
     3'                  ',I3,' FLUX COMPONENTS PER NODE',/)
      DO 30  ISEG = 1,NSEG
C-->  READ BOUNDARY NODES
      READ (5,5010)  (LNODE(L),L=1,LBN)
 5010 FORMAT ( (16I5) )
C-->  READ BOUNDARY FLUX
      READ (5,5020)  ((FLUX(K,IS),IS=1,NG),K=1,LBN)
 5020 FORMAT ( (8F10.4) )
      WRITE (6,5030) ISEG,(LNODE(L),L=1,LBN)
 5030 FORMAT ((12I8) )
      DO 5  L = 1,LBN
      IF ( LNODE(L).LT.1 .OR. LNODE(L).GT.M )  GOTO 40
    5 WRITE (6,5040)   (FLUX(L,IS),IS=1,NG)
 5040 FORMAT ( (6F15.4) )
C     EXTRACT COORDINATES
      CALL  ELCORD (M,LBN,NSPACE,X,COORD,LNODE)
C-->  CALCULATE BOUNDARY FLUX MATRICES (PROB DEPENDENT)
      CALL  BFLUX (FLUX,COORD,LBN,NSPACE,NFLUX,
     1             NG,C,S,IOPT)
      IF ( IOPT.EQ.0 )  GO TO 30
C     INSERT BOUNDARY FLUX MATRICES INTO SYSTEM EQ
      CALL  INDXEL (LBN,NFLUX,NG,LNODE,INDEX)
      IF ( IOPT.EQ.1 .OR. IOPT.EQ.3 )
     1     CALL STORCL (NDFREE,NFLUX,INDEX,C,CC)
      IF ( IOPT.EQ.2 .OR. IOPT.EQ.3 )
     1     CALL STORSQ (NDFREE,MAXBAN,NFLUX,INDEX,S,SS)
   30 CONTINUE
C     CLEAR ARRAY C AND S FOR LATER USE
      CALL ZEROA (NFLUX,C)
      ISEG = NFLUX*NFLUX
      CALL ZEROA (ISEG,S)
      RETURN
   40 WRITE (6,50)
   50 FORMAT (' INVALID NODE NUMBER IN INFLUX')
      STOP
      END
```

Subject and Author Index

Abel, J. F., 232, 245, 267
Abscissas, 94, 98, 102
Akin, J. E., 84, 124, 125, 148, 212, 217, 276, 277, 281
Allan, T., 205
Area coordinates, 102, 109
Arya, S. K., 212
Assembly of equations, 7, 8, 15, 22, 27, 129–149
Axisymmetric integrals, 81, 281, 326
Aziz, A. K., 2

Bandwidth, 8, 9, 39, 49, 163
Barlow, J. 106
Bathe, K. H., 46, 95, 117, 170, 232, 317, 331
Becker, E. B., 126
Berztiss, A. T., 21
Bettess, P., 126
Bibliography, 345
Blackburn, W. S., 124
Briggs, J. M., 333, 337
Bookkeeping, 5, 15, 131, 135, 136
Boolean assembly, 15, 144–148
Body forces, 232
Boundary conditions, 8, 153–164, 184
Boundary codes, 6, 11, 32, 35, 142, 191
Boundary flux, 54, 55, 246, 250, 267, 273, 287
Boundary matrices, 54, 267, 272, 287

Campbell, J., 106, 161
Carey, G. F., 267, 269
Carpenter, C. J., 126
Choleski factorization, 166
Christie, I., 127
Chung, T. J., 267, 281, 317
Cohen, M., 105
Column height, 46, 48, 49
Condensation (Static), 59–61
Conduction, 183, 245–255
Connectivity, 6, 38
Constitutive matrix, 232
Constraint equations, 30, 33, 153–164
Contours, 84–89, 177, 179, 247
Convection, 183
Convective coefficient, 184
Convergence criteria, 4
Cook, R. D., 105, 161, 232
Coordinate transformation, 225
Crank-nicolson, 323
Cubic element, 119, 218, 305
Curiskis, J. I., 164
Current density, 280

Daly, P., 276
Data generation, 10, 295–315
Data requirements, 32, 33, 34
Data storage, 21
De Boor, C. E., 2
Deflection, 201
Degree of freedom number, 5, 43, 129
Derivative
  local, 67
  global, 70, 73
  nodal parameters 218, 226
Desai, C. S., 232, 245, 267
De Vries, G., 1
Diagonalized matrices, 323–326, 330
Differential equation, 212–224
Direct assembly, 129–144
Direction cosines, 194, 225
Displacements, 5, 204
Dynamic applications, 331–344
Dynamic storage, 21

Eigenvalues, 105
Element degree, 18
Element incidences, 6, 38, 142

Element matrices, 28, 29, 51, 58, 185, 194, 206, 212
Element properties, 33, 41
Element thickness, 232
Essential condition, 184
Euler integration, 320, 323
Exact integrals, 78, 83, 84, 92, 93, 225

Factorial function, 78, 83
Fenves, S. J., 333
Film thickness, 205, 211
Flow chart, 25–31
Flow, viscous, 255–258, 284
Flux components, 54, 255, 275
Forces, 204, 206, 229, 245
Frame, 225–229
Frontal method, 148–151
Functional, 2, 14

Galerkin criterion, 3
Gallagher, R. H., 232
Gamma function, 78
Gartling, D., 126
Gauss factorization, 166, 167
Gauss-Choleski factorization, 166, 167
Gaussian quadrature, 94–100, 246, 275
Global derivatives, 70, 73
Gray, W. H., 84
Griffiths, D. F., 346
Gupta, A. K., 117

Half-bandwidth, 8, 39, 140, 155, 165
Haroun, M., 105
Hayashi, H., 205
Heat conduction, 184, 245, 328
Heat convection, 34, 184
Heat generation, 245, 286
Heat loss, 186, 190
Henshell, R. D., 22, 123
Hexahedra, 111, 248, 285
Hinton, E., 1, 105, 106
Huebner, K. H., 205
Hughes, T. J. R., 105, 127
Hyperbolic equations, 331–344

Inclined roller, 154, 202
Initial strain, 194
Internal nodes, 60
Internal quantities, 59
Interpolate, 16
Interpolation functions, 4, 7, 15, 71, 107–127, 301, 302
Inviscid flow, 281
Irons, B. M., 262
Isoparametric elements, 65–90, 249, 261, 282, 295

Jacobian, 69, 72, 79, 90, 101, 123

Lagrange interpolation, 107, 112
Laplace equation, 259, 264, 299
Least squares, 4, 212, 217
Lewis, R. W., 320
Linear acceleration, 333
Linear elements
  hexahedra, 111, 288
  line, 15, 184, 206, 212
  quadrilateral, 66, 245, 277
  triangle, 11, 232, 256
Linear velocity, 321
Liu, W. K., 317
Load capacity, 206, 211
Load vector, 7, 53
Local coordinates, 65, 82, 110, 111, 113
Local derivatives, 67, 110, 111, 113
Lowther, D. A., 126
Lubrication, 205–211
Lumped matrices, 324
Lynn, P. P., 212

Magnetic flux density, 280, 281
Malkus, D. S., 105
Mass matrix, 325, 337
Martin, H. C., 267, 269
Material number, 231, 240
Meek, J. L., 193, 195, 231
Mesh generation, 10, 295–316
M.H.D. Plasma, 280–283
Minimal integration, 105
Mitchell, A. R., 127
Moan, T., 106
Modulus of elasticity, 204, 225
Moment, 228
Myers, G. E., 56, 251, 286, 318, 323, 326

Natural boundary conditions, 3, 14, 184

Newmark, N. M., 332
Nodal
  boundary condition code, 6, 11, 32, 142, 191
  constraints, 6, 16, 33, 153–164
  coordinates, 6, 32, 191, 260
  displacements, 193, 232
  forces, 204, 229, 245
  moment, 226, 229
  pressure, 205, 255
  properties, 33
  rotations, 226
  temperature, 184, 245, 285
  thickness, 205, 211
  velocity, 205, 211, 255
Norrie, D. H., 1
Numerical integration, 63, 75, 95–106, 249, 288

Oden, J. T., 2
Optimal points, 16
Output, 175
Owen, D. R. J., 22, 295

PAFEC, 22, 295
Parabolic equations, 318
Pardue, R. M., 149
Park, K. C., 317
Pascal triangle, 18, 112
Patch test, 261–266
Penalty, 160–162
Plane frame, 225–229
Plane stress, 232–245
Plane truss, 193–204
Plasma, 280–283
Poisson equation, 231, 245, 285
Poisson's ratio, 232
Post-solution calculations, 9, 31, 180, 181, 256, 262
Potential flow, 259, 275
Pressure, 5, 205, 255

Quadratic elements
  line, 116
  triangle, 113, 305
  quadrilateral, 107, 259, 305
Quadratic functional 14, 144
Quadrilaterial element, 66, 107, 110, 117, 128, 245, 259, 282, 305

Radau quadratures, 101–104
Rank, 105, 106
Razzague, A., 261
Reactions, 156
Reddi, M. M., 205
Reddy, J. N., 2
Reduced integration, 105

Schechter, R. S., 255
Selective integration, 105, 106
Serendipity elements, 107, 112
Shaw, K. G., 123
Shear modulus, 232
Silvester, P. P., 126
Simplex elements, 77–84
Singularity element, 122, 276–279
Skyline, 46–49
Slider bearing, 205–211
Solution techniques, 9, 165–175
Stiffness matrix, 7, 51, 232
Storage, 21
Strain-displacement, 232
Stream, 259, 280
Stress, 194
Stress–strain law, 232
Structural analysis, 193, 225, 232
Surface tractions, 234
System equations, 7, 13, 15, 129–149, 165–175

Taylor, R. L., 22, 46, 170
Temperature, 184, 190, 245, 286
Thermal conductivity, 184, 245, 285
Thermal expansion, 194, 204
Thermal strain, 194
Thickness, 205, 234, 244
Topology, 6
Transient applications, 317–344
Transition elements, 115, 119, 128
Triangular elements, 78, 101, 113, 305
Truss, 193–203

Unit triangle, 82, 102–104
Ural, 0, 232, 233, 239

Valliappan, S., 164
Variational form, 2, 14, 281
Variable properties, 41, 43
Vector potential, 280
Velocity, 5, 205, 255, 259

Velocity potential, 259–275
Viscosity, 205, 255

Wada, S., 205
Waveguide, 276–279
Weaver, W., 225
Weighted residual, 3, 7, 212, 217
Weights, 94, 101
Whiteman, J. R., 1, 117, 277
Wilson, E. L., 46, 95, 117, 170, 232, 317, 331, 333
Wooten, J. W., 281
Wyatt, E. A., 126

Zienkiewicz, O. C., 22, 102, 105, 127, 193, 194, 245, 256, 285, 317

Subroutine Index

APLYBC, 23, 26, 30, 156, *A*-356
ASYMBL, 23, 26, 131, 134

BANMLT, 195, 202
BANSUB, 136
BFLUX, 34, 54, 241, 250, 272
BTDB, 243
BTDIAB, 244

CALPRT, 74, 186
CCOUNT, 32, *A*-362
CEQBAN, 30, 163
CHANGE, 224
CONDSE, 60
CONTUR, 177, 179, 247
CONVRT, 135
CORECT, 26

DCHECK, 68, 72, 127
DEGPAR, 34, *A*-365
DERCU, 218
DERIV, 29, 76, 87, 128
DER2CU, 219
DER3T, 113
DER4Q, 68, 70, 71, 128, 246
DER412, 128
DER6T, 113
DER8H, 111
DER8Q, 108, 110, 259
DIRECT, 335, 344

ELBAND, 37, 39, 46
ELCOL, 28, 52, 198, 208, 214, 237, 249, 289
ELCORD, 28, 38, 41
ELFRE, 28, 31, 44
ELHIGH, 47
ELPOST, 28, 53, 186, 188, 198, 209, 237, 257
ELPRTY, 28, 41, 43
ELSQ, 28, 52, 186, 188, 197, 207, 213, 219, 227, 235, 248, 260, 287
EULER, 320, 321, 326, 327, 344

FACTOR, 23, 26, 170, 173
FMONE, 287, 292
FORCER, 327, 337
FOURPT, 311

GAUSCO, 95, 97
GAUS2D, 95, 97, 247
GDERIV, 29, 70, 73, 75, 246
GENELM, 23, 27, 54, 58, 131

IJQUAD, 304
IJTRI, 304
INCEQ, 26, 32, 35, *A*-362
INDXEL, 27, 31, 44, 49, 129
INDXPT, 43, 49, 129, 155
INEL, 311
INFLUX, 26, 34, *A*-366
INPROP, 26, *A*-363
INPTS, 309
INPUT, 26, 35, *A*-361
INVDET, 29, 72, 76
INVECT, 26, 34, 35, *A*-365
ISOPAR, 76, 282, 284
I2BY2, 73, 75, 246
I3BY3, 73

JACOB, 29, 69, 70, 72, 75, 246

LAPL8H, 288
LAPL8Q, 261
LCONTC, 87
LCOUNT, 304, 313
LNODES, 27, 31, 37, 38
LPTPRT, 28, 41, 43

MATPRT, 234, 240, 286

MATWRT, 240
MAXMIN, 176, 178
MESH2D, 303, 307
MMULT, 87, 199, 209, 219, 227, 236, 238, 258
MODBAN, 156
MODEL, 22, 26, *A*-355
MODFUL, 30, 155, 159
MODFY1, 30, 156, 159
MODFY2, 156
MODFY3, 156
MOD2FL, 196
MSMULT, 236
MTMULT, 227, 236

NGRAND, 29, 76, 284

OMITP, 304, 312

PCOUNT, 304, 312
PENLTY, 30, 162
PENMOD, 30, 162
PLASMA, 282
PLOTIJ, 304, 314
PLTSET, 273
POST, 23, 180, 181
POSTEL, 31, 53, 180, 186, 189, 195, 199, 209, 238, 257, 262
PTCODE, 32, 38, *A*-361
PTCORD, 41
PT1, 26, *A*-359
PT2, 26, *A*-359
PT3, 26, *A*-360

REVISE, 155

SCHECK, 67, 71, 127
SET, 26, *A*-358
SHAPE, 29, 76, 87, 114, 128
SHPCU, 218
SHP3T, 112
SHP4Q, 67, 71, 74, 128, 246
SHP412, 118, 119
SHP6T, 113
SHP8H, 111
SHP8Q, 108, 110, 259
SINGLR, 122, 125, 276
SKYDIA, 46, 48
SKYFAC, 170, 174
SKYHI, 47
SKYSOL, 170, 174
SKYSTR, 131, 134
SKYSUB, 135
SNORMV, 287, 292
SOLVE, 23, 26, 170, 173
START, 220
STORCL, 27, 118, 130, 132
STORSQ, 27, 130, 132
STRFUL, 130, 133
SYSBAN, 26, 37, 46, 155
SYMTRI, 127

TRGEOM, 92

VECT2D, 274

WRTELM, 23, 26, 176, 178
WRTPT, 23, 26, 176, 177

ZEROA, 26, 29
ZEROI, 26